Personality in Nonhuman Animals

Jennifer Vonk · Alexander Weiss
Stan A. Kuczaj
Editors

Personality in Nonhuman Animals

Springer

Editors
Jennifer Vonk
Department of Psychology
Oakland University
Rochester, MI
USA

Stan A. Kuczaj
Department of Psychology
University of Southern Mississippi
Hattiesburg, MS
USA

Alexander Weiss
Psychology, School of Philosophy, Psychology and Language Sciences
University of Edinburgh
Edinburgh
UK

ISBN 978-3-319-59299-2 ISBN 978-3-319-59300-5 (eBook)
DOI 10.1007/978-3-319-59300-5

Library of Congress Control Number: 2017941479

Printed on acid-free paper

This Springer imprint is published by Springer Nature
The registered company is Springer International Publishing AG
The registered company address is: Gewerbestrasse 11, 6330 Cham, Switzerland

In Memoriam: Stan A. Kuczaj

On April 14, 2016, the scientific community lost a beloved colleague and friend, Dr. Stan A. Kuczaj. One of Stan's principal research interests was animal personality, but he did not begin his research career with this topic. After a somewhat later than usual start in academia, Stan earned a Ph.D. in Child Psychology in 1976 from the University of Minnesota where his focus was on language development. In 1989, he was invited to serve as a Visiting Professor at the University of Hawaii's Kewalo Basin Marine Mammal Laboratory in order to collaborate with Dr. Louis Herman on studies of dolphin language and syntax. Inspired by the incredible cognitive abilities of the dolphins, he founded the Marine Mammal Behavior and Cognition Laboratory at the University of Southern Mississippi in 1996 and shifted his research efforts to comparative psychology, and specifically marine mammals.

Stan was a prolific scholar and exceptional collaborator. He authored nearly 150 publications and was frequently invited to present at conferences and colloquia at other universities. His research with dolphins was highlighted in five different documentaries, and featured as the cover article in *National Geographic*. His remarkable success in the field of psychology resulted in a legacy of more than 50 master's- and doctoral-level students working in a variety of fields. The work of some of those students is represented in this volume. Given his enormous contributions to the field of marine mammal behavior and cognition, and the fact that his laboratory was one of the few places where enthusiastic students could be trained to carry on his work, the loss of Stan will affect the field for many years to come.

Animal personality was a special interest to Stan, although I am not sure when he first became interested in the subject. When I arrived to his lab in 2004, there was some talk of a former student being interested in the subject, but that nothing had been formally researched. I remember when Stan brought me into his office and asked if I would be interested in studying dolphin personality—I was not enthused. I came into the program wanting to study dolphin language, and admittedly thought that dolphin personality sounded too anthropomorphic. However, after reading Gosling's (2001) review of animal personality (at Stan's request), I quickly changed my feelings toward the subject. Stan and I then devised our first dolphin personality questionnaire based on the Five Factor Model and published the first ever article on dolphin personality (Highfill and Kuczaj 2007). This project launched our long-term collaboration examining personality in a variety of species. Stan would ask everyone he knew that studied animals to complete a personality questionnaire. Being the well-liked guy that he was, everyone would agree! I currently have boxes and boxes full of these questionnaires in my office!

I would venture to say that animal personality was one of Stan's favorite research topics. He witnessed the increasing interest over the years and would remark how animal personality often came up in casual conversations at conferences. He believed the field was growing and that was very exciting for him. Nearly 10 years after his first publication on dolphin personality, he had gone on to publish several papers and author many presentations on the topic. At the time of Stan's death, several of his current graduate students were focusing on animal personality for their master's and doctoral work. Stan believed it was important to stop thinking of members of a species as being carbon copies of each other and to focus on individual differences within species. Indeed, at Stan's urging, Michael Beran and I co-edited a special issue on individual differences (Beran and Highfill 2011) for the *International Journal of Comparative Psychology*, which Stan was chief editor of, in 2011. He realized the value in recognizing that personality greatly influences animal behavior and cognition.

The publication of this volume attests to his enthusiasm for the subject. Stan believed that the more we understand about animal personality, the better we understand animal behavior. His ultimate goal was for animal personality research to reach all species (big and small), so that we could study both the commonalities and differences across species. I believe this volume brings us closer to his goal and I hope that research in the field of animal personality will continue to flourish.

Lauren Highfill

References

Beran, M. J., & Highfill, L. E. (2011). Paying more attention to what (some) nonhuman animals and (some) humans can do: An introduction to the special issue on individual differences in comparative psychology. *International Journal of Comparative Psychology, 24*, 1–3.

Gosling, S. D. (2001). From mice to men: What can we learn about personality from animal research? *Psychological Bulletin, 127*, 45–86.

Highfill, L. E., & Kuczaj II, S. A. (2007). Do bottlenose dolphins (*Tursiops truncatus*) have distinct and stable personalities? *Aquatic Mammals, 33*, 380.

Preface

Stan was Chair of the Psychology Department at the University of Southern Mississippi in 2005, so it was he who offered me my first tenure-track position. I was eager to accept the offer with the promise of being able to collaborate with Stan on studies of sea lions housed at the time at the Marine Life Oceanarium in Gulfport, MS. These plans never came to fruition because of the catastrophic damage of Hurricane Katrina, which devastated the MS coast within one month of the start date of my new faculty position. Stan led our department through this tumultuous time when research was put on the back burner and the focus was on repairing lives and maintaining our campus on the coast. We remained friends and colleagues, but never had the chance to collaborate until we were reunited through our commitment to Division 6 of the American Psychological Association (APA), which is now the Society for Behavioral Neuroscience and Comparative Psychology, thanks to the work of previous president David Washburn, who co-authors the introduction to this volume. Stan was the president of the division/society at the time of his passing, and he and I were working closely together to help further its mission. In working together on the Division 6 program for the 2015 meeting of the APA, Stan agreed to organize a session on personality in nonhumans in conjunction with the Society for Personality and Social Psychology (Division 8 of the APA). It was in this session that we met Alex Weiss, who graciously agreed to participate in this session. Out of the lively discussion that arose during this session, the idea for this volume was born. Stan and I readily embraced the opportunity to collaborate on a project, and so it is bittersweet that this volume has come to fruition in his absence. I am grateful to the support and assistance of my co-editor, Alex Weiss, who has been instrumental in bringing this project to life. It is our hope that this volume, along with the journal *Animal Behavior and Cognition* that Stan founded, will serve as a long-lasting testament to the impact that Stan has made on this field in general, and on his colleagues more personally.

Rochester, USA — Jennifer Vonk

Contents

Part I Introduction

1 **A History of Animal Personality Research** 3
Will Whitham and David A. Washburn

Part II Models and Perspectives in the Study of Personality

2 **Exploring Factor Space (and Other Adventures) with the Hominoid Personality Questionnaire** 19
Alexander Weiss

3 **The Interpersonal Circumplex: A Complementary Approach for Understanding Animal Personality** 39
Virgil Zeigler-Hill and Lauren Highfill

4 **The Quantitative and Molecular Genetics of Individual Differences in Animal Personality** 55
Veronika N. Laine and Kees van Oers

5 **Personality from the Perspective of Behavioral Ecology** 73
Jon E. Brommer and Barbara Class

Part III Personality in Various Taxonomic Groups

6 **Feeling Fishy: Trait Differences in Zebrafish (*Danio Rerio*)** 111
Kanza M. Khan and David J. Echevarria

7 **Personality in Elasmobranchs with a Focus on Sharks: Early Evidence, Challenges, and Future Directions** 129
J.S. Finger, F. Dhellemmes and T.L. Guttridge

8 **Personality and Individuality in Reptile Behavior** 153
R. Mark Waters, Bonnie B. Bowers and Gordon M. Burghardt

9 **Personality in Swine** 185
Kristina M. Horback

10 **Personality in Dogs** 205
Jamie L. Fratkin

11 **Felid Personality and Its Implications** 225
Marieke Cassia Gartner

12 **What Do We Want to Know About Personality in Marine Mammals?** ... 237
Erin E. Frick, Amber J. de Vere and Stan A. Kuczaj

Part IV Applications of the Study of Nonhuman Personality

13 **Individual Differences in Nonhuman Animals: Examining Boredom, Curiosity, and Creativity** ... 257
Malin K. Lilley, Stan A. Kuczaj and Deirdre B. Yeater

14 **Phenotype Management: An Inclusive Framework for Supporting Individuals' Contributions to Conservation Populations** ... 277
Jason V. Watters, Samantha Bremner-Harrison and David M. Powell

15 **You Are What You Eat: The Interplay Between Animal Personality and Foraging Ecology** ... 295
Sandra M. Troxell-Smith and Valentina S.A. Mella

16 **Applications of Research in Nonhuman Animal Personality** ... 307
Carly Loyer and James Ha

Index ... 319

Contributors

Bonnie B. Bowers Department of Psychology, Hollins University, Roanoke, VA, USA

Samantha Bremner-Harrison Nottingham Trent University, Nottingham, UK

Jon E. Brommer Department of Biology, University of Turku, Turku, Finland

Gordon M. Burghardt Departments of Psychology and Ecology & Evolutionary Biology, University of Tennessee, Knoxville, TN, USA

Barbara Class Department of Biology, University of Turku, Turku, Finland

Amber J. de Vere Department of Psychology, University of Southern Mississippi, Hattiesburg, MS, USA

F. Dhellemmes Leibniz-Institute of Freshwater Ecology and Inland Fisheries, Berlin, Germany

David J. Echevarria Department of Psychology, The University of Southern Mississippi, Hattiesburg, MS, USA

J.S. Finger Faculty of Life Sciences, Humboldt Universität zu Berlin, Berlin, Germany

Jamie L. Fratkin Department of Psychology, The University of Texas at Austin, Austin, TX, USA

Erin E. Frick Department of Psychology, University of Southern Mississippi, Hattiesburg, MS, USA

Marieke Cassia Gartner Philadelphia Zoo, Philadelphia, PA, USA

T.L. Guttridge Bimini Biological Field Station Foundation, Bimini, Bahamas

James Ha Department of Psychology, University of Washington, Seattle, WA, USA

Lauren Highfill Eckerd College, St. Petersburg, FL, USA

Kristina M. Horback Department of Animal Science, University of California, Davis, Davis, CA, USA

Kanza M. Khan Department of Psychology, The University of Southern Mississippi, Hattiesburg, MS, USA

Stan A. Kuczaj Department of Psychology, University of Southern Mississippi, Hattiesburg, MS, USA

Veronika N. Laine Department of Animal Ecology, Netherlands Institute of Ecology (NIOO-KNAW), Wageningen, The Netherlands

Malin K. Lilley The University of Southern Mississippi, Hattiesburg, MS, USA

Carly Loyer Department of Psychology, University of Washington, Seattle, WA, USA

Valentina S.A. Mella School of Life and Environmental Sciences, The University of Sydney, Sydney, NSW, Australia

David M. Powell Wildlife Conservation Society, Bronx Zoo, New York, NY, USA

Sandra M. Troxell-Smith Department of Biological Sciences, University of Illinois at Chicago, Chicago, IL, USA

Kees van Oers Department of Animal Ecology, Netherlands Institute of Ecology (NIOO-KNAW), Wageningen, The Netherlands

David A. Washburn Georgia State University, Atlanta, GA, USA

R. Mark Waters Ohio University Eastern Campus, St. Clairsville, OH, USA

Jason V. Watters San Francisco Zoological Society, San Francisco, CA, USA

Alexander Weiss Department of Psychology, School of Philosophy, Psychology and Language Sciences, The University of Edinburgh, Edinburgh, UK

Will Whitham Georgia State University, Atlanta, GA, USA

Deirdre B. Yeater Sacred Heart University, Fairfield, CT, USA

Virgil Zeigler-Hill Oakland University, Rochester, MI, USA

Part I
Introduction

A History of Animal Personality Research

1

Will Whitham and David A. Washburn

Abstract

Though the study of animal personality has experienced explosive growth in the last 20 years, its history can be traced to the early days of comparative psychology. Early descriptions of nonhuman animals as sociable or fearful and the like have been progressively replaced across the years with systematic coding of behavioral patterns across multiple dimensions of temperament, in a fashion (and with results) similar to the way the topic is studied in humans. This chapter will explore the researchers, laboratories, and methodologies of animal personality research as it evolved from a methodologically impermeable curiosity to an increasingly important determinant of individual and species variability in behavior.

For much of the twentieth century, animal personality was comparative psychology's moon: clearly visible, obvious to any observer, yet decidedly impossible to bring into the laboratory for study. For as long as there has been a comparative psychology, researchers have described, privately or professionally, the individual character of subject animals. Yet these researchers lacked any accepted framework by which they could sensibly and consistently measure individual differences in temperament or personality. The study of human personality was relatively difficult and contentious in its own right (as indicated, for example, by the longstanding person–situation debate; e.g., Kenrick and Funder 1988); not surprisingly, the methodological and philosophical challenges to the study of (nonhuman) animal personality were even more formidable.

The history of the comparative study of personality is not one of gradual progressions, of an accumulation of observations that manifest into functional theory. Neither was there an emergence of eminent animal personality theorists and dedicated research programs at each step of a coherent process. The history is disjointed; one of fits and starts, as largely isolated scientists combatted Watsonian dogma and a pre-paradigmatic lack of direction in an attempt to build a study of animal personality from the ground up. This chapter will cover the different

W. Whitham (✉) · D.A. Washburn
Georgia State University, Atlanta, GA, USA
e-mail: wwhitham1@student.gsu.edu

D.A. Washburn
e-mail: dwashburn@gsu.edu

J. Vonk et al. (eds.), *Personality in Nonhuman Animals*,
DOI 10.1007/978-3-319-59300-5_1

approaches to animal personality that were taken up to the turn of the twenty-first century, at which point the disjointedness began to cohere into the field of animal personality research that will be explored by the other chapters of this volume. After initial consideration of early comparative authors, this chapter will be organized as a series of extended case studies of the scholars and laboratories that fought a strong, resistant tide in their attempts to codify the vast individual differences in temperament manifest in nonhuman animal (henceforth "animal") life. Special consideration will be given to the process by which these parties innovated new methods and analyses to present a way forward for the study of animal personality.

Early Expressions

Even as the study of human personality was in its infancy, and the empirical study of animal personality was nonexistent, there was some rudimentary acknowledgement of meaningful individual differences in animals. As early as the late nineteenth century, there existed a tendency for early comparative psychologists to describe their subjects as individuals rather than as mere representatives of the norms of the species. British polymath and comparative psychologist Leonard Trelawny Hobhouse reported the results of his comparative work in such a fashion (Hobhouse 1915). A cat subject, for example, was "a sociable creature, who follows his friends about in the half dog-like way that some cats have" (p. 155); a chimpanzee had "an extremely retiring and unsociable disposition" (p. 235). The first investigation of rhesus monkeys in a psychology laboratory by A. J. Kinnaman likewise blended rigorous scholarship with informal discussion of the two monkeys' individual temperaments (Kinnaman 1902a, b). His monkeys were inferred to have expressed "triumph, villainy, jealousy, anger and risibility… in the manner and speed of limb and body movement, or in the viscera of the observers who read into them a feeling like that which they suppose they would have under similar circumstances" (p. 106). It may be that these examples are more the result of stylistic convention than empirical positions on the animal mind. These early scholars did not explicitly study animal personality, nor even necessarily regard such study as legitimate. Yet the language used to describe the dispositions, behavioral tendencies, and temperamental traits of their participant animals is noteworthy, particularly in light of the shifts in psychology that would come with the advent of Watsonian behaviorism in Western psychology years later.

Constitutional Differences and Functional Disturbances (1927)

Ivan Pavlov was perhaps the first researcher to incorporate temperament into animal research in a work often translated as "Constitutional Differences and Functional Disturbances: Experimental Neuroses" (Pavlov 1966). Over the course of his research on digestion and conditioned reflexes, Pavlov observed individual peculiarities in his famous dogs. Some animals that were selected as subjects for their attentive demeanor and enthusiasm were found to fall asleep rapidly during the experimental procedure. Others were fearful, cowering at any unexpected noise and remaining highly vigilant in behavior for the duration of experimental testing. Although Pavlov did not formally incorporate these classes of animals (translated as "temperaments", "types", or "constitutional differences", and similar to the laid-back versus uptight distinction in macaques studied in recent years by Suomi and colleagues, discussed below) into his research program, he nevertheless provided a basic framework for categorizing the temperaments of the dogs he studied in his lab.

In keeping with the physiological timbre of his work, Pavlov initially classified his animals into broad groupings based on what he understood to be the qualities of their nervous systems. The highly enthusiastic dogs that he chose for his first experiments could be understood as the product of a highly excitatory nervous system. In addition to their manic behavior, animals of this type required demonstrably more stimulation in order to maintain attention to the experimental procedure. If the dogs did not receive novel stimuli at a rapid pace, their excitatory predispositions gave way to drowsiness and rest. This type contrasted sharply with another class of dogs observed by Pavlov that he viewed as having a much more equilibrated and inhibited nervous system. These often fearful dogs were quick to engage in specific motoric behaviors—shrinking to the floor, fleeing with tucked tail—but this lack of inhibition was not a more general character of the animal. Rather, the animals remained active across a variety of experimental settings and were capable of substantial inhibitory activity that was demanded by the experimental setting. Pavlov found that most dogs could be explained in terms of two dichotomous dimensions: their tendency to exhibit moderate or extreme excitation and their tendency to exhibit moderate or extreme inhibition.

Pavlov recognized that his animals fell into temperamental categories besides these as well, and conjectured that 24 or more types of nervous system may be described. Most commonly he adopted a framework with four principal nervous system types, in homage to Galen's temperaments (e.g., Stelmack and Stalikas 1991). The rapidly excited and rapidly inhibited animals described first were classified sanguine, and the more fearful and measured animals classified as melancholic. More uncommonly, Pavlov encountered animals of phlegmatic temperament. These dogs were extremely restrained in their behaviors, seeming disinterested and neither friendly nor hostile, yet capable of extreme excitation when the inhibited restraint was upset. Finally, choleric animals could be described by the inconsistency of their inhibitory responses.

Application of these human temperament constructs to dogs, even informally, is notable. Pavlov himself would perhaps not agree with this recognition, as he shifted fluidly between discussions of the mental activity of dogs and humans throughout his writing. He suggested that he felt that the empirical definition of stable, heritable nervous system types would be forthcoming. Instead, the particular way in which Pavlov framed his observations of the individual character of his dogs would become a rarity for most of the twentieth century (see Burdina and Melikhova 1961 for one return to the analysis of animal nervous system types).

A Behavior Rating Scale for Young Chimpanzees (1938)

In 1938, the Yale Laboratories of Primate Biology and Meredith Crawford published the first empirical exploration of animal personality in the form of "A Behavior Rating Scale for Young Chimpanzees" (Crawford 1938). Crawford's motivation for the exploration was clear from the outset: "Only a few days' work with a group of chimpanzees is sufficient to impress the observer with the vast differences between particular animals" (p. 79). And like Pavlov's proposed research into animal temperament, the scale was both an empirical investigation into observed individual differences and a pragmatic pursuit. The ability to capture the particular behavior patterns of an animal would likely be relevant and useful for the handling, caretaking, and research participation of the animal.

The paper describes the development and application of the rating scale from its conception. In the first version of the scale, six raters were asked to rate chimpanzees on 44 multiple-choice items divided into five groups: interactions with humans, interactions with other chimpanzees, behavior in experiments, individual characteristics, and trait ratings. The items that made up the scale took many forms, from behavioral indices (e.g., "Amount of motor activity displayed", "Amount of masturbation") to introspected states (e.g., "Apparent confidence in observer", "Desire to please observer") to assessment of more general qualities of the animal (e.g., "Intelligence", "Friendliness"). Notably, and with credit to the empiricism of Crawford, items were then assessed for both their reliability and validity.

Raters had less direct contact with the adult chimpanzees of the colony, and thus were unable to use earliest versions of the rating scale to assess these animals confidently. The use of multiple-choice options on the scale was found to be inadequate, and raters more easily rated each animal's score for an item on a continuous line between two extremes. For example, an animal might be rated as halfway between silent and boisterous on a measure of "Noisiness". Raters also ascribed a score to their own confidence in rating each animal on each item. Items were removed from the scale for a few reasons. Items related to sexual activity were particularly subject to the hormonal cycles of individual animals, and thus removed from the scale. Unreliable items were removed, and highly intercorrelated items were collapsed together.

When the scale was used to assess the chimpanzees in two consecutive years, an additional measure of reliability, the test–retest reliability of the ratings, was measured. On each measure of reliability, the items of the final scale are high, above .7 in most cases by each measure. This indicates not only the potential usefulness of the measure, but also the stability of individual characteristics as chimpanzees develop year-after-year. The items that were most and least reliable, however, are telling. The items that were more behavioral (e.g., amount of motor activity) tended to be least reliable, whereas those that required more inference by the rater (e.g., desire to please) were the most reliable. This can be interpreted as evidence that what the rating scale primarily measured was the raters' shared anthropomorphizations of the animals, compounded by the fact that the raters, as caretakers, almost certainly discussed their general impressions of individual animals on a regular basis. Crawford acknowledged as much, and also suggested that the raters may have been consciously or unconsciously attempting to match the ratings of animals on second assessment to what they reported on the initial assessment (pp. 85–86).

In the final version of the behavior rating scale, the animals were meaningfully rated on 22 items measured on a continuous scale. This acted as the first quantitative index of animal personality, and Crawford described his research as a preliminary investigation into chimpanzee personality types that acted as a proof of concept for what is possible in the study of animal personality. Moreover, the level of detail in Crawford's methods and analyses is instructive given the novelty of using such a scale to assess animal personality, and, indeed, the novelty of assessing animal personality empirically in any form. His concerns with using a behavioral rating scale with sophisticated observers would become a constant thread of animal personality research into the twenty-first century (see Gosling 2001 for review). Crawford's questions were the same as those that motivate twenty-first-century trait-rating studies: Do reliable differences between animals' item ratings indicate individual differences in psychical constructs? Do reliable intercorrelations between items indicate the presence of reliable trait groupings?

In the introduction to the rating scale, Crawford was adamant: "So different from that of every other animal, and so consistent with itself is the behavior of each ape, that one cannot escape the conclusion that every chimpanzee must possess a distinct personality" (p. 79). Robert Yerkes, the founder and director of the Yale Laboratories of Primate Biology, agreed. In his 1925 book *Almost Human*, Yerkes wrote that "[Apes] are so highly individualized and they so quickly make a place for themselves in one's world of social relations that it is entirely inadequate to describe them merely by type, or as gibbons, orangs, or chimpanzees" (p. 52). Alas, this zeal for this exploration of animal personality did not translate into progress in the field more broadly: publications on the subject would not become any less scarce in the ensuing decades (Freeman and Gosling 2010). The innovation of Crawford's rating scale thus became a mere footnote to his distinguished career in military psychology (for more details, see Benjamin et al. 2002).

The Relationship Between Emotionality and Various Other Salients of Behavior in the Rat (1940)

Billingslea (1941) published a short investigation on personality variation of different rat strains, although this fact is heavily disguised by the behaviorist language of the author. Strains of rats bred for their 'emotionality', as indexed by the frequency of their urination and defecation in an open field, were then tested on additional measures of the rats' five "salients of individuality" (p. 69). An activity salient was measured by the number of turns each rat made on a rodent wheel in its home cage. Problem-solving ability was measured by how well the animals learned to complete two tasks. In one task the rats were required to learn to tear through a paper barrier to reach food, and in the other, the animals were required to learn to use their paws to reach food that could not be reached with their mouths. Aggression was measured both by the amount of fighting an animal participated in and by the animals' reactions to being "attacked" by a jet of air blown by the experimenter. Timidness–savageness was measured by the animals' reaction to miniature versions of the open-field test and by the animals' responses to experimenters. Finally, neuroses were indexed in the same way as emotionality: the open-field test. Billingslea's preliminary analysis toward this purpose reported that emotional rats were less aggressive and neurotic while being more timid and active than the non-emotional rats.

Although Billingslea divided rats into emotional and non-emotional groups for his analyses, his stated goal was to better understand how the measured dimensions combine to make up the specific behavior profile of an individual animal. The personality and individuality of the ubiquitous white rat is rarely considered, and it is perhaps assumed by experimenters that the homogeneity that rat colonies are designed to maintain must eliminate individual variation of this kind. Billingslea's modest investigation into this question at a time in which animal personality research was all but nonexistent is truly exceptional.

Temperament in Chimpanzees (1949)

Upon assuming the directorship of the Yale Laboratories, Karl Lashley promised a program of research on "Individual Differences in Temperament" (King and Weiss 2011). Yet the execution of this proposed research program was inconsistent. The duty largely fell to Lashley's former doctoral student Dr. Donald Hebb, a brilliant physiologist and psychologist with no experience with chimpanzees nor desire to study them (Beach 1987). In spite of Hebb's initial reservations about the project, his 1949 "Temperament in Chimpanzees" is an elegant contribution to the early study of animal personality.

Hebb's investigation was conducted by observing individual chimpanzees' behaviors toward human- and object-based stimuli. The first set of observations of chimpanzee behaviors toward humans was of each chimpanzee's behaviors toward the caretaker during the animal's daily, midday meal. The second and third observations were of the chimpanzees' responses to two human confederates, one who played the role of a timid subordinate and one who played the role of a fearless dominant. Object-based observations were similarly subdivided. Some objects were highly salient and primate-related (e.g., a chimp skull, a stuffed spider monkey), whereas others were not (e.g., representations of dogs or snakes). In addition, chimpanzees were given pictures of familiar individuals and strangers, and a board that they could use to enact events outside of their cages.

The behaviors elicited by the animals across these situations were coded; however, Hebb found little to analyze when using only the individual behavior ratings. It was only when these behaviors were collapsed together into larger categories that meaningful interpretations could be made. Hebb's broad categories included friendly behaviors, aggressive behaviors, quasi-aggressive behavior, avoidance, and unresponsiveness. Like Crawford before him, Hebb had a keen interest in determining the reliability of the chimpanzees' scores on these categories as both a measure of the appropriateness of his methodology and as a measure of the stability of chimpanzees' behavior ratings over time. As with Crawford's behavior rating scale, Hebb's measure was highly reliable when animals were observed in similar circumstances some months after the initial testing. Even more notable is the degree to which the animals' scores were stable: Approximately 8 years after the animals' first exposure to the object test, the apes' fear ratings to a different set of objects were remarkably similar to those of their first exposure.

Hebb would appear to have found the animal personality jackpot: a reliable measure of a variety of meaningful individual differences within a species that are stable across (at a minimum, a significant portion of) the lifespan. His post-hoc assignment of behaviors into human-like behavioral categories was, by his own admission (p. 197), something of an overreach. And yet, at that point in the history of animal personality research, such assignment was the only sensible way to analyze the data. Without the structure of a human-like framework for investigating individual differences in behavioral suites, "the investigator is left with an indigestible mass of facts without relation to one another, and with little value for the prediction of more complex aspects of behavior" (Hebb 1949, p. 196). [Of course, this remains one position in the ongoing contemporary debate regarding the merits of anthropomorphism in comparative research (e.g., Burghardt 2006).] On issues of emotion and temperament in animals, Hebb was pragmatic rather than dogmatic: "Whatever the anthropomorphic terminology may seem to imply about conscious states in the chimpanzee, it provides *an intelligible and practical guide to behavior*" (italics original, Hebb 1946, p. 88).

Nevertheless, animal personality research lacked anyone to follow that guide. Hebb left his position at Yerkes Laboratories before his animal personality research was published, and though he is ascribed as saying that "five years studying temperament in chimpanzees taught him more about human behavior than he learned in any other five years except his first" (paraphrased by Beach 1987, p. 187), he was not to return to Yerkes nor animal personality research. No researcher would immediately take up the mantle.

The Emotions Profile Index (1966, 1973, 1978)

A unique set of contributions to animal personality research comes from the collaborations of Peter Buirski, Robert Plutchik, and Henry Kellerman. From positions at the John Jay College of Criminal Justice at the City University of New York, Albert Einstein College of Medicine, and the Postgraduate Center for Mental Health, respectively, the authors developed and adapted a personality rating scale for use with humans, olive baboons, chimpanzees, and dolphins (Buirski et al. 1973, 1978; Kellerman 1966; Plutchik and Kellerman 1974). Whether by the unorthodox background of the authors or the changing landscape of psychology, work using the Emotions Profile Index (EPI) succeeded in pushing the boundaries of animal personality research.

Buirski, Plutchik, and Kellerman were unafraid of assuming, explicitly, that the same personality constructs applied equally well to all animal species. Indeed, the Emotions Profile Index was originally designed for use with humans on the basis of a general theory of human emotion advanced by Plutchik and Kellerman (1974). The central tenet of the theory was that all human personality is constructed from eight basic emotional states, the frequency and intensity of which can be quantified and understood as an individual personality profile. This profile might then be used as a diagnostic tool in clinical practice, counseling, education, or the workplace.

The version of the EPI for humans is a series of pairwise choices between two descriptive terms. Twelve terms were used in the scale, and all possible pairs of these 12 terms are included in the EPI for a total of 66 pairwise choices. These 12 terms (adventurous, affectionate, brooding, cautious, gloomy, impulsive, obedient, quarrelsome, resentful, self-conscious, shy, and sociable) were mapped onto the eight basic emotional states (fear, anger, joy, sadness, acceptance, disgust, expectancy, surprise), presenting the individual who administered the EPI with an instant profile of an individual's personality.

To explore just how basic or fundamental the basic emotions that make up the EPI are, the same rating scale was then used by experienced observers to rate the personalities of animals. The first animals tested in such a way were dolphins housed at John Lilly's Communication Research Institute (Kellerman 1966). From a sample of only three dolphins, Kellerman identified both individual and species differences in the EPIs of the animals. Dolphins more generally, and two of Kellerman's subjects in particular, were understood to be playful, accepting, joyful, and primarily occupied with pleasure-seeking. More exceptional animals, like Kellerman's final dolphin subject, exhibited a wider range of personality expression. The normative pleasure-seeking may have been in conflict with more anxious or fearful personality dimensions. As can perhaps be deduced from these descriptions, Kellerman's interpretation of the dolphins' scores was distinctly psychoanalytic. Many unconscious motivations and states were presumed to combine so as to form the observed traits and behaviors on which the EPI is based, and the root cause of emotions and personality might, for any subject, be difficult to discern. For the most anxious animals, Kellerman recommended supportive psychotherapy.

The second use of the EPI with animals involved a wild troop of olive baboons (Buirski et al. 1973). Compared with the previous use of the EPI with dolphins, this application to baboons was greatly expanded. Three observers, rather than one, completed EPIs for the seven animals that made up the baboon troop, and those ratings were then compared with observed behavioral markers of grooming and dominance. EPI scores were reliable across raters and were variable across animals (i.e., reliable individual differences were observed), taken by the authors as evidence for the validity of the EPI as a measure of animal personality. Animals that were scored as most sullen and jealous and least fearful (i.e., the dominant monkeys) were groomed the most by other animals. More subordinate animals were profiled as more affectionate and fearful. Although the authors noted that further analyses could not be made sensibly without adequate normative values for baboon EPI scores, they did offer that baboon scores were highly similar to those of humans for five of the eight emotional indices. Baboons, in comparison to humans, were more sociable and accepting.

The third use of the EPI with animals was with another wild primate troop: the groups of wild chimpanzees studied by Jane Goodall at the Gombe Stream National Park (Buirski et al. 1978). This application of the EPI, with seven raters of 23 animals, was another expansion of the method, and reliability scores remained high despite the greater numbers of raters and animals. And unlike the scores of dolphins and baboons, the scores of the chimpanzees were described with consideration to the specific life history and dynamics of the group. For example, one chimpanzee male with a physical impediment was nevertheless quite aggressive and impulsive when in the company of his dominant brother. Exhibition of these specific personality traits would have likely been significantly muted when the crippled chimpanzee was in a less supportive social environment. Another instructive example from the troop comes from the mother–daughter pair of Passion and Pom (Buirski and Plutchik 1991). The pair was highly deviant behaviorally, engaging in infanticide and cannibalism, and the aggressive, uncontrolled EPI profile of Passion contrasts sharply with female chimpanzee norms. More generally, chimpanzee scores and their relations to dominance rankings were similar to those of baboons. Dominant animals were aggressive and impulsive, whereas subordinates were more timid and sociable.

Issues with the use of the EPI with human subjects, much less with animals, are not difficult to identify. Applications of the EPI to animals were anthropomorphic in the extreme, explicitly assuming that biological similarities and common evolutionary heritage of humans and animals yielded common capacities for emotionality and personality. The EPI itself was only completed by a small number of raters (as few as one in the case of Kellerman's investigation of dolphin personality) and for a small number of animals. Inter-rater reliability was high throughout (correlating the EPI scores of human adults, male baboons, and male chimpanzees yielded a correlation of 0.9), and the three primate species, graded on the exact same rating scale, present a strikingly similar range of individual personality profiles. Yet it can never be confidently stated that similarities within and between different species' EPI scores were the result of shared personality constructs or a demonstration of how consistently human raters can infer human-like mental states from a diversity of animal behavior.

The influence of the EPI cannot be denied, and the productivity of Buirski, Plutchik, and Kellerman makes the EPI among the most-used rating methods in comparative psychology (Gosling 2001). Although the EPI offered a standardized, cross-species method for personality research, something the field had never before boasted, fundamental issues with the scale's validity certainly reflected, and may well have contributed to, some bad habits of animal personality researchers and the generally poor reputation of the field that have been corrected only in recent decades. Though the EPI offered a way forward for the study of animal personality, it seems fortunate that other methodologies were developed concurrently.

Personality in Monkeys: Factor Analyses of Rhesus Social Behavior (1973)

In contrast to the approach of finding ways to measure individual differences on classes of behavior that seem important for theoretical or even face-valid reasons, some researchers have employed a strategy of casting a wide net of measures and looking subsequently for underlying patterns like personality factors or latent variables. Factor analysis, a statistical sorting technique often applied in the study of human personality, was not used in animal personality research until the publication of "Personality in Monkeys: Factor Analyses of Rhesus Social Behavior" (Chamove et al. 1972), although Van Hooff (1970) did use principal component analysis for a similar purpose to understand social behaviors of captive chimpanzees. This innovation emerged from the collaboration of clinical psychologist Arnold Chamove and two giants of psychology, personality theorist Hans Eysenck and comparative psychologist Harry Harlow, at Harlow's primate research facility. Ten behaviors of 168 juvenile rhesus macaques were recorded as the animals were exposed to different social and experimental settings including interactions with three groupmates, exposure to an infant conspecific, and exposure to a docile adult male conspecific. The behaviors were coded by the duration in which the monkey engaged in them, and included social and nonsocial versions of play, fear, and avoidance behaviors.

Once analyzed, the animals' behaviors during their interactions with three groupmates clearly loaded onto three factors: hostile, fearful, and social behaviors. A possible analogy between these factors and Eysenck's three major personality factors (neuroticism-stability, extraversion-introversion, and psychoticism) was dutifully noted by the authors (Eysenck and Eysenck 1968). They were also careful to note that similarities in the factor loadings of different species may be illusory. In spite of the superficial similarity in the factors, it was concluded that "It would be premature to seek to prove the identity of the factors in these different species; no acceptable method exists at the moment for any such proof" (Chamove et al. 1972; p. 502). Critically, animal personality research was regarded as worth pursuing, and even as a construct that was both critically important and strikingly underrepresented in comparative psychology. Such a line of inquiry was not impossible; the interpretive framework was simply incomplete.

Madingley Questionnaire (1978–Present)

A landmark series of studies on animal personality was published by Joan Stevenson-Hinde and colleagues beginning in the latter part of the 1970s (Stevenson-Hinde et al. 1980a, b; Stevenson-Hinde and Zunz 1978). Her work on individual differences in rhesus macaques carried with it no assumption that all animals experienced the same basic emotions, as in work with the EPI. Stevenson-Hinde was trained as an experimental psychologist in the Skinnerian tradition, and this training, combined with her interest in variables outside of the immediate stimulus environment and an eye to developing theories of human personality, allowed for the development of a new tool for the study of animal personality (Stevenson-Hinde and Hinde 2011).

In truth, Stevenson-Hinde's work is not different in kind from what has already been reviewed. The development of the scale that would eventually become the Madingley Questionnaire was strikingly similar to that of Crawford's behavior rating scale, with terminology from Sheldon's *Scale for Temperament* (1942), and analysis via factor analyses of the kind used by Chamove, Eysenck, and Harlow (see previous section; Stevenson-Hinde and Hinde 2011). Nevertheless, the combined effects of her commitment to her scale (she used it across multiple years and multiple publications) and its relative adaptability (the experimenter is afforded a certain degree of flexibility in changing the scale) presented a substantive contribution to animal personality research (Stevenson-Hinde and Hinde 2011). Initial versions of Stevenson-Hinde's rating scale included 33 items that were to be rated on a seven-point scale by three observers. This version of the scale loaded the ratings of the rhesus macaques into two factors, one a spectrum from "Confident" to "Fearful" and the other, a spectrum from "Active" to "Slow". Later versions removed unreliable items from the initial set and replaced them with ones of a more social nature. The new set of items loaded onto a third factor and included items related to sociality on a spectrum from "Sociable" to "Solitary". These ratings and loadings were highly consistent over multiple years for adult animals, with juveniles' scores more irregular.

A novel and important approach of Stevenson-Hinde was the standardization of individual animals' scores relative to population means for each component. In this way, she could derive a truly individual profile for each of the animals based on their specific deviations from the norms of the species. Whether it was for this unique element or another reason, Stevenson-Hinde is regarded as a pioneer in the study of personality research (Gosling et al. 2003). Indeed, so impactful were her early works that they are sometimes erroneously cited as being the first studies of animal personality (e.g., Clarke and Boinski 1995). Although it is impossible to say with any certainty what precipitated the boom in animal personality research that occurred at the conclusion of the twentieth century and into the twenty-first century, Stevenson-Hinde offered an attractive corpus of scholarship that synthesized the strongest ideas from previous animal personality research just as a generation of animal personality research began. Her influence cannot be understated, and the Madingley questionnaire remains an invaluable tool for the study of animal personality (e.g., Freeman et al. 2013).

Suomi's Laid-Back, Uptight, and Jumpy Monkeys (1989–Present)

As noted above, some of the earliest attempts to describe individual differences in animal personality were focused on the organisms' responsiveness to novel or threatening stimuli. Assessments of individual differences in fearfulness, timidity, anxiety, shyness, or hypervigilance in response to such situations appear to be stable within individuals and across contexts, and predictive with respect to a wide range of behavioral outcomes. Over the last three decades, Steve Suomi and his collaborators have examined the biological basis and social–behavioral consequences of variations in behavioral reactivity as a personality trait in monkeys (e.g., Higley and Suomi 1989; Schneider et al. 1991; Suomi 2001; Suomi et al. 2011). Suomi and his colleagues have focused on three subgroups of monkeys within these variations in reactivity. About 20% of the monkeys respond consistently and characteristically to environmental stressors with fear, stress, and avoidance. These animals, dubbed "uptight" (or fearful, or anxious) by Suomi's team, respond behaviorally and physiologically with stress responses to even mild challenges—the kinds of things that would produce curiosity and exploration in other ("laid-back") members of the same species and group. The third group of monkeys of interest to these researchers is the 5–10% of animals that are reliably impulsive, showing poor behavioral inhibition in a wide range of environments. Unlike the uptight and laid-back animals, these monkeys are much more likely to behave in inappropriate and maladaptive ways, for instance by moving repeatedly between a dominant male and a desired food, or between a high-ranking mother and her infant. These aggressive, jumpy monkeys also show neurobiological markers that correspond to the stable behavioral patterns. Suomi and collaborators have also shown that the traits that underlie these three groups are highly heritable but modifiable by early experience, that the distributions of these groups are consistent across laboratory and naturalistic settings, and that the behavioral types correspond to adjective-rating assessments of personality like those discussed in the previous section (Bolig et al. 1992). Further, the temperament differences recorded in these rhesus monkeys appear to map onto variations that are observed in human children (i.e., with parallels between uptight monkeys and anxious or fearful children, and between jumpy monkeys and aggressive children). Without suggesting that reactivity and behavioral inhibition capture all of the variability in personality differences for animals, Suomi and colleagues have shown powerful predictive relations between this trait and a wide range of important social and behavioral competencies and problems such as affiliation or social isolation, reproductive success or sexual impairment, and longevity versus premature death from violence (Higley et al. 2011).

The Five-Factor Model Plus Dominance in Chimpanzee Personality (1997–Present)

The final case study included in this review is the Five Factor Model + Dominance for chimpanzees (FFM + D; King and Figueredo 1997). In most ways, the work of King and Figueredo reflected the natural evolution of the various adjective-based personality ratings that preceded them. For example, Figueredo et al. (1995) had compared the personality structure of macaques and zebra finches using a modified version of the Madingley Questionnaire. Like the research with the EPI and earlier versions of the Madingley Questionnaire discussed above, the FFM + D was based on rating scales originally designed for use with humans. In this case, the model was Goldberg's (1990) taxonomy of the "Big-Five" factor framework, in which human personality can be measured with adjectives selected to reflect five latent factors or dimensions: Surgency (Extraversion/Introversion), Agreeableness, Emotional Stability (or its reverse, Neuroticism), Intellect (Openness), and Conscientiousness. King and Figueredo (1997) selected 40 adjectives from Goldberg's inventory (and added three words: clumsy, autistic, manipulative) for use with chimpanzees. A total of 100 chimpanzees from 12 zoological parks were rated for each adjective on a 7-point scale by a total of 53 raters (averaging about four raters per chimpanzee, with high observed inter-rater reliability). Principal axis factor analysis of these ratings revealed six factors. The first and largest of the factors, accounting for about 21% of the variability, did not correspond to any of the factors in either Goldberg's (1990) analyses or the FFM more generally. The highest-loaded adjectives on this factor were dominant and submissive, and all of the adjectives that loaded on this factor suggested a Dominance dimension. The other five factors revealed in the King and Figueredo analysis corresponded rather directly with the

Big Five from Goldberg and many other studies (see, for example, reviews by Digman 1990; McRae 2009). The second factor included variables like solitary, active, playful, sociable, and friendly, and was interpreted as Surgency (Extraversion). The third factor (including impulsive, defiant, reckless, erratic) suggested Dependability or Conscientiousness, whereas the fourth factor (including sympathetic, helpful, sensitive, protective) was interpreted as Agreeableness. Stable, excitable, and unemotional appeared to form an Emotional Stability factor, whereas inventive and inquisitive loaded together in a factor interpreted as Openness. More than 72% of the total variability was accounted for by the six-factor solution.

The general correspondence between the Big Five personality traits and factors two through six of the King and Figueredo data was impressive, although the authors acknowledged that some of the items had their strongest loadings for the chimpanzees in a different factor than what Goldberg (1990) found for humans. Nevertheless, the report provided impressive evidence both for the existence of apparent personality traits in these animals, and also for the utility of assessment strategies for understanding animal personality that are comparable to those used for humans.

A Recent History of Animal Personality Research

The animal personality area has been definitively reviewed and summarized by Samuel Gosling and his collaborators (e.g., Freeman and Gosling 2010; Gosling 2001, 2008; Gosling and John 1999; Jones and Gosling 2005). These reviews illustrate the commonalities and differences in findings about animal personality across studies. They also serve to document the tremendous increase in interest in the topic in the last two decades, at least to the degree that scholarly output is an indicator of interest. Figure 1.1 updates the publication counts reported by Freeman and Gosling (2010), including conservative projections of the number of empirical, theoretical, and review publications that should be in print by the end of the present decade. To highlight the recent surge of interest, note that of the 468 articles and chapters that have been published to date on primate personality, 75% have appeared in the two decades since King and Figueredo (1997). It seems likely that PsycINFO will index more articles and chapters on animal personality (with animals as the subject population) between 2010 and 2019 than in all of the previous decades combined.

In addition to chronicling the change in interest in the topic, the highly cited reviews of animal personality research by Gosling (2001) and Gosling and John (1999) likely contributed to the start of the surge in interest. Attempts to synthesize the contributions of scholars like Crawford and Hebb indicate a self-awareness about the field that animal personality research had previously lacked; that is, there seems to be a tacit acknowledgment that although no single work of those authors had birthed a multi-faceted program of animal personality research, the combined efforts across investigators might have done just that. The reviews also inevitably evaluated the current climate in which questions of animal personality might be studied and found it to be quite a bit more forgiving than in Crawford's or Hebb's time. Gone was much of the behaviorist dogma that may have otherwise stifled the growth of animal personality research, without abandoning the methodological behaviorism that elevates the enterprise from speculation to science. The language of psychology had loosened, and loosened enough that the exploration of the individual characters of a diverse array of animals was no longer verboten.

The methodological innovations discussed in the present chapter remain important: the EPI, Madingley Questionnaire, and Chimpanzee Personality Questionnaire were used in two-thirds of the animal personality studies reviewed by Freeman et al. (2011). More broadly, the general use of subjective personality ratings (ideally anchored by specific behaviors) followed by analysis of factor structure has become conventional, with the specific assessment methods and latent variables being continually refined and described

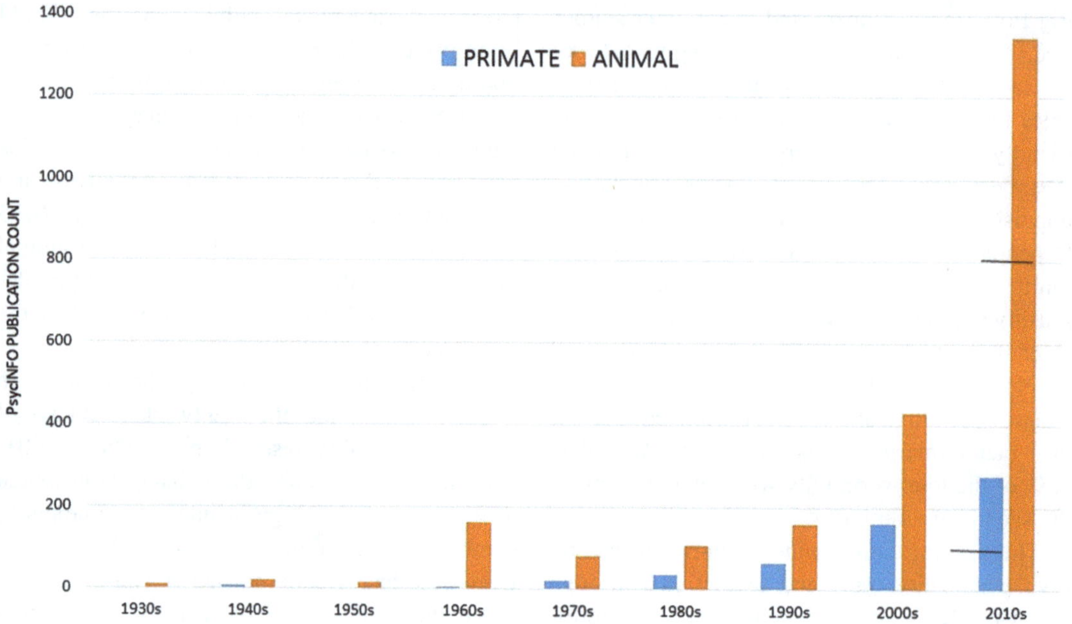

Fig. 1.1 PsycINFO publication counts by decade for "primate and personality" or "animal and personality" (population = animals). The *bars* in the 2010s column represent projections for the 2010–2019 decade, based on publications to date (marked by the *horizontal lines* in this column)

(e.g., Freeman et al. 2013). Nonhuman primates remain popular subjects of these studies, but personality research is being conducted with a growing number of other animal species. Within the last year, for example, there have been published reports of personality research with fish, birds, spiders, lizards, and wild boars, among other species. As is the case for other topics within comparative cognition, there has been a surge of research on personality of canines—bringing the field full circle, in a sense, from the early musings of Pavlov discussed above. Although there remains much to learn about the nature, causes, and implications of individual and species differences in personality among nonhuman animals, the field is much more empirically grounded and much more theoretically cohesive, compared to the broad early descriptions of savage rats (Utsurikawa 1917), gregarious pigeons (Taylor 1932), negative dogs (Pavlov and Petrova 1934), unsociable cats (Romanes 1912), and confident chimpanzees (Yerkes 1939).

That said, this overview of the history of the research area serves to highlight that there remain concerns and problems within the field, even after these many decades of research on personality in animals, and in the midst of this tremendous surge in interest and scholarly activity. The challenges that characterize other areas of inquiry within comparative cognition—from anthropomorphism to uncontrolled sources of error—plague animal personality researchers as well, who must also wrestle with issues (e.g., the circularity of inferring traits that presumably explain behavior from subjective ratings of similar behaviors) faced by scientists who study personality in humans. In part because of this, it could be claimed that the hundreds of discrete steps the field has taken since the 1970s have advanced our knowledge less than those two or three leaps reflected in the seminal studies reviewed in this chapter. The other contributions to this volume seek to remedy this, illuminating the relation between evolution, heredity, biology, experience, development, cognition, social variables, temperament, and behavior.

It seems that comparative psychology is at the brink of a big change, in which individual differences become as interesting and important as group (e.g., species, or mother-reared versus nursery reared, or laboratory-housed versus free-range) similarities and differences have been. Future histories of comparative psychology will indicate whether this statement is true; but if individual differences are indeed to become an increasing focus of comparative psychology, it seems likely that the struggles and successes that have characterized the study of animal personality will lead the way.

Acknowledgements Preparation of this chapter was supported by Grant HD-060563 from the National Institute of Child Health and Human Development, and by Georgia State University. For more information, contact the authors at wwhitham1@student.gsu.edu or dwashburn@gsu.edu.

References

Beach, F. (1987). Donald Olding Hebb. *American Psychologist, 42,* 186–187. doi:10.4249/scholarpedia.3719

Benjamin, L. T. J., McKay, C. L., Osborn, W. C., & Prophet, W. W. (2002). Meredith P. Crawford (1910–2002). *American Psychologist, 57,* 980–982. doi:10.1037//0003-066X.57.11.980

Billingslea, F. Y. (1941). The relationship between emotionality and various other salients of behavior in the rat. *Journal of Comparative Psychology, 31,* 69–77.

Bolig, R., Price, C. S., O'Neill, P. L., & Suomi, S. J. (1992). Subjective assessment of reactivity level and personality traits of rhesus monkeys. *International Journal of Primatology, 13,* 287–306.

Buirski, P., Kellerman, H., Plutchik, R., Weininger, R., & Buirski, N. (1973). A field study of emotions, dominance, and social behavior in a group of baboons (*Papio anubis*). *Primates, 14,* 67–78.

Buirski, P., & Plutchik, R. (1991). Measurement of deviant behavior in a gombe chimpanzee: Relation to later behavior. *Primates, 32,* 207–211.

Buirski, P., Plutchik, R., & Kellerman, H. (1978). Sex differences, dominance, and personality in the chimpanzee. *Animal Behaviour, 26,* 123–129.

Burdina, V. N., & Melikhova, E. F. (1961). The importance of the type of the nervous system in the body's response to injurious factors. *Bulletin of Experimental Biology and Medicine, 51,* 413–417. doi:10.1007/BF00779524

Burghardt, G. M. (2006). Critical anthropomorphism, uncritical anthropocentrism, and naïve nominalism. *Comparative Cognition and Behavior Reviews, 2,* 136–138. doi:10.3819/ccbr.2008.20009

Chamove, A. S., Eysenck, H. J., & Harlow, H. F. (1972). Personality in monkeys: Factor analyses of rhesus social behavior. *Quarterly Journal of Experimental Psychology, 24,* 496–504.

Clarke, A. S., & Boinski, S. (1995). Temperament in nonhuman primates. *American Journal of Primatology, 37,* 103–125.

Crawford, M. P. (1938). A behavior rating scale for young chimpanzees. *Journal of Comparative Psychology, 26,* 79–92. doi:10.1037/h0054503

Digman, J. M. (1990). Personality structure: Emergence of the five-factor model. *Annual Review of Psychology, 41,* 417–440. doi:10.1146/annurev.ps.41.020190.002221

Eysenck, S. B. G., & Eysenck, H. J. (1968). The measurement of psychoticism: A study of factor stability and reliability. *British Journal of Social and Clinical Psychology, 7,* 286–294.

Figueredo, A. J., Cox, R. L., & Rhine, R. J. (1995). A generalizability analysis of subjective personality assessments in the stumptail macaque and the zebra finch. *Multivariate Behavioral Research, 30,* 167–197. doi:10.1207/s15327906mbr3002

Freeman, H. D., Brosnan, S. F., Hopper, L. M., Lambeth, S. P., Schapiro, S. J., & Gosling, S. D. (2013). Developing a comprehensive and comparative questionnaire for measuring personality in chimpanzees using a simultaneous top-down/bottom-up design. *American Journal of Primatology, 75,* 1042–1053.

Freeman, H. D., & Gosling, S. D. (2010). Personality in nonhuman primates: A review and evaluation of past research. *American Journal of Primatology, 72,* 653–671. doi:10.1002/ajp.20833

Freeman, H. D., Gosling, S. D., & Schapiro, S. J. (2011). Comparison of methods for assessing personality in nonhuman primates. *Personality and Temperament in Nonhuman Primates* (pp. 17–40). New York: Springer.

Goldberg, L. R. (1990). An alternative "description of personality": The big-five factor structure. *Personality Processes and Individual Differences, 59,* 1216–1229. doi:10.1037/0022-3514.59.6.1216

Gosling, S. D. (2001). From mice to men: What can we learn about personality from animal research? *Psychological Bulletin, 127,* 45–86. doi:10.1037//0033-2909.127.1.45

Gosling, S. D. (2008). Personality in non-human animals. *Social and Personality Psychology Compass, 2,* 985–1001. doi:10.1111/j.1751-9004.2008.00087.x

Gosling, S. D., & John, O. P. (1999). Personality dimensions in nonhuman animals: A cross-species review. *Current Directions in Psychological Science, 8,* 69–75. doi:10.1111/1467-8721.00017

Gosling, S. D., Kwan, V. S. Y., & John, O. P. (2003). A dog's got personality: A cross-species comparative approach to personality judgments in dogs and humans. *Journal of Personality and Social Psychology, 85,* 1161–1169. doi:10.1037/0022-3514.85.6.1161

Hebb, D. O. (1946). Emotion in man and animal: An analysis of the intuitive processes of recognition. *Psychological Review, 53,* 88–106. doi:10.1037/h0063033

Hebb, D. O. (1949). Temperament in chimpanzees: I. Method of analysis. *Journal of Comparative and Physiological Psychology, 42,* 192–206.
Higley, J. D., & Suomi, S. J. (1989). Temperamental reactivity in non-human primates. In G. A. Kohnstamm, J. E. Bates, & M. K. Rothbart (Eds.), *Temperament in childhood* (pp. 153–167). Oxford: Wiley.
Higley, J. D., Suomi, S. J., & Chaffin, A. C. (2011). Impulsivity and aggression as personality traits in nonhuman primates. In A. Weiss, J. E. King, L. Murray, A. Weiss, J. E. King, & L. Murray (Eds.), *Personality and temperament in nonhuman primates* (pp. 257–283). New York, NY: Springer. doi:10.1007/978-1-4614-0176-6_10
Hobhouse, L. T. (1915). *Mind in evolution*. London: Macmillan and Co.
Jones, A. C., & Gosling, S. D. (2005). Temperament and personality in dogs (*Canis familiaris*): A review and evaluation of past research. *Applied Animal Behaviour Science, 95,* 1–53. doi:10.1016/j.applanim.2005.04.008
Kellerman, H. (1966). The emotional behavior of dolphins, *Tursiops truncatus*: Implications for psychoanalysis. *International Mental Health Research Newsletter, 8*(1), 3–7.
Kenrick, D. T., & Funder, D. C. (1988). Profiting from controversy: Lessons from the person-situation debate. *American Psychologist, 43,* 23–34.
King, J. E., & Figueredo, A. J. (1997). The five-factor model plus dominance in chimpanzee personality. *Journal of Research in Personality, 31,* 257–271. doi:10.1006/jrpe.1997.2179
King, J. E., & Weiss, A. (2011). Personality from the perspective of a primatologist. In A. Weiss, E. J. King, & L. Murray (Eds.), *Personality and temperament in nonhuman primates* (pp. 77–99). New York, NY: Springer. doi:10.1007/978-1-4614-0176-6_4
Kinnaman, A. J. (1902a). Mental life of two macacus rhesus monkeys in captivity I. *The American Journal of Psychology, 13,* 98–148.
Kinnaman, A. J. (1902b). Mental life of two macacus rhesus monkeys in captivity II. *The American Journal of Psychology, 13,* 173–218.
McCrae, R. R. (2009). The five-factor model of personality traits: Consensus and controversy. In P. J. Corr & G. Matthews (Eds.), *The Cambridge handbook of personality psychology* (pp. 148–161). New York, NY: Cambridge University Press. doi:10.1017/CBO9780511596544.012
Pavlov, I. (1966). Constitutional differences and functional disturbances: Experimental neuroses. In M. Kaplan (Ed.), *Essential Works of Pavlov* (pp. 261–267). New York: Bantam Books.
Pavlov, I. P., & Petrova, D. R. (1934). A contribution to the physiology of the hypnotic state of dogs. *Journal of Personality, 2,* 189–200.
Plutchik, R., & Kellerman, H. (1974). *Emotions profile index manual*. Torrance, CA: Western Psychological Services.
Romanes, G. J. (1912). *Animal intelligence*. New York, NY: Appleton & Company.
Schneider, M. L., Moore, C. F., Suomi, S. J., & Champoux, M. (1991). Laboratory assessment of temperament and environmental enrichment in rhesus monkey infants (*Macaca mulatta*). *American Journal of Primatology, 25,* 137–155. doi:10.1002/ajp.1350250302
Stelmack, R. M., & Stalikas, A. (1991). Galen and the humour theory of temperament. *Personality and Individual Differences, 12,* 255–263.
Stevenson-Hinde, J., & Hinde, C. A. (2011). Individual characteristics: Weaving psychological and ethological approaches. In A. Weiss, E. J. King, & L. Murray (Eds.), *Personality and temperament in nonhuman primates* (pp. 3–14). New York, NY: Springer.
Stevenson-Hinde, J., Stillwell-Barnes, R., & Zunz, M. (1980a). Individual differences in young rhesus monkeys: Consistency and change. *Primates, 21,* 498–509.
Stevenson-Hinde, J., Stillwell-Barnes, R., & Zunz, M. (1980b). Subjective assessment of rhesus monkeys over four successive years. *Primates, 21,* 66–82.
Stevenson-Hinde, J., & Zunz, M. (1978). Subjective assessment of individual rhesus monkeys. *Primates, 19,* 473–482.
Suomi, S. J. (2001). Socialization in primates. In J.G. Borkowski., S.L. Ramey., & M. Bristol-Power (Eds.), *Parenting and the child's world: Influences on academic, intellectual, and social-emotional development* (pp. 265–282). Mahwah, NJ: Lawrence Erlbaum Associates.
Suomi, S. J., Chaffin, A. C., & Higley, J. D. (2011). Reactivity and behavioral inhibition as personality traits in nonhuman primates. In *Personality and temperament in nonhuman primates* (pp. 285–311). Berlin: Springer.
Taylor, W. S. (1932). The gregariousness of pigeons. *Comparative Psychology, 13*(1), 127–131.
Utsurikawa, N. (1917). Temperamental differences between outbred and inbred strains of the albino rat. *Journal of Animal Behavior, 7,* 111–129.
Van Hooff, J. A. R. A. M. (1970). A component analysis of the structure of the social behaviour of a semi-captive chimpanzee group. *Experientia, 26,* 549–550.
Yerkes, R. M. (1925). *Almost human*. New York: Century Co.
Yerkes, R. M. (1939). The life history and personality of the chimpanzee. *The American Naturalist, 73,* 97–112.

Part II

Models and Perspectives in the Study of Personality

Exploring Factor Space (and Other Adventures) with the Hominoid Personality Questionnaire

2

Alexander Weiss

Abstract

In this chapter, I describe a research tradition for investigating the evolutionary origins of the personalities of different species of primates, including humans. To that end, I describe the tradition's origins and methods, set out its assumptions, and evaluate its ability to solve empirical problems related to personality. In doing so, I will clear up misconceptions and assuage critics of the approaches that define this research tradition. After surveying the literature that originated from this research tradition, I conclude that it solves many problems as well as other research traditions do. I then identify its limitations and the need for further studies of wild populations, and propose a new direction for the study of personality and behavior.

Finally: It was stated at the outset, that this system would not be here, and at once, perfected. You cannot but plainly see that I have kept my word. But I now leave my cetological system standing thus unfinished, even as the great Cathedral of Cologne was left, with the crane still standing upon the top of the uncompleted tower. For small erections may be finished by their first architects; grand ones, true ones, ever leave the cope-stone to posterity. God keep me from ever completing anything. This whole book is but a draught—nay, but the draught of a draught. Oh, Time, Strength, Cash, and Patience!

—*Herman Melville,* Moby-Dick; or, The Whale, *1851*

Although the study of personality in nonhuman animals is not new (see Whitham and Washburn, this volume), it has, in the past 20 years or so, migrated from the fringes of the psychological and biological sciences into the mainstream. The genetic and evolutionary forces that maintain variation in personality traits, the limited plasticity of behavior, personality development, and the covariation of personality traits are now recognized by many as key to understanding the

A. Weiss (✉)
Department of Psychology, School of Philosophy, Psychology and Language Sciences, The University of Edinburgh, Edinburgh, UK
e-mail: alex.weiss@ed.ac.uk

J. Vonk et al. (eds.), *Personality in Nonhuman Animals*,
DOI 10.1007/978-3-319-59300-5_2

evolution of animal behavior. Reviews of work in these and other areas are widely available (e.g., Dingemanse and Wolf 2010; Réale et al. 2007; Sih et al. 2004; Weiss and Adams 2013), including in the present volume. You will excuse me then, I hope, for taking a different tack for, although I will discuss some of the literature, I have set as my goal to describe what I think of as the research habits that have guided my work on personality structure in nonhuman primates. To this end, I will explore the origins of these habits, what this work does and does not assume, the thorny question of anthropomorphism, what empirical problems have been solved, and what problems remain. I will conclude by discussing how methodological and statistical advances that are benefiting human personality research may benefit the study of personality and behavior in nonhuman primates and other species.

Before moving on to this exercise in introspection, it will benefit the reader if I define what I mean by "research habits" and also explain why I am taking this unconventional approach in writing this chapter. Research habits here refer to how I collect animal personality data and, more importantly, what I do with those data; that is, how I analyze them and interpret the results after they have been collected. What I have just described may sound like a paradigm (Kuhn 1970), and the work that I do and similar work on personality ratings by others, has been labeled such in the earlier chapter by Whitham and Washburn. However, my research habits (I cannot speak for others who conduct similar research) are not unthinking and automatic, but follow from pragmatic constraints, accidents of history, learning from experience, and otherwise. In this sense, and this is arguably true for research in psychology more generally (Leahey 1992), my research habits are probably better characterized as part of a research tradition (Laudan 1977) than as a Kuhnian paradigm. In other words, these research habits have been adopted as part of an approach to studying the question of animal personality because they lead to the solution of certain empirical problems.

Having spelled out what I mean by research habits, the question remains as to why I am taking the time to lay them out in this chapter. The overarching goal of my research on personality is to develop a taxonomy of personality structure in nonhuman primates, mammals, and other species for which these methods are appropriate. Doing so, in my view, is crucial to establishing the functional bases of personality traits. However, comparing species requires comparable methods. As such, I would like this chapter to serve as a guide for future comparative studies of personality structure, whether carried out by me and my colleagues or by others, including those who come along after I am dead, no longer able to carry out research, or have moved on to other things, whichever comes first. That way, if there is a need to deviate from these methods, for example to address a reviewer comment or because of a particular characteristic of the species or sample under study, or a wish to pursue some more exploratory analyses, these deviations or explorations will be clear. Another purpose that I hope this chapter will serve is to inform colleagues, including young investigators and often reviewers, why either I or my colleagues have decided to pursue a particular analysis or interpret results in one way instead of another. In doing so, I hope that this chapter will bring forth discussions that enable me and others to cultivate research habits that are more effective at solving problems concerning personality evolution.

Origins

Speaking broadly, the habits in question involve obtaining ratings on a large number of personality traits (typically 40 or more), and then using data reduction to find out what traits cluster together. As is clear from the introduction to this volume and elsewhere (Stevenson-Hinde and Hinde 2011), there is nothing novel or innovative about assessing animal personality using questionnaires. My own work began in 1997 or 1998, at around the time when animal personality research was being increasingly recognized as serious scientific business. I was a Ph.D. student and Jim King, who would soon become my Ph.

D. supervisor, and A. J. Figueredo, who was my Ph.D. supervisor at the time, recently published a paper describing their study of 100 zoo-housed chimpanzees (*Pan troglodytes*) that had been rated on 43 trait descriptive adjectives by keepers, volunteers, and researchers (King and Figueredo 1997).

Around the time that Jim King and A. J. Figueredo were about to study chimpanzee personality, there was a growing consensus that human personality could be described by five dimensions collectively known as "the Big Five" or "Five-Factor Model" (Digman 1990; Goldberg 1990; McCrae and John 1992). As such, when developing what would be known as the Chimpanzee Personality Questionnaire or CPQ, Jim King sampled 41 items from Goldberg's (1990) paper describing the taxonomy of the human Big Five. Jim King's reason for doing so was to determine the extent to which the Five-Factor Model domains were present in chimpanzees, a finding that would lend support to the Big Five or Five-Factor Model; and to ensure that the questionnaire items measured traits characterizing not just behavioral differences among chimpanzees, but also differences in emotional reactivity and stability, interactions with conspecifics, and differences in cognitive domains, such as self-control and curiosity (James E. King, Personal Communication).

To achieve these goals, when selecting trait descriptive adjectives, Jim King made sure to select adjectives from as many of the clusters or facets that defined each of the five human personality domains—neuroticism, extraversion, openness to experience, agreeableness, and conscientiousness (King and Figueredo 1997). In addition, when selecting trait descriptive adjectives, he made sure that the adjectives would be applicable to chimpanzees and did his best to sample positive and negative markers for a given domain (King and Figueredo 1997). The top panel of Table 2.1 shows that he was fairly successful in achieving these goals. Unfortunately, as is also clear from the top panel of Table 2.1, it proved difficult or impossible to find trait descriptive adjectives, such as "unimaginative," to take an example from Goldberg's taxonomy, related to the low pole of openness, that is, being "closed to experience." As a result, the only openness markers represented the positive pole of this domain. The questionnaire also included two trait descriptive adjectives ("clumsy" and "autistic") based on descriptions of chimpanzee behavior. Finally, Jim King came up with one to three sentences for each of the 43 trait descriptive adjectives. These sentences defined each item in terms of chimpanzee behavior, but in a manner consistent with the common dictionary definition. For example, the description for the item "fearful" is: "Subject reacts excessively to real or imagined threats by displaying behaviors such as screaming, grimacing, running away or other signs of anxiety or distress." Including these sentences was important as it reduced the degrees of freedom that raters can use when deciding what a trait means and where, on the scale for a trait, to place a particular chimpanzee.

In the end, King and Figueredo (1997) found that the chimpanzee personality traits defined six mostly uncorrelated factors. Remarkably, they did find evidence for five factors that were analogs or homologues of the five human domains. The other factor they found was the first and largest factor and comprised of trait markers of multiple domains in such a way to suggest that it was related to competitive prowess or dominance. They thus named this factor "dominance." Importantly, they found that the consistency between raters was on par with what had been found in studies of personality in humans (Costa and McCrae 1992; McCrae and Costa 1987; 1989; all cited in King and Figueredo 1997).

My work in this heady milieu began with the suggestion that I study the heritability of the CPQ domains in a dataset consisting of the original 100 chimpanzees and 45 additional individuals. I found considerable evidence that at least the dominance domain was heritable; none of the personality variation appeared to be attributable to shared zoo environments (Weiss et al. 2000). The heritability findings were later supported by quantitative and molecular genetic studies of orangutans, *Pongo spp.* (Adams et al. 2012), chimpanzees (Hopkins et al. 2012; Latzman et al. 2015a, b; Wilson et al. 2016), bonobos, *Pan*

Table 2.1 Origins of the 54 traits comprising the Hominoid Personality Questionnaire

	Location in Human Five-Factor Model									
	Neuroticism		Extraversion		Openness		Agreeableness		Conscientiousness	
	+	−	+	−	+	−	+	−	+	−
CPQ traits										
Dominant			■							
Submissive				■						
Dependent/follower	■									
Independent		■								
Fearful	■									
Decisive									■	
Timid				■						
Cautious									■	
Intelligent					■					
Persistent									■	
Bullying								■		
Stingy/greedy								■		
Solitary				■						
Lazy										■
Active			■							
Playful			■							
Sociable			■							
Depressed				■						
Friendly							■			
Affectionate							■			
Imitative					■					
Impulsive			■							
Defiant								■		
Reckless										■
Erratic										■
Irritable								■		
Predictable									■	
Aggressive								■		
Jealous	■									
Disorganized										■
Sympathetic							■			
Helpful							■			
Sensitive							■			
Protective							■			
Gentle							■			
Stable		■								
Excitable	■									
Unemotional		■								
Inventive					■					
Inquisitive					■					
Manipulative								■		
n traits	4	3	5	4	4	0	7	6	4	4
OPQ add-ons										
Anxious	■									
Vulnerable	■									
Cool		■								
Curious					■					
Conventional						■				
n traits	6	4	5	4	5	1	7	6	4	4

Table 2.1 continued

	Location in Human Five-Factor Model									
	Neuroticism		Extraversion		Openness		Agreeableness		Conscientiousness	
	+	–	+	–	+	–	+	–	+	–
HPQ add-ons										
Thoughtless										■
Distractible										■
Quitting										■
Individualistic					■					
Innovative					■					
Unperceptive						■				
n traits	6	4	5	4	7	2	7	6	4	7

Note CPQ traits are found in the original Chimpanzee Personality Questionnaire (King and Figueredo 1997). OPQ add-ons are traits found in the Orangutan Personality Questionnaire (Weiss et al. 2006). HPQ add-ons are traits found in the Hominoid Personality Questionnaire (Weiss et al. 2009). Filled boxes under "+" indicate that the item is a marker of the positive pole of the human personality domain. Filled boxes under "–" indicate that the item is a marker of the negative pole of the human personality domain

paniscus (Staes et al. 2016), rhesus macaques, *Macaca mulatta* (Brent et al. 2014), and of non-primates (van Oers et al. 2005; Van Oers and Sinn 2013).

Shortly afterwards, I addressed whether the association between the chimpanzee dominance domain and a measure of subjective well-being described in a paper by King and Landau (2003), who developed and validated this measure, was attributable to shared genes. For me, this study was more exciting than my original heritability study. It marked the first time that I used chimpanzee data to address a question about individual differences in humans, namely why do people who are emotionally stable and extraverted report being happier than people who are neurotic and introverted? Although I focused on dominance, because it was the only domain in my previous study that I found to be heritable, this new study had the ability to inform the human debate. This was because several loadings on dominance (see Table 2.1 in King and Figueredo 1997) were associated with high extraversion (e.g., the positive loading of "assertive") and low neuroticism (e.g., the negative loading of "fearful"), and so dominance captured trait variance related to higher subjective well-being in humans. The analyses were conducted using data from 128 chimpanzees on which we had personality and subjective well-being data, and the results were stark. Variation in dominance and subjective well-being were both attributable to the same additive genetic effects, which accounted for about 60% of the variation in dominance and 40% of the variation in subjective well-being (Weiss et al. 2002).

At about the same time that I was watching the maximum likelihood models of this study converge, I began to collect my own data for a side project on orangutan personality. Perhaps because I was working on my doctoral research, I decided to take the easy road and more or less used the same personality questionnaire as was used in the previous chimpanzee studies. In fact, initially, I made only two changes to the CPQ. The first was to add three items—"anxious", "vulnerable", and "cool"—related to the Five-Factor Model neuroticism domain; the first two items were based on two neuroticism facets (Costa and McCrae 1992); the last had no specific origins in human or animal personality studies, but its descriptive sentence, that is, "Subject seems unaffected by emotions and is usually undisturbed, assured, and calm." clearly makes it a marker of the low pole of neuroticism (Weiss et al. 2006). The second change was to add two items—"curious" and "conventional"—related to the Five-Factor Model openness domain; these items were based on another Five-Factor Model questionnaire (McCrae and Costa 1985) and were chosen to represent high and low openness, respectively.

However, this questionnaire, the Orangutan Personality Questionnaire, was short-lived, for at a late stage in that study, I created six more items (Weiss et al. 2009). These additional trait adjectives were derived from a bipolar adjective measure of the Five-Factor Model (McCrae and Costa 1985) and included "thoughtless," "distractible," and "quitting," which were related to the low pole of conscientiousness, "individualistic" and "innovative," which were related to the high pole of openness, and "unperceptive," which was related to the low pole of openness. I christened the revised questionnaire the "Hominoid Personality Questionnaire" (or HPQ), a name that I now admit was rather grandiose given that I was studying only chimpanzees and orangutans.[1]

The preceding paragraph naturally leads to the question of why I augmented the questionnaire with those particular items instead of leaving the questionnaire alone or choosing a different set of items? The decision to add the first five items followed the results of a study by Jim King, myself, and Kay Farmer. In this study, which was published many years later (King et al. 2005), we used targeted orthogonal Procrustes rotation (McCrae et al. 1996) to compare the personality factors obtained (i.e., the personality structure) from ratings of 74 zoo-housed chimpanzees that were not part of King and Figueredo's (1997) study and those obtained from ratings of 43 chimpanzees living in a naturalistic sanctuary in the Republic of the Congo. To our delight, the dominance, extraversion, conscientiousness, and agreeableness domains were identified in both samples, and there was considerable similarity across all of the factor loadings. On the other hand, neuroticism and openness domains clearly comparable to those found in zoo-housed chimpanzees were not found in their sanctuary-housed counterparts. However, the factor loadings from the targeted rotation suggested that those domains were present (see Table 5 in King et al. 2005); a follow-up analysis found similar results when comparing the personality structure found in the 74 zoo-housed chimpanzees to that found in the 100 zoo-housed chimpanzees from King and Figueredo's original study (King, et al. 2005, pp. 401–402).[2]

The findings of our study of chimpanzees in zoos and in the naturalistic sanctuary led us to suspect that, although chimpanzee variants of neuroticism and openness are present, these domains were not as robust because neuroticism was represented by only three items and openness was represented by only two items. In hindsight, I suspect that neuroticism and openness were represented by so few items because King and Figueredo did not anticipate that so many items representing these traits would cluster into the broad dominance domain. Our interpretation of the results was later vindicated by two studies of chimpanzees, each using a different questionnaire, and developed based on different principles, that found evidence for neuroticism and openness domains (Dutton 2008; Freeman et al. 2013). However, at the time, Jim King and I pursued what we thought was the simplest solution, which was to increase the number of items, especially for openness, that might tap these domains.

Turning to the decision to include three new items for conscientiousness, this was motivated by two observations. The first was that, at the time, conscientiousness domains had reportedly been found only in humans and in chimpanzees, suggesting that this domain evolved recently (Gosling and John 1999, p. 71). The second was the surprising finding that orangutans did not appear to possess a conscientiousness domain. Instead, orangutan personality structure was defined by extraversion, dominance, neuroticism,

[1] Readers can go to http://extras.springer.com/2011/978-1-4614-0175-9/ to obtain the HPQ for chimpanzees, orangutans, and rhesus macaques at. Filenames prefixed by "weiss" are English-language versions of the HPQ. Filenames prefixed by "weissmurayama" are Japanese-language versions of the HPQ. The HPQ is also available in Dutch, German, Chinese, French, and Spanish. Until I develop a website, please contact me if you wish to obtain any of these other versions or the LaTeX code should you wish to adapt the HPQ for your own purposes.

[2] Similar studies of the chimpanzee personality conducted after the HPQ was developed that used the original 43 CPQ items yielded similar results (Weiss et al. 2009, 2007).

and agreeableness domains, which resembled those found in chimpanzees, and an intellect domain, which combined aspects of openness and conscientiousness. In the resulting paper, published 2 years later, we speculated that this finding, the absence of a distinct conscientiousness domain, may be attributable to the fact that the OPQ did not include items suitable for measuring conscientiousness in a semi solitary species (Weiss, et al. 2006, p. 508). As noted in the paper, we thought that a straightforward way to test this possibility would be to include items related to less social aspects of conscientiousness in our questionnaire, which is exactly what Jim King and I did when developing the HPQ. In addition, although we did not raise the possibility in the paper, we thought that a straightforward way to test Gosling and John's hypothesis about the recent emergence of conscientiousness would be to use the HPQ to assess other species of primates. I will describe these studies later.

Throughout this period and up to 2007, when I was analyzing data that I collected with Miho Inoue-Murayama on chimpanzees living in Japan, I adopted the remaining research habits that flavor this research tradition. Because this period of time included my 3 years of postdoctoral training, studying human personality with Paul Costa, I adopted some habits from human personality research. Since that time, I have been surprised by the extent to which some of these habits, which I continue to adhere to and teach, have caused me and my students' grief. I hope that by describing them and their rationale in the next section will spare myself, my students, and future researchers from further suffering.

Habits

So to put the question plainly, how do I analyze HPQ data on some new species? The first step is to estimate the interrater reliabilities of the HPQ items. To do so, like King and Figueredo (1997), I use two of the intraclass correlation coefficients (*ICC*s) described in Shrout and Fleiss (1979). The first, *ICC*(3,1), is an estimate of how reliable an individual rater's rating is likely to be, and therefore can be used to compare reliabilities across studies that differ in how many raters are used. The second, *ICC*(3,*k*), is an estimate of the reliability of the average (mean) score across raters (*k* refers to the number of raters per target). Because later analyses are conducted after aggregating scores across raters, *ICC*(3,*k*) is informative with respect to the reliability of the measures used in these analyses. To compute these, *ICC*s requires treating raters as fixed effects. This makes sense because, in studies of nonhuman animals housed in zoos, research facilities, sanctuaries, and even in human homes, the raters have expertise in working with nonhuman animals and/or know these animals better than anybody else. They are not a random sample drawn from the entire universe of possible raters. One question that often arises when estimating interrater reliabilities in samples derived from multiple facilities, as is often the case in these studies, is whether to obtain variance components with or without taking into account the fact that targets (and raters) are nested within facilities. Although I began by taking this nesting into account (Weiss et al. 2006, 2009), I no longer do so (e.g., Morton et al. 2013). This change in habit came about following an insight: so long as each animal and each rater have a unique ID, when zookeepers and other personnel at a facility rate only the animals at their facility, the nesting is already taken into account. Of course, if some animals have lived in and have been rated at two or more facilities, one should take nesting into account.

The second step is to screen out items with interrater reliabilities that are zero or negative. I do not test whether the interrater reliabilities are significantly different from zero or use some higher interrater reliability as a cut-off point. This is a shockingly liberal criterion to some, and it has been the source of criticism by reviewers. So why use such a liberal cut-point? The main reason for being such 'wide-eyed liberals' when it comes to interrater reliabilities is that intraclass correlations are ratios of the true score variance, which comes from the subject, to the total variance, which comprises the subject variance and the error variance, the latter being the subject $\times$ rater interaction (Shrout and Fleiss 1979). As such, it is possible that these low interrater reliabilities indicate that the trait assessed by an item is rare, i.e., something seen infrequently in few animals, and not that it is unimportant (see Stevenson-Hinde and Hinde 2011 for a discussion). An example taken from another field illustrates this point. Researchers in genetic epidemiology often estimate the heritability of various diseases. Heritability estimates are ratios of the amount of variation attributable to genetic effects over the total amount of variation, and so resemble interrater reliabilities. In fact, both are special cases within generalizability theory (Shavelson et al. 1989). The heritability of stomach cancer is ~1% (Czene et al. 2002). The authors of the paper did not take this to mean that genetic variation does not contribute to this trait. In fact, this study, which comprised ~9.6 million people (the whole of Sweden, in fact), found that the heritability was statistically significant. Thus, because of the fact that, without very large sample sizes, it is impossible to determine whether low interrater reliabilities are statistically significant, it is probably best to retain these items until they are shown to be otherwise and to not throw out possibly meaningful traits at this early stage. Another reason to accept such low interrater reliabilities is that, if they are random error, they are unlikely to load onto any domains derived via principal components analysis or factor analysis in a meaningful way. A third reason is that a key determinant for how many subjects are needed in principal components analyses or factor analyses to obtain a stable structure is the ratio of items to factors: the higher the ratio, the fewer subjects are needed (see Table 2.1 in MacCallum et al. 1999). As such, setting a low cut-point ensures that principal components analyses and factor analyses of the HPQ and related questionnaires will reveal stable structures.

Another habit deserving of mention and explanation is our use of principal components analysis as our primary means of data reduction. This practice is (rightly I suspect) frowned upon by psychometricians because principal components analysis does not reduce the dataset by modeling the correlations between variables as being caused by one or more latent variables, but by finding linear composites of the variables in the analysis (Widaman 2007). I began using principal components analysis based on advice I received from a senior researcher at a poster session in 1999. It became a habit after I began using an additional method to determine how many factors or components were present in a sample. Specifically, I initially used a scree plot and my judgement of whether factors or components "made sense" to determine how many dimensions to extract. Of course, this widespread practice is somewhat subjective, and has led to disagreements about the number of human personality domains (see the exchange in Eysenck 1992; Zuckerman et al. 1991). To get around this subjectivity, I began to use parallel analysis in addition to these other methods to determine the number of dimensions. In parallel analysis, one conducts factor analyses or principal components analyses on several sets of randomly generated datasets that are identical to the real dataset in terms of the number of items and the number of subjects (Horn 1965). The eigenvalues, which indicate how much variance each factor or component accounts for, derived from the real data are then compared to the distribution of eigenvalues generated from the randomly generated data (Horn 1965). Factors or components with eigenvalues that fall below a cut-off (usually the 95th percentile) are discarded. Early on I noticed that, parallel analysis, conducted using code provided by O'Connor (2000), when applied to factor analysis, typically recommended one or two more factors than did the scree plot. These factors were also not interpretable, for instance, they included only a single item. This did not happen when I used principal components analyses. Little did I know then that I was not the first person to observe this phenomenon, nor that it was attributable to a specific method for extracting dimensions in factor analysis (Buja and Eyuboglu 1992). Since discovering that, using the fa.parallel function in the psych package (Revelle 2015) and setting the option "SMC" (estimate communalities by using squared multiple correlations) to "False," I can avoid the problem of trivial factors, I have begun leaning toward factor analysis as my primary method for data reduction. However, I plan to still report the results of principal components analyses, which, in any event, tend to be similar to those derived from factor analyses (Velicer 1977).

My final habit deserving of mention concerns how I interpret the personality domains, that is, how do I decide what they mean and what to label them? To do so, I first apply an orthogonal (usually varimax) and oblique (usually promax) rotation to the factors or components. In a promax rotation, factors or components are allowed to correlate. If the correlations between factors or components are high and/or lead to loadings on one or more factors or components that differ from loadings derived from a varimax rotation, where the correlations are fixed at zero, I interpret the results of the promax rotation. I otherwise interpret the results of the varimax rotation. In all of the HPQ-based studies that I have conducted to date, the mean of the absolute correlations between factors or components has been low (less than 0.2). Moreover, the factors or components that result from these different rotations are close to identical. Nonetheless, I now present or provide the results of both rotations. Next, depending on the study species, when interpreting the factors or components and assigning labels to them, I rely on all available information, including, but not limited to, what is known about the species' socioecology and the personality structures of closely related species (e.g., Weiss et al. 2015) or of species where there is evidence of convergent evolution in behavior (e.g., Morton et al. 2013). Also, where available, correlations of the personality domains with behaviors are another source of information that I use to label personality domains (e.g., Morton et al. 2013). Finally, as a principle, I avoid labels that are too specific and, whenever possible, use the same label for factors or components that are similar to those of other species, and especially other closely related species. I follow these principles because they acknowledge the fact that personality dimensions and similar constructs do not have clearly defined boundaries (Adams et al. 2015; King and Weiss 2011) and because I want to avoid the confusion generated in a literature where many different terms are used to refer to what is possibly the same construct.

After interpreting and labeling factors, I estimate the interrater and, if data are available, retest reliabilities of these domains. I then turn my attention to addressing the questions described in other chapters within this volume. Before describing the results of attempts of my research tradition to solve empirical problems, I will spell out what this research tradition does and does not assume, which I hope will clarify some misconceptions.

Assumptions and Non-assumptions

The first assumption is that the factors or components derived from the HPQ (or any other set of traits, regardless of how measured) are biologically relevant and can be used to better understand the functions of those traits. It follows, then, that the factors or components revealed in rating studies are believed to not be implicit theories of personality based on folk psychology, artifacts of the questionnaire, such as the semantic similarity of the items (see Kenrick and Funder 1988 for a summary of these criticisms) or, because we are studying animal personality, anthropomorphic projections (Uher 2011).

This assumption and its antecedents, and especially that concerning anthropomorphic projection, has been used as a bludgeon by reviewers and, more gently (and genteelly), and usually over drinks, by other researchers, to criticize animal personality research based on ratings. From what I can gather, this criticism stems from two sources. The first is the HPQ's origins in the Big Five or Five-Factor Model traditions of human personality research. The second is an aversion to studying animal behavior by means other than observation that is held by ethologists and learning theorists. This aversion is often framed in terms of preferring measures that are "objective" to those that are "subjective." The origins of this aversion can be traced to the founders of modern ethology and learning theory who reacted against the relative lack of rigor of their predecessors (Wynne 2005).

However, inspection of the bases of this criticism suggests that this criticism is lacking. Let us examine the concerns raised by the origin of the HPQ and related instruments. First, there appears to be a lack of awareness or a deliberate ignoring of the reasons for the HPQ's origins and the fact that care was taken to insure that it was appropriate for assessing nonhuman primate personality (both described earlier in this chapter).

Second, large studies of human personality find that the personality factors measured by questionnaires are found in many human cultures (Eysenck and Eysenck 1985; Ion et al. 2017; McCrae et al. 2005; Schmitt et al. 2007) and that the factors or components that emerge from questionnaires are attributable to common genetic influences on sets of traits and not implicit biases about the associations between traits (McCrae et al. 2001; Yamagata et al. 2006) or their semantic similarity (Rowe 1982). Closer to the theme of this chapter, a study of dogs, *Canis familiaris* (Kwan et al. 2008), and a study of chimpanzees and orangutans (Weiss et al. 2012) found no evidence that anthropomorphic projection adversely influenced ratings.

Third, studies using the CPQ and the HPQ have not yielded the Five-Factor Model, even in our closest animal relations, chimpanzees and bonobos (King and Figueredo 1997; Weiss et al. 2009; Weiss et al. 2007; Weiss et al. 2015). A study of spotted hyenas (*Crocuta crocuta*) that used an adaptation of a questionnaire based on a human personality scale (the Big Five Mini-markers) also failed to find the human Big Five (Gosling 1998).

Fourth, like other human observations, behavioral codings are susceptible to error and bias. For example, the meaning of behaviors can be misinterpreted (Carter et al. 2012a, b). The distinction between the objective and subjective is therefore a false dichotomy. This statement should not be controversial. Researchers in animal behavior acknowledge that their measures are not without bias or error and take steps to mitigate the problem, such as video recording subjects, asking independent observers to code videos, and computing interobserver reliabilities. There have also recently been recommendations for further steps that researchers should take to avoid misinterpreting behavioral traits (Carter et al. 2012a, b).

Given that interrater or interobserver reliability estimates reflect the degree to which traits or behaviors are easy or difficult to assess, it should be possible to determine whether one approach is more "objective" than another. I do not have data in hand that would enable me to compare behavioral observations and ratings directly. However, analyses of interrater reliability estimates reported in a study of crab-eating macaques (*M. fascicularis*) measured at two time points (Uher et al. 2013a, b) and from a study of brown capuchin monkeys (*Sapajus apella*) measured once (Uher and Visalberghi 2016) are informative. In both studies, the researchers measured personality traits via so-called "trait-adjective items" or so-called "behavior-descriptive verb items." The behavior-descriptive verb items, although not being direct observations of behavior, were generated from behavioral observations and so were designed to be "less colloquial" than trait adjectives (Uher et al. 2013a, b, p. 429). To highlight the differences between these types of items, for example, impulsiveness in crab-eating macaques was measured by the trait-adjective item "*Name* is impulsive." (Uher 2016d, p. 2) and by the behavior-descriptive verb item "When he/she does not like something, *Name* shakes trees or jumps on or slaps others." (Uher 2016c, p. 2). The results of my analyses did not support the view that behavioral observations are more objective than trait ratings. Briefly, for crab-eating macaques at time 1, the mean *ICC* (3,1) of behavior-descriptive verb ratings (0.43) was higher than that of trait-adjective ratings (0.39), but Welch's t test revealed that the difference was not significant, $t_{32.279} = -0.72$, $p =$ 0.48 (Uher 2016c, d). For the macaque ratings at time 2, the mean *ICC*(3,1) of behavior-descriptive verb ratings (0.40) was lower than that of trait-adjective ratings (0.43), but this difference was also not significant, $t_{33.651} = 0.50$, $p = 0.62$

(Uher 2016c, d). Finally, for brown capuchin monkeys, the mean *ICC*(3,1) of behavior-descriptive verb ratings (0.39) was also lower than that of trait-adjective ratings (0.43), and, once again, this difference was not significant, $t_{37.922} = 0.62$, $p = 0.54$ (Uher 2016a, b). The code and *ICC* data for these analyses can be found at https://github.com/alexweissuk/uher_icc_comparison.

Another assumption of the research tradition that I described is that similar behaviors in closely related species serve similar functions and have similar genetic, physiological, and developmental underpinnings. This assumption is central to research that seeks to understand animal behavior, and has yielded insights into, the evolutionary bases of, among other things, human culture (Whiten et al. 1999).

In addition to holding these (and probably) other assumptions, there are things that this tradition does not assume. For one, it does not assume that personality traits or domains serve similar functions, or have similar genetic, physiological, and developmental bases, even in closely related species. In other words, we do not assume that the same traits or domains in different species are homologues. After all, if we did, research on a species would stop after determining that species' personality structure, and the research tradition would be a descriptive enterprise (Braithwaite 1968). What we have instead is research examining this question, which has yielded encouraging results with respect to whether similar personality domains in closely related species are, in fact, homologues.

In describing these studies and their results I will first return to the findings on personality domains and subjective well-being. To recap, there is a trend across primate species, including humans, for more emotionally stable and social individuals to experience higher levels of subjective well-being compared to those who are less emotionally stable and less social. Furthermore, in chimpanzees, this association is largely attributable to common genetic effects (Weiss et al. 2002). Starting in 2008, studies of human twins (Weiss et al. 2008) and siblings (Hahn et al. 2013), and a study of interrelated orangutans (Adams et al. 2012) all found that personality and measures related to subjective well-being share common genetic underpinnings. Finally, in humans, polygenic scores (weighed sums indicating how many single nucleotide polymorphisms for a trait an individual has) for extraversion were associated with higher life satisfaction and subjective well-being and polygenic scores for neuroticism were associated with lower subjective well-being (Weiss et al. 2016).

Additional evidence comes from neurophysiological and molecular genetic studies. MRI studies of chimpanzees rated on the CPQ and a related questionnaire identified associations between areas of the chimpanzee brain and personality domains that are similar to the associations expected based on human findings (Blatchley and Hopkins 2010; Latzman et al. 2015a, b). Concerning molecular genetic evidence, CPQ and HPQ studies of arginine vasopressin 1a receptor polymorphisms and personality in chimpanzees (Hopkins et al. 2012; Wilson et al. 2016) and in bonobos (Staes et al. 2016) are consistent with human findings (Bachner-Melman et al. 2005; Pappa et al. 2016) and point to an association between arginine vasopressin 1a receptor polymorphisms and socially appropriate behaviors.

Studies of development also support the likelihood that similar personality domains are homologous. Specifically, with some exceptions, which I will discuss in the next section, at least in the case of chimpanzees (King et al. 2008) and orangutans (Weiss and King 2015), sex and age differences in similar domains are similar to one another and to sex differences (Costa et al. 2001) and age differences (Roberts et al. 2008) found in comparable human personality domains.

Some things that this research tradition does not assume have not been studied. Although I will not present an exhaustive catalog here, some are worth noting. For one, the research tradition that I described does not assume that different domains or traits in distantly related species have different functions. In other words, different personality domains in different species could serve the same function in those two species. Second, this work does not assume that personality traits are isomorphic with behaviors, that is, there is no assumption that each trait that makes up a personality domain corresponds to a single behavior. A third matter concerns personality structure. Specifically, there is no assumption as to what the personality structure of a given species will be or should be.

Finally, this research tradition does not assume that the labels or even definitions of the components or factors are immutable. Instead, they are better characterized as hypotheses. Data collected in the future on revised versions of the HPQ, behavioral observations, or on the genetics, physiology, and development of these domains, will test these hypotheses, and some revisions will no doubt be necessary. One area in which this might happen would involve the domains related to competitive prowess that have been identified in nonhuman primates but not in humans. In particular, to recognize species differences in social organization, my collaborators and I have labeled domains similar to the chimpanzee dominance "assertiveness" in brown capuchin monkeys (Morton et al. 2013) and in bonobos (Weiss et al. 2015). Should future studies not support this distinction by, for example, showing that all such domains underlie similar behaviors and are related to genes with similar functions, then, depending on these findings, it would make sense to rename the capuchin and bonobo domains, the chimpanzee domains, or both.

Evaluating This Research Tradition

Having discussed what this research tradition does and does not assume, it is worth asking whether it has been successful in solving empirical problems important to personality research. This is an important question as research traditions stand or fall on this criteria, and even a technically sophisticated research tradition will be abandoned or changed if it cannot solve the problems in its domain (Laudan 1977).[3]

When it comes to characterizing species' personalities, a straightforward question is whether the structure of the traits measured using the above-described approach resembles the structure of traits measured using other approaches. As I noted earlier, questionnaires developed based on other conceptions of personality (e.g., Dutton 2008) or by other approaches (Freeman et al. 2013) yield chimpanzee personality structures similar to those found using the CPQ and HPQ. More striking evidence comes from a study of brown capuchin monkeys rated on trait-adjective items and behavior-descriptive verb items, the latter, recall, being identified from observations of naturally occurring behaviors (Uher and Visalberghi 2016). A factor analysis of these ratings yielded personality domains similar to those identified from principal components analysis or factor analysis of HPQ ratings (Morton, et al. 2013).

Other evidence suggests that scales such as the CPQ, OPQ, and HPQ characterize biologically meaningful relations between traits. The first comes from a factor analysis of ratings of crab-eating macaques on trait adjective descriptors and behavioral verb descriptors identified from observations of naturally occurring behaviors. The analysis yielded dimensions labeled playful-active-curious, aggressive-competitive, prosocial-gregarious, and assertive-nonanxious (Uher et al. 2013a, b, p. 657). There has not yet been, at least to my knowledge, a study of crab-eating macaques using the HPQ. However, crab-eating macaques are a relatively despotic macaque species (Thierry 2000) and, in an earlier paper, my colleagues and I reported on the personality structures of Japanese (*M. fuscata*), Assamese (*M. assamensis*), Barbary macaques

[3]The question of whether this research tradition is as good, better, or poorer at solving conceptual problems (Laudan 1977) than some other research tradition is beyond the scope of this chapter.

(*M. sylvanus*), Tonkean macaques (*M. tonkeana*), and crested (*M. nigra*) macaques, all rated using the HPQ (Adams et al. 2015). One of the species investigated, Assamese macaques, have a level of despotism similar to that of crab-eating macaques (Thierry 2000). Analysis of the HPQ ratings found personality domains labeled confidence, activity, openness, friendliness, and opportunism (Adams, et al. 2015), and so, with the exception of the separate activity and openness domains, the Assamese macaque structure was similar to that found in crab-eating macaques by Uher (2013b). Finally, a study of Hanuman langurs (*Semnopithecus entellus*) compared the personality structure based on ratings on an offshoot of the CPQ to that based on behavioral observations, and found them to be similar (Konečná et al. 2008).

Moreover, as I noted in detail earlier, in addition to identifying similar personality structure to other methods, the research tradition described here has led to greater understanding of the genetic and neurophysiological mechanisms that underpin personality in chimpanzees and bonobos. Likewise, as also discussed in detail earlier in this chapter, this research found that the associations between personality traits and affect described in humans, nonhuman primates, and in other species likely reflect their genetic architecture.

Moving to the ability to solve other problems, studies of sex and age differences in chimpanzees and in orangutans have contributed to the understanding of the evolutionary origins of these differences. In addition, these studies have informed debates concerning whether biological or sociocultural processes drive age differences in human personality.

Personality sex differences actually differ some between chimpanzees, humans, and orangutans. Cross-cultural studies of humans have found that, although the magnitude of the difference varies across countries, compared to men, women are higher in neuroticism, extraversion, openness, agreeableness, and conscientiousness (Costa et al. 2001). In chimpanzees, however, the only sex differences consistent with the human pattern were for conscientiousness and agreeableness; males and females did not differ in extraversion or in openness and males were higher in neuroticism than females (King et al. 2008). This finding together with findings suggesting that an activity facet of extraversion was higher in males than in females, and so also differing from the human pattern, and a pattern of age differences suggesting that males but not females exhibit a prolonged period of aggression, led us to hypothesize that chimpanzee personality differences partly reflect the higher levels of aggression in male chimpanzees than in humans (King et al. 2008). Later findings that, in orangutans, sex differences in neuroticism were like those in humans (Weiss and King 2015), supported this hypothesis as male orangutans do not display the same intensity of aggression that is found in male chimpanzees.

With respect to age differences, among humans, lifetime changes in personality suggest that individuals become more mature as they age, with individuals declining in neuroticism, extraversion, and openness, but increasing in agreeableness and conscientiousness (Roberts et al. 2008). This appears to be a universal phenomenon, having been found in several cultures (Bleidorn et al. 2013; McCrae et al. 2005). One explanation for these trajectories is that they are products of biological maturation (McCrae and Costa 2003). An alternative explanation for these trajectories is that they are products of social and cultural forces, such as the need for increased responsibility when entering the workforce or increased desire for caretaking when having children (Roberts et al. 2005). A study of chimpanzee personality found that the pattern of age differences for the five human-like domains is conserved, although, among males, there is a less consistent decline of neuroticism and extraversion (King et al. 2008). This finding, too, was attributed to the pattern of aggression among male chimpanzees. Moreover, among orangutans, extraversion, neuroticism, and agreeableness, domains that orangutans share with humans and chimpanzees, the age differences were like those of humans except for agreeableness, which was lower in older individuals (Weiss and King

2015). Finally, both studies found that, despite differences in how their societies are structured, after adjusting for the different rate of development in humans and the two great ape species, the magnitudes of age differences were similar across all three species (King et al. 2008; Weiss and King 2015).

The findings concerning age differences in two species that differ in many ways from humans are more consistent with what one would expect if the development of personality resembled that of height and other physical characteristics than what one would expect if human sociocultural forces were responsible. However, this research also highlights the possible functions of some of these traits in these species. As noted above, the pattern of development for extraversion and neuroticism in male chimpanzees is consistent with the higher aggression noted in male chimpanzees. Furthermore, the decline in orangutan agreeableness suggests that the increases found in humans and chimpanzees may be adaptive for individuals that live in large groups of, at best, distantly related individuals (Weiss and King 2015). In other words, among species that are solitary or semisolitary such as orangutans (Galdikas 1985), there is no selection for age-related rises in agreeableness.

Studying personality and reproductive success among captive individuals is problematic because, in these settings, reproduction is controlled so as to prevent inbreeding and overpopulation and to maximize the genetic diversity of the species (see Watters et al., this volume). The HPQ has also only seldom been deployed to study wild populations, and these studies have, for the most part, been cross-sectional. As such, there is no definitive answer to whether these personality domains are related to reproductive success. However, studies have examined the association between personality domains assessed using the HPQ and related measures, and dominance rank, an important asset in the social lives of nonhuman primates that is associated with fitness outcomes, including reproductive success (see, e.g., Pusey et al. 1997).

Evidence that the HPQ and related measures capture precursors of dominance rank includes the ubiquity of domains related to dominance rank in nonhuman primates that the HPQ, like other rating scales, unveils. Further evidence includes findings that the social organization of nonhuman primate species is reflected in the make-up of personality domains related to dominance rank. For example, a study of six macaque species revealed that the degree of despotism that characterizes each species is associated with the structure of personality traits related to aggression (e.g., "bullying") and social competence (e.g., "independent") in those species (Adams et al. 2015). There is some suggestion for a similar phenomenon among African apes. Specifically, if one examines the cardinal CPQ, OPQ, or HPQ traits associated with dominance rank, i.e., "dominant" and "submissive," one sees that they are more strongly related to the dominance-like domains in captive chimpanzees (King and Figueredo 1997; Weiss et al. 2009, 2007) and captive western lowland gorillas, *Gorilla gorilla gorilla* (Schaefer and Steklis 2014), than they are in captive bonobos (Weiss, et al. 2015) or in wild Virunga mountain gorillas, *G. beringei beringei* (Eckardt et al. 2015).

Moreover, consistent with the fact that, unlike other great apes, female bonobos exhibit so-called "partial female dominance" (see, e.g., Furuichi 2011), female bonobos score higher than males in assertiveness (Staes et al. in press) whereas in chimpanzees (King et al. 2008; Weiss et al. 2009, 2007),[4] orangutans (Weiss and King 2015), and mountain gorillas (Eckardt et al. 2015), males score higher. The previously discussed study of spotted hyenas, another species in which females are higher ranking than males, also found that females were more assertive (Gosling 1998).

This research tradition has also identified direct evidence of the role of dominance-like personality domains. A study of provisioned, semi-free-ranging Barbary macaques that assessed personality using an offshoot of the CPQ found that the temporal stability of dominance

[4]In the Weiss et al. 2009 study, the sex effect was not significant ($p = .0694$), but the direction of the effect was comparable to what had been found in the other studies.

rank was attributable to the personality domain, confidence (Konečná et al. 2012). In addition, among wild Virunga mountain gorillas, the dominance domain was associated with dominance strength and spending less time staring at other gorillas (Eckardt et al. 2015), and in zoo- and laboratory-housed brown capuchin monkeys, individuals who were higher in assertiveness were groomed more often, less often solitary, and less likely to be targets of aggression (Morton et al. 2013). Similarly, among bonobos, individuals higher in assertiveness were groomed more often, less likely to be targets of aggression, and less likely to retreat from agonistic encounters (Staes et al. in press).

One area that perhaps illustrates the effectiveness of this research tradition more than any other is its ability to uncover the phylogenetic origins of personality traits (Gosling and Graybeal 2007). The best example comes from work on conscientiousness. As I previously mentioned, an early review noted that, of the species studied, conscientiousness was found only in chimpanzees. Based on this, the authors of the review reasonably hypothesized that conscientiousness emerged relatively recently in African apes (Gosling and John 1999). However, with the exception of King and Figueredo's 1997 study of chimpanzees, the personality measures used in the other studies cited by Gosling and John did not include many traits related to self-control, behavioral predictability, persistence, and other markers of conscientiousness. Fortunately, testing these hypotheses is a simple matter of assessing multiple related species using the HPQ. Doing so revealed that conscientiousness and/or a related trait, attentiveness, exists in captive western lowland gorillas (Schaefer and Steklis 2014), captive bonobos (Weiss et al. 2015), and possibly exists in wild Virunga mountain gorillas (Eckardt et al. 2015). Using the HPQ to unearth the phylogenetic origins of domains like conscientiousness or attentiveness has also revealed an attentiveness domain in brown capuchin monkeys (Morton et al. 2013) and, using a modified version of the HPQ, in common marmosets, *Callithrix jacchus* (Iwanicki and Lehmann 2015), but not in Old World monkeys (Adams et al. 2015; Konečná et al. 2008; Konečná et al. 2012).[5] These studies therefore indicate that, conscientiousness-like domains emerged in the African apes, but also independently emerged in some species of platyrrhines. Moreover, because conscientiousness was found in common marmosets, there is a need to re-evaluate whether domains like conscientiousness emerge only in species that are cognitively and socially complex (King and Weiss 2011).

All in all, then, these studies have uncovered personality domains that appear to be species-specific adaptations, highlight the functional significance of traits, and allow one to address questions about their proximate and ultimate origins. In a real sense, then, in the case of personality, this research tradition is capable of addressing Tinbergen's four questions about animal behavior (2005).

Conclusion

Before I conclude I want to stress that, as I have noted elsewhere (Weiss and Adams 2013) I do not think of this research tradition as fixed or the only way to study animal personality. It will change with the advent of problems that its present form has difficulties solving and other research traditions are bound to be equally good at solving problems. As I stated from the outset, I merely wanted to describe my own tradition and to evaluate it. Others will probably differ in their judgement.

Ultimately, the approach used should be suitable for the questions researchers seek to address. For example, Brent et al. (2014), Sussman et al. (2014) have, to good ends, used ethological approaches to address questions concerning the genetics, evolution, and development of macaque personality. It is thus important to remember that it would be foolhardy to use the HPQ without modifications (or at all) to address questions about species other than haplorrhine primates (see Koski 2011 and

[5]In my mind, whether orangutans have something like a conscientiousness or attentiveness domain is unresolved.

references therein). Moreover, given recent findings on wild bonobos (Garai et al. 2016), it is unclear whether HPQ traits are defined by the same structure in wild and captive populations (see also Gurven et al. 2013 for the case of humans). As such, until more is known, or a revision of the HPQ addresses any discrepancies, it would be premature to draw strong conclusions about species differences when they are confounded by whether the samples live in the wild, are semi-free-ranging, or are housed in captivity.

Finally, the study of Hanuman langurs found very strong correlations between personality domains and behavioral indices generated from the raw behavioral observation data (see Table A1 in Konečná, et al. 2008). Similarly, recent research on humans has shown machine learning algorithms can be used to measure personality via individuals' Facebook "Likes" (Youyou et al. 2015). These findings suggest that each questionnaire item or personality domain does not necessarily correspond to a single discrete behavior. They also recommend the possible gains that can be made by using machine learning approaches to study associations between personality and behavior provided the sample sizes are large enough.

As I noted at the beginning of this chapter, this research tradition has met with criticism, especially with regards to its origins in the traditions of human personality research. However, the history of animal behavior includes prominent examples of successful research traditions that directly sprung from human psychology (e.g., Harding et al. 2004). As such, the admonishment from some that human personality research is an inappropriate starting place for studying the personalities of other members of our primate lineage strikes me as not just shortsighted, but downright fanciful.

References

Adams, M. J., King, J. E., & Weiss, A. (2012). The majority of genetic variation in orangutan personality and subjective-well being is nonadditive. *Behavior Genetics, 42,* 675–686. doi:10.1007/s10519-012-9537-y

Adams, M. J., Majolo, B., Ostner, J., Schuelke, O., De Marco, A., Thierry, B., et al. (2015). Personality structure and social style in macaques. *Journal of Personality and Social Psychology, 109,* 338–353. doi:10.1037/pspp0000041

Bachner-Melman, R., Zohar, A. H., Bacon-Shnoor, N., Elizur, Y., Nemanov, L., Gritsenko, I., et al. (2005). Link between vasopressin receptor AVPR1A promoter region microsatellites and measures of social behavior in humans. *Journal of Individual Differences, 26,* 2–10. doi:10.1027/1614-0001.26.1.2

Blatchley, B. J., & Hopkins, W. D. (2010). Subgenual cingulate cortex and personality in chimpanzees (*Pan troglodytes*). *Cognitive, Affective and Behavioral Neuroscience, 10,* 414–421. doi:10.3758/CABN.10.3.414

Bleidorn, W., Klimstra, T. A., Denissen, J. J. A., Rentfrow, P. J., Potter, J., & Gosling, S. D. (2013). Personality maturation around the world: A cross-cultural examination of social-investment theory. *Psychological Science*. doi:10.1177/0956797613498396

Braithwaite, R. B. (1968). *Scientific explanation: A study of the function of theory, probability, and law in science*. Cambridge: Cambridge University Press.

Brent, L. J. N., Semple, S., MacLarnon, A., Ruiz-Lambides, A., Gonzalez-Martinez, J., & Platt, M. J. (2014). Personality traits in rhesus macaques (*Macaca mulatta*) are heritable but do not predict reproductive output. *International Journal of Primatology, 35,* 188–209. doi:10.1007/s10764-013-9724-6

Buja, A., & Eyuboglu, N. (1992). Remarks on parallel analysis. *Multivariate Behavioral Research, 27,* 509–540. doi:10.1207/s15327906mbr2704_2

Carter, A. J., Feeney, W. E., Marshall, H. H., Cowlishaw, G., & Heinsohn, R. (2012a). Animal personality: What are behavioural ecologists measuring? *Biological Reviews, 88,* 465–475. doi:10.1111/brv.12007

Carter, A. J., Marshall, H. H., Heinsohn, R., & Cowlishaw, G. (2012b). How not to measure boldness: Demonstration of a jingle fallacy in a wild social primate. *Animal Behaviour, 84,* 603–609. doi:10.1016/j.anbehav.2012.06.015

Costa, P. T., Jr., & McCrae, R. R. (1992). *Revised NEO Personality Inventory (NEO-PI-R) and NEO Five-Factor Inventory (NEO-FFI) professional manual*. Odessa, FL: Psychological Assessment Resources.

Costa, P. T., Jr., Terracciano, A., & McCrae, R. R. (2001). Gender differences in personality traits across cultures: Robust and surprising findings. *Journal of Personality and Social Psychology, 81,* 322–331. doi:10.1037//0022-3514.81.2.322

Czene, K., Lichtenstein, P., & Hemminki, K. (2002). Environmental and heritable causes of cancer among 9.6 million individuals in the Swedish family-cancer database. *International Journal of Cancer, 99,* 260–266. doi:10.1002/ijc.10332

Digman, J. M. (1990). Personality structure: Emergence of the Five-Factor Model. *Annual Review of Psychology, 41,* 417–440. doi:10.1146/annurev.ps.41.020190.002221

Dingemanse, N. J., & Wolf, M. (2010). Recent models for adaptive personality differences: a review. *Philosophical Transactions of the Royal Society B Biological Sciences, 365,* 3947–3958. doi:10.1098/rstb.2010.0221

Dutton, D. M. (2008). Subjective assessment of chimpanzee (*Pan troglodytes*) personality: Reliability and stability of trait ratings. *Primates, 49,* 253–259. doi:10.1007/s10329-008-0094-1

Eckardt, W., Steklis, H. D., Steklis, N. G., Fletcher, A. W., Stoinski, T. S., & Weiss, A. (2015). Personality dimensions and their behavioral correlates in wild Virunga mountain gorillas (*Gorilla beringei beringei*). *Journal of Comparative Psychology, 129,* 26–41. doi:10.1037/a0038370

Eysenck, H. J. (1992). Four ways five factors are not basic. *Personality and Individual Differences, 13,* 667–673. doi:10.1016/0191-8869(92)90237-J

Eysenck, H. J., & Eysenck, M. W. (1985). *Personality and individual differences*. New York: Plenum Press.

Freeman, H. D., Brosnan, S. F., Hopper, L. M., Lambeth, S. P., Schapiro, S. J., & Gosling, S. D. (2013). Developing a comprehensive and comparative questionnaire for measuring personality in chimpanzees using a simultaneous top–down/bottom–up design. *American Journal of Primatology, 75,* 1042–1053. doi:10.1002/ajp.22168

Furuichi, T. (2011). Female contributions to the peaceful nature of bonobo society. *Evolutionary Anthropology, 20,* 131–142. doi:10.1002/evan.20308

Galdikas, B. M. F. (1985). Orangutan sociality at Tanjung-Puting. *American Journal of Primatology, 9,* 101–119. doi:10.1002/ajp.1350090204

Garai, C., Weiss, A., Arnaud, C. M., & Furuichi, T. (2016). Personality in wild bonobos (*Pan paniscus*). *American Journal of Primatology, 78,* 1178–1189. doi:10.1002/ajp.22573

Goldberg, L. R. (1990). An alternative "description of personality": The Big-Five factor structure. *Journal of Personality and Social Psychology, 59,* 1216–1229. doi:10.1037/0022-3514.59.6.1216

Gosling, S. D. (1998). Personality dimensions in spotted hyenas (*Crocuta crocuta*). *Journal of Comparative Psychology, 112,* 107–118. doi:10.1037/0735-7036.112.2.107

Gosling, S. D., & Graybeal, A. (2007). Tree thinking: A new paradigm for integrating comparative data in psychology. *The Journal of General Psychology, 134,* 259–277. doi:10.3200/GENP.134.2.259-278

Gosling, S. D., & John, O. P. (1999). Personality dimensions in nonhuman animals: A cross-species review. *Current Directions in Psychological Science, 8,* 69–75. doi:10.1111/1467-8721.00017

Gurven, M., von Rueden, C., Massenkoff, M., Kaplan, H., & Lero Vie, M. (2013). How universal is the Big Five? Testing the five-factor model of personality variation among forager-farmers in the Bolivian Amazon. *Journal of Personality and Social Psychology, 104,* 354–370. doi:10.1037/a0030841

Hahn, E., Johnson, W., & Spinath, F. M. (2013). Beyond the heritability of life satisfaction: The roles of personality and twin-specific influences. *Journal of Research in Personality, 47,* 757–767. doi:10.1016/j.jrp.2013.07.003

Harding, E. J., Paul, E. S., & Mendl, M. (2004). Animal behaviour: Cognitive bias and affective state. *Nature, 2004,* 312. doi:10.1038/427312a

Hopkins, W. D., Donaldson, Z. R., & Young, L. J. (2012). A polymorphic indel containing the RS3 microsatellite in the 5′ flanking region of the vasopressin V1a receptor gene is associated with chimpanzee (*Pan troglodytes*) personality. *Genes, Brain, and Behavior, 11,* 552–558. doi:10.1111/j.1601-183X.2012.00799.x

Horn, J. L. (1965). A rationale and test for the number of factors in factor analysis. *Psychometrika, 30,* 179–185. doi:10.1007/BF02289447

Ion, A., Iliescu, D., Aldhafri, S., Rana, N., Ratanadilok, K., Widyanti, A., et al. (2017). A cross-cultural analysis of personality structure through the lens of the HEXACO model. *Journal of Personality Assessment, 99,* 25–34. doi:10.1080/00223891.2016.1187155

Iwanicki, S., & Lehmann, J. (2015). Behavioral and trait rating assessments of personality in common marmosets (*Callithrix jacchus*). *Journal of Comparative Psychology, 129,* 205–217. doi:10.1037/a0039318

Kenrick, D. T., & Funder, D. C. (1988). Profiting From controversy: Lessons from the person-situation debate. *American Psychologist, 43,* 23–34. doi:10.1037//0003-066X.43.1.23

King, J. E., & Figueredo, A. J. (1997). The Five-Factor Model plus dominance in chimpanzee personality. *Journal of Research in Personality, 31,* 257–271. doi:10.1006/jrpe.1997.2179

King, J. E., & Landau, V. I. (2003). Can chimpanzee (*Pan troglodytes*) happiness be estimated by human raters? *Journal of Research in Personality, 37,* 1–15. doi:10.1016/S0092-6566(02)00527-5

King, J. E., & Weiss, A. (2011). Personality from the perspective of a primatologist. In A. Weiss, J. E. King, & L. Murray (Eds.), *Personality and temperament in nonhuman primates* (pp. 77–99). New York: Springer.

King, J. E., Weiss, A., & Farmer, K. H. (2005). A chimpanzee (*Pan troglodytes*) analogue of cross-national generalization of personality structure: Zoological parks and an African sanctuary. *Journal of Personality, 73,* 389–410. doi:10.1111/j.1467-6494.2005.00313.x

King, J. E., Weiss, A., & Sisco, M. M. (2008). Aping humans: Age and sex effects in chimpanzee (*Pan troglodytes*) and human (*Homo sapiens*) personality. *Journal of Comparative Psychology, 122,* 418–427. doi:10.1037/a0013125

Konečná, M., Lhota, S., Weiss, A., Urbánek, T., Adamová, T., & Pluháček, J. (2008). Personality in free-ranging Hanuman langur (*Semnopithecus entellus*) males: Subjective ratings and recorded behavior. *Journal of Comparative Psychology, 122,* 379–389. doi:10.1037/a0012625

Konečná, M., Weiss, A., Lhota, S., & Wallner, B. (2012). Personality in Barbary macaques (*Macaca sylvanus*): Temporal stability and social rank. *Journal of Research in Personality, 46,* 581–590. doi:10.1016/j.jrp.2012.06.004

Koski, S. E. (2011). How to measure animal personality and why does It matter? Integrating the psychological and biological approaches to animal personality. In M. Inoue-Murayama, S. Kawamura, & A. Weiss (Eds.), *From genes to animal behavior: Social structures, personalities, communication by color* (pp. 115–136). Tokyo: Springer.

Kuhn, T. S. (1970). *The structure of scientific revolutions* (2nd ed.). Chicago: University of Chicago Press.

Kwan, V. S. Y., Gosling, S. D., & John, O. P. (2008). Anthropomorphism as a special case of social perception: A cross-species social relations model analysis of humans and dogs. *Social Cognition, 26,* 129–142. doi:10.1521/soco.2008.26.2.129

Latzman, R. D., Freeman, H. D., Schapiro, S. J., & Hopkins, W. D. (2015a). The contribution of genetics and early rearing experiences to hierarchical personality dimensions in chimpanzees (*Pan troglodytes*). *Journal of Personality and Social Psychology, 109,* 889–900. doi:10.1037/pspp0000040

Latzman, R. D., Hecht, L. K., Freeman, H. D., Schapiro, S. J., & Hopkins, W. D. (2015b). Neuroanatomical correlates of personality in chimpanzees (*Pan troglodytes*): Associations between personality and frontal cortex. *Neuroimage, 123,* 63–71. doi:10.1016/j.neuroimage.2015.08.041

Laudan, L. (1977). *Progress and its problems: Towards a theory of scientific growth.* London: Routledge & Kegan Paul.

Leahey, T. H. (1992). The mythical revolutions of American psychology. *American Psychologist, 47,* 308–318. doi:10.1037/0003-066X.47.2.308

MacCallum, R. C., Widaman, K. F., Zhang, S., & Hong, S. (1999). Sample size in factor analysis. *Psychological Methods, 4,* 84–99. doi:10.1037/1082-989X.4.1.84

McCrae, R. R., & Costa, P. T., Jr. (1985). Updating Norman's "adequate taxonomy": Intelligence and personality dimensions in natural language and in questionnaires. *Journal of Personality and Social Psychology, 49,* 710–721. doi:10.1037/0022-3514.49.3.710

McCrae, R. R., & Costa, P. T., Jr. (1987). Validation of the five-factor model of personality across instruments and observers. *Journal of Personality and Social Psychology, 52,* 81–90. doi:10.1037/0022-3514.52.1.81

McCrae, R. R., & Costa, P. T., Jr. (1989). Different points of view: Self reports and ratings in the assessment of personality. In J. P. Forgas & J. M. Innes (Eds.), *Recent advances in social psychology: An international perspective* (pp. 429–439). New York: Elsevier.

McCrae, R. R., & Costa, P. T., Jr. (2003). *Personality in adulthood: A Five-Factor Theory perspective.* New York, NY: Guilford Press.

McCrae, R. R., Jang, K. L., Livesley, W. J., Riemann, R., & Angleitner, A. (2001). Sources of structure: Genetic, environmental, and artifactual influences on the covariation of personality traits. *Journal of Personality, 69,* 511–535. doi:10.1111/1467-6494.694154

McCrae, R. R., & John, O. P. (1992). An introduction to the Five-Factor Model and its applications. *Journal of Personality, 60,* 175–215. doi:10.1111/j.1467-6494.1992.tb00970.x

McCrae, R. R., Terracciano, A., & 78 Members of the Personality Profiles of Cultures Project. (2005a). Universal features of personality traits from the observer's perspective: Data from 50 cultures. *Journal of Personality and Social Psychology, 88,* 547–561. doi:10.1037/0022-3514.88.3.547

McCrae, R. R., Terracciano, A., & 79 Members of the Personality Profiles of Cultures Project. (2005b). Personality profiles of cultures: Aggregate personality traits. *Journal of Personality and Social Psychology, 89,* 407–425. doi:10.1037/0022-3514.89.3.407

McCrae, R. R., Zonderman, A. B., Bond, M. H., Costa, P. T., Jr., & Paunonen, S. V. (1996). Evaluating replicability of factors in the revised NEO personality inventory: Confirmatory factor analysis versus Procrustes rotation. *Journal of Personality and Social Psychology, 70,* 552–566. doi:10.1037/0022-3514.70.3.552

Morton, F. B., Lee, P. C., Buchanan-Smith, H. M., Brosnan, S. F., Thierry, B., Paukner, A., et al. (2013). Personality structure in brown capuchin monkeys (*Sapajus apella*): Comparisons with chimpanzees (*Pan troglodytes*), orangutans (*Pongo spp.*), and rhesus macaques (*Macaca mulatta*). *Journal of Comparative Psychology, 127,* 282–298. doi:10.1037/a0031723

O'Connor, B. P. (2000). SPSS and SAS programs for determining the number of components using parallel analysis and Velicer's MAP test. *Behavior Research Methods, Instruments, and Computers, 32,* 396–402. doi:10.3758/BF03200807

Pappa, I., St Pourcain, B., Benke, K., Cavadino, A., Hakulinen, C., Nivard, M. G., et al. (2016). A genome-wide approach to children's aggressive behavior: The EAGLE consortium. *American Journal of Medical Genetics Part B: Neuropsychiatric Genetics, 171,* 562–572. doi:10.1002/ajmg.b.32333

Pusey, A. E., Williams, J., & Goodall, J. (1997). The influence of dominance rank on the reproductive success of female chimpanzees. *Science, 277,* 828–831. doi:10.1126/science.277.5327.828

Réale, D., Reader, S. M., Sol, D., McDougall, P. T., & Dingemanse, N. J. (2007). Integrating animal temperament within ecology and evolution. *Biological Reviews, 82,* 291–318. doi:10.1111/j.1469-185X.2007.00010.x

Revelle, W. (2015). psych: Procedures for personality and psychological research (Version 1.2.12). Evanston, IL: Northwestern University. Retrieved from http://personality-project.org/r/psych.manual.pdf

Roberts, B. W., Wood, D., & Caspi, A. (2008). The development of personality traits in adulthood. In O. P. John (Ed.), *Handbook of personality: Theory and research* (3rd ed., pp. 375–398). New York, NY: Guilford Press.

Roberts, B. W., Wood, D., & Smith, J. L. (2005). Evaluating Five Factor Theory and social investment perspectives on personality trait development. *Journal of Research in Personality, 39,* 166–184. doi:10.1016/j.jrp.2004.08.002

Rosenthal, R., & Fode, K. L. (1963). The effect of experimenter bias on the performance of the albino-rat. *Behavioral Science, 8,* 183–189. doi:10.1002/bs.3830080302

Rosenthal, R., & Lawson, R. (1964). A longitudinal-study of the effects of experimenter bias on the operant learning of laboratory rats. *Journal of Psychiatric Research, 2,* 61–72. doi:10.1016/0022-3956(64)90003-2

Rowe, D. C. (1982). Monozygotic twin cross-correlations as a validation of personality structure: A test of the semantic bias hypothesis. *Journal of Personality and Social Psychology, 43,* 1072–1079. doi:10.1037/0022-3514.43.5.1072

Schaefer, S. A., & Steklis, H. D. (2014). Personality and subjective well-being in captive male western lowland gorillas living in bachelor groups. *American Journal of Primatology, 76,* 879–889. doi:10.1002/ajp.22275

Schmitt, D. P., Allik, J., McCrae, R. R., Benet-Martinez, V., Alcalay, L., Ault, L., et al. (2007). The geographic distribution of big five personality traits: Patterns and profiles of human self-description across 56 nations. *Journal of Cross-Cultural Psychology, 38,* 173–212. doi:10.1177/0022022106297299

Shavelson, R. J., Webb, N. M., & Rowley, G. L. (1989). Generalizability theory. *American Psychologist, 44,* 922–932. doi:10.1037/0003-066X.44.6.922

Shrout, P. E., & Fleiss, J. L. (1979). Intraclass correlations: Uses in assessing rater reliability. *Psychological Bulletin, 86,* 420–428. doi:10.1037/0033-2909.86.2.420

Sih, A., Bell, A. M., Johnson, J. C., & Ziemba, R. E. (2004). Behavioral syndromes: An integrative overview. *Quarterly Review of Biology, 79,* 241–277. doi:10.1086/422893

Staes, N., Eens, M., Weiss, A., & Stevens, J. M. G. (2017). Bonobo personality: age and sex differences and links with behavior and dominance. In: B. Hare & S. Yamamoto (Eds.), *Bonobos: Unique in mind, brain, and behavior.* Oxford: Oxford University Press.

Staes, N., Weiss, A., Helsen, P., Korody, M., Eens, M., & Stevens, J. M. G. (2016). Bonobo personality traits are heritable and associated with vasopressin receptor gene 1a variation. *Scientific Reports, 6,* 38193. doi:10.1038/srep38193

Stevenson-Hinde, J., & Hinde, C. A. (2011). Individual characteristics: Weaving psychological and ethological approaches. In A. Weiss, J. E. King, & L. Murray (Eds.), *Personality and temperament in nonhuman primates* (pp. 3–14). New York: Springer.

Sussman, A. F., Mates, E. A., Ha, J. C., Bentson, K. L., & Crockett, C. M. (2014). Tenure in current captive settings and age predict personality changes in adult pigtailed macaques. *Animal Behaviour, 89,* 23–30. doi:10.1016/j.anbehav.2013.12.009

Thierry, B. (2000). Covariation of conflict management patterns across macaque species. In F. Aureli & F. B. M. de Waal (Eds.), *Natural conflict resolution* (pp. 106–128). Berkeley, CA: University of California Press.

Tinbergen, N. (2005). On aims and methods of Ethology (Reprinted from Zeitschrift fur Tierpsychologie, vol 20, pg 410, 1963). *Animal Biology, 55,* 297–321. doi:10.1163/157075605774840941

Uher, J. (2011). Personality in nonhuman primates: What can we learn from human personality psychology? In A. Weiss, J. E. King, & L. Murray (Eds.), *Personality and temperament in nonhuman primates* (pp. 41–76). New York: Springer.

Uher, J. (2016a). *Capuchin personality inventory—Behaviour-descriptive verb items (CPI-BV).* Retrieved from http://www.primate-personality.net/questionnaires/CPI-BV_capuchin_personality-inventory_EN.pdf

Uher, J. (2016b). *Capuchin personality inventory—Trait-adjective items (CPI-TA).* Retrieved from http://www.primate-personality.net/questionnaires/CPI-TA_capuchin_personality-inventory_EN.pdf

Uher, J. (2016c). *Macaque personality inventory for captive individuals (MPIc-BV)—English.* Retrieved from http://www.primate-personality.net/questionnaires/MPIc-BV_macaque_personality-inventory_EN.pdf

Uher, J. (2016d). *Macaque personality inventory for captive individuals (MPIc-TA)—English.* Retrieved from http://www.primate-personality.net/questionnaires/MPIc-TA_macaque_personality-inventory_EN.pdf

Uher, J., Addessi, E., & Visalberghi, E. (2013a). Contextualised behavioural measurements of personality differences obtained in behavioural tests and social observations in adult capuchin monkeys (*Cebus apella*). *Journal of Research in Personality, 47,* 427–444. doi:10.1016/j.jrp.2013.01.013

Uher, J., & Visalberghi, E. (2016). Observations versus assessments of personality: A five-method multi-species study reveals numerous biases in ratings and methodological limitations of standardised assessments. *Journal of Research in Personality, 61,* 61–79. doi:10.1016/j.jrp.2016.02.003

Uher, J., Werner, C. S., & Gosselt, K. (2013b). From observations of individual behaviour to social representations of personality: Developmental pathways, attribution biases, and limitations of questionnaire methods. *Journal of Research in Personality, 47,* 647–667. doi:10.1016/j.jrp.2013.03.00

van Oers, K., de Jong, G., van Noordwijk, A. J., Kempenaers, B., & Drent, P. J. (2005). Contribution of genetics to animal personalities: A review of case studies. *Behaviour, 142,* 1191–1212. doi:10.1163/156853905774539364

Van Oers, K., & Sinn, D. L. (2013). Quantitative and molecular genetics of animal personality. In C. Carere & D. Maestripieri (Eds.), *Animal personalities: Behavior, physiology, and evolution.* Chicago: University of Chicago Press.

Velicer, W. F. (1977). An empirical comparison of the similarity of principal component, image, and factor patterns. *Multivariate Behavioral Research, 12,* 3–22. doi:10.1207/s15327906mbr1201_1

Weiss, A., & Adams, M. J. (2013). Differential behavioral ecology. In C. Carere & D. Maestripieri (Eds.), *Animal personalities: Behavior, physiology and evolution* (pp. 96–123). Chicago, IL: University of Chicago Press.

Weiss, A., Baselmans, B. M. L., Hofer, E., Yang, J., Okbay, A., Lind, P. A., et al. (2016). Personality polygenes, positive affect, and life satisfaction. *Twin Research and Human Genetics, 19,* 407–417. doi:10.1017/thg.2016.65

Weiss, A., Bates, T. C., & Luciano, M. (2008). Happiness is a personal(ity) thing: The genetics of personality and well-being in a representative sample. *Psychological Science, 19,* 205–210. doi:10.1111/j.1467-9280.2008.02068.x

Weiss, A., Inoue-Murayama, M., Hong, K.-W., Inoue, E., Udono, S., Ochiai, T., et al. (2009). Assessing chimpanzee personality and subjective well-being in Japan. *American Journal of Primatology, 71,* 283–292. doi:10.1002/ajp.20649

Weiss, A., Inoue-Murayama, M., King, J. E., Adams, M. J., & Matsuzawa, T. (2012). All too human? Chimpanzee and orang-utan personalities are not anthropomorphic projections. *Animal Behaviour, 83,* 1355–1365. doi:10.1016/j.anbehav.2012.02.024

Weiss, A., & King, J. E. (2006). Searching for the genetic and environmental contributions to personality and happiness in chimpanzees (*Pan troglodytes*). In T. Canli (Ed.), *Biology of Personality and Individual Differences*. New York: Guilford Press.

Weiss, A., & King, J. E. (2015). Great ape origins of personality maturation and sex differences: A study of orangutans and chimpanzees. *Journal of Personality and Social Psychology, 108,* 648–664. doi:10.1037/pspp0000022

Weiss, A., King, J. E., & Enns, R. M. (2002). Subjective well-being is heritable and genetically correlated with dominance in chimpanzees (*Pan troglodytes*). *Journal of Personality and Social Psychology, 83,* 1141–1149. doi:10.1037//0022-3514.83.5.1141

Weiss, A., King, J. E., & Figueredo, A. J. (2000). The heritability of personality factors in chimpanzees (*Pan troglodytes*). *Behavior Genetics, 30,* 213–221. doi:10.1023/A:1001966224914

Weiss, A., King, J. E., & Hopkins, W. D. (2007). A cross-setting study of chimpanzee (*Pan troglodytes*) personality structure and development: Zoological parks and Yerkes National Primate Research Center. *American Journal of Primatology, 69,* 1264–1277. doi:10.1002/ajp.20428

Weiss, A., King, J. E., & Perkins, L. (2006). Personality and subjective well-being in orangutans (*Pongo pygmaeus* and *Pongo abelii*). *Journal of Personality and Social Psychology, 90,* 501–511. doi:10.1037/0022-3514.90.3.501

Weiss, A., Staes, N., Pereboom, J. J. M., Inoue-Murayama, M., Stevens, J. M. G., & Eens, M. (2015). Personality in bonobos. *Psychological Science, 26,* 1430–1439. doi:10.1177/0956797615589933

Whiten, A., Goodall, J., McGrew, W. C., Nishida, T., Reynolds, V., Sugiyama, Y., et al. (1999). Cultures in chimpanzees. *Nature, 399,* 682–685. doi:10.1038/21415

Widaman, K. F. (2007). Common factors versus components: Principals and principles, errors and misconceptions. In R. Cudek & R. C. MacCallum (Eds.), *Factor analysis at 100: Historical developments and future directions* (pp. 177–203). Mahwah, NJ: Lawrence Erlbaum Associates.

Wilson, V. A. D., Weiss, A., Morimura, N., Idani, G., Matsuzawa, T., & Inoue-Murayama, M. (2016). Chimpanzee personality and the arginine vasopressin receptor 1a genotype. *Behavior Genetics*. doi:10.1007/s10519-016-9822-2

Wynne, C. D. L. (2005). The emperor's new anthropomorphism. *The Behavior Analyst Today, 6,* 151–154. doi:10.1037/h0100066

Yamagata, S., Suzuki, A., Ando, J., Ono, Y., Kijima, N., Yoshimura, K., et al. (2006). Is the genetic structure of human personality universal? A cross-cultural twin study from North America, Europe, and Asia. *Journal of Personality and Social Psychology, 90,* 987–998. doi:10.1037/0022-3514.90.6.987

Youyou, W., Kosinski, M., & Stillwell, D. (2015). Computer-based personality judgments are more accurate than those made by humans. *Proceedings of the National Academy of Sciences of the United States of America, 112,* 1036–1040. doi:10.1073/pnas.1418680112

Zuckerman, M., Kuhlman, D. M., Thornquist, M., & Kiers, H. (1991). Five (or three) robust questionnaire scale factors of personality without culture. *Personality and Individual Differences, 12,* 929–941. doi:10.1016/0191-8869(91)90182-B

The Interpersonal Circumplex: A Complementary Approach for Understanding Animal Personality

3

Virgil Zeigler-Hill and Lauren Highfill

Abstract

This chapter reviews the benefits of adopting the interpersonal circumplex as a supplement to current approaches for understanding animal personality. The interpersonal circumplex is a representation of social behavior that is organized in a circular fashion as it is defined by a two-coordinate system consisting of agency (as the vertical axis) and communion (as the horizontal axis). The interpersonal circumplex was developed to describe and understand the social behavior of humans. We believe that the interpersonal circumplex may be helpful in gaining a better understanding of animal personality as well, because it may have implications for (1) conceptualizing the structure of animal personality and (2) accounting for the role of social behavior in animal personality. We review the limited research that has been conducted using the interpersonal circumplex in animals and suggest directions for future research.

Keywords

Interpersonal · Circumplex · Big Five · Five-Factor · Complementarity · Agency · Communion

Research concerning animal personality has the potential to address essential questions concerning the genetic, neurochemical, physiological, developmental, and environmental bases of personality within and across species (Weinstein et al. 2008). Despite the tremendous potential of animal personality research, there is still a great deal of uncertainty and debate regarding how to best conceptualize it. A common approach to this problem has been to adapt models of human personality traits for use with animals (e.g., Freeman et al. 2013; Freeman and Gosling 2010; Weinstein et al. 2008). The basic idea underlying this approach is that researchers interested in animal personality can use the large body of accumulated

V. Zeigler-Hill (✉)
Oakland University, Rochester, MI, USA
e-mail: zeiglerh@oakland.edu

L. Highfill
Eckerd College, St. Petersburg, FL, USA

J. Vonk et al. (eds.), *Personality in Nonhuman Animals*,
DOI 10.1007/978-3-319-59300-5_3

research concerning human personality as a starting point for understanding animal personality (e.g., Gosling and John 1999; King and Figueredo 1997). The focus on personality traits—for both humans and nonhuman species—is consistent with the idea that these traits are connected with how organisms process information about their social environments and often have implications for the motives and goals that organisms adopt (e.g., McAdams 1995).

Researchers concerned with human personality have attempted to develop a common dimensional structure for personality to organize recognized traits into a manageable framework, guide future research, and facilitate communication between researchers. Early attempts to accomplish this goal led to a proliferation of trait taxonomies from a variety of theoretical perspectives (see Pervin and John 1999, for a review). By the 1930s, there were so many competing taxonomies that researchers often had difficulty deciding how to conceptualize personality traits. It was around this time that Allport (e.g., Allport 1937; Allport and Odbert 1936) began advocating for the lexical approach to describing personality that had been introduced by Sir Francis Galton (1884). The lexical approach to personality is based on the assumption that the vast majority of personality traits would have become encoded in natural language such that a careful examination of language should reveal the most commonly used—and most useful—descriptors of personality. There were various attempts to develop a trait taxonomy using this lexical approach that often involved scouring dictionaries and thesauri for words that have been used to describe personality and attempting to find an underlying factor structure to organize these descriptors (e.g., Allport and Odbert 1936; Borgatta 1964; Cattell 1946; Fiske 1949; Norman 1963, 1967; Tupes and Christal 1961). This work was essentially ignored for decades until it was "rediscovered" by researchers in the 1980s (e.g., Digman and Takemoto-Chock 1981; Goldberg 1981, 1982; John et al. 1988; McCrae and Costa 1982). By the 1990s, a broad consensus had been reached that human personality traits could be best described by five broad traits referred to as the Big Five dimensions of personality (e.g., Goldberg 1993; John 1990) or the Five-Factor Model of personality (e.g., Costa and McCrae 1992; Digman 1990). The Big Five dimensions consist of extraversion, agreeableness, neuroticism, conscientiousness, and openness.

The development of the Big Five has provided personality researchers with a shared descriptive taxonomy that captures a number of personality descriptors used in natural language and allows researchers to more effectively communicate about these concepts (see John and Srivastava 1999, for a review). Despite the advantages associated with the Big Five, this model has certainly not escaped criticism. First, the Big Five personality dimensions do not capture all personality traits. Rather, these dimensions simply represent a broad level of abstraction regarding personality (John and Srivastava 1999). It is commonly recognized that more specific personality characteristics—sometimes referred to as "facets"—are contained within each broad personality dimension (e.g., the extraversion dimension contains facets such as warmth, gregariousness, and assertiveness; Costa and McCrae 1992). In addition, other personality traits are not directly captured by this model. For example, narcissism may reflect a blend of high levels of extraversion and low levels of agreeableness (Paulhus 2001) but narcissism includes features that are not captured by the Big Five personality dimensions (e.g., lack of humility; Lee and Ashton 2005). Second, the Big Five personality dimensions are widely accepted but there is still some disagreement regarding the exact nature of these dimensions (e.g., Block 1995; Eysenck 1992; McAdams 1992; Pervin 1994). For example, there has been considerable debate regarding whether openness captures "openness to experience" or "general intellect" (see Nusbaum and Silvia 2011, for a review). Further, it has been argued that the Big Five personality dimensions have failed to account for a sixth personality dimension referred to as honesty-humility (Ashton et al. 2004). Third, there are concerns about whether the Big Five dimensions are really independent of each other (e.g., Block 1995; Digman 1997; Eysenck 1992). For example, Digman

(1997) analyzed personality trait ratings from a variety of studies that utilized both self-reports and observer reports and found consistent evidence for two higher order factors that he referred to using the intentionally nondescript labels of *alpha* (agreeableness, conscientiousness, and emotional stability) and *beta* (extraversion and openness). However, it has been argued that the correlations between these traits may simply be artifacts reflecting response styles rather than capturing substantive connections between these traits (e.g., Ferguson et al. 2013; McCrae et al. 2008; Revelle and Wilt 2013). Fourth, the theoretical underpinnings of the Big Five personality dimensions have been criticized for being relatively weak (e.g., Loevinger 1994; McAdams 1992). That is, the original factor-analytic derivations of the Big Five personality dimensions were based on the assumption that everyday language could be used to capture the structure of personality rather than these investigations being guided by a more substantial theory that would have led to a priori predictions concerning which personality traits would emerge from this process.

Despite its weaknesses, many researchers interested in animal personality have used adaptations of the Big Five when studying nonhuman species (see Weinstein et al. 2008, for a review). This approach has been at least moderately successful in helping researchers understand animal personality. For example, it appears that as many as three of the Big Five dimensions (i.e., extraversion, agreeableness, and neuroticism) emerge consistently from research employing a wide array of species, which suggests that these may be useful constructs when considering animal personality (e.g., Gosling 2001; Gosling and John 1999). However, there are clear limitations associated with applying the Big Five dimensions to nonhuman species given that it was specifically developed for understanding human personality (e.g., Freeman et al. 2013). One limitation is that relying too heavily on the Big Five dimensions may lead researchers to consider personality traits that are not relevant to the target species (e.g., is openness important for understanding the personality of opossums?). Another limitation of focusing on the Big Five dimensions is that it may result in researchers failing to acknowledge the importance of species-specific traits (e.g., dominance has been found to be particularly important for understanding the personality structure of chimpanzees; Weiss et al. 2007).

The primary purpose of this chapter is to suggest that researchers who are interested in gaining an even clearer understanding of animal personality may benefit from supplementing their use of traits —whether based on the Big Five personality dimensions or other theoretical perspectives—with the interpersonal circumplex (Benjamin 1974; Carson 1969; Horowitz 2004; Kiesler 1983, 1996; Leary 1957[1]; Wiggins 1979; see Gurtman 2009, for a review). The interpersonal circumplex is a model that was originally developed to describe and understand the social behavior of humans. We believe that the application of the interpersonal circumplex to animal personality may help researchers gain a better understanding of animal personality because the interpersonal circumplex may have implications for (1) conceptualizing the structure of animal personality and (2) accounting for the role of social behavior in animal personality. Before we examine the potential benefits of using the interpersonal circumplex to gain a better understanding of animal personality, it will be necessary for us to review how the use of the interpersonal circumplex has shaped the understanding of personality in humans.

The Interpersonal Circumplex in Humans

One area of personality that has consistently attracted a great deal of theoretical and empirical attention has been social behavior (e.g., Carson

[1]In addition to his pioneering research concerning the interpersonal circumplex, Timothy Leary is also known for his controversial work concerning the therapeutic potential of psychedelic drugs (i.e., LSD and psilocybin) during the 1960s. Leary also popularized a number of catchphrases during the 1960s and 1970s (e.g., "turn on, tune in, drop out," "think for yourself and question authority") and was once referred to as "the most dangerous man in America" by President Nixon for some of his actions (e.g., a prison escape that drew international attention).

1969; Leary 1957; Sullivan 1953; Wiggins 1991; Wiggins and Pincus 1992). The emphasis placed on social behavior when considering personality is not surprising given the importance of the social context in which humans live. This interpersonal perspective has revealed two underlying dimensions of personality that are often referred to as *agency* (i.e., the tendency to display power, mastery, and self-assertion rather than weakness, failure, and submission) and *communion* (i.e., the tendency to engage in behaviors connected with intimacy, union, and solidarity rather than remoteness, hostility, and separation; Wiggins and Pincus 1992). These terms are derived from the work of Bakan (1966) who identified agency and communion as the "fundamental modalities" of human existence. These two pervasive aspects of social behavior have emerged in theories across numerous disciplines under many different names (see Hogan and Blickle, in press, for a review). For example, Adler (1939) referred to these dimensions as *superiority striving* (agency) and *social interest* (communion). Hogan's (1982) socioanalytic theory posited *getting ahead* (agency) and *getting along* (communion) as two basic strategies adopted by humans in their social interactions. In evolutionary psychology, Buss (2015) has argued that successful reproduction depends on *navigating status hierarchies* (agency) as well as *forming coalitions and alliances* (communion). In anthropology, Redfield (1960) observed that social groups depend on members *getting a living* (agency) and *living together* (communion). In sociology, Parsons and Bales (1955) argued that human groups depend on the completion of tasks related to *group survival* (agency) and *socio-emotional tasks* (communion). McAdams (1988, 2001) found that the stories people develop about their identities center around two basic themes that he calls *power* (agency) and *intimacy* (communion). The central idea underlying social exchange theory (Foa and Foa 1980) is that the exchange of *status* (agency) and *love* (communion) is at the core of all social interactions. Taken together, these various approaches suggest that agency and communion may play a central role in social behavior.

Despite the emergence of agency and communion as the primary components of social behavior in various models, the clearest and most comprehensive depiction of these fundamental dimensions of social behavior is the interpersonal circumplex (Carson 1969, 1989; Horowitz 2004; Kiesler 1983, 1996; Leary 1957; Wiggins 1979; see Gurtman 2009, for a review). The interpersonal circumplex is a two-dimensional representation of social behavior that is organized in a circular fashion with no clear beginning or end. This model is defined by a two-coordinate system that is represented by vertical (agency) and horizontal (communion) axes (see Fig. 3.1). The interpersonal circumplex integrates agency and communion in an attempt to offer a relatively comprehensive model of social behavior (Gurtman 2009; Horowitz et al. 2006; McAdams et al. 1996; Wiggins 1991; Wiggins and Trapnell 1996). The agency dimension captures behaviors that emphasize the self as a distinct unit by focusing on one's own individual influence, control, or mastery over the self, other people, and the environment. The agency dimension is often linked with ideas connected to autonomy, dominance, power, status, control, self-protection, self-assertion, and self-expansion. In contrast, the communion dimension captures behaviors that emphasize connections by enhancing participation in unions with other people. The communion dimension captures ideas linked with intimacy, sociability, love, affiliation, union, and friendliness.

The interpersonal circumplex has its origins in the "interpersonal circle" that was developed by the Kaiser Research Group in the 1950s as a tool for the interpersonal diagnosis of personality (e.g., Freedman et al. 1951; Leary 1957). Wiggins (1979) is believed to have coined the term "interpersonal circumplex" when he merged the Kaiser Research Group's interpersonal circle with the "circumplex" model (i.e., a set of variables that have a particular pattern of correlations that suggests a circular structure) that was developed by Guttman (1954). The two-dimensional structure of the interpersonal circumplex suggests that all interpersonal variables (e.g., behaviors, needs, values, problems) can be represented as a blend of agency and communion, depending on the

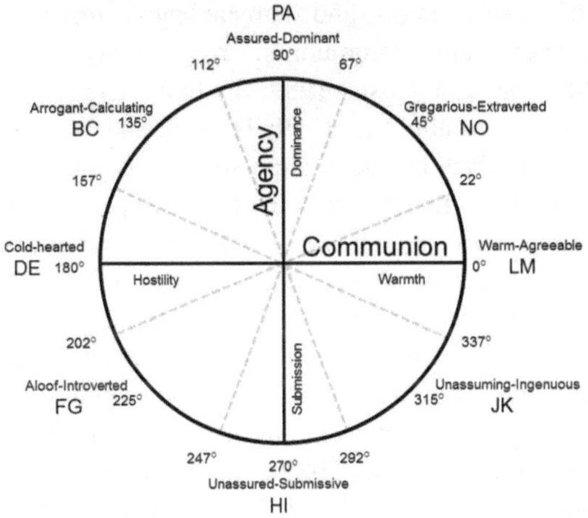

Fig. 3.1 The interpersonal circumplex consists of eight octants (labeled around the perimeter). These octants are blends of agency and communion (labeled along the vertical and horizontal axes)

location of that variable within the space defined by the interpersonal circumplex (Gurtman and Pincus 2003; Kiesler 1983, 1996; Wiggins 1982). More specifically, the interpersonal circumplex is a continuous curve, which implies that there is a uniform distribution of variables around the center of the circumplex resulting in a circular organization of interpersonal traits (Gurtman and Pincus 2003). This circular arrangement provides clear connections between various social behaviors by characterizing each behavior as a blend of agency and communion.

Wiggins (1979) argued that the interpersonal circumplex may serve as the basis for a *circular taxonomy* for the interpersonal domain because all interpersonal traits are believed to be contained within the circumplex. Although there is no fixed set of interpersonal "traits" because the interpersonal circumplex may be separated into segments of varying breadth (e.g., quarters, eighths, sixteenths), the circumplex is most often divided into eight sectors referred to as *octants* that represent various combinations of agency and communion (e.g., aloof-introverted behavior is a blend of submissiveness and hostility; Gurtman and Pincus 2003; Kiesler 1983, 1996; Wiggins 1982). The octants are alphabetically labeled in a counterclockwise direction around the circumplex at 45° intervals originating at the positive horizontal axis: Assured-Dominant (PA; 90°), Arrogant-Calculating (BC; 135°), Cold-hearted (DE; 180°), Aloof-Introverted (FG; 225°), Unassured-Submissive (HI; 270°), Unassuming-Ingenuous (JK; 315°), Warm-Agreeable (LM; 0°), and Gregarious-Extraverted (NO; 45°). The degrees that are located around the perimeter of the circumplex indicate the boundaries and midpoint of each octant. The labels for each octant consist of two letters (e.g., PA) that are derived from the original division of the interpersonal circumplex into 16 sectors labeled "A" through "P" (Freedman et al. 1951).

The interpersonal circumplex represents social behavior as being evenly distributed around the orthogonal axes of agency and communion with the circular structure implying certain connections and conflicts between interpersonal traits (Plutchik 1997). One consequence of this circular—and continuous—arrangement is that each interpersonal trait has both neighboring traits as well as an opposing trait (Plutchik 1980). The proximity of the octants corresponds to their interpersonal similarity such that similarity decreases as the distance between the octants increases (i.e., adjacent octants are more similar than those that are more distant). That is, social behaviors that are more similar to each other are closer together on the circumplex. For example, aloof-introverted (FG) behavior is more similar to unassured-submissive (HI) behavior than it is to unassuming-ingenuous (JK) behavior. In addition, social behaviors that are on opposite sides of the interpersonal circumplex (i.e., separated by 180° as shown in Fig. 3.1) are considered to be opposing traits (e.g., aloof-introverted [FG] opposes gregarious-extraverted [NO]).

It is important to note that the interpersonal circumplex is not the only type of circumplex model (Plutchik and Conte 1997). Circumplex models have been developed in various areas including the structure of emotions (e.g., Plutchik 1997; Yik and Russell 2004), vocational interests (e.g., Holland 1973; Tracey 2000; Tracey and Rounds 1996), cognitive abilities (Lorr 1997), and brand personality in marketing (Sweeney and Brandon 2006). Wiggins and Trobst (1997) argued that a model must demonstrate a circumplex structure (i.e., the proximity of circumplex elements should correspond to their similarity) as well as a plausible interpersonal rationale in order to be considered an *interpersonal circumplex*. Most interpersonal circumplex models use the ideas of agency and communion introduced by Bakan (1966) as the basis for their interpersonal formulations (i.e., these models view all social behavior as blends of agency and communion). These models have focused on various aspects of interpersonal behavior including interpersonal *styles*, interpersonal *needs*, interpersonal *values*, and interpersonal *problems*. The interpersonal circumplex has been used to improve our understanding of a wide variety of phenomena including basic personality dimensions (e.g., Barford et al. 2015; Hofstee et al. 1992), darker aspects of personality (e.g., Ruiz et al. 2001; Southard et al. 2015), and self-esteem (e.g., Zeigler-Hill 2006, 2010; Zeigler-Hill et al. 2011).

Implications of the Interpersonal Circumplex for the Structure of Animal Personality

Although the interpersonal circumplex and the Big Five personality dimensions were developed in different research contexts and tend to be used by different researchers, our proposal that animal personality research would benefit from employing the interpersonal circumplex does not conflict with the tendency for animal personality researchers to use models that emphasize a relatively simple personality structure in their work. For example, the interpersonal circumplex and the Big Five are actually complementary perspectives rather than competing perspectives despite their apparent differences (McCrae and Costa 1989; Trapnell and Wiggins 1990). The complementary relationship between the Big Five model and the interpersonal circumplex stems from the fact that the Big Five model contains representations of the underlying dimensions of the interpersonal circumplex such that extraversion and agreeableness represent agency and communion, respectively (e.g., McCrae and Costa 1989; Plutchik and Conte 1997; Schmidt et al. 1999; Trapnell and Wiggins 1990). That is, the interpersonal circumplex is essentially nested within the Big Five personality dimensions, which is a more comprehensive model because it includes interpersonal behaviors captured by extraversion and agreeableness as well as intrapsychic characteristics (e.g., affective and motivational factors) captured by neuroticism, conscientiousness, and openness (McCrae and Costa 1989). As noted by McCrae and Costa (1989), "The five-factor model provides a larger framework in which to orient and interpret the circumplex, and the interpersonal circle provides a useful elaboration about aspects of two of the five factors" (p. 593). However, the connections between the Big Five and the interpersonal circumplex are not as simple as they appear on the surface because extraversion and agreeableness differ by approximately 30°–45° from the dimensions of agency and communion. This indicates that these dimensions are clearly related to each other but that extraversion and agreeableness are not simply synonyms for agency and communion, respectively (i.e., the two models have slightly different factor rotations; Hopwood 2010; Lorr and Strack 1990; McCrae and Costa 1989; Wiggins and Pincus 1992, 1994). The difference in the rotations of these two models is depicted in Fig. 3.2.

Despite their complementary natures, there are important differences between the interpersonal circumplex and the Big Five personality dimensions. For one, the Big Five dimensions are at least somewhat atheoretical having been developed using a factor-analytic approach to identify a simple structure among personality descriptors (McAdams 1992). In contrast, the interpersonal circumplex has a strong theoretical foundation because it is based on interpersonal theory (e.g., Sullivan 1953) and its circular structure acknowledges the meaningful connections between traits, which may be blurred or distorted by factor-analytic models that attempt to extract relatively distinct personality

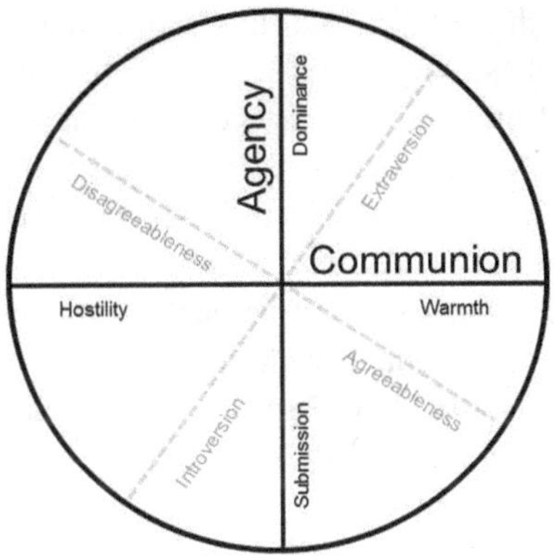

Fig. 3.2 An illustration of the overlap between the Interpersonal Circumplex and the Big Five dimensions of extraversion and agreeableness. The two models have slightly different factor rotations such that extraversion and agreeableness differ by approximately 30°–45° from the dimensions of agency and communion

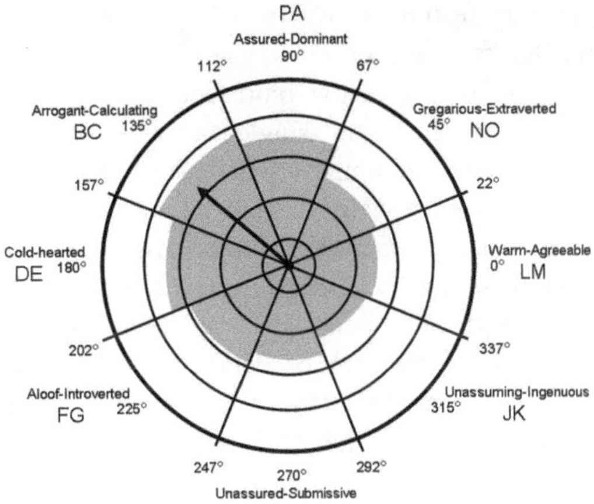

Fig. 3.3 A circumplex profile based on owner ratings of their pet cats. The pattern of this profile shows that cat owners view their pets as displaying social behavior that is best described as arrogant-calculating (i.e., a blend of hostility and dominance)

dimensions (e.g., Gurtman 1997). As a result, the Big Five personality dimensions may fail to capture some of the complex information that exists in personality (Sweeney and Brandon 2006). For example, a Big Five personality dimension profile is often depicted by showing an individual's score for each dimension in a linear fashion that is not arranged in a psychometrically meaningful way but is simply based on established convention (e.g., researchers often arrange the profile in the following order: Neuroticism, Extraversion, Openness, Agreeableness, and Conscientiousness).

Also, in contrast to the Big Five, the structural features of profiles based on the interpersonal circumplex are meaningful (Gurtman 2009). Perhaps more importantly, profiles based on the interpersonal circumplex have been shown to provide important information about animal personality (Woodward and Bauer 2007; Zeigler-Hill and Highfill 2010). As an example, Fig. 3.3 shows an average *circular profile* based on owner ratings of their cats (Zeigler-Hill and Highfill 2010). The circular profile presents the average scores for cats on each of the eight octants of the circumplex. As with the interpersonal circumplex, the circular arrangement of this profile means that it has no beginning or end. The ordering of the octant scores on this profile is meaningful because it mirrors the arrangement of the octants on the circumplex (e.g., more similar octants are located more closely together on the profile). The pattern of this profile shows that the shaded region—which represents higher scores for each octant as it radiates out from the origin—peaks in the arrogant-calculating (BC) octant. This peak captures the predominant trend in the profile and shows that pet owners view their cats as displaying social behavior that is best described as a blend of hostility and dominance (i.e., arrogant and calculating). It is also important to note the corresponding distortion of the profile on the opposing side of the circumplex (i.e., the unassuming-ingenuous [JK] octant) that shows a low score (i.e., cat owners do not typically view the social behavior of their pets as being characterized by a blend of submission and warmth).

An even more concise method for summarizing social behavior with the interpersonal circumplex is to use the *vector* method, which is simply the point—or location—within the interpersonal circumplex that captures the standing of an individual on the agency and communion axes (i.e., the coordinates within the space defined by these axes). This vector summary feature is symbolized by an arrow that radiates from the origin of the circumplex. The direction of the arrow represents *angular displacement* (i.e., the single location on the interpersonal circumplex that best characterizes the behavioral style of these pets) and the length of the arrow represents *amplitude* (i.e., the strength of the behavioral style of these pets). Angular displacement (i.e., the angle of the vector) is a measure of central tendency that captures the predominant social behavior of the organism that is being evaluated (Gurtman and Balakrishnan 1998). In contrast, amplitude (i.e., vector length) indicates the extent to which interpersonal content was observed, such that longer vector lengths are indicative of profiles with stronger interpersonal content. The vector length for example concerning how cat owners viewed their pets indicates a well-defined profile with adequate interpersonal content (see Gurtman and Balakrishnan 1998, for additional information concerning interpersonal profiles).

Another potential benefit to employing the interpersonal circumplex when studying animal personality is that the dimensions of agency and communion may be more easily observed than some intrapsychic personality dimensions. This is important because animal personality research tends to utilize rating or coding approaches to assess animal personality (Archer 1973; Gosling 2001; Highfill et al. 2010; Horback, this volume; Manteca and Deag 1993; Watters and Powell 2012). These methods rely on observers being able to understand the personalities of animals based on their behaviors. The rating method most often relies on observers who are familiar enough with the animal to describe their personality traits based on their cumulative impression of the animal (e.g., their own past experiences with the animal or their observations of the animal interacting with conspecifics). The rating approach relies on the impressions that observers have formed of the animals they are rating. In contrast, the coding technique involves observation and recording of specific animal behaviors in either naturalistic settings or testing situations that are designed to reveal certain personality traits (e.g., exposing animals to novel objects and recording their behaviors). Both the rating and coding approaches are easier to use when they are focused on behaviors that are more amenable to observation despite the fact that there is at least some element of subjectivity in both methods. The social behaviors involved in the interpersonal circumplex could be captured using either the rating or coding method.

Two studies have used the rating method to examine whether the interpersonal circumplex could be applied to animals (Woodward and Bauer 2007; Zeigler-Hill and Highfill 2010). The results of both studies that have examined the interpersonal circumplex with animals suggest that the circular structure of the interpersonal traits was maintained, which supports the use of this model with household pets (i.e., cats and dogs). This is important because there are often limitations associated with the use of human personality models for nonhuman species that focus on intrapsychic dimensions that may not be applicable to animals (e.g., conscientiousness; Gosling 2001). In contrast, the social behaviors described by the interpersonal circumplex appear to be relatively well-suited for describing the behavioral styles of cats and dogs. These studies will hopefully serve as the foundation for the interpersonal circumplex to be applied to a much wider variety of species in future studies using both the rating method and the coding method. That is, we hope that the interpersonal circumplex will be used in a manner that is at least somewhat similar to how the Emotions Profile Index has been employed across species (see Whitman and Washburn, this volume, for a review). Although there are only two studies that have examined the interpersonal circumplex in animals, it is important to note that traits consistent with the dimensions of agency and communion have been frequently identified across species (e.g., dominance, sociability; Gosling 1998, 2001; King and Figueredo 1997; Sapolsky and Ray 1989; Weinstein et al. 2008). For example, personality traits such as dominance have been shown to emerge alongside Big Five dimensions (e.g., extraversion, agreeableness) in certain species (e.g., chimpanzees; Weiss et al. 2007). The use of the interpersonal circumplex in animal personality research may also facilitate within- and cross-species comparisons in social behavior because it provides researchers with a common language.

Implications of the Interpersonal Circumplex for the Role of Social Behavior in Animal Personality

The focus of the interpersonal circumplex on social behavior distinguishes this approach from other models of personality. The basic premise of the interpersonal perspective is that personality functioning is concerned with what an individual *does* rather than with what an individual *is* (Hopwood et al. 2013). That is, social behavior is what allows us to understand the personality features of individuals. Interpersonal *complementarity* is the most basic dynamic pattern in interpersonal theory (i.e., an interaction between individuals in which the social behavior of an individual evokes or elicits a particular response from his or her interaction partner; Carson 1969; Horowitz et al. 2006; Kiesler 1983; Leary 1957; Sadler et al. 2009; Sadler and Woody 2003). The idea underlying complementarity is that it occurs when there is a "match" between the social motives of two individuals who are engaged in an interaction (see Markey and Markey 2007 or Tracey 2004, for extended discussions of complementarity). The model of complementarity proposed by Leary (1957)—and extended by Carson (1969)—defines complementarity as reciprocal (or opposing) social behaviors on the agency dimension along with corresponding (or similar) social behaviors on the communion dimension. Carson (1969) proposed that complementarity can be viewed as a social exchange of status (agency) and affiliation (communion). Reciprocal behavior on the agency dimension means that dominance exhibited by one individual will pull for—or elicit—submissive behavior from the other individual (just as submissive behavior pulls for dominance). Correspondence on the communion dimension means that warm behavior will pull for warmth (just as hostile behavior will pull for hostility). For

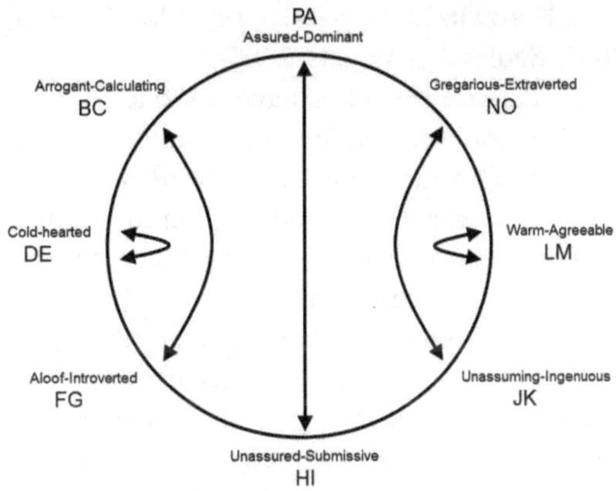

Fig. 3.4 The model of complementarity suggests reciprocal (or opposite) styles on the agency dimension but corresponding (or similar) styles on the communion dimension

example, dominant interpersonal behavior (e.g., "I am the leader of this group") communicates a bid for status (e.g., "I am stronger/better/smarter than you") that impacts others in the social environment and will either elicit a complementary response (e.g., "Yes, you are the leader and I will follow you") or a non-complementary response (e.g., "No, you are not the leader and I will not follow you"). An illustration of complementarity is provided in Fig. 3.4.

Kiesler (1996) argued that complementary social relationships are more stable and satisfying because the social behaviors of the individual are in harmony with the social behaviors—and social position—of his or her interaction partner. This means that complementarity is more than a description of social behavior because it is one part of a dynamic "field-regulatory" system (Gurtman 2009; Pincus and Gurtman 2006). For example, two male gorillas who are housed together in a zoo will most likely develop a more stable relationship if the dominance displays of one male are greeted by submissive behaviors, rather than defiance or challenge, from the other male. Similarly, a chimpanzee that offers an affectionate touch to a conspecific will most likely be more satisfied when that gesture is met with warmth rather than hostility. The predictions concerning complementarity have been examined in a large number of studies in humans (e.g., Horowitz 2004; Kiesler 1996; Markey et al. 2003; Sadler and Woody 2003).

To date, few studies have considered interpersonal complementarity in situations involving nonhuman species (e.g., Woodward and Bauer 2007; Zeigler-Hill and Highfill 2010). Zeigler-Hill and Highfill (2010) found that pet owners were more satisfied with their pets when there was correspondence between their own interpersonal styles and their perceptions of the social behaviors of their pets on the communion dimension. That is, both dog owners and cat owners were more satisfied with their pets when they believed their pets exhibited a level of warmth that was similar to their own. Although this pattern emerged for both dogs and cats, the association was significantly stronger for cats. This may suggest that cat owners are especially sensitive to the perceived warmth of their pets. This association is especially interesting considering that cats were perceived as being less warm than dogs in general and there was more variability in the perceptions of cat owners toward their pets than was reported among dog owners. In terms of agency, Zeigler-Hill and Highfill (2010) found that cat owners were more satisfied with their pets when they perceived them to engage in complementary social behavior. For example, dominant cat owners reported more positive attitudes toward their pets when they believed their pets were submissive. However, complementarity on the agency dimension did not emerge as a significant predictor of positive attitudes toward dogs (e.g., dominant dog owners were no more likely to report positive attitudes toward their pets when they perceived them as submissive). This suggests that agency may have different associations with the attitudes that cat owners and dog owners have toward their pets.

Although the existing studies concerning complementarity in animals have been encouraging, the results in humans have been inconsistent (see Horowitz et al. 2006, for a review). One pattern that has emerged is that the principle of complementarity tends to be confirmed in cooperative interactions but not in competitive interactions. For example, "friendly dominant" behavior from one person often leads to "friendly submissive" behavior from the other. However, the principle of complementarity does not seem to hold as consistently for situations that involve eliciting behaviors from the other side of the interpersonal circumplex which concerns hostility (i.e., low levels of communion). Orford (1986) provided a review of the literature which showed that people often respond to "hostile dominant" behavior with their own "hostile dominant" behavior instead of the theoretically expected "hostile submissive" behavior. Further, Orford (1986) also found that people frequently respond to "hostile submissive" behavior with "friendly dominant" behavior instead of the theoretically expected "hostile dominant" behavior.

These contradictory results have led to some alternative versions of the interpersonal circumplex. One alternative model proposed by Horowitz et al. (2006) argues that the negative pole of communion is "indifference" rather than "hostility." This change has implications for the meaning of many interpersonal behaviors. For example, the BC octant which captures "arrogant-calculating" behavior in the traditional circumplex (i.e., a blend on hostility and dominance) is believed to capture a blend of indifference and dominance in this alternative model. Other changes in this alternative model include complementarity being conceptualized as the response that would satisfy the motive underlying the eliciting behavior, the recognition that social behavior may differ with regard to its ambiguity, and the idea that a failure to experience complementarity will likely lead to negative affect. Future research concerning complementarity in animals should consider alternative interpersonal circumplex models—such as the one provided by Horowitz et al. (2006)—as well as the traditional model. Of course, it is quite possible that none of these interpersonal circumplex models will map onto nonhuman social behavior very well. Rather, the interpersonal circumplex may require additional modifications in order to adequately describe the social behavior of animals (e.g., reconsidering the label "warmth" for the positive pole of the communion dimension). It is also possible that the interpersonal circumplex will be most useful when it is used in conjunction with other models of animal personality that capture dimensions beyond agency and communion such as the Emotions Profile Index (Buirski et al. 1973).

The use of the interpersonal circumplex to understand the complementarity of behavioral styles among animals could lead to several useful applications. For example, the previous results concerning complementarity between pet owners and their pets (Woodward and Bauer 2007; Zeigler-Hill and Highfill 2010) suggest that the interpersonal circumplex may be useful for improving the match between pets and their owners. The American Society for the Prevention of Cruelty to Animals (ASPCA) has used various programs at animal shelters across the United States in an effort to increase adoption rates and decrease return rates. One of these programs was the "Meet Your Match" program that assigned personality labels to all of the adoptable pets (e.g., "constant companion") and the potential adopters also provided information about their own personalities. This information allowed the program to make suggestions concerning potential matches. We believe that the interpersonal circumplex may be a useful tool for enhancing programs such as this by providing additional information concerning compatibility.

Another potential application of the interpersonal circumplex with animals would be to improve decision-making at zoos, sanctuaries, and farms concerning the grouping of these animals. For example, if zookeepers knew the dominance levels of each individual animal, then they could consider the possibility of keeping highly dominant animals apart in order to minimize conflict and improve the welfare of the group (Cassinello and Pieters 2000; Cristol 1995; Digby and Kahlenberg 2002; Sinn and Moltschaniwskyj 2005). In addition, the use of the interpersonal circumplex may be helpful for animal management techniques such as breeding and reintroduction programs. It is important to note, however, that further research would be needed in order to determine whether the interpersonal circumplex could be applied to non-domesticated animals in contexts other than households as well as determining whether this model could be used to predict the behaviors of animals with their conspecifics.

Conclusion

Research concerning animals has the potential to address many fundamental questions concerning personality in different species (Weinstein et al. 2008). For example, animal personality research may elucidate the evolutionary processes that have shaped the development of personality as well as the forces that maintain individual differences (see Dingemanse and Reale 2005, for a review). We believe that researchers who are interested in gaining a clearer understanding of animal personality may benefit from supplementing their current approaches with the interpersonal circumplex. We believe that the interpersonal circumplex model may be beneficial for understanding the structure of animal personality and accounting for the role of social behavior in animal personality. Agency and communion—which are the two central dimensions of the interpersonal circumplex—are likely to play an important role in the personality structures of most social animals. Agency is likely to be an important personality trait across a variety of species because many animal groups have some sort of status hierarchy which may have implications for fitness (e.g., better options for mates, food, and other resources). The pervasiveness of communion is most likely tied to the fact that the ability to form social relationships is crucial to the survival of many organisms. For example, even relatively solitary animals such as bears have to be able to engage in short-term mating interactions in order to procreate. In addition to the benefits that the interpersonal circumplex may have for our understanding of animal personality, it may also have implications for understanding the connections that the social behaviors of nonhuman species have with fitness-related consequences such as mating, reproduction, health, and survival.

References

Adler, A. (1939). *Social interest*. New York, NY: Plenum.

Allport, G. W. (1937). *Personality: A psychological interpretation*. New York, NY: Holt.

Allport, G. W., & Odbert, H. S. (1936). Trait names: A psycho-lexical study. *Psychological Monographs, 47*, (1, Whole No. 211).

Archer, J. (1973). Tests for emotionality in rats and mice: A review. *Animal Behavior, 21*, 205–235.

Ashton, M. C., Lee, K., Perugini, M., Szarota, P., de Vries, R. E., Di Blas, L., et al. (2004). A six-factor structure of personality-descriptive adjectives: Solutions from psycholexical studies in seven languages. *Journal of Personality and Social Psychology, 86*, 356–366.

Bakan, D. (1966). *The duality of human existence: Isolation and communication in Western man*. Boston, MA: Beacon Press.

Barford, K. A., Zhao, K., & Smillie, L. D. (2015). Mapping the interpersonal domain: Translating between the Big Five, HEXACO, and Interpersonal Circumplex. *Personality and Individual Differences, 86*, 232–237.

Benjamin, L. S. (1974). Structural analysis of social behavior. *Psychological Review, 81*, 392–425.

Block, J. (1995). A contrarian view of the five-factor approach to personality description. *Psychological Bulletin, 117*, 187–215.

Borgatta, E. F. (1964). The structure of personality characteristics. *Behavioral Science, 12*, 8–17.

Buirski, P., Kellerman, H., Plutchik, R., Weininger, R., & Buirski, N. (1973). A field study of emotions, dominance, and social behavior in a group of baboons (*Papio anubis*). *Primates, 14*, 67–78.

Buss, D. M. (2015). *Evolutionary psychology: The new science of the mind* (5th ed.). Boston, MA: Pearson.

Carson, R. C. (1969). *Interaction concepts of personality*. Chicago, IL: Aldine.

Carson, R. C. (1989). Personality. *Annual Review of Psychology, 40*, 227–248.

Cassinello, J., & Pieters, I. (2000). Multi-male captive groups of endangered dama gazelle: Social rank, aggression, and enclosure effects. *Zoo Biology, 19*, 121–129.

Cattell, R. B. (1946). *The description and measurement of personality*. Yonkers, NY: World Books.

Costa, P. T., & McCrae, R. R. (1992). *The NEO-PI-R professional manual*. Odessa, FL: Psychological Assessment Resources.

Cristoll, D. A. (1995). Costs of switching social groups for dominant and subordinate dark-eyed juncos (*Junco hyemalis*). *Behavioral Ecology and Sociobiology, 37*, 93–101.

Digby, L. J., & Kahlenberg, S. M. (2002). Female dominance in blue-eyed black lemurs (*Eulemur macaco flavifrons*). *Primates, 43*, 191–199.

Digman, J. M. (1990). Personality structure: Emergence of the five-factor model. *Annual Review of Psychology, 41*, 417–440.

Digman, J. M. (1997). Higher-order factors of the big five. *Journal of Personality and Social Psychology, 73*, 1246–1256.

Digman, J. M., & Takemoto-Chock, N. K. (1981). Factors in the natural language of personality: Re-analysis, comparison, and interpretation of six major studies. *Multivariate Behavioral Research, 16*, 149–170.

Dingemanse, N. J., & Réale, D. (2005). Natural selection and animal personality. *Behaviour, 142*, 1159–1184.

Eysenck, H. J. (1992). Four ways five factors are not basic. *Personality and Individual Differences, 13*, 667–673.

Ferguson, E., Chamorro-Premuzic, T., Pickering, A., & Weiss, A. (2013). Five into one doesn't go: A critique of the general factor of personality. In T. Chamorro-Premuzic, S. von Stumm, & A. Furnham (Eds.), *The Wiley-Blackwell handbook of individual differences* (pp. 162–186). Chichester: Wiley-Blackwell.

Fiske, D. W. (1949). Consistency of the factorial structures of personality ratings from different sources. *Journal of Abnormal and Social Psychology, 44*, 329–344.

Foa, E. B., & Foa, U. G. (1980). Resource theory of social exchange. In J. W. Thibaut, J. T. Spence, & R. C. Carson (Eds.), *Contemporary topics in social psychology* (pp. 99–131). Morristown, NJ: General Learning Press.

Freedman, M. B., Leary, T. F., Ossario, A. G., & Coffey, H. S. (1951). The interpersonal dimension of personality. *Journal of Personality, 20*, 143–161.

Freeman, H., Brosnan, S., Hopper, L., Lambeth, S., Schapiro, S., & Gosling, S. (2013). Developing a comprehensive and comparative questionnaire for measuring personality in chimpanzees using a simultaneous top-down/bottom-up design. *American Journal of Primatology, 75*, 1042–1053.

Freeman, H. D., & Gosling, S. D. (2010). Personality in nonhuman primates: A review and evaluation of past research. *American Journal of Primatology, 72*, 653–671.

Galton, F. (1884). Measurement of character. *Fortnightly Review, 36*, 179–185.

Goldberg, L. R. (1981). Language and individual differences: The search for universals in personality lexicons. In L. Wheeler (Ed.), *Review of personality and social psychology* (Vol. 2, pp. 141–165). Beverly Hills, CA: Sage.

Goldberg, L. R. (1982). From Ace to Zombie: Some explorations in the language of personality. In C. D. Speilberger & J. N. Butcher (Eds.), *Advances in personality assessment* (Vol. 1, pp. 203–234). Hillsdale, NJ: Erlbaum.

Goldberg, L. R. (1993). The structure of phenotypic personality traits. *American Psychologist, 48*, 26–34.

Gosling, S. D. (1998). Personality dimensions in spotted hyenas (*Crocuta crocuta*). *Journal of Comparative Psychology, 112,* 107–118.

Gosling, S. D. (2001). From mice to men: What can we learn about personality from animal research? *Psychological Bulletin, 127,* 45–86.

Gosling, S. D., & John, O. P. (1999). Personality dimensions in non-human animals: A cross-species review. *Current Directions in Psychological Science, 8,* 69–75.

Gurtman, M. B. (1997). Studying personality traits: The circular way. In R. Plutchik & H. R. Conte (Eds.), *Circumplex models of personality and emotions*. Washington, DC: American Psychological Association.

Gurtman, M. B. (2009). Exploring personality with the interpersonal circumplex. *Social and Personality Psychology Compass, 3,* 601–619.

Gurtman, M. B., & Balakrishnan, J. D. (1998). Circular measurement redux: The analysis and interpretation of interpersonal circle profiles. *Clinical Psychology: Science and Practice, 5,* 344–360.

Gurtman, M. B., & Pincus, A. L. (2003). The circumplex model: Methods and research applications. In J. Schinka & W. Velicer (Eds.), *Handbook of psychology: Research methods in psychology* (Vol. 2, pp. 407–428). Hoboken, NJ: Wiley.

Guttman, L. (1954). A new approach to factor analysis: The radex. In P. F. Lazarsfeld (Ed.), *Mathematical thinking in the social sciences* (pp. 258–348). Glencoe, IL: Free Press.

Highfill, L., Hanbury, D., Kristiansen, R., Kuczaj, S., & Watson, S. (2010). Rating vs. coding in animal personality research. *Zoo Biology, 29,* 509–516.

Hofstee, W. K., De Raad, B., & Goldberg, L. R. (1992). Integration of the Big Five and circumplex approaches to trait structure. *Journal of Personality and Social Psychology, 63,* 146–163.

Hogan, R. (1982). A socioanalytic theory of personality. In M. M. Page (Ed.), *Nebraska symposium on motivation 1982. Personality: Current theory and research* (pp. 55–89). Lincoln, NE: University of Nebraska Press.

Hogan, R., & Blickle, G. (in press). Socioanalytic theory: Basic concepts, supporting evidence, and practical implications. In V. Zeigler-Hill & T. K. Shackelford (Eds.), *The SAGE handbook of personality and individual differences*. Thousand Oaks, CA: Sage.

Holland, J. L. (1973). *Making vocational choices: A theory of careers*. Englewood Cliffs, NJ: Prentice-Hall.

Hopwood, C. J. (2010). An interpersonal perspective on the personality assessment process. *Journal of Personality Assessment, 92,* 471–479.

Hopwood, C. J., Wright, A. G. C., Ansell, E. B., & Pincus, A. L. (2013). The interpersonal core of personality pathology. *Journal of Personality Disorders, 27,* 270–295.

Horowitz, L. M. (2004). *Interpersonal foundations of psychopathology*. Washington, DC: American Psychological Association.

Horowitz, L. M., Wilson, K. R., Turan, B., Zolotsev, P., Constantino, M. J., & Henderson, L. (2006). How interpersonal motives clarify the meaning of interpersonal behavior: A revised circumplex model. *Personality and Social Psychology Review, 10,* 67–86.

John, O. P. (1990). The "Big Five" factor taxonomy: Dimensions of personality in the natural language and in questionnaires. In L. A. Pervin (Ed.), *Handbook of personality: Theory and research* (pp. 66–100). New York, NY: Guilford.

John, O. P., Angleitner, A., & Ostendorf, F. (1988). The lexical approach to personality: A historical review of trait taxonomic research. *European Journal of Personality, 2,* 171–205.

John, O. P., & Srivastava, S. (1999). The Big Five trait taxonomy: History, measurement, and theoretical perspectives. In L. A. Pervin & O. P. John (Eds.), *Handbook of personality: Theory and research* (2nd ed., pp. 102–138). New York, NY: Guilford.

Kiesler, D. J. (1983). The 1982 interpersonal circle: A taxonomy for complementarity in human transactions. *Psychological Review, 90,* 185–214.

Kiesler, D. J. (1996). *Contemporary interpersonal theory and research*. New York, NY: Wiley.

King, J. E., & Figueredo, A. J. (1997). The five-factor model plus dominance in chimpanzee personality. *Journal of Research in Personality, 31,* 257–271.

Leary, T. F. (1957). *Interpersonal diagnosis of personality: A functional theory and methodology for personality evaluation*. New York, NY: Ronald.

Lee, K., & Ashton, M. C. (2005). Psychopathy, Machiavellianism, and narcissism in the Five-Factor Model and the HEXACO model of personality structure. *Personality and Individual Differences, 38,* 1571–1582.

Loevinger, J. (1994). Has psychology lost its conscience? *Journal of Personality Assessment, 62,* 2–8.

Lorr, M. (1997). The circumplex model applied to interpersonal behavior, affect and psychotic syndromes. In R. Plutchik & H. R. Conte (Eds.), *Circumplex models of personality and emotions* (pp. 47–56). Washington, DC: American Psychological Association.

Lorr, M., & Strack, S. (1990). Wiggins interpersonal adjective scales: A dimensional view. *Personality and Individual Differences, 11,* 423–425.

Manteca, X., & Deag, J. M. (1993). Individual differences in temperament if domestic animals: A review of methodology. *Animal Welfare, 2,* 247–268.

Markey, P. M., Funder, D. C., & Ozer, D. J. (2003). Complementarity of interpersonal behaviors in dyadic interactions. *Personality and Social Psychology Bulletin, 29,* 1082–1090.

Markey, P. M., & Markey, C. N. (2007). Romantic ideals, romantic obtainment, and relationship experiences: The complementarity of interpersonal traits among romantic partners. *Journal of Social and Personal Relationships, 24,* 517–533.

McAdams, D. P. (1988). *Power, intimacy, and the life story*. New York, NY: Guilford.

McAdams, D. P. (1992). The five factor model in personality: A critical appraisal. *Journal of Personality, 60,* 329–361.

McAdams, D. P. (1995). What do we know when we know a person? *Journal of Personality, 63,* 365–396.

McAdams, D. P. (2001). The psychology of life stories. *Review of General Psychology, 5,* 100–122.

McAdams, D. P., Hoffman, B. J., Mansfield, E. D., & Day, R. (1996). Themes of agency and communion in significant autobiographical scenes. *Journal of Personality, 64,* 339–377.

McCrae, R. R., & Costa, P. T. (1982). Self-concept and the stability of personality: Cross-sectional comparisons of self-reports and ratings. *Journal of Personality and Social Psychology, 43,* 1282–1292.

McCrae, R. R., & Costa, P. T. (1989). The structure of interpersonal traits: Wiggins' circumplex and the five factor model. *Journal of Personality and Social Psychology, 56,* 586–595.

McCrae, R. R., Yamagata, S., Jang, K. L., Riemann, R., Ando, J., Ono, Y., et al. (2008). Substance and artifact in the higher-order factors of the Big Five. *Journal of Personality and Social Psychology, 95,* 442–455.

Norman, W. (1963). Toward an adequate taxonomy of personality attributes: Replicated factor structure in peer nomination personality types. *Journal of Abnormal and Social Psychology, 66,* 574–583.

Norman, W. (1967). *2800 personality trait descriptors: Normative operating characteristics for a university population.* Ann Arbor, MI: University of Michigan.

Nusbaum, E. C., & Silvia, P. J. (2011). Are openness and intellect distinct aspects of openness to experience? A test of the O/I model. *Personality and Individual Differences, 51,* 571–574.

Orford, J. (1986). The rules of interpersonal complementarity: Does hostility beget hostility and dominance, submission? *Psychological Review, 93,* 365–377.

Parsons, T., & Bales, R. F. (1955). *Family, socialization and interaction process.* Glencoe, IL: Free Press.

Paulhus, D. L. (2001). Normal narcissism: Two minimalist accounts. *Psychological Inquiry, 12,* 228–230.

Pervin, L. A. (1994). A critical analysis of current trait theory. *Psychological Inquiry, 5,* 103–113.

Pervin, L. A., & John, O. P. (1999). *Handbook of personality: Theory and research* (2nd ed.). New York, NY: Guilford Press.

Pincus, A. L., & Gurtman, M. B. (2006). Interpersonal theory and the interpersonal circumplex: Evolving perspectives on normal and abnormal personality. In S. Strack (Ed.), *Differentiating normal and abnormal personality* (2nd ed., pp. 83–111). New York, NY: Springer.

Plutchik, R. (1980). *Emotion: A psychoevolutionary synthesis.* New York, NY: Harper and Row.

Plutchik, R. (1997). The circumplex as a general model of the structure of emotions and personality. In R. Plutchik & H. R. Conte (Eds.), *Circumplex models of personality and emotions* (pp. 17–46). Washington, DC: American Psychological Association.

Plutchik, R., & Conte, H. R. (1997). *Circumplex models of personality and emotions.* Washington, DC: American Psychological Association.

Redfield, R. (1960). How society operates. In H. L. Shapiro (Ed.), *Man, culture, and society* (pp. 345–368). New York, NY: Oxford University Press.

Revelle, W., & Wilt, J. (2013). The general factor of personality: A general critique. *Journal of Research in Personality, 47,* 493–504.

Ruiz, J. M., Smith, T. W., & Rhodewalt, F. (2001). Distinguishing narcissism and hostility: Similarities and differences in interpersonal circumplex and five-factor correlates. *Journal of Personality Assessment, 76,* 537–555.

Sadler, P., Ethier, N., Gunn, G. R., Duong, D., & Woody, E. (2009). Are we on the same wavelength? Interpersonal complementarity as shared cyclical patterns during interactions. *Journal of Personality and Social Psychology, 97,* 1005–1020.

Sadler, P., & Woody, E. (2003). Is who you are who you're talking to? Interpersonal style and complementarily in mixed-sex interactions. *Journal of Personality and Social Psychology, 84,* 80–95.

Sapolsky, R. M., & Ray, J. C. (1989). Styles of dominance and their endocrine correlates among wild olive baboons (*Papio anubis*). *American Journal of Primatology, 18,* 1–13.

Schmidt, J. A., Wagner, C. C., & Kiesler, D. J. (1999). Covert reactions to Big Five personality traits: The Impact Message Inventory and the NEO-PI-R. *European Journal of Psychological Assessment, 15,* 221–232.

Sinn, D. L., & Moltschaniwskyj, N. A. (2005). Personality traits in dumpling squid (*Euprymna tasmanica*): Context-specific traits and their correlation with biological characteristics. *Journal of Comparative Psychology, 119,* 99–110.

Southard, A. C., Noser, A. E., Pollock, N. C., Mercer, S. H., & Zeigler-Hill, V. (2015). The interpersonal nature of dark personality features. *Journal of Social and Clinical Psychology, 34,* 555–586.

Sullivan, H. S. (1953). *The interpersonal theory of psychiatry.* New York, NY: Norton.

Sweeney, J. C., & Brandon, C. (2006). Brand personality: Exploring the potential to move from factor analytical to circumplex models. *Psychology and Marketing, 23,* 639–663.

Tracey, T. J. G. (2000). Analysis of circumplex models. In H. E. A. Tinsley & S. D. Brown (Eds.), *Handbook of applied multivariate statistics and mathematical modeling* (pp. 641–664). San Diego, CA: Academic Press.

Tracey, T. J. G., & Rounds, J. (1996). The spherical representation of vocational interests. *Journal of Vocational Behavior, 48,* 3–41.

Trapnell, P. D., & Wiggins, J. S. (1990). Extension of the Interpersonal Adjective Scales to include the Big Five dimensions of personality. *Journal of Personality and Social Psychology, 59,* 781–790.

Tupes, E. C, & Christal, R. E. (1961). *Recurrent personality factors based on trait ratings* (Technical Report ASD-TR-61-97). Lackland Air Force Base, TX: U.S. Air Force.

Tracey, T. J. G. (2004). Levels of interpersonal complementarity: A simplex representation. *Personality and Social Psychology Bulletin, 30*, 1211–1225.

Watters, J. V., & Powell, D. M. (2012). Measuring animal personality for use in population management in zoos: Suggested methods and rationale. *Zoo Biology, 31,* 1–12.

Weinstein, T. A. R., Capitanio, J. P., & Gosling, S. D. (2008). Personality in animals. In O. P. John, R. W. Robins, & L. A. Pervin (Eds.), *Handbook of personality: Theory and research* (3rd ed., pp. 328–348). New York, NY: Guilford.

Weiss, A., King, J. E., & Hopkins, W. D. (2007). A cross-setting study of chimpanzee (Pan troglodytes) personality structure and development: Zoological parks and Yerkes National Primate Research Center. *American Journal of Primatology, 69,* 1264–1277.

Wiggins, J. S. (1979). A psychological taxonomy of trait-descriptive terms: The interpersonal domain. *Journal of Personality and Social Psychology, 37,* 395–412.

Wiggins, J. S. (1982). Circumplex models of interpersonal behavior in clinical psychology. In P. C. Kendall & J. N. Butcher (Eds.), *Handbook of research methods in clinical psychology* (pp. 183–221). New York, NY: Wiley.

Wiggins, J. S. (1991). Agency and communion as conceptual coordinates for the understanding and measurement of interpersonal behavior. In W. Grove & D. Cicchetti (Eds.), *Thinking clearly about psychology: Essays in honor of Paul E. Meehl: Vol. 2, Personality and psychopathology* (pp. 89–113). Minneapolis, MN: University of Minnesota Press.

Wiggins, J. S., & Pincus, A. L. (1992). Personality: Structure and assessment. *Annual Review of Psychology, 43,* 473–504.

Wiggins, J. S., & Pincus, A. L. (1994). Personality structure and the structure of personality disorders. In P. T. Costa Jr. & T. A. Widiger (Eds.), *Personality disorders and the five-factor model of personality* (pp. 73–93). Washington, DC: American Psychological Association.

Wiggins, J. S., & Trapnell, P. D. (1996). A dyadic-interactional perspective on the five-factor model. In J. S. Wiggins (Ed.), *The five-factor model of personality: Theoretical Perspectives* (pp. 88–162). New York, NY: Guilford.

Wiggins, J. S., & Trobst, K. K. (1997). When is a circumplex an 'interpersonal circumplex'? The case of supportive actions. In R. Plutchik & H. R. Conte (Eds.), *Circumplex models of personality and emotions* (pp. 57–80). Washington, DC: American Psychological Association.

Woodward, L. E., & Bauer, A. L. (2007). People and their pets: A relational perspective on interpersonal complementarity and attachment in companion animal owners. *Society and Animals, 15,* 169–189.

Yik, M., & Russell, J. (2004). On the relationship between circumplexes: Affect and Wiggins' IAS. *Multivariate Behavioral Research, 39,* 203–230.

Zeigler-Hill, V. (2006). Contingent self-esteem and the interpersonal circumplex: The interpersonal pursuit of self-esteem. *Personality and Individual Differences, 40,* 713–723.

Zeigler-Hill, V. (2010). The interpersonal nature of self-esteem: Do different measures of self-esteem possess similar interpersonal content? *Journal of Research in Personality, 44,* 22–30.

Zeigler-Hill, V., Clark, C. B., & Beckman, T. E. (2011). Fragile self-esteem and the interpersonal circumplex: Are feelings of self-worth associated with interpersonal style? *Self and Identity, 10,* 509–536.

Zeigler-Hill, V., & Highfill, L. (2010). Applying the interpersonal circumplex to the behavioral styles of dogs and cats. *Applied Animal Behaviour Science, 124,* 104–112.

4 The Quantitative and Molecular Genetics of Individual Differences in Animal Personality

Veronika N. Laine and Kees van Oers

Abstract

One of the main goals in current personality research is to identify genes behind the measured behavioral variations. This is important in order to study how, under the influence of the environment, gene expression changes are translated into the observed phenotypes. The advances, especially in genomic technologies, have made it possible to identify genetic loci behind these variations, also concerning non-model species. In this chapter, we will describe the role and relevance of quantitative and molecular genetic approaches in explaining the existence and maintenance of variation in animal personality. We here will provide (1) a timely review on the papers published on this topic, (2) an overview of the current situation and progress, and (3) a view on the likely new avenues the field will take.

Introduction

It has now been established that, just as in humans, individuals of a wide range of animal species express variation in personality. Animal personality describes between-individual differences in behaviors that are consistent over context and time (Carere and Maestripieri 2013). Often, these personality traits correlate and thereby form suites of traits, so-called behavioral syndromes (Sih et al. 2004). Whereas the phenomenon of individual consistency in behavioral traits was described before (e.g., Burtt 1973; Williams et al. 1962), there has been a large increase in papers that attempt to explain the causes and consequences of variation in personality traits over the past three decades (see for overview: Carere and Maestripieri 2013). The study of animal personality has now grown to be a multidisciplinary field, which unites scientists who use various approaches to study animal behavior and work in fields ranging from comparative psychology to neuroscience and evolutionary biology.

V.N. Laine (✉) · K. van Oers
Department of Animal Ecology, Netherlands Institute of Ecology (NIOO-KNAW), P.O. Box 50, 6700 AB Wageningen, The Netherlands
e-mail: V.Laine@nioo.knaw.nl

K. van Oers
e-mail: K.vanOers@nioo.knaw.nl

J. Vonk et al. (eds.), *Personality in Nonhuman Animals*,
DOI 10.1007/978-3-319-59300-5_4

When trying to explain the origin of variation in personality traits an immediate question is to what extent variation in personality traits can be explained by genes, the environment, and the interaction between the two. Where genetic studies will provide information on the genomic structure and mechanisms, possible constraints, and the potential for natural selection to act on different levels of variation (van Oers and Sinn 2013), such information is valid only when tested along axes of relevant environmental variation. Already from early studies, it was obvious that both genes and the environment play important roles in describing variation in behaviors, but their relative contribution and importance has been a point of discussion since then (see e.g., Groothuis and Trillmich 2011; van Oers et al. 2005). In more recent years this discussion has changed into a more interactive one, in which it is recognized that the absolute role for genes and environment cannot be separated and the integration rather than the contrast between the two will yield new insights (Groothuis and Trillmich 2011; van Oers and Sinn 2011).

In this chapter, we will describe the role and relevance of quantitative and molecular genetic approaches in explaining the existence and maintenance of variation in animal personality. Although several reviews and some book chapters have been written on this topic (Dochtermann and Roff 2010; Dochtermann et al. 2014; van Oers et al. 2005; van Oers and Mueller 2010; van Oers and Sinn 2011, 2013), we here will provide (1) a timely review on the papers published on this topic, (2) an overview of the current situation and progress, and (3) the likely new avenues the field will take.

Theory of Genetic Influence of Animal Personality

The identification of a heritable factor is the essential starting point for all evolutionary research on any behavioral trait (Boake et al. 2002). Part of the phenotypic variation that is observed in behavioral traits in animals is transmitted across generations (Stirling et al. 2002). The presence of some level of heritability, where resemblance can be observed among relatives across or within generations, has been identified in many studies, regardless of the mechanism that may be responsible for this resemblance. Traditionally, "heritable variation", "genetic variation", and "genomic variation" were envisioned as interchangeable terms. Genetic variation due to genomic polymorphisms was viewed as the basis for individual differences in the expression of behaviors, and phenotypic plasticity, i.e., the residual irreversible and reversible phenotypic variance, was assigned to the interaction of the expression of the genes with the environment, due to, for example, development and learning (Arnold et al. 2007; Brydges et al. 2008; Quinn and Cresswell 2005). Selection was thereby expected to be acting only on the additive genetic component, reflected in the so-called narrow-sense heritability estimate. This is in contrast to the broad-sense heritability, which, apart from the additive genetic component, includes other components that may contribute to resemblance among relatives, like maternal effects, non-additive effects, or epistasis, the interaction between genes (Lynch and Walsh 1998). Recent ideas build on the observation that transgenerational inheritance may be caused by processes other than genetic polymorphisms (Danchin et al. 2011) and this view has consequences for how we expect measures of heritability and genomic differences among individuals to be associated with one another. In this chapter we therefore make a clear distinction between quantitative genetics, which may not explicitly take into account the genomic and epigenetic mechanisms responsible for the heritable component, and molecular genetics, in which an intrinsic interest exists in explaining direct causes for changes in gene expression on the genomic level (see below).

Personality as a Polygenic Trait

As with many other quantitative traits, many loci with small effects are expected to be involved in animal personality variation. Although this theory is accepted by the majority of researchers, hardly any direct proof for it exists. To our knowledge, only one study specifically tested the polygenetic nature of one personality trait in a wild species (Santure et al. 2015) by combining a chromosome partitioning approach (Robinson et al. 2013) and a calculation of the number of loci contributing to the variation (Guan and Stephens 2011). Both a positive association between the amount of variation explained and the size of a chromosome, and an estimation that hundreds of loci were identified that contributed to the variation in exploratory behavior, point to the fact that indeed such traits are influenced by many genes of small effect distributed throughout the genome (Santure et al. 2015). These loci are expected to play roles at different organizational levels. There will be sets of genes that will affect behavioral trait variation per se. Polymorphisms in such genes are likely to affect the amount or structure of proteins influencing behavior in a direct way. Whereas, on the other hand, there will be genes involved in the plasticity a trait expresses (Dingemanse and Wolf 2013). A more direct cause for the quantitative nature of behavioral traits is the fact that some loci will be involved in the pathway, but will not have genomic variation among individuals (Visser et al. 2010). Such genes may, for example, be genes that are essential in the pathway, but may only be switched on or off. These genes can therefore not account for genomic variation among individuals, i.e., will not bear any polymorphisms, and thus, cannot be identified using conventional molecular genetic approaches (e.g., Zou and Zeng 2009). However, they can be identified using expression studies, because the expression levels of these genes may be important drivers of a plastic expression of a trait (Brem and Kruglyak 2005).

Non-additive Genetic Effects

Where additive genetic effects describe the independent effects of loci on the trait, in cases when genes interact, non-additive effects may affect the expression and inheritance of such traits. Substantial non-additive genetic effects may point to the stabilizing nature of selection acting on personality traits (Penke et al. 2007). Studies have shown that non-additive genetic variance may contribute to substantial amounts of the heritability of behavioral traits in general (Meffert and Hicks 2002) and personality traits specifically (Adams et al. 2012; Jordan et al. 2007; van Oers et al. 2004), indicating that this might be a general feature of personality traits, explaining some of the variation in heritability estimates (van Oers and Sinn 2013). Where the presence of such non-additive genetic effects can be incorporated in quantitative genetic models (see below), studies that test for these effects are still rare. The exact origin of non-additive effects is often difficult to assess and simulations have shown that complex patterns of, for example, epistasis may apparently be there, while the effect of the genes is purely additive (Eaves and Verhulst 2014).

Distinction Between Quantitative and Molecular Genetics

Here we want to make a distinction between two approaches that can be used to study heritable variation in traits, including behavioral traits. On the one hand, studies use information on resemblance among individuals to study inheritance patterns of behavioral traits across generations (quantitative genetics). On the other hand, studies can investigate the architecture and function of genomic structures such as genes at a molecular level (molecular genetics). This distinction in methods is important for identifying the level of questions one wants to answer and the implications of the results from these studies.

Quantitative Genetics of Animal Personality

The goal in quantitative genetic studies is to identify the fraction of variation in a phenotype that is attributable to a heritable component. Quantitative traits, such as the majority of behavioral traits, are characterized by a continuous distribution, likely caused by polymorphisms in many loci, the expressions of which, and therefore the phenotypes, are responsive to environmental fluctuations.

Genetic studies in captivity often rely on artificial selection of specific traits (e.g., Drent et al. 2003; van Oortmerssen and Bakker 1981; van Oers et al. 2004) or crosses between breeds (Laine et al. 2014; van Oers et al. 2014). The so-called narrow-sense heritability derived from such experiments may indicate the evolutionary potential of the traits at a specific time point, given the additive genetic variation of the parental population. Whereas these studies have given us much insight into the genetic mechanisms and the standing genetic variation in certain populations at certain time points, validation and experimental studies are needed in natural populations to assess how these findings relate to variation under natural conditions.

A quantitative genetic analysis in a natural population requires that individuals are measured for their behavior and that information on the relatedness among those individuals is available (Falconer and Mackay 1996; Lynch and Walsh 1998; Kruuk et al. 2008). This information is then used to construct statistical models that integrate phenotypic and relatedness information, so-called animal models. Animal models use pedigrees to estimate genetic parameters (Kruuk 2004), such as the additive genetic variance (V_A) and the heritability (h^2), the latter being the ratio of V_A to the total phenotypic variance (VP). Animal models were originally designed to determine the genetic merit of individuals by calculating the genetic variance components of mainly farm animals for animal breeders (Henderson 1984) to emphasize specific traits, but were later adopted by quantitative geneticists working on pedigreed natural populations (Kruuk 2004; Lynch and Walsh 1998), and are now also used in personality studies in captive (Careau et al. 2011; van Oers et al. 2004) and natural populations (Brommer and Kluen 2012; Dingemanse and Dochtermann 2013; Quinn et al. 2009; Taylor et al. 2012).

Quantitative genetic studies in natural populations typically find that between 20 and 50% of the phenotypic variation in animal personality traits can be attributable to a heritable component (for reviews see Dochtermann et al. 2014; Van Oers et al. 2005; Van Oers and Sinn 2013). In a recent example, Petelle et al. (2015) estimated the additive genetic variance and covariance of four personality traits, docility, sociability, activity, and exploratory behavior, in a wild population of yellow-bellied marmots (*Marmota flaviventris*). Using behavioral information from individuals that were part of marker-based pedigree spanning 11 years, they found that about 10–15% of the total phenotypic variation could be explained by an additive genetic component and around 40% of the between-individual variance could be explained by resemblance between relatives. However, in a recent meta-analysis including 70 measures of both heritability and repeatability in 10 study systems, Dochtermann et al. (2014) tested the degree to which the repeatable part of behavioral variation could be attributed to additive genetic variation. They found that approximately 52% of the consistent part of animal personality variation was the result of additive genetic variation. This meta-analysis shows that the heritability of personality is higher than previously estimated, which might demonstrate that heritable differences are likely to be a major contributor to variation in animal personality as well as support the phenotypic gambit: that evolutionary inferences drawn from repeatability estimates may often be justified. Furthermore, while additive genetic variation is a primary contributor to personality, considerable variation remains to be explained. The animal model approach also allows for other components of the phenotypic variance to be modeled, such as common environment or maternal effects, often requiring structured experiments and/or multiple measures of the same or related individuals. Brommer and Kluen (2012), for example, conducted a 3-year study of reciprocal cross-fostering within a pedigreed blue tit (*Cyanistes caeruleus*) population. Heritability estimates of handling aggression, breathing rate, and docility confirmed those found in non-experimental studies (h^2 = 16–29%), but they were strongly affected by a "nest-of-rearing" component, indicating that the phenotypic similarity of relatives may also be due to common environment effects.

More recently, quantitative genetic methods have been applied to assess more than only the heritable component in personality traits. Two important components of personality research are (1) the multivariate nature of personality traits and (2) the assumption that individuals are consistent over time and context. To integrate these aspects into animal models, both genetic correlations (Dochtermann and Roff 2010) and the genetic basis of phenotypic plasticity (Dingemanse et al. 2009) can be integrated into these animal models (Adams et al. 2012).

Animal Personality Trait Correlations

Where individual consistency is an important feature of animal personality, the observed phenotypic correlations often referred to as behavioral syndrome (Sih et al. 2004) is another aspect. Where the genetic structure underlying a trait will provide information on the evolvability of a single trait, traits are not expected to be expressed independently. Nevertheless, estimations of genetic correlations are still rare (but see Dochtermann and Roff 2010; Roff 1996; van Oers et al. 2005). This lack of estimates of genetic correlations calls for the need to view and analyze traits in a multivariate way. Quantitative genetic methods allow for the characterization of this so-called multivariate behavioral phenotype by estimating within, among, and independent structures (Dochtermann and Roff 2010). In Brommer and Kluen's (2012) study on blue tits, for example, it was specifically tested whether phenotypic correlations provided a correct description of the genetic correlations. They found that the phenotypic and genetic correlations indeed went in the same direction and were roughly of the same size when looking within cohorts. In a follow-up study, however, they concluded that this genetic correlation could disappear during development due to age-specific genetic effects (Class and Brommer 2015). Moreover, in yellow-bellied marmots (*Marmota flaviventris*), researchers found that only one of the phenotypic correlations between the traits could be explained by a positive genetic correlation, indicating that even when traits are correlated on a phenotypic level, different mechanisms might be affecting the expression of these traits (Petelle et al. 2015). This is important because carryover effects (described by Sih et al. 2004) are a valid problem only if the two measures of the same trait in different contexts are indeed measures of the same trait, i.e., that the genetic correlation is approaching 1. If there is no genetic correlation between these measures of the same trait, these measures should be seen as different traits, and selection could be acting independently on them. Second, if seemingly different traits in the same or different contexts have some overlap in the genomic structure, the evolution of these traits is not independent. This may lead to temporary constraints in the independent evolution of these traits (but see Roff et al. 1996).

Molecular Genetics of Animal Personality

Knowledge of the molecular genetic basis of personality traits provides information on the expression and regulation of genes at the molecular level. Moreover, genes provide the foundation for the nervous system to express behaviors through the production of polypeptides. Knowing the structure and function of the molecular mechanisms in personality traits will help us in answering questions regarding the origin of the variation in personality traits, the expected responses to environmental variation, and the possible constraints there are in expressing behavioral variation. Identifying genetic variants in the genome will also elucidate the micro-evolutionary processes and historical selection behind personality traits, and will eventually help explain the maintenance of variation in personality (see also Bell and

Aubin-Horth 2010). The rapid speed at which molecular genetic tools have been developed over the last decade has, in particular, helped to provide a more profound knowledge of the mechanisms by means of integrating quantitative genetics and molecular biology.

Identifying the biological mechanisms of variation in behavior can help us understand the genetic composition, characteristics of the nervous system, physiological state, developmental history, environmental factors, and biochemical reactions an organism has and experiences during its lifetime (Anholt and Mackay 2009; Krebs and Davies 1997). Because of the complexity of personality and the multigenic architecture and high sensitivity to environmental influence, unraveling the genetic background of behavior is challenging (Anholt and Mackay 2004; Bendesky and Bargmann 2011; Flint 2003; Flint and Mott 2001).

The main resource for genetic variation is mutation. Often mutations are rare and neutral/deleterious but sometimes natural selection favors random mutations in the genome. If this mutation affects genes that control an organism's behavior, this will cause the behavioral patterns to be altered. Genes that affect behavior can have two kinds of effects. First, some genes influence the manifestation of the behavior, i.e., the gene affects a trait in which there is little or no variation in the population, and the trait is not environmentally determined. This can be studied by mutagenesis approaches: targeted, gene-trap or ENU (*N*-ethyl-*N*-nitrosourea) mutagenesis, as it has been done, for example, in behavioral studies of mice (Belknap et al. 2001; Bucan and Abel 2002; Vitaterna et al. 2006). In addition to these mutagenesis approaches and traditional transgenesis technologies, there are new genome editing techniques such as CRISPR-Cas-based RNA-guided DNA endonucleases (CRISPR-Cas) that allow genomes to be modified by targeted insertion, removal, or replacement of genomic DNA areas (reviewed by Kratochwil and Meyer 2014; Singh et al. 2015). In behavioral studies CRISPR-Cas has been used in mice. Swiech et al. (2014) perturbed multiple genes in the mouse brain in vivo using the CRISPR-Cas method. These authors characterized the effects of genome modifications in postmitotic neurons using biochemical, genetic, electrophysiological, and behavioral readouts and found in behavioral tests that the modification affected the memory of the mice. Their study successfully demonstrated that the CRISPR-Cas method can be used in the studies of gene function in the brain.

For personality traits, a more likely case is, when a particular set of genes contributes to the behavioral variation. With quantitative trait loci (QTL) mapping approaches (see examples below) it is possible to identify and estimate the contributions of these genes to the observed phenotypic variance (Boake et al. 2002). Any QTL analysis assumes that the behavior is variable within the mapping population and that the effects of the QTL can be monitored by creating mutations, and this is possible especially in model organisms. However, it is often only possible to rely on the mutations that have occurred naturally and are manifesting in natural populations.

Behavioral genetics studies pose many problems that need to be resolved. One such problem in personality studies is that the replication of the behavioral measurements can be difficult. If the measured personality trait is relying on unreliable, non-replicated measures, this might give spurious results in, e.g., genetic association studies. However, some behaviors, like the reaction to novelty, can be measured only once. Another challenge is the small effect size (strength of the relationship between an individual genetic variant and personality) in association studies. Effect size determines the power of the study and it is dependent on the sample size. In many wild animal QTL studies, the sample sizes are in the hundreds (Slate 2013) and the expectation is that increasing the sample size would help the QTL detection. This expectation has been proven correct in model organisms and in humans by Flint and Mackay (2009). However, even if the studies have small effect sizes, they can be useful. A single-nucleotide polymorphism (SNP) with a small effect may reveal the underlying biological pathways and especially when many of these small effect SNP studies are combined in meta-analyses, significant pathways may be revealed. For example, Ripke et al. (2013) used a multi-stage genome-wide association study (GWAS) that combined actual GWA analysis with a meta-analysis in a single study to reveal new risk loci for schizophrenia.

In this chapter we regard studies on wild species as studies in which the individuals under study are descended from recently sampled non-domesticated individuals. In contrast, genetic model species (i.e., model to human behavior) are typically reared in the laboratory for many generations. Most behavioral genetic studies have been conducted on such genetic model organisms, which include mice, Drosophila, and, more recently, domesticated animals as subjects (Fitzpatrick et al. 2005; Inoue-Murayama 2009). New models have also been emerging especially for behavioral studies. One of them being honeybees, which have provided valuable insights into social behavior (Fitzpatrick et al. 2005). Furthermore, the three-spined stickleback (*Gasterosteus aculeatus*) is a well-known model species for evolutionary biology and it has a large amount of variation in behavior as well as morphology and physiology (Bell and Foster 1994; Bell and Stamps 2004). Finally, the great tit, *Parus major*, a wild passerine bird has proved to be an excellent species for ecological, evolutionary, and behavioral studies (Laine et al. 2016). Despite the potential ecological and evolutionary significance of wild populations, and the fact that more variation in behavior is seen in the natural setting, the underlying genetic basis of behaviors and especially personality traits has rarely been studied in free-living populations. However, there are methods available currently to change the situation. First, a "candidate gene approach" can easily be used in wild populations (Fidler et al. 2007; Fitzpatrick et al. 2005). Second, the most recent developments in genomic techniques enable genome-wide approaches in QTL analyses to be undertaken in almost any species. The lack of genome sequence information has been one of the main limitations of conducting both candidate gene and QTL studies in non-model organisms but, thanks to technological developments, more and more whole genomes for species have become available. In the following paragraphs we present examples from both candidate gene and QTL approaches.

Candidate Gene Approaches

In the candidate gene approach the assumption is that the trait is determined to a significant extent by a smaller number of genes, whose effects are large enough to be detected. Many recent studies that use a candidate gene approach have demonstrated that the knowledge of gene-trait associations in "model" organisms such as humans, domestic fowl, or mice can be utilized for studying the genetic basis of behavioral variation in wild animal populations (Fidler et al. 2007; Fitzpatrick et al. 2005). In addition to taking candidate genes from model organism studies, genes can be selected based on an a priori knowledge of the biochemical pathway in the trait in question. For example, a zebrafish model was used successfully to identify many candidate genes and pathways to regulating aggression in fish, combining gene expression profiling, behavioral analyses, and pharmacological manipulations (Filby et al. 2010).

In a candidate gene approach wild individuals can be categorized into behavioral types without measuring the actual behavior. In addition, candidate gene studies require little prior sequence information and are therefore well-suited for behavioral genetics analyses in natural populations of non-model species (Fitzpatrick et al. 2005). Earlier studies demonstrated that molecular markers located within or near candidate genes can be a useful resource for the identification of genes associated with adaptive phenotypic divergence, especially when there are no other genomic resources available (Shikano et al. 2010; Tonteri et al. 2010). In addition, markers closely linked to functionally important genes are useful in the construction of comparative genetic maps, in which they can be used as comparative anchor-tagged sequence loci (Lyons et al. 1997).

In humans, among the most studied candidate genes for personality variation are various polymorphisms within the dopamine receptor D4 (*DRD4*) gene and the serotonin transporter gene, solute carrier family 6 member 4 (*SLC6A4/SERT*) (Delvecchio et al. 2016; Savitz and Ramesar 2004). Polymorphisms within the *DRD4* gene have been found to account for about 3% of the variation in novelty seeking in humans (Munafò et al. 2008). There are a series of polymorphic regions that might affect the expression or function of serotonin transporter (Ozaki et al. 2003; Wendland et al. 2006). The most studied polymorphism is the 5-HTTLPR, which is a functional polymorphism within the promoter sequence of the serotonin transporter gene (Heils et al. 1995), and it has been linked to variety of behaviors and behavioral disorders, including impulsivity, mood, and anxiety (Eley and Plomin 1997; Gordon and Hen 2004). However, the evidence for an association is inconsistent when slightly different measures of the trait (harm avoidance, neuroticism, etc.) are used (Munafò et al. 2009). Nevertheless, *SLC6A4* may play a role in anxiety and other affective disorders, but its effects might be subtle, for example, on amygdala activation, which has been associated with depression and anxiety (Murphy et al. 2013).

In addition to humans, an association between exploratory behavior and the *DRD4* homologue has been detected in many species, for example, in apes (Shimada et al. 2004), dogs (Ito et al. 2004), fish (Boehmler et al. 2007), and birds (Fidler et al. 2007). In the latter study, Fidler et al. (2007) found that great tits artificially selected for divergent levels of exploratory behavior differed in the allele frequency of an exonic SNP in the *DRD4* gene. The association was confirmed in a natural population, where the levels of exploratory behavior differed for great tits with different genotypes (Fidler et al. 2007). However, the *DRD4* studies done with great tits also highlight concerns regarding candidate gene studies: (unaccounted) population structuring leading to possible false-positive associations especially in the case of complex traits like personality. When the association was tested across samples of four great tit populations, the association was significant in only one with an estimated effect size of around 5% (Korsten et al. 2010). This could indicate that the *DRD4* polymorphism is linked to the functional variant in some but not all populations, or that the association is dependent on the environment or other population-specific characteristics (Korsten et al. 2010; Mueller et al. 2013).

Another popular gene that has been used in personality studies is the monoamine oxidase A (*MAO-A*), polymorphisms of which have been linked to antisocial behavior and aggression in humans and mice (Nelson and Trainor 2007; Scott et al. 2008; Shih and Thompson 1999). One well-known example comes from human studies where the interaction between the *MAO-A* genotype and the rearing environment affected aggressive behavior (Caspi et al. 2002). This study showed that children carrying the short form of the *MAO-A* promoter gene, which decreases *MAO-A* activity, are more likely to develop antisocial problems when exposed to abusive home environments. This environmental effect has less effect on individuals carrying the long form of the promoter. Later studies supported these results by showing that the low activity *MAO-A* in combination with traumatic early life events predisposes individuals toward aggressive behavior (Frazetto et al. 2007; Kim-Cohen et al. 2006) and in a recent review of Weeland et al. (2015) they describe ways to conduct and compare gene–environment interactions in mental disorders. Other genes with possible effects on variation in personality include the dopamine receptor D2 gene, the serotonin receptor genes 5-HT2c and HTR2a, and the tyrosine hydroxylase gene (reviewed in Savitz and Ramesar 2004).

In addition to population structuring affecting the candidate gene studies, there are limitations to the use of candidate gene studies for animal personality. Where this approach directly tests the effects of mutations located in specific genes for their association to the trait of interest, it can permit only the identification of genes that have been, in some way, related to the trait previously. The candidate gene approach, therefore, cannot reveal the association of totally new mutations, unlike genome-wide approaches, such as linkage mapping or genome-wide association studies. Moreover, because candidate gene studies are often conducted with relatively small samples, only polymorphisms that are relatively abundant in the population can be tested. Furthermore, it is likely that many genes are involved in epistatic interactions. The effects of single polymorphism or gene associations might therefore be a bad indicator for the ultimate role of the genes for variation in trait expression. Thus, the results from candidate gene studies need to be validated by testing for these associations in several populations and species, and also by adding gene expression and genome-wide association approaches to ascertain causal links between the polymorphism and the behavior.

Genome-Wide QTL Approaches

Another approach to assess which genomic regions underlie variation in personality is a more bottom-up approach in which no a priori expectation about genes related to measured traits is needed using QTL analyses. This approach is more suitable for studies of behavior because of the multigenic architecture of behavioral traits. These genome-wide QTL analyses can be divided into two groups: linkage-based studies and association mapping studies.

Linkage-Based Mapping (QTL Mapping)

In linkage-based mapping, also known as QTL mapping, the aim is to identify genetic markers that predict the phenotype, by linking pedigreed populations with individual genotypic data consisting of genetic markers that are evenly distributed over the whole genome (Slate 2005). This analysis method is based on pedigree information (one or many pedigrees from natural populations or crossing experiments) and the mapping populations consist either of inbred line crosses, crosses between outbred populations, or natural populations with known pedigree structure (Lynch and Walsh 1998; Slate 2005).

The predictive markers are close to the causal loci, and so the predictive markers and the causal loci tend to segregate together. Most predictive markers are expected to be in close proximity to the causal locus because the probability of recombination breaking the linkage increases with physical distance. Linkage-based mapping permits the determination of genomic regions that are involved in explaining heritable phenotypic variation and this is then used to predict the genetic architecture of the measured trait (Erickson et al. 2004). Often the result is a candidate chromosomal region linked to variation in the phenotype, which covers several dozens to hundreds of genes. This region can be narrowed by haplotype sharing or the identification of more polymorphic markers at these specific sites using so-called fine mapping. The candidate genes are identified from these areas for further genetic studies with the aim of identifying loci that are of major importance for the trait variation.

Many QTL mapping studies on animal personality traits have been conducted with rodents using controlled crosses between lines and strains (e.g., Gershenfeld et al. 1997; Hovatta and Barlow 2008). Additionally, personality has been found to be associated with animal well-being in farm animals (Christiansen and Forkman 2007; Horback, this volume; Koolhaas et al. 2001; Rodenburg et al. 2008), leading to behavior genetic studies identifying QTL for behavioral traits in livestock (Haskell et al. 2014). For example, in a study by Gutierrez-Gil et al. (2008) 29 QTL regions were identified in a cross between two cattle populations measured for several temperament-related behaviors. In total, each region explained only a small fraction of the phenotypic variation, ranging from 4 to 8%. However, the most notable candidate gene found in one of the regions was the *DRD4* gene, showing that *DRD4* is probably an important gene in cattle personality traits.

In 2005, Slate published a review of the prospects for QTL studies in natural populations. This paper has good guidelines for what techniques can be used to perform QTL studies with non-model organisms, especially in the context of ecological and evolutionary issues. The biggest issue for most species for which personality data have been collected in natural populations is the lack of pedigree information and the lack of a sufficient number of markers to be able to construct genetic maps. It is therefore possible that some QTL mapping studies on behavior did not result in the detection of any QTL and thus remained unpublished, even in the presence of heritable variation. Because modern meta-analytical methods could overcome these biases, there is an urge to also publish studies when no significant QTLs are found. However, genomic polymorphism data for high-quality genetic maps are now becoming available for many non-model species. A whole-genome linkage map of the zebra finch (*Taeniopygia guttata*) based on about 2000 SNP markers has been constructed (Backström et al. 2010), while tens of thousands of SNPs have been used to construct the great tit linkage map (Van Bers et al. 2010; van Oers et al. 2014) to name just two dominant species used in animal personality research.

Nevertheless, association studies in wild populations based on these numbers of markers show only limited success. In a QTL study on personality in wild bighorn sheep (*Ovis canadensis*), Poissant et al. (2013) identified two suggestive QTLs using a linkage analysis based on 238 microsatellite loci genotyped in 310 pedigreed individuals, but no regions exceeding genome-wide significance were detected. A study on exploratory behavior in great tits in two independent replicate populations that used both linkage and association mapping methods also failed to reveal any significant regions associated with this often investigated personality trait (Santure et al. 2015), even though sample sizes were larger than in the wild bighorn sheep study (1000 individuals and 5500 SNPs).

To increase the power of these QTL studies, natural variation can be magnified by artificial selection or artificial crosses, by, for example, F2-cross designs. In Laine et al. (2014), nine-spined stickleback (*Pungitius pungitius*), three behavioral measurements were QTL mapped using a F2-cross design between marine and pond populations. Many QTL regions were located in the linkage map, which supports the polygenic regulation of behavior. In one of the regions they found QTLs influencing different behavioral traits. This finding suggests that the genetic factors influencing one behavior may have pleiotropic effects on other behaviors or that genetic factors influencing different behaviors cluster into this linkage group. QTL mapping of schooling behavior has also been successfully used in three-spined sticklebacks using two types of crosses: benthic × marine backcrosses and an F2 intercross, which revealed new candidate regions for behavioral differences in sticklebacks (Greenwood et al. 2015). Another example of QTL mapping and cross design comes from the study of anti-predatory behavior in rainbow trout (*Oncorhynchus mykiss,* Christensen et al. 2014). Through using clonal rainbow trout lines derived from either wild or hatchery populations, they identified several behaviors that varied between clonal lines and they found QTLs for behavioral and size traits.

Association Mapping

Association mapping, also known as genome-wide association study (GWAS), is also based on recombination, but the recombination used in this method is historical. GWAS uses population-based data with abundant recombination history and individuals are genotyped with thousands of SNPs using microarrays or chip sequencing to capture most of the genetic variation in the studied population (Risch 2000). With this method, the entire genome is covered and the aim is to identify the genes with potentially known functions that are causing variation in the trait of interest (Bush and Moore 2012). The significant genomic areas or candidate SNPs can be used for further fine mapping or in experimental studies. The power to detect QTL at an intermediate frequency is comparable between linkage and GWA studies. However, in GWA studies, allele frequencies can be more extreme and this means that increasing sample sizes and also marker densities increases the number of QTL detected and this decreases the average effect sizes (Mackay et al. 2009).

The vast majority of personality GWA studies use human participants because of the high numbers of individuals and markers needed for the analyses (De Moor et al. 2012). The first GWA study on the five major human personality factors, neuroticism, extraversion, agreeableness, openness to experience, and conscientiousness, used a sample of 3972 individuals from an isolated population on Sardinia, Italy that was genotyped with 362 129 SNPs (Terracciano et al. 2008). A few of the most promising SNPs that were identified were successfully replicated in two independent samples, one from USA and the other from the Netherlands. The authors highlight two important conclusions that can be drawn from this pioneering study. First, personality traits are influenced by many genes that each explain only a small proportion of variation (1–2%), and these polymorphisms are detected only when sample sizes are large enough. Second, genetic effects are most probably found when specific phenotypes are measured, rather than when pooled into broader factors or using principal component analysis (Terracciano et al. 2008). This suggests that personality traits, like boldness or risk-taking behaviors, should be broken down into smaller units of individual behaviors. The same outcome could also indicate that specific genes play a role in determining variation in single personality trait, but additional genes modify the correlations among traits. This issue points to one of the challenges in the behavior genetics of personality: the definition and quantification of behavior (Sokolowski 2001). We expect that more association studies will be possible in natural systems in the near future, because with decreasing sequencing and genotyping costs and the invention of new genomic and statistical tools, these methods will become available also for non-model species as in Santure et al. (2015).

Pleiotropy

Generally, no genes have been identified as solely affecting behavior, and "behavioral genes" therefore do not directly control behavior. These genes instead influence the development and function of behavior by affecting the development of the nervous system that is related to adult behavior (Sokolowski 2001). In addition, these genes affect behavior, physiology, and morphology. Genes, therefore, are often responsible for many distinct and unrelated phenotypes, a phenomenon called pleiotropy (Stearns 2010). Pleiotropic effects in behavior have especially been studied in insects. In the fruit fly (*Drosophila melanogaster*) system where mutants are relatively easy to produce, researchers discovered that producing mutants in candidate genes for behavior had pleiotropic effects in both development and behavior, or in several behaviors [reviewed by Anholt and Mackay (2004) and Sokolowski (2001)]. Furthermore, in honeybees (*Apis mellifera*) there have been demonstrations of pleiotropy in the context of reproduction and social behavior (reviewed by Page et al. 2012). But examples extend also to vertebrates. In the guppy (*Poecilia reticulata*), lines selected for large or for small brains, were known to differ in cognitive ability (Kotrschal et al. 2014). When testing individuals from these selected lines for three personality traits, the authors found that large-brained animals explored an open field faster and stayed in the open more compared to individuals from the small-brained line. Large-brained animals also secreted less cortisol in a stressful situation and were slower to feed from a novel food source, altogether pointing toward a more reactive personality type in large-brained animals, indicating pleiotropic effects between brain development and behavior.

In a combined QTL and eQTL study on a range of behavioral traits in 294 German Landrace piglets (*Sus domesticus*), the most evident gene found was *PER1*, which has known physiological implications for circadian rhythms, and was shown to associate with variation in a backtest trait, a validated personality and coping trait in pigs (Ponsuksili et al. 2015), indicating that this gene may have a role for both timing and personality traits. Another striking example of pleiotropic effects between physiology and behavior comes from the melanocortin system studies. It has been shown that there is a widespread relationship between melanin-based coloration and other phenotypic traits in vertebrates (reviewed by Ducrest et al. 2008). One of the most famous examples of pleiotropism comes from the domestication of wild silver foxes (*Vulpes vulpes*). The Russian geneticist Belyaev found that, when selecting for tame foxes, many other features were also changed such as droopy ears, a smaller skull size, bigger litter sizes, a faster reproductive cycle, and a piebald fur pattern (white patches, depigmentation, Trut et al. 1997; Trut 1999). More recent examples of this relationship include Mafli et al. (2011) who found that darker shell coloration predicted greater aggressiveness and boldness in Hermann's tortoises (*Eurotestudo boettgeri*). Genetic studies should therefore not only concentrate on the most popular genes studied in behavioral genetics but also widen the search to other gene networks as well.

Future Research

Research on the genetic mechanism of personality differences saw a transition from testing effects of candidate gene variation toward whole-genome methods in molecular genetics. For quantitative genetics it changed from describing heritability components in unmanipulated mid-term pedigrees to multivariate experimental setups. Now there is a great need for integrating such molecular and quantitative genetic methods (Bell and Dochtermann 2015). For a truly integrative framework, experiments should be planned that allow variance partitioning at the level of both trait variation and trait plasticity. A main challenge will be to link the heritable variation, or even the variance components responsible for heritable variation, to genomic mechanisms. Moreover, the conceptual frameworks developed in quantitative and molecular genetics should be integrated, to develop testable hypotheses.

Because genes and environment cannot be separated (see e.g., Groothuis and Trillmich 2011; van Oers et al. 2005), future research also needs to focus on integrating genetic research with epigenetic research. In addition to additive genetic variation and dominance genetic effects, heritable personality differences can arise from transgenerational epigenetic effects, i.e., all transgenerational processes that contribute to the non-genetic determination of the phenotype, such as paternal/maternal effects and permanent environmental effects (Youngson and Whitelaw 2008). The genomic mechanisms responsible for such transgenerational effects are largely unknown, although processes such as DNA methylation of cytosines, small RNA's, and the modification of chromatin proteins are strong candidates (Moore 2015). Therefore, epigenetic mechanisms may play a major role in the expression and heritability of animal personality traits (Ledón-Rettig et al. 2012). First, studies that investigate the DNA methylation in animals indicate that DNA methylation might be a mechanism involved in mediating transgenerational inheritance. In a study of great tits in the Netherlands, birds from a line selected for fast exploratory behavior (Drent et al. 2003) have been found to have increased levels of DNA methylation in the promoter region of *DRD4* (Verhulst et al. 2016). This study was confirmed by results in a wild population in Spain, where methylation at a single CpG dinucleotide in the *DRD4* promoter region was related to exploration score in urban great tits (Riyahi et al. 2017). These results support the idea that some of the effects found for *DRD4* may be mediated by DNA methylation (Verhulst et al. 2016). The next steps should focus on investigating the contribution of epigenetic processes, such as DNA methylation for personality traits in natural systems and its subsequent role for understanding the ecology and evolution of behavioral consistency.

Where current studies focus on finding allelic differences among individuals that are associated with individual differences in behavior, in future studies we will be moving toward more concrete functional links between genomic variation, gene expression, and behavior. A very nice example of this in a recent study is a large-scale genetical genomics analysis of chicken (*Gallus gallus*) brains from individuals differing in open-field behavior (Johnson et al. 2016). By combining information on genome-behavior associations with associations between gene expression and behavior, the authors identified 10 putative loci that correlated with both behavior (bQTL) and gene expression (eQTL). Four of these loci were formerly linked to bipolar disorder, major depressive disorder, and schizophrenia in the mouse Heterogeneous Stock anxiety data set and human GWAS data sets. Assuming that most adaptive and functional variation lies in the coding sequence of genes underestimates the role of regulatory mechanisms (Groothuis and Trillmich 2011; Rasmuson 2009). Studying gene expression with gene polymorphisms rather than a single one in isolation will be more likely to lead to further understanding the consistency and plasticity of personality traits.

The validation of the findings of association studies will be mitigated by new tools that are increasingly becoming available. Tools, that allow for experimental modification of genomes, even in unconventional model species such as wild species, make personality genetics research more and more experimental. One example of such a tool is the above-mentioned CRISPR, which has many advantages for behavioral studies. For one, this method allows many mutations to be introduced in parallel, which is especially advantageous for multigenic behavioral traits. In addition, CRISPR can be applied to any species.

Conclusion

In the past two decades, research on the genetical background of animal personality traits focused on mapping the quantitative and molecular structure of personality traits, trait correlations, and plasticity. We should now move toward more integration among fields of research. Animal personality offers a unique opportunity to not only combine quantitative genetics with molecular genetics, but it also offers the greatest diversity in processes to be studied in concordance, in single organisms. Animal personality traits as labile quantitative traits are playgrounds for studying intra-individual, inter-individual, and inter-trait associations. Personality will therefore help us understand the complexity of organisms at different levels. Genetic studies may thereby function as a multidisciplinary link between research fields in achieving a more holistic understanding how organisms react to their changing environments.

References

Adams, M. J., King, J. E., & Weiss, A. (2012). The majority of genetic variation in orangutan personality and subjective well-being is nonadditive. *Behavior Genetics, 42,* 675–686. doi:10.1007/s10519-012-9537-y

Anholt, R. R., & Mackay, T. F. (2009). *Principles of behavioral genetics*. Amsterdam: Academic.

Anholt, R. R., & Mackay, T. F. (2004). Quantitative genetic analyses of complex behaviors in Drosophila. *Nature Reviews Genetics, 5,* 838–849. doi:10.1038/nrg1472

Arnold, K. E., Ramsay, S. L., Donaldson, C., & Adam, A. (2007). Parental prey selection affects risk-taking behavior and spatial learning in avian offspring. *Proceedings of the Royal Society B: Biological Sciences, 274,* 2563–2569. doi:10.1098/rspb.2007.0687

Backström, N., Forstmeier, W., Schielzeth, H., Mellenius, H., Nam, K., Bolund, E., et al. (2010). The recombination landscape of the zebra finch *Taeniopygia guttata* genome. *Genome Research, 20,* 485–495. doi:10.1101/gr.101410.109

Bell, A. M., & Aubin-Horth, N. (2010). What can whole genome expression data tell us about the ecology and evolution of personality? *Philosophical Transactions of the Royal Society B: Biological Sciences, 365,* 4001–4012. doi:10.1098/rstb.2010.0185

Bell, A. M., & Stamps, J. A. (2004). Development of behavioral differences between individuals and populations of sticklebacks, *Gasterosteus aculeatus*. *Animal Behavior, 68,* 1339–1348. doi:10.1016/j.anbehav.2004.05.007

Bell, A. M., & Dochtermann, N. A. (2015). Integrating molecular mechanisms into quantitative genetics to understand consistent individual differences in behavior. *Current Opinion in Behavioral Sciences, 6,* 111–114. doi:10.1016/j.cobeha.2015.10.014

Bell, M. A., & Foster, S. A. (1994). *The evolutionary biology of the threespine stickleback*. Oxford: Oxford University Press.

Belknap, J. K., Hitzemann, R., Crabbe, J. C., Phillips, T. J., Buck, K. J., & Williams, R. W. (2001). QTL analysis and genomewide mutagenesis in mice: Complementary genetic approaches to the dissection of complex traits. *Behavior Genetics, 31,* 5–15. doi:10.1023/A:1010249607128

Bendesky, A., & Bargmann, C. I. (2011). Genetic contributions to behavioral diversity at the gene–environment interface. *Nature Reviews Genetics*. doi:10.1038/nrg3065

Bers, N. E., van Oers, K., Kerstens, H. H., Dibbits, B. W., Crooijmans, R. P., Visser, M. E., et al. (2010). Genome-wide SNP detection in the great tit Parus major using high throughput sequencing. *Molecular Ecology, 19,* 89–99. doi:10.1111/j.1365-294x.2009.04486.x

Boake, C., Arnold, S., Breden, F., Meffert, L., Ritchie, M., Taylor, B., et al. (2002). Genetic tools for studying adaptation and the evolution of behavior. *The American Naturalist, 160*(S6), S143–S159. doi:10.1086/342902

Boehmler, W., Carr, T., Thisse, C., Thisse, B., Canfield, V. A., & Levenson, R. (2007). D4 Dopamine receptor genes of zebrafish and effects of the antipsychotic clozapine on larval swimming behavior. *Genes, Brain and Behavior, 6,* 155–166. doi:10.1111/j.1601-183x.2006.00243.x

Brem, R. B., & Kruglyak, L. (2005). The landscape of genetic complexity across 5,700 gene expression traits in yeast. *Proceedings of the National Academy of Sciences, 102,* 1572–1577. doi:10.1073/pnas.0408709102

Brommer, J. E., & Kluen, E. (2012). Exploring the genetics of nestling personality traits in a wild passerine bird: Testing the phenotypic gambit. *Ecology and Evolution, 2,* 3032–3044. doi:10.1002/ece3.412

Brydges, N. M., Colegrave, N., Heathcote, R. J., & Braithwaite, V. A. (2008). Habitat stability and predation pressure affect temperament behaviors in populations of three-spined sticklebacks. *Journal of Animal Ecology, 77,* 229–235. doi:10.1111/j.1365-2656.2007.01343.x

Bućan, M., & Abel, T. (2002). The Mouse: Genetics meets behavior. *Nature Reviews Genetics, 3,* 114–123. doi:10.1038/nrg728

Burtt, H. E., & Giltz, M. L. (1973). Personality as a variable in the behavior of birds. *Journal of Science, 73,* 65–82.

Bush, W. S., Moore, J. H., Lewitter, F., & Kann, M. (2012). Chapter 11: Genome-Wide Association Studies. *PLoS Computational Biology, 8*(12), e1002822.

Careau, V., Thomas, D., Pelletier, F., Turki, L., Landry, F., Garant, D., et al. (2011). Genetic correlation between resting metabolic rate and exploratory behavior in deer mice (*Peromyscus maniculatus*). *Journal of Evolutionary Biology, 24,* 2153–2163. doi:10.1111/j.1420-9101.2011.02344.x

Carere, C., & Maestripieri, D. (2013). *Animal personalities: Behavior, physiology, and evolution*. Chicago: The University of Chicago Press.

Caspi, A. (2002). Role of genotype in the cycle of violence in maltreated children. *Science, 297,* 851–854. doi:10.1126/science.1072290

Christensen, K. A., Brunelli, J. P., Wheeler, P. A., & Thorgaard, G. H. (2014). Antipredator behavior QTL: Differences in rainbow trout clonal lines derived from wild and hatchery populations. *Behavior Genetics, 44,* 535–546. doi:10.1007/s10519-014-9663-9

Christiansen, S. B., & Forkman, B. (2007). Assessment of animal welfare in a veterinary context—a call for ethologists. *Applied Animal Behavior Science, 106*(4), 203–220. doi:10.1016/j.applanim.2007.01.004

Class, B., & Brommer, J. E. (2015). A strong genetic correlation underlying a behavioral syndrome disappears during development because of genotype–age interactions. *Proceedings of the Royal Society B: Biological Sciences, 282,* 20142777. doi:10.1098/rspb.2014.2777

Danchin, É., Charmantier, A., Champagne, F. A., Mesoudi, A., Pujol, B., & Blanchet, S. (2011). Beyond DNA: Integrating inclusive inheritance into an extended theory of evolution. *Nature Reviews Genetics, 12,* 475–486. doi:10.1038/nrg3028

Delvecchio, G., Bellani, M., Altamura, A. C., & Brambilla, P. (2016). The association between the serotonin and dopamine neurotransmitters and personality traits. *Epidemiology and Psychiatric Sciences, 25,* 109–112. doi:10.1017/s2045796015001146

Moor, D., Costa, P. T., Terracciano, A., Krueger, R. F., de Geus, E. J. C., Toshiko, T., et al. (2012). Meta-analysis of genome-wide association studies for personality. *Molecular Psychiatry, 17,* 337–349. doi:10.1038/mp.2010.128

Dingemanse, N. J., Plas, F. V., Wright, J., Reale, D., Schrama, M., Roff, D. A., et al. (2009). Individual experience and evolutionary history of predation affect expression of heritable variation in fish personality and morphology. *Proceedings of the Royal Society B: Biological Sciences, 276,* 1285–1293. doi:10.1098/rspb.2008.1555

Dingemanse, N. J., & Wolf, M. (2013). Between-individual differences in behavioral plasticity within populations: Causes and consequences. *Animal Behavior, 85,* 1031–1039. doi:10.1016/j.anbehav.2012.12.032

Dingemanse, N. J., & Dochtermann, N. A. (2013). Quantifying individual variation in behavior: Mixed-effect modelling approaches. *Journal of Animal Ecology, 82*(1), 39–54. doi:10.1111/1365-2656.12013

Dochtermann, N. A., & Roff, D. A. (2010). Applying a quantitative genetics framework to behavioral syndrome research. *Philosophical Transactions of the Royal Society B: Biological Sciences, 365,* 4013–4020. doi:10.1098/rstb.2010.0129

Dochtermann, N. A., Schwab, T., & Sih, A. (2015). The contribution of additive genetic variation to

personality variation: Heritability of personality. *Proceedings of the Royal Society B: Biological Sciences, 282,* 2014–2201. doi:10.1098/rspb.2014.2201
Drent, P. J., van Oers, K., & Noordwijk, A. J. (2003). Realized heritability of personalities in the great tit (*Parus major*). *Proceedings of the Royal Society B: Biological Sciences, 270,* 45–51. doi:10.1098/rspb.2002.2168
Ducrest, A., Keller, L., & Roulin, A. (2008). Pleiotropy in the melanocortin system, coloration and behavioral syndromes. *Trends in Ecology & Evolution, 23,* 502–510. doi:10.1016/j.tree.2008.06.001
Eaves, L., & Verhulst, B. (2014). Problems and pit-falls in testing for G × E and epistasis in candidate gene studies of human behavior. *Behavior Genetics, 44,* 578–590. doi:10.1007/s10519-014-9674-6
Eley, T. C., & Plomin, R. (1997). Genetic analyses of emotionality. *Current Opinion in Neurobiology, 7,* 279–284. doi:10.1016/s0959-4388(97)80017-7
Erickson, D. L., Fenster, C. B., Stenoien, H. K., & Price, D. (2004). Quantitative trait locus analyses and the study of evolutionary process. *Molecular Ecology, 13,* 2505–2522. doi:10.1111/j.1365-294x.2004.02254.x
Falconer, D., & Mackay, T. F. C. (1996). *Introduction to quantitative genetics* (4th ed.). Harlow: Longman.
Fidler, A. E., van Oers, K., Drent, P. J., Kuhn, S., Mueller, J. C., & Kempenaers, B. (2007). Drd4 gene polymorphisms are associated with personality variation in a passerine bird. *Proceedings of the Royal Society B: Biological Sciences, 274*(1619), 1685–1691. doi:10.1098/rspb.2007.0337
Filby, A. L., Paull, G. C., Hickmore, T. F., & Tyler, C. R. (2010). Unravelling the neurophysiological basis of aggression in a fish model. *BMC Genomics, 11,* 498. doi:10.1186/1471-2164-11-498
Fitzpatrick, M., Benshahar, Y., Smid, H., Vet, L., Robinson, G., & Sokolowski, M. (2005). Candidate genes for behavioral ecology. *Trends in Ecology & Evolution, 20,* 96–104. doi:10.1016/j.tree.2004.11.017
Flint, J., & Mackay, T. F. (2009). Genetic architecture of quantitative traits in mice, flies, and humans. *Genome Research, 19,* 723–733. doi:10.1101/gr.086660.108
Flint, J. (2003). Analysis of quantitative trait loci that influence animal behavior. *Journal of Neurobiology, 54,* 46–77. doi:10.1002/neu.10161
Flint, J., & Mott, R. (2001). Finding the molecular basis of quantitative traits: Successes and pitfalls. *Nature Reviews Genetics, 2,* 437–445. doi:10.1038/35076585
Frazzetto, G., Lorenzo, G. D., Carola, V., Proietti, L., Sokolowska, E., Siracusano, A., et al. (2007). Early trauma and increased risk for physical aggression during adulthood: The moderating role of MAOA genotype. *PLoS ONE, 2,* e486. doi:10.1371/journal.pone.0000486
Gershenfeld, H. K., & Paul, S. M. (1997). Mapping quantitative trait loci for fear-like behaviors in mice. *Genomics, 46,* 1–8. doi:10.1006/geno.1997.5002
Gordon, J. A., & Hen, R. (2004). Genetic approaches to the study of anxiety. *Annual Review of Neuroscience, 27,* 193–222. doi:10.1146/annurev.neuro.27.070203.144212
Greenwood, A. K., Ardekani, R., Mccann, S. R., Dubin, M. E., Sullivan, A., Bensussen, S., et al. (2015). Genetic mapping of natural variation in schooling tendency in the threespine stickleback. *Genes|Genomes|Genetics, 5,* 761–769. doi:10.1534/g3.114.016519
Groothuis, T. G., & Trillmich, F. (2011). Unfolding personalities: The importance of studying ontogeny. *Developmental Psychobiology, 53,* 641–655. doi:10.1002/dev.20574
Guan, Y., & Stephens, M. (2011). Bayesian variable selection regression for genome-wide association studies and other large-scale problems. *The Annals of Applied Statistics, 5,* 1780–1815. doi:10.1214/11-aoas455
Gutierrez-Gil, B., Ball, N., Burton, D., Haskell, M., Williams, J. L., & Wiener, P. (2008). Identification of quantitative trait loci affecting cattle temperament. *Journal of Heredity, 99,* 629–638. doi:10.1093/jhered/esn060
Haskell, M. J., Simm, G., & Turner, S. P. (2014). Genetic selection for temperament traits in dairy and beef cattle. *Frontiers in Genetics, 5,* 368. doi:10.3389/fgene.2014.00368
Heils, A., Teufel, A., Petri, S., Seemann, M., Bengel, D., Balling, U., et al. (1995). Functional promoter and polyadenylation site mapping of the human serotonin (5-HT) transporter gene. *Journal of Neural Transmission, 102,* 247–254. doi:10.1007/bf01281159
Henderson, C. R. (1984). *Applications of linear models in animal breeding*. Guelph, ON: University of Guelph.
Hovatta, I., & Barlow, C. (2008). Molecular genetics of anxiety in mice and men. *Annals of Medicine, 40,* 92–109. doi:10.1080/07853890701747096
Inoue-Murayama, M. (2009). Genetic polymorphism as a background of animal behavior. *Animal Science Journal, 80,* 113–120. doi:10.1111/j.1740-0929.2008.00623.x
Ito, H., Nara, H., Inoue-Murayama, M., Shimada, M. K., Koshimura, A., Ueda, Y., et al. (2004). Allele frequency distribution of the canine dopamine receptor d4 gene exon iii and i in 23 breeds. *Journal of Veterinary Medical Science, 66,* 815–820. doi:10.1292/jvms.66.815
Johnsson, M., Williams, M. J., Jensen, P., & Wright, D. (2016). Genetical genomics of behavior: A novel chicken genomic model for anxiety behavior. *Genetics, 202,* 327–340. doi:10.1534/genetics.115.179010
Jordan, K. W., Carbone, M., Yamamoto, A., Morgan, T. J., & Mackay, T. F. (2007). Quantitative genomics of locomotor behavior in *Drosophila melanogaster*. *Genome Biology, 8,* R172. doi:10.1186/gb-2007-8-8-r172
Kim-Cohen, J., Caspi, A., Taylor, A., Williams, B., Newcombe, R., Craig, I. W., et al. (2006). MAOA, maltreatment, and gene–environment interaction predicting children's mental health: New evidence and a

meta-analysis. *Molecular Psychiatry, 11,* 903–913. doi:10.1038/sj.mp.4001851

Koolhaas, J. M., De Boer, S. F., Buwalda, B., Van der Vegt, B. J., Carere, C., & Groothuis, A. G. G. (2001). How and why coping systems vary among individuals. In D. M. Broom (Ed.), *Coping with challenge: Welfare in animals including humans* (pp. 197–209). Dahlem: Dahlem University Press.

Korsten, P., Mueller, J. C., Hermannstädter, C., Bouwman, K. M., Dingemanse, N. J., Drent, P. J., et al. (2010). Association between DRD4 gene polymorphism and personality variation in great tits: A test across four wild populations. *Molecular Ecology, 19,* 832–843. doi:10.1111/j.1365-294x.2009.04518.x

Kotrschal, A., Lievens, E. J., Dahlbom, J., Bundsen, A., Semenova, S., Sundvik, M., et al. (2014). Artificial selection on relative brain size reveals a positive genetic correlation between brain size and proactive personality in the guppy. *Evolution, 68,* 1139–1149. doi:10.1111/evo.12341

Kratochwil, C. F., & Meyer, A. (2014). Closing the genotype–phenotype gap: Emerging technologies for evolutionary genetics in ecological model vertebrate systems. *BioEssays, 37,* 213–226. doi:10.1002/bies.201400142

Krebs, J. R., & Davies, N. B. (1997). *Behavioural ecology: An evolutionary approach.* Wiley-Blackwell.

Kruuk, L. E. (2004). Estimating genetic parameters in natural populations using the 'animal model'. *Philosophical Transactions of the Royal Society B: Biological Sciences, 359,* 873–890. doi:10.1098/rstb.2003.1437

Kruuk, L. E., Slate, J., & Wilson, A. J. (2008). New answers for old questions: The evolutionary quantitative genetics of wild animal populations. *Annual Review of Ecology Evolution and Systematics, 39,* 525–548. doi:10.1146/annurev.ecolsys.39.110707.173542

Laine, V. N., Gossmann, T. I., Schachtschneider, K. M., Garroway, C. J., Madsen, O., Verhoeven, K. J., et al. (2016). Evolutionary signals of selection on cognition from the great tit genome and methylome. *Nature Communications, 7,* 10474. doi:10.1038/ncomms10474

Laine, V. N., Herczeg, G., Shikano, T., Vilkki, J., & Merilä, J. (2014). QTL analysis of behavior in nine-spined sticklebacks (*Pungitius pungitius*). *Behavior Genetics, 44,* 77–88. doi:10.1007/s10519-013-9624-8

Ledon-Rettig, C. C., Richards, C. L., & Martin, L. B. (2012). Epigenetics for behavioral ecologists. *Behavioral Ecology, 24,* 311–324. doi:10.1093/beheco/ars145

Lynch, M., & Walsh, B. (1998). *Genetics and analysis of quantitative traits.* Sunderland, MA: Sinauer.

Lyons, L. A., Laughlin, T. F., Copeland, N. G., Jenkins, N. A., Womack, J. E., & O'brien, S. J. (1997). Comparative anchor tagged sequences (CATS) for integrative mapping of mammalian genomes. *Nature Genetics, 15,* 47–56. doi:10.1038/ng0197-47

Mackay, T. F., Stone, E. A., & Ayroles, J. F. (2009). The genetics of quantitative traits: Challenges and prospects. *Nature Reviews Genetics, 10,* 565–577. doi:10.1038/nrg2612

Mafli, A., Wakamatsu, K., & Roulin, A. (2011). Melanin-based coloration predicts aggressiveness and boldness in captive eastern Hermann's tortoises. *Animal Behavior, 81,* 859–863. doi:10.1016/j.anbehav.2011.01.025

Meffert, L., Hicks, S., & Regan, J. (2002). Nonadditive genetic effects in animal behavior. *The American Naturalist, 160*(S6), S198–S213. doi:10.1086/342896

Moore, D. S. (2015). *The developing genome: An introduction to behavioral epigenetics.* Oxford: Oxford University Press.

Mueller, J. C., Korsten, P., Hermannstaedter, C., Feulner, T., Dingemanse, N. J., Matthysen, E., et al. (2013). Haplotype structure, adaptive history and associations with exploratory behavior of the DRD4 gene region in four great tit (*Parus major*) populations. *Molecular Ecology, 22,* 2797–2809. doi:10.1111/mec.12282

Munafò, M. R., Freimer, N. B., Ng, W., Ophoff, R., Veijola, J., Miettunen, J., et al. (2009). 5-HTTLPR genotype and anxiety-related personality traits: A meta-analysis and new data. *American Journal of Medical Genetics Part B: Neuropsychiatric Genetics, 150B*(2), 271–281. doi:10.1002/ajmg.b.30808

Munafò, M. R., Yalcin, B., Willis-Owen, S. A., & Flint, J. (2008). Association of the Dopamine D4 receptor (DRD4) gene and approach-related personality traits: Meta-Analysis and new cata. *Biological Psychiatry, 63,* 197–206. doi:10.1016/j.biopsych.2007.04.006

Murphy, S. E., Norbury, R., Godlewska, B. R., Cowen, P. J., Mannie, Z. M., Harmer, C. J., et al. (2013). The effect of the serotonin transporter polymorphism (5-HTTLPR) on amygdala function: A meta-analysis. *Molecular Psychiatry, 18,* 512–520. doi:10.1038/mp.2012.19

Nelson, R. J., & Trainor, B. C. (2007). Neural mechanisms of aggression. *Nature Reviews Neuroscience, 8,* 536–546. doi:10.1038/nrn2174

van Oers, K., Drent, P. J., Jong, G. D., & Noordwijk, A. J. (2004). Additive and nonadditive genetic variation in avian personality traits. *Heredity, 93,* 496–503. doi:10.1038/sj.hdy.6800530

van Oers, K., & Mueller, J. C. (2010). Evolutionary genomics of animal personality. *Philosophical Transactions of the Royal Society B: Biological Sciences, 365,* 3991–4000. doi:10.1098/rstb.2010.0178

van Oers, K., Santure, A. W., Cauwer, I. D., Bers, N. E., Crooijmans, R. P., Sheldon, B. C., et al. (2014). Replicated high-density genetic maps of two great tit populations reveal fine-scale genomic departures from sex-equal recombination rates. *Heredity, 112*(3), 307–316. doi:10.1038/hdy.2013.107

van Oers, K., Jong, G. D., Noordwijk, A. V. K., & Drent, P. (2005). Contribution of genetics to the study of animal personalities: A review of case studies. *Behavior, 142,* 1185–1206. doi:10.1163/156853905774539364 van Oers, K., & Sinn, D. L. (2013). Quantitative and molecular genetics of animal personality. In C. Carere & D. Maestripieri (Eds.), *Animal personalities: Behavior, physiology, and evolution* (pp. 148–200). Chicago: University of Chicago Press.

van Oers, K., & Sinn, D. L. (2011). Toward a basis for the phenotypic gambit: Advances in the evolutionary genetics of animal personality. In *From genes to animal behavior primatology monographs* (pp. 165–183). doi:10.1007/978-4-431-53892-9_7

Oortmerssen, G. A., & Bakker, T. C. (1981). Artificial selection for short and long attack latencies in wild *Mus musculus domesticus*. *Behavior Genetics, 11,* 115–126. doi:10.1007/bf01065622

Ozaki, N., Goldman, D., Kaye, W. H., Plotnicov, K., Greenberg, B. D., Lappalainen, J., et al. (2003). Serotonin transporter missense mutation associated with a complex neuropsychiatric phenotype. *Molecular Psychiatry, 8,* 933–936. doi:10.1038/sj.mp.4001365

Page, R. E., Rueppell, O., & Amdam, G. V. (2012). Genetics of reproduction and regulation of honeybee (*Apis melliferal.*) social behavior. *Annual Review of Genetics, 46,* 97–119. doi:10.1146/annurev-genet-110711-155610

Penke, L., Denissen, J. J., & Miller, G. F. (2007). The evolutionary genetics of personality. *European Journal of Personality, 21,* 549–587. doi:10.1002/per.629

Petelle, M. B., Martin, J. G., & Blumstein, D. T. (2015). Heritability and genetic correlations of personality traits in a wild population of yellow-bellied marmots (*Marmota flaviventris*). *Journal of Evolutionary Biology, 28,* 1840–1848. doi:10.1111/jeb.12700

Poissant, J., Réale, D., Martin, J., Festa-Bianchet, M., & Coltman, D. (2013). A quantitative trait locus analysis of personality in wild bighorn sheep. *Ecology and Evolution, 3,* 474–481. doi:10.1002/ece3.468

Ponsuksili, S., Zebunke, M., Murani, E., Trakooljul, N., Krieter, J., Puppe, B., et al. (2015). Integrated Genome-wide association and hypothalamus eQTL studies indicate a link between the circadian rhythm-related gene PER1 and coping behavior. *Scientific Reports, 5,* 16264. doi:10.1038/srep16264

Quinn, J. L., Patrick, S. C., Bouwhuis, S., Wilkin, T. A., & Sheldon, B. C. (2009). Heterogeneous selection on a heritable temperament trait in a variable environment. *Journal of Animal Ecology, 78,* 1203–1215. doi:10.1111/j.1365-2656.2009.01585.x

Quinn, J., & Cresswell, W. (2005). Personality, anti-predation behavior and behavioral plasticity in the chaffinch *Fringilla coelebs*. *Behavior, 142,* 1377–1402. doi:10.1163/156853905774539391

Rasmuson, M. (2009). Old and new ideas about genes and behavior. *Hereditas, 146,* 198–203. doi:10.1111/j.1601-5223.2009.02138.x

Ripke, S., O'Dushlaine, C., Chambert, K., Moran, J. L., Kahler, A. K., Akterin, S., et al. (2013). Genome-wide association analysis identifies 13 new risk loci for schizophrenia. *Nature Genetics, 45,* 1150–1159. doi:10.1038/ng.2742

Risch, N. J. (2000). Searching for genetic determinants in the new millennium. *Nature, 405,* 847–856. doi:10.1038/35015718

Riyahi, S., Björklund, M., Mateos-Gonzalez, F., & Senar, J. C. (2017). Personality and urbanization: Behavioral traits and DRD4 SNP830 polymorphisms in great tits in Barcelona city. *Journal of Ethology*. doi:10.1007/s10164-016-0496-2

Robinson, M. R., Santure, A. W., Decauwer, I., Sheldon, B. C., & Slate, J. (2013). Partitioning of genetic variation across the genome using multimarker methods in a wild bird population. *Molecular Ecology, 22,* 3963–3980. doi:10.1111/mec.12375

Rodenburg, T. B., Komen, H., Ellen, E. D., Uitdehaag, K. A., & Arendonk, J. A. (2008). Selection method and early-life history affect behavioral development, feather pecking and cannibalism in laying hens: A review. *Applied Animal Behavior Science, 110,* 217–228. doi:10.1016/j.applanim.2007.09.009

Roff, D. A. (1996). The evolution of genetic correlations: An analysis of patterns. *Evolution, 50,* 1392. doi:10.2307/2410877

Santure, A. W., Poissant, J., Cauwer, I. D., van Oers, K., Robinson, M. R., Quinn, J. L., et al. (2015). Replicated analysis of the genetic architecture of quantitative traits in two wild great tit populations. *Molecular Ecology, 24,* 6148–6162. doi:10.1111/mec.13452

Savitz, J. B., & Ramesar, R. S. (2004). Genetic variants implicated in personality: A review of the more promising candidates. *American Journal of Medical Genetics, 131B*(1), 20–32. doi:10.1002/ajmg.b.20155

Scott, A. L., Bortolato, M., Chen, K., & Shih, J. C. (2008). Novel monoamine oxidase A knock out mice with human-like spontaneous mutation. *NeuroReport, 19,* 739–743. doi:10.1097/wnr.0b013e3282fd6e88

Shih, J., & Thompson, R. (1999). Monoamine oxidase in neuropsychiatry and behavior. *The American Journal of Human Genetics, 65,* 593–598. doi:10.1086/302562

Shikano, T., Ramadevi, J., Shimada, Y., & Merilä, J. (2010). Utility of sequenced genomes for microsatellite marker development in non-model organisms: A case study of functionally important genes in nine-spined sticklebacks (*Pungitius pungitius*). *BMC Genomics, 11,* 334. doi:10.1186/1471-2164-11-334

Shimada, M. K., Inoue-Murayama, M., Ueda, Y., Maejima, M., Murayama, Y., Takenaka, O., et al. (2004). Polymorphism in the second intron of dopamine receptor D4 gene in humans and apes. *Biochemical and Biophysical Research Communications, 316,* 1186–1190. doi:10.1016/j.bbrc.2004.03.006

Sih, A., Bell, A., & Johnson, J. (2004). Behavioral syndromes: An ecological and evolutionary overview. *Trends in Ecology & Evolution, 19,* 372–378. doi:10.1016/j.tree.2004.04.009

Singh, P., Schimenti, J. C., & Bolcun-Filas, E. (2015). A mouse geneticist's practical guide to CRISPR applications. *Genetics, 199*(1), 1–15. doi:10.1534/genetics.114.169771

Slate, J. (2013). From beavis to beak color: A simulation study to examine how much QTL mapping can reveal about the genetic architecture of quantitative traits. *Evolution*. doi:10.1111/evo.12060

Slate, J. (2005). INVITED REVIEW: Quantitative trait locus mapping in natural populations—progress, caveats and future directions. *Molecular Ecology, 14,* 363–379. doi:10.1111/j.1365-294x.2004.02378.x

Sokolowski, M. B. (2001). Drosophila: Genetics meets behavior. *Nature Reviews Genetics, 2,* 879–890. doi:10.1038/35098592

Stearns, F. W. (2010). One hundred years of pleiotropy: A retrospective. *Genetics, 186,* 767–773. doi:10.1534/genetics.110.122549

Stirling, D. G., Réale, D., & Roff, D. A. (2002). Selection, structure and the heritability of behavior. *Journal of Evolutionary Biology, 15,* 277–289. doi:10.1046/j.1420-9101.2002.00389.x

Swiech, L., Heidenreich, M., Banerjee, A., Habib, N., Li, Y., Trombetta, J., et al. (2014). In vivo interrogation of gene function in the mammalian brain using CRISPR-Cas9. *Nature Biotechnology, 33,* 102–106. doi:10.1038/nbt.3055

Taylor, R. W., Boon, A. K., Dantzer, B., Réale, D., Humphries, M. M., Boutin, S., et al. (2012). Low heritabilities, but genetic and maternal correlations between red squirrel behaviors. *Journal of Evolutionary Biology, 25,* 614–624. doi:10.1111/j.1420-9101.2012.02456.x

Terracciano, A., Sanna, S., Uda, M., Deiana, B., Usala, G., Busonero, F., et al. (2008). Genome-wide association scan for five major dimensions of personality. *Molecular Psychiatry, 15,* 647–656. doi:10.1038/mp.2008.113

Tonteri, A., Vasemägi, A., Lumme, J., & Primmer, C. R. (2010). Beyond MHC: Signals of elevated selection pressure on Atlantic salmon (*Salmo salar*) immune-relevant loci. *Molecular Ecology, 19,* 1273–1282. doi:10.1111/j.1365-294x.2010.04573.x

Trut, L. N., Iliushina, I. Z., Prasolova, L. A., & Kim, A. A. (1997). The hooded allele and selection of wild Norway rats *Rattus norvegicus* for behavior. *Genetika, 33,* 1156–1161. [Translation in English available in: Russian Journal of Genetics, 1997. 33, 983–989].

Trut, L. (1999). Early canid domestication: The farm-fox experiment. *American Scientist, 87,* 160. doi:10.1511/1999.20.813

Verhulst, E. C., Mateman, A. C., Zwier, M. V., Caro, S. P., Verhoeven, K. J., & van Oers, K. (2016). Evidence from pyrosequencing indicates that natural variation in animal personality is associated with DRD4 DNA methylation. *Molecular Ecology, 25,* 1801–1811. doi:10.1111/mec.13519

Visser, M. E., Caro, S. P., van Oers, K., Schaper, S. V., & Helm, B. (2010). Phenology, seasonal timing and circannual rhythms: Towards a unified framework. *Philosophical Transactions of the Royal Society B: Biological Sciences, 365,* 3113–3127. doi:10.1098/rstb.2010.0111

Vitaterna, M. H., Pinto, L. H., & Takahashi, J. S. (2006). Large-scale mutagenesis and phenotypic screens for the nervous system and behavior in mice. *Trends in Neurosciences, 29,* 233–240. doi:10.1016/j.tins.2006.02.006

Weeland, J., Overbeek, G., Castro, B. O., & Matthys, W. (2015). Underlying mechanisms of gene–environment interactions in externalizing behavior: A systematic review and search for theoretical mechanisms. *Clinical Child and Family Psychology Review, 18,* 413–442. doi:10.1007/s10567-015-0196-4

Wendland, J. R., Martin, B. J., Kruse, M. R., Lesch, K., & Murphy, D. L. (2006). Simultaneous genotyping of four functional loci of human SLC6A4, with a reappraisal of 5-HTTLPR and rs25531. *Molecular Psychiatry, 11,* 224–226. doi:10.1038/sj.mp.4001789

Williams, R. J., Pelton, R. B., & Siegel, F. L. (1962). Individuality as exhibited by inbred animals; its implications for human behavior. *Proceedings of the National Academy of Sciences, 48,* 1461–1466. doi:10.1073/pnas.48.8.1461

Youngson, N. A., & Whitelaw, E. (2008). Transgenerational epigenetic effects. *Annual Review of Genomics and Human Genetics, 9,* 233–257. doi:10.1146/annurev.genom.9.081307.164445

Zou, W., & Zeng, Z. (2009). Multiple interval mapping for gene expression QTL analysis. *Genetica, 137,* 125–134. doi:10.1007/s10709-009-9365-z

Personality from the Perspective of Behavioral Ecology

5

Jon E. Brommer and Barbara Class

Abstract

Behavioral ecologists consider behaviors that show significant between-individual variation as aspects of personality. When multiple aspects of personality covary, these are viewed as a behavioral syndrome. Meta-analyses have demonstrated that behaviors typically are repeatable and that behavioral syndromes are common across a wide variety of taxa. The core interest in behavioral ecology is to understand why such between-individual differences in behavior arise and how they are maintained. We present in this chapter an overview of two inter-connected research avenues: evolutionary quantitative genetics and individual optimization theories. We outline the basic premises of these approaches and summarize what empirical studies have demonstrated thus far. We emphasize the increasing recognition of the hierarchical nature of aspects of personality and behavioral syndromes in behavioral ecology, as well as the plasticity of personality and behavioral syndromes with respect to environmental conditions and to age. We present an overview of insights derived from explicit incorporation of between-individual variation in behavioral plasticity into current personality research in behavioral ecology, which emphasizes how personality is likely to be less consistent across individuals than originally perceived.

Behavioral Ecology

Behavioral ecology aims to understand why animals behave the way they do under ecologically relevant conditions. The question "why" is typically understood in terms of Tinbergen's (1963) recognition of this question's four aspects, although with an emphasis on understanding behavior from its ultimate

J.E. Brommer (✉) · B. Class
Department of Biology, University of Turku, Turku, Finland
e-mail: jon.brommer@utu.fi

B. Class
e-mail: barbara.class@utu.fi

J. Vonk et al. (eds.), *Personality in Nonhuman Animals*,
DOI 10.1007/978-3-319-59300-5_5

(evolutionary) functional perspective. As a consequence, a behavioral ecologist is typically interested in the fitness costs and benefits that a behavior imparts to the individual displaying this behavior. This type of thinking is central to a major theoretical framework underlying behavioral ecology: individual optimization. Under individual optimization, one expects the behavior adopted by an individual to be the behavior maximizing the individual's fitness. Because of this perspective, behavior is viewed as concrete and directly quantifiable: A behavior is hence a measurable trait of an individual similar to other traits, such as body size, coloration, and number of offspring. Behaviors are either quantified by direct observation (e.g., rate at which a parent feeds its offspring), by experimental measurement within the ecological setting (e.g., latency to attack a stuffed conspecific simulating territorial intrusion), by measurement of a behavior in a completely artificial setting (e.g., the number of exploratory flights a bird makes in a short period of time directly after release in a closed room), or by a combination of these approaches (Carter et al. 2013). Much of behavioral ecology is based on studying wild populations, but certain studies in semi-captive populations or laboratory populations are also considered as behavioral ecology. Irrespective of how the behavior is quantified, a behavioral ecologist's interest is to relate this behavior to performance of the individual within its ecological context. The focus on quantifying behavior, and from those measures distill what can be considered as aspects of personality, characterizes what has been termed the "biological perspective." This approach differs therein from the "psychological perspective," where aspects of personality of a subject are predefined and these aspects are rated by a person close to the subject (Koski 2011). The field of behavioral ecology is considered to be still developing its approach to define and measure personality (Carter et al. 2013).

In this chapter, we distinguish two partly overlapping research avenues; one based on evolutionary quantitative genetics, and one based on individual optimization and related behavioral ecological theories. Behavioral ecologists have become enthused about the study of personality only about a decade or so ago, and there is hence a large amount of recent work published and several recent reviews of personality from different behavioral ecological perspectives (e.g., David and Dall 2016; Dingemanse and Wolf 2010; Carere and Maestripieri 2013; Réale et al. 2010; Réale et al. 2007; Sih et al. 2004). Because of this, our objective here is to provide a non-exhaustive overview of personality research in this field. We have further structured this chapter such that we initially discuss insights when the focus is on one measured behavior, from both the perspective of evolutionary quantitative genetics and of individual optimization theory. In the second part, we extend our overview to the perspective where multiple behaviors are considered simultaneously.

Personality in Behavioral Ecology

Personality, also called "animal personality," "temperament," or "coping style" in behavioral ecology, refers to the phenomenon that behaviors tend to be repeatable (Bell et al. 2009). That is, if we measure the behavior *beh* of individual *i* (1,2,3, …, n_i) at trial *t* (1,2,3,… n_t), we may, in the simplest case, assume that

$$beh_{it} = \mu_{beh} + \mu_F + ind_i + \varepsilon_{it}, \qquad (5.1)$$

where μ_{beh} is the mean behavior over all individuals measured, μ_F denotes deviation in the mean due to fixed effects (e.g., age of the individual), ind_i is the individual-specific deviation from the overall fixed-effect mean, and ε_{it} the residual deviation. These two deviations are assumed to be random draws made from a normal distribution with a mean of zero and a certain variance σ^2, which we notate here as $ind \sim N(0, \sigma^2_{ind})$, and $\varepsilon \sim N(0, \sigma^2_{res})$. In general, other assumptions could also apply, for example heterogeneous residual errors. If we measure the same behavior repeatedly for a number of individuals, then Eq. (5.1) describes a linear mixed model where the ID of the individual codes a random effect, which can be solved using standard software to yield estimates of σ^2_{ind} and σ^2_{res}, and hence of repeatability (Falconer and MacKay 1996) defined as

$$R = \frac{\sigma^2_{ind}}{\sigma^2_{ind} + \sigma^2_{res}} = \frac{\sigma^2_{ind}}{\sigma^2_{phen}}, \qquad (5.2)$$

where σ^2_{phen} denotes the so-called phenotypic variance (conditional on the fixed effects included in the model), which is the estimated variance at the phenotypic level for, in this case, behavior *beh*. Repeatability as defined here (Eq. 5.2) is thus the proportion of phenotypic variance (conditional on fixed effects in the model) explained by between-individual variance (cf. Falconer and MacKay 1996). In the formulation above, repeatability is derived from a mixed model. Because the inclusion of significant fixed effects explains, by definition, part of the residual and, under some conditions, individual-specific variance, the calculation of repeatability is specific to the model structure used. Thus, careful comparison of repeatabilities across studies is warranted (cf. Bell et al. 2009). Repeatability is equivalent to the intraclass correlation (*ICC*) commonly used in other fields. Repeatability can be calculated also for behaviors that are not Gaussian distributed as long as the behavior is linear on some scale (e.g., a logit scale) and the proper influence of that scaling on residual variances is included. For a thorough overview of repeatability and its quantification in various contexts, we refer the reader to Nakagawa and Schielzeth (2010).

The central feature in the verbal definitions of personality for behavioral ecologists includes the property "consistency" across time and situations (cf. Réale et al. 2007). Consistency is typically deemed present when there is a nonzero repeatability R (Eq. 5.2) of measures of a behavior repeatedly quantified for a set of individuals during multiple trials. There is confusion over the terms "(within-individual) consistency" and "between-individual variation" in the literature (for an overview, see Japyassú and Malange 2014). We here relate these terms to the realization that high repeatability may stem from reduced residual variance σ^2_{res}(i.e., increased within-individual consistency) and/or increased between-individual variance σ^2_{ind}. Nevertheless, only nonzero between-individual variation σ^2_{ind} is evidence of a nonzero repeatability in the focal behavior, irrespective of the amount of residual variance σ^2_{res}. Cleasby et al. (2015) provide a detailed overview and discussion of approaches to quantify behavioral consistency. Even more confusing is the plethora of terms used in the literature for a repeatable behavior, including but not restricted to labeling the focal behavior "personality," "a personality trait," or giving it a predefined conceptual term (e.g., a repeatable measure of behavior is labeled "boldness"). Indeed, a seminal paper in animal personality research in behavioral ecology (Réale et al. 2007) explicitly argued that behaviors could be mapped onto an analogy of the human personality, the "Big Five" or "Five-Factor Model" (McCrae and Costa 1999). Within the behavioral ecology literature, there are examples where aspects of personality (e.g., boldness) are a priori assumed to be quantified by (typically one) measured behavior (e.g., flight initiation distance), as well as examples where authors consistently refer to the measured behavior as such, without imposing conceptual inferences with respect to aspects of personality. In this chapter, we view personality as a concept, and adhere to the notion that a focal, measured behavior is simply termed as such; finding that the focal behavior is repeatable constitutes evidence of the existence of personality in the study population, and the focal behavior could be considered an aspect of personality. Thus, we are also hesitant to label specific measures of behaviors according to some "established" description of personality when the behavior is shown to be repeatable, but—as noted above—there are other approaches in behavioral ecology (cf. Carter et al. 2013), and this view, hence, primarily reflects our stance on this issue and not necessarily the general consensus in the field. A summary of studies in recent (since 2010) issues in the journal

Behavioral Ecology and Sociobiology (see Table 5.1) provides an overview both of the range of organisms studied but also of the types of behaviors studied and of the naming convention used. All of these studies calculated the repeatability (see Table 5.1) and considered this evidence that the behavior studied was an aspect of personality.

The approach to personality research in behavioral ecology, as described above, follows traditional approaches in evolutionary biology, and its statistical framework in evolutionary quantitative genetics (Réale et al. 2007). Indeed, the behavioral ecological perspective on personality is explicitly an evolutionary one. In general, if we consider the phenotype as anything we can measure on an individual, then for evolution of this phenotype to occur, it is not sufficient that there simply is variation in the phenotype. Phenotypic variation across individuals must also be caused by something that is inherited by the next generation from the parental generation (Falconer and MacKay 1996). As a minimal requirement, the phenotype must be repeatable (Falconer and MacKay 1996), because non-repeatable phenotypic variation is equivalent to "noise" and/or measurement error. Thus, a necessary although not sufficient requirement for evolution to occur is the presence of σ^2_{ind}; there is variation across individuals in their individual-specific value *ind* [Eq. (5.1)]. Clearly, assessment of whether a phenotype is repeatable is a common-sense and generic approach to identify traits that are amenable for further analyses, and its implementation when it comes to behavior is therefore not exceptionally novel or unusual. In fact, it is rather surprising that interest in repeatable behaviors only boomed in the last decade or so (although repeatable behaviors as such were documented earlier, see references and discussion in Réale et al. 2007).

Variance partitioning, as described above, is central to quantitative genetics and the quantitative genetics toolbox hence is a natural one for behavioral ecology (Dochtermann and Roff 2010). Quantitative genetics is, of course, primarily interested in the genetic level, and uses a statistical approach to further partition σ^2_{ind} into its putative causal elements, where, for the simplest quantitative genetic description of behavior *beh*

$$beh_{it} = \mu_{beh} + \mu_F + (a_i + pe_i) + \varepsilon_{it}, \qquad (5.3)$$

where the bracketed terms a_i denotes the breeding value and pe_i the permanent environment of individual i and other terms as in Eq. (5.1). Equation (5.3) is a simplified description of the genetics as it ignores genetic dominance, epistasis, and other interactions within the genome, but is in practice commonly used (Lynch and Walsh 1998). Again, the standard statistical assumption of identical and independent distribution is applied and $a \sim N(0, \sigma^2_A)$ and $pe \sim N(0, \sigma^2_{PE})$, where σ^2_A and σ^2_{PE} are the additive-genetic and permanent environmental variances respectively. A breeding value of an individual can be heuristically understood as the summation of all genes' effect on the phenotype. In the absence of any other effect, an individual's phenotype would be its breeding value. The permanent environment is the effect that is not due to additive genes but that is conserved across the repeated records of the individual. The permanent environmental effect truly is a potpourri. An intuitive example of a permanent environmental effect is maternal effects (when not modeled explicitly); because each individual has one mother, any effect this mother has on its phenotype is maintained (i.e., permanent) across repeated measures of the phenotype. Other examples include consistent spatial differences in resource availability between individuals. However, genetic dominance (when not modeled explicitly) may appear as a permanent environmental effect, too.

Our emphasis in this chapter is not on the genetics of personality, which is discussed elsewhere (see Laine and Van Oers, this volume). Nevertheless, we wish to point out that the bracketed term in Eq. (5.3) is identical to the individual-specific value *ind* in Eq. (5.1). Thus, under the (simplified) quantitative genetic framework above, variation between individuals

Table 5.1 Overview of the kind of organisms, behaviors and terminology for the traits investigated in behavioral ecology, as well as their repeatability

Study	Common name	Species	Behavior name	Description	Repeatability
Lichtenstein et al. (2016)	Western black widow spider	*Lactrodectus hesperus*	Boldness	Response to a puff of air (retreat or not)	0.15
Michelangeli et al. (2016)	Delicate skink	*Lampropholis delicata*	Sociability	Time spent basking in proximity to conspecific	0.34
DiRienzo et al. (2016)	Field cricket	*Gryllus integer*	Boldness	Latency to emerge head from the vial	0.17
Yuen et al. (2016)	African striped mouse	*Rhabdomys pumilio*	Exploration	Latency to contact novel object	0.77
Mell et al. (2016)	Viviparous lizard	*Zootoca vivipara*	Risk-taking (boldness)	Time head hidden after a simulated predator attack	0.52
Wexler et al. (2016)	Flour beetle	*Tribolium castaneum*	Edge preference	Proportion of time spent in the periphery of the arena	0.86
Toscano and Monaco (2015)	Mud crab	*Panopeus herbstii*	Activity	Proportion of observations where crabs were observed active (/20 total observations)	0.231
Careau et al. (2015)	Chipmunk	*Tamias striatus*	Docility	Seconds spent immobile in handling-bag test	0.258
Rödel et al. (2014)	European rabbit	*Oryctolagus cuniculus*	Offensive agonsistic behavior	The frequency (interactions/h) of events when the focal animal was observed chasing another animal	0.226
Stein and Bell (2015)	Three-spined stickleback	*Gasterosteus aculeatus*	Bites	Number of bites towards intruder	0.14
Boulton et al. (2014)	Sheepshead swordstail	*Xiphophorus birchmanni*	Activity	Percentage of time moving at a minimum 1.5 cm/s (%)	0.193
Toscano et al. (2014)	Mud crab	*Panopeus herbstii*	Refuge use	Proportion of the 30 observations where crabs were completely in the oyster shell refuge	0.173
Kluen et al. (2014)	Blue tit	*Cyanistes caeruleus*	Handling aggression	Aggression score during handling (1–5)	0.4
Wilson et al. (2013)	Sheepshead swordstail	*Xiphophorus birchmanni*	No. attacks	Number of attacks on opponent (characterised by a sudden forward acceleration towards the opponent that may or may not result in contact) (sqrt)	0.442
Fitzsimmons and Bertram (2013)	Field cricket	*Gryllus veletis*	Aggressiveness	Sum of all aggression scores (1–4)*time performed	0.15
Alcalay et al. (2014)	Ant lion	*Myrmeleon hyalinus*	Displacement to distance	Delta of the first and last coordinates along the track length divided by the distance length	0.22

(continued)

Table 5.1 (continued)

Study	Common name	Species	Behavior name	Description	Repeatability
Cordes et al. (2013)	Lesser wax moth	*Achroia grisella*	Silence response	Duration of silence response to bat signal	0.3
Lapierre et al. (2011)	Song sparrow	*Melospiza melodia*	Sharing strategy	Spearman's correlation coefficient between a given male's output of each song type and that song type's neighborhood sharing coefficient, across all the songs within the male's repertoire	0.38
Koski (2011)	Chimpanzee	*Pan troglodytes*	Submission	Frequency of submissive behaviors (crouch, pant grunt)	0.34
Pruitt et al. (2011)	Western black widow spider	*Latrodectus hesperus*	Retreat darkness	Females' response towards light intensity: distance from the darkest edge of the Nalgene enclosure after a 24-h settlement period	0.47
Grim et al. (2014)	Blackbird	*Turdus merula*	Egg ejection	Eject or not a non-mimetic egg	0.7
Stoffer et al. (2015)	Wolf spider	*Schizocosa ocreata*	Female selectivity	Cumulative number of receptivity displays to the courting male with large tufts/male with small tufts	0.235

Compilation is based on a search for articles containing the term "repeatab*" published from 2010 onwards in the journal *Behavioral Ecology and Sociobiology*. The search returned 22 studies quantifying 1–16 behaviors each (total of 89 behaviors) of which we randomly selected one behavior per article. The overall mean repeatability of all these behaviors was 38%

in behavior may be caused by additive-genetic and/or permanent environmental effects. Clearly, for evolution to occur, the phenotype must have a nonzero heritability h^2, defined $h^2 = \sigma_A^2/\sigma_{phen}^2$, the theoretical upper estimate of which is the repeatability (Eq. 5.2) (Falconer and MacKay 1996). In general, quantitative geneticists have a variety of approaches to estimate the variances associated with the terms in Eq. (5.3), some of which are amenable for use in behavioral ecology (Wilson et al. 2010a, b). Whereas the heritability of behavior is fairly modest (about 14%), comparative analyses of published studies do show that additive-genetic variance σ_A^2 constitutes about 52% of the between individual variance σ_{ind}^2 in behaviors (Dochtermann et al. 2015). Thus, genetic factors are typically responsible for a sizeable fraction of the repeatability underlying animal personality. Hence, the evidence to date supports a hierarchical view where genes determine most of the variation between individuals and between-individual variation is a pronounced (although not massive) proportion of the phenotypic variance in behaviors (see Fig. 5.1).

Quantitative genetics, and, in particular, its variance partitioning approach are viewed as valuable analytical tools also outside behavioral ecology, for example in studying epidemiology (Worth et al. 2014). Behaviors that are repeatable but not heritable may be associated with differential performance of individuals, but because such behavioral variation is not inherited, any fitness differences associated with it will not cause evolution (Fisher 1958). Nevertheless, behavioral ecologists are not necessarily interested in whether, or to what extent, behavior is heritable. Rather, the focus is on identifying repeatable behaviors as the hallmark of aspects of

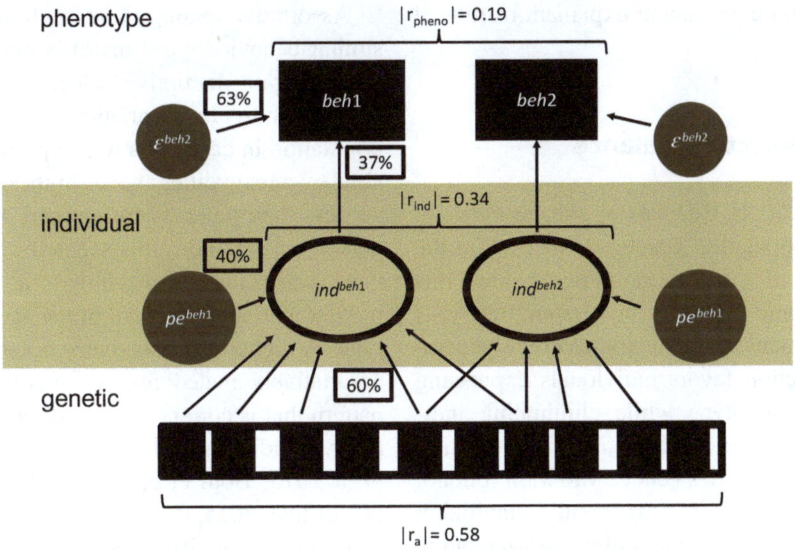

Fig. 5.1 A schematic summary of the hierarchical view of personality following from the variance partitioning perspective. On the genetic level, there is the genome with many genes (*black boxes*), some of which are involved in determining intrinsic propensity for one or more behaviors expressed by the individual. In this schematic example, genes are assumed to pleiotropically affect the behaviors, but other mechanisms may apply. The sum of the genetic effects is the breeding value *a*, Eq. (5.3). A defining feature of the behavioral ecology view of personality is the interest on the individual level (in *light gray*), here characterized as individual-specific values (*ind*) for two behaviors (*beh*1 and *beh*2), schematically presented to be influenced by the underlying genes and any other effects which are permanently associated with the individual, permanent environmental effect *pe*. Within boxes, the approximate variance explained by one hierarchical level to the next is given. Thus, 60% of variance in individual-specific values for a behavior is expected to be caused by genetic differences (Dochtermann et al. 2015). The individual-specific effect is expected to explain about 30% of variance in the behavior when measured on the phenotypic level (Bell et al. 2009). Expected magnitude of the correlations are given for the genetic level (r_a, Dochtermann 2011), individual-specific level (r_{ind}) and phenotypic level (r_{pheno}, Brommer and Class 2017)

personality and to explore the fitness consequences (costs and benefits) of such behaviors. From this perspective, a genetic underpinning of the focal behavior (Réale et al. 2007) is but one pathway for causing differences between individuals.

The Maintenance of Personality from an Evolutionary Quantitative Genetic Perspective

A focal endeavor of behavioral ecology is to study the selective consequences of variation in behavior. Meta-analysis shows that aspects of personality are under selection (Smith and Blumstein 2008). Whenever there is selection acting, one is faced with a need to explain how genetic and, as a corollary, between-individual, variation is maintained. This is because selection, barring disruptive selection, implies there is an optimal behavior or, when selection is directional, higher fitness for extreme values of the focal behavior. Why then do we still find personality? The explanations provided by evolutionary quantitative genetics for why personality (i.e., between-individual variation in behavior) persists are not unique to personality research, but are applicable to other traits. Evolutionary ecologists often discuss these processes in the light of “evolutionary stasis,” which refers to the phenomenon that traits under selection appear to not show micro-evolutionary changes (e.g., Merilä et al. 2001). In this section, we discuss some of the more prominent explanations related to personality.

Mutation–Selection Balance

Mutation–selection balance is one mechanism invoked by population geneticists to explain the maintenance of genetic variation. Under this scenario, genetic variation in personality exists because of mutations occurring at every generation and selection favors individuals expressing the optimal phenotype while eliminating individuals deviating from this optimum. Stabilizing selection thus decreases genetic variation (but not totally in the case of "apparent" stabilizing selection) but this variation will be restored at the next generation thanks to new mutations (Lande 1976; Nettle 2006; Penke et al. 2007; Zhang and Hill 2005). Importantly, because polygenic traits, such as behaviors, involve many genes, these traits may have a reasonable opportunity to accumulate mutations although the mutation rate per se is very low. Hence, one may attain a balance between the erosion of variation by (relatively mild) selection and the generation of new variation. The mutation-selection mechanism was supported for personality traits in humans (Verweij et al. 2012) but, to our knowledge, has not been empirically tested in other species.

Disruptive Selection and Sexual Selection

Genetic variation in aspects of personality within a population can also be maintained by selection if behavioral types at the extreme ends of a continuum have a higher fitness than intermediate behavioral types (disruptive selection, Lynch and Walsh 1998). Evidence for disruptive survival selection in behavioral traits is however scarce, but has been documented for adult exploratory behavior in eastern chipmunks (Bergeron et al. 2013) and antipredator behavior in garter snakes (Brodie 1992).

Assortative mating (when individuals that are similar behavior types mate) is one process that can generate disruptive selection and maintain between-individual variation in behavior in a population in case pairs where partners that have similar personalities have higher reproductive success than pairs where partners have different personalities. Under this scenario, all behavioral types can achieve equal fitness as long as they partner with an individual of the same behavioral type. A higher reproductive success of pairs that assortatively mated for personality is in fact a pattern that is commonly found, both under laboratory conditions and in the wild (Ariyomo and Watt 2013; Both et al. 2005; Class et al. 2014; David et al. 2015; Gabriel and Black 2012; Harris and Siefferman 2014; Kralj-Fišer et al. 2013; Schuett et al. 2011; Sinn et al. 2006; Spoon et al. 2006). Finally, the genetic variation underlying personality can be maintained by sexual selection if bold and shy males achieve similar paternity rates. In great tits for instance, more explorative males were found to sire more extra-pair young but fewer within-pair young with no difference in total paternity compared to shy males, hence reflecting the existence of personality-dependent mating tactics (Patrick et al. 2012).

Sexual Antagonistic Selection

Personality can be maintained by sexual antagonistic selection when the same behavioral trait, which is positively genetically correlated across sexes, is selected in opposite directions in males and females (Rice and Chippindale 2001). Empirical studies found sexual antagonistic selection in great tit exploration behavior (Dingemanse et al. 2004), and in sociality in the comb-footed spider (Pruitt and Riechert 2009a).

Balancing Selection

Fluctuations of environmental factors in time and/or space cause antagonistic selection pressures over time/space that maintain between-individual variation in behavior (Dingemanse and Réale 2013; Koolhaas et al. 2007; Nettle 2006; Penke et al. 2007). For instance, predation risk can vary in time (fluctuations in predator density) or space (some areas are less accessible to predators than others). Shy individuals are then predicted to be more likely to survive than bold individuals during years with high predator density and in unprotected areas, whereas bold individuals are predicted to be more likely to survive than shy individuals during years with lower predator density and in protected areas. Overall, both shy and bold behavioral types have a similar fitness and are maintained in the population. This hypothesis is supported by studies in wild and captive populations showing that fluctuations in food abundance (Both et al. 2005; Dingemanse et al. 2004; Kontiainen et al. 2009; Quinn et al. 2009; Le Coeur et al. 2015; Montiglio et al. 2014; Vetter et al. 2016), density (Cote et al. 2008; Le Galliard et al. 2015; Nicolaus et al. 2016; Quinn et al. 2009), predation risk (Réale and Festa-Bianchet 2003) and environmental variability in time (Réale et al. 2009; Taylor et al. 2014) or space (Monestier et al. 2015) can act to maintain variation in personalities by alternatively selecting different personality types over time/space, leading to a balance where one behavioral type never consistently outperforms the other behavioral type(s).

From an evolutionary point of view, negative frequency-dependent selection is another type of balancing selection where the fitness of individuals expressing a (heritable) behavior decreases with an increasing frequency of individuals expressing the same behavior in the population. As a result, genetic variation underlying personality is maintained because the different phenotypes achieve equal fitness payoffs at equilibrium. Empirical evidence for this mechanism maintaining behavioral variation, however, remains scarce and only a few empirical studies demonstrated negative frequency-dependent selection on heritable behavior (Fitzpatrick et al. 2007; Kralj-Fišer and Schneider 2012; Pruitt and Riechert 2009b).

Genotype-Environment or Genotype-Age Interaction

Aspects of personality may be under consistent directional selection, but the individual or genotype expressing the extreme personality may differ due to interactions with the environment or age. Thus, selection will tend to favor different individuals or genotypes as the environment changes or individuals age. Further on in this chapter we discuss plasticity of personality and conclude that there is strong evidence for between-individual variation in the plasticity of aspects of personality in response to environmental contexts and age.

Theory Explaining the Evolution of Repeatable Behavior (Individual Optimization)

We have thus far taken a data-driven view of personality, and of efforts linking it to selection and evolution. In this section, we provide an overview of how behavioral ecology theory explains the existence of personality. Behavioral ecology theory is strongly based on the concept of individual optimization (Krebs and Davies 2009). Individual optimization assumes that natural selection has proceeded such that a behavior displayed by an individual will be at the fitness optimum for the focal individual, and the individual is thus constantly balancing the fitness costs and benefits (Houston and McNamara 1999). Classical behavioral ecology terminology for this phenomenon uses a (perhaps somewhat confusing) semantic, where individuals are said to "decide" which "action," out of the set of all alternative actions, to express. The decision in this case does not imply a conscious action of a calculating animal mind, but is instead a shorthand formulation for natural selection having selected genes for a specific trait expression in a specific situation. A classic example concerns optimal reproductive decisions regarding clutch size and seasonal timing of reproduction (Daan and Tinbergen 1997).

In general, many optimization theories recognize that fitness payoffs arise from a game-theoretical situation where a focal individual's fitness is dependent on what the other individuals are doing (Brommer 2000). Individual optimization is, furthermore, a phenotypic framework. It is naïve to the underlying genetics, and, hence, assumes implicitly that any required genetic mechanism will evolve and that any genetic constraint will be removed during long-term evolution (Weissing 1996). The exception to the latter are the trade-offs, which posit that decisions leading to an increase in one aspect of fitness (e.g., produce a higher clutch size, or foraging success) also entail fitness costs (e.g., a reduction in survival). The trade-offs underlying the costs and benefits in optimization theory are fundamental ones, typically involving the resource allocation dilemma. Given that each individual has finite resources, it can allocate each unit of these resources to "A," "B," "C," etc. (including different behaviors), but each unit of resource can be spent only once. Individual optimization theory is powerful because it allows testable predictions to be generated. For example, if an individual produces an optimal clutch size, then both experimental increases and decreases of clutch size should result in lower fitness (Daan and Tinbergen 1997). In general, individual optimization theory has led behavioral ecology to focus on quantification and experimentation.

The existence of repeatable behaviors is, at first glance, diametrically opposite to the concept of individual optimization (Sih et al. 2004). Clearly, if each individual in a population can facultatively adjust its behavior, one would expect all individuals to behave optimally under any condition. As a consequence, there would not be any between-individual variation and thus no personality. In the behavioral ecology literature, this viewpoint is sometimes expressed as an assumption that theory predicts the mean (a single optimum), and that variation around this mean is noise. This sentiment, however, underappreciates individual optimization theory and forgets that the focus of the theory has always been on the individual level. For example, why do we see such variation in nature in reproductive output across individuals in a population? Individual optimization theory argues that this is because each individual makes an optimal decision on its reproductive output under the costs and benefits it expects to be subjected to in its specific situation, or—more generally, given its state (Houston and McNamara 1999). Thus, the short answer of individual optimization theory to the question "Why do animals show between-individual variation in behavior?" is that each individual expresses the behavior that is optimal given the individual's state. Evolutionary explanations along this line for the existence of consistent individual differences have been well reviewed (Dall et al. 2004; Dingemanse and Réale 2013; Dingemanse and Wolf 2010; Réale et al. 2007, 2010; Sih et al. 2004; Wolf and Weissing 2010; Wolf et al. 2013). In this section, we cover some of the basics of this theory, and relate it to examples of empirical work.

Stable State Distribution and Feedback Loops

Individuals vary in their state, defined as any feature that affects the costs and benefits of an individual's actions (e.g., energy reserves, metabolism, predation risk, age, information state, social rank, etc.; see Wolf et al. 2013) and individuals can flexibly adjust their behavior to their state to maximize their fitness. Hence, between-individual variation in behavior is constant whenever between-individual differences in state are maintained (stable state; Wolf et al. 2013). Especially when dealing with behaviors, however, it is clear that states can be affected by individuals' behaviors (through resource acquisition or learning for instance). Hence, even small initial differences in state between individuals can be reinforced through so-called positive feedbacks between behavior and state (Luttbeg and Sih 2010; Sih et al. 2015; Sih and Bell 2008; Wolf et al. 2008). Such positive feedbacks thus have the potential to generate consistent between-individual variation.

How likely are mechanisms for positive feedbacks between behavior and state in nature?

Some mechanisms are argued to always generate positive feedback. These mechanisms are, for instance, learning (by being bold, individuals increase their information that allows them to be bolder), state-dependent safety (boldness decreases risks of predation by increasing conditions that favors boldness), the winner/loser effect (individuals more likely to win a fight when they won the previous fight). Positive feedbacks can also be mediated by physiological characteristics of the individual, such as metabolism or hormones. Furthermore, aspects related to the extrinsic environment where the individuals are living can generate positive feedbacks. For example, when parasites are prevalent in the environment, a bold individual may be more likely to be infected, which increases its energetic needs and hence increases its boldness to assure it acquires the additional resources needed. Finally, the social environment may create strong positive feedbacks: an individual's aggression elicits aggression by others, and vice versa (Sih et al. 2015).

Not all mechanisms, however, necessarily give rise to positive feedback loops. The same mechanism may, under other assumed pathways, produce negative feedbacks loops, which will erode differences in state between individuals and thus eliminate the differences in behavior between individuals. One example of feedback that may either create positive or negative feedback is the asset protection principle (Wolf et al. 2007). Assuming that risk-taking increases individuals' immediate fitness, individuals can optimize their lifetime fitness by adjusting their level of risk-taking to their assets (residual reproductive value or RRV). An individual has a high RRV when it has a high probability to survive to the next breeding period(s) and/or when it has a high probability of reproducing in the next breeding period(s). For instance, individuals that have a high RRV are predicted to take fewer risks than individuals that have a lower RRV, because the former have more to lose than the latter (Clark 1994; Wolf et al. 2007). However, the asset protection principle will explain only long-term differences in individual behavior if behavior has little or negative effect on the individual's assets (Luttbeg and Sih 2010; McElreath et al. 2007; Sih and Bell 2008; Sih et al. 2015). This is because positive effects of risk-taking on the individual's RRV create a negative feedback process: By taking risks, an individual increases its RRV, which makes it less likely to express risky behavior. As a consequence, when risk-taking behavior positively affects assets, the resulting feedback process will erode any initial differences in assets across individuals, and hence the behavior of all individuals is expected to be similar, unless risk-taking behavior has small or negative effects on assets.

A fair amount of empirical work has explored whether the required negative relationship between risk-taking behavior and assets are found in nature. We here divided empirical work on the asset protection principle to date into three categories. First, we considered studies investigating age-specific changes in behavior. Individuals are predicted to take more risks and favor their current reproduction as they age, because (all else being equal) older individuals are expected to have lower residual reproductive value than younger ones. Although this age effect itself does not explain between-individual variation in behavior, findings of a positive relationship between boldness, reproductive investment, and age are taken to support the asset protection principle. Most (62%, 18/29) of the empirical studies between risk-taking behavior and age supported the asset protection principle by finding clear or partial support for the expected relationship (see Table 5.2). A second category is formed by studies testing whether individuals show lower risk-taking behavior when they are in better somatic condition, under the assumption that individuals in good condition have a higher residual reproductive value. This prediction received full or partial support by 58% (15/26) studies (see Table 5.2). Third, there are studies testing whether risk-taking behavior is more likely to be expressed by individuals showing greater parental care under the assumption that individuals investing heavily in current reproduction presumably are experiencing a reduced potential for future reproduction. That is, parents who invest heavily in current reproduction are assumed to pay the costs for this investment through a lowered probability to survive or reproduce in the future and therefore have fewer assets than parents making a lower investment in current reproduction. This third predicted relationship received full or partial support by 79% (11/14) of studies (see Table 5.2).

While there is thus empirical evidence in support of the negative relationship between putative assets and risk-taking behavior required by theory, the evidence is not particularly strong. In fact, simple tests of the estimates compiled imply that, only in the last category, is there some statistical support for a nonrandom probability of the required relationship (see Table 5.2). Thus, to the extent that the metrics used in the studies summarized in Table 5.2 critically asses the asset protection principle, we can conclude that both negative associations and positive associations between risk-taking behavior and assets are likely. A second issue is that the majority of the studies are correlational. Thus, they describe the association between behavior and the putative asset as found in the cross-section of individuals in the study population. Correlational studies do not inform us how an individual has changed its behavior when its assets have changed. The latter relationship clearly is the critical one with respect to the theory. To date, only one study tested for the effect of experimentally altered RRV on boldness, and found support for the asset protection principle (Nicolaus et al. 2012).

Apart from the asset protection principle, behavioral ecologists have invoked several putative mechanisms with the potential to generate feedback loops between state and risk-taking behavior. Risk-taking is a generic term for any behavior that helps individuals acquire more resources at the cost of increasing the risks of being predated. We list what we see as the main hypothesized mechanisms and examples of empirical studies supporting these mechanisms or not (see Table 5.3). The first mechanism, starvation avoidance, is a process that causes individuals with low energy reserves to forage more. Foraging more entails risks. In contrast, individuals that have sufficient energy reserves take fewer risks by needing to forage less, but—as a consequence—they deplete their energy reserves. In this situation, the feedback is negative and the individual's state and personality will hence converge. However, this feedback can become positive in some situations, for example if individuals that have a higher condition lower their predation risk by foraging in pairs, which allows them to continue foraging (Sih et al. 2015).

A positive relationship between metabolism and personality can generate a positive feedback between state and behavior. A high metabolism increases energetic needs and thus can increase the motivation to feed (and hence increases risk-taking during foraging). In turn, an increase in food intake increases mass, which increases metabolism (Biro and Stamps 2008, 2010; Careau et al. 2008; Sih et al. 2015). There have been an increasing number of studies on the link between metabolic rate and behavior supporting the "performance model" (positive relationship

Table 5.2 Overview of empirical studies testing for the asset protection principle

Asset trait	Required relationship with risk-taking	Support	Partial support	No support	Opposite	*N*	Test *P*
Age	Positive	13	5	4	7	29	0.2
Traits linked to condition	Negative	13	2	6	5	26	0.4
Parental care	Positive	6	5	0	3	14	0.03

The direction of the relationship is the one where risk-taking behavior is negatively associated with assets (Residual Reproductive Value) as required by theoretical models. Only behavior related to risk-taking is included. In total we used 69 traits reported in 37 studies. For each category, a chi-square test is performed contrasting the number of studies finding full or partial support versus studies finding no or opposite results

Table 5.3 An overview of the kind of mechanisms investigated by empirical studies that have the potential to generate feedbacks between state and behavior (as postulated to exist in theoretical models)

Mechanism	More risk taken by individual with	Feedback assumed	Example studies
Starvation avoidance	Less reserves	Negative	*Supported*: Dosmann et al. (2015), Niemelä et al. (2013), Hall et al. (2015), Mishra et al. (2011) *Not supported*: Dammhahn and Almeling (2012), Wilson et al. (2010b), David et al. (2012), Andersson and Höglund, (2012) (opposite)
Metabolism	High Basal Metabolic Rate (BMR), higher energy requirements	Positive	*Supported*: Mathot et al. (2015), Huntingford et al. (2010), McKenzie et al. (2015) *Not supported*: Bijleveld et al. (2014) (experimental, opposite results); Careau et al. (2015) (opposite), Royauté et al. (2015), Le Galliard et al. (2013)
State-dependent safety	More anti-predatory defenses	Positive	*Supported*: Ahlgren et al. (2015), Hulthén et al. (2013), Kuo et al. (2015), Briffa and Twyman (2011) (experimental) *Not supported*: De Winter et al. (2016) (opposite results)

For each mechanism, we denote the expected relationship between risk-taking behavior and the focal trait of the mechanism. We denote the direction of the feedback that most studies assume is present and provide references to studies which serve as examples

between costly behaviors and basal metabolic rate, see Mathot and Dingemanse 2015). However, the relationship between metabolism and behavior might depend on the behavior tested, the way in which basal metabolic rate is measured and whether it is actually linked with energetic constraints (Mathot and Dingemanse 2015). Finally, state-dependent safety can generate a positive feedback loop between state and behavior because bolder individuals are predicted to have better antipredator defenses and hence can take more risks while foraging, which, in turn, increases their state and provides better antipredator defenses.

Empirical evidence for these different mechanisms is mixed (see examples in Table 5.3). As for the asset protection hypothesis, most studies investigating feedbacks focus on correlations between state and behavior, and hence do not test for causal relationships between these. Thus, they, at best, provide partial support for the existence of feedback loop mechanisms. A problematic issue of the theory is that there is little a priori knowledge of which type of feedback is operational in any given system. Thus, there is little guidance provided to motivate empirical workers to focus on establishing causal links. Perhaps as a consequence, studies experimentally investigating state-behavior feedback loops remain scarce although a call has been made for empiricists to test theoretical predictions from the models (Sih et al. 2015). Clearly, there is also no reason to expect that only one mechanism is acting. Indeed, Luttbeg and Sih (2010) showed that three state-dependent mechanisms can act simultaneously to maintain variation in behavior and state. These mechanisms are asset protection, starvation avoidance (smaller individuals take more risks, low RRV take more risks, negative feedback), and state-dependent safety (larger individuals are bolder, positive feedback). In their model, however, the maintenance of behavioral and state variation depends on the relative strength of the positive feedback compared to the negative feedbacks. Both these feedback processes depend, furthermore, on the levels of risk and resource availability (Luttbeg and Sih 2010; Sih et al. 2015). At present, therefore, numerous mechanisms and putative feedback processes have been developed and are being worked on (see Sih et al. 2015), without there necessarily being a strong overarching theme. It is, in any case, likely that system specific factors will come into play, as we are dealing with (i) massive differences in the organisms studied, and (ii) systems exposed to ecologically relevant interactions including large spatiotemporal variation in resource availability as well in predation pressures.

Negative Frequency-Dependence

Negative frequency-dependence is another phenotypic mechanism that can maintain variation in behavior and is common, e.g., in social species (Dall et al. 2004). Under this scenario the fitness payoffs of an individual's tactic negatively depend on the frequency of the same tactic and the alternative tactic in the population. This mechanism mainly concerns social behaviors and famous examples of negative frequency-dependence are the hawk-dove game or producer-scrounger situation (Maynard Smith 1982). In a hawk-dove game for example, an individual involved in a dyadic interaction (aggression or cooperation) benefits from adopting a tactic different from its partner's tactic. As a result, both tactics coexist at equilibrium in the population. Simulation studies showed that under this scenario, variation can arise even when initial differences in state between individuals are small or absent (Wolf and McNamara 2012, Wolf et al. 2007). Negative frequency-dependent selection can be associated with positive frequency-dependent selection in the rock paper scissor game in which more than two tactics coexist (Sinervo and Lively 1996). However, an important critique is that negative frequency-dependence gives rise to variation, but the process does not explain consistency because individuals will switch between tactics across repeated interactions (Wolf and McNamara 2012). The evolution of consistency has been explained by social responsiveness (Dall et al. 2004; Johnstone 2001; Johnstone and Manica 2011; McNamara et al. 2008; Wolf et al. 2011), which can itself be negative frequency-dependent (Wolf et al. 2008). Predictions from this "social niche specialization hypothesis" were supported in social spiders, where behavioral repeatability is reinforced by repeated social interactions and is higher in groups of familiar individuals than in groups of unfamiliar individuals (Laskowski and Pruitt 2014; Modlmeier et al. 2014). However, this hypothesis was not supported in sticklebacks (Laskowski and Bell 2014) or meerkats (Carter et al. 2014). Finally, we note that behavioral consistency can be explained by other positive feedbacks, which we did not consider in much detail, such as learning (Rosenzweig and Bennett 1996; Tinker et al. 2009; Wolf et al. 2008).

Behavioral Plasticity

Thus far we have been considering single behaviors as aspects of personality, emphasizing their between-individual variation and, thus, consistency. Consistency, however, does not imply that aspects of personality are constant over longer time periods or across different contexts. Indeed, behaviors are labile, which means that individuals adjust to variation in their internal state (age, assets) or external environment (predation risk, temperature, conspecifics). Therefore, individuals can have a repeatable behavior and be plastic at the same time and there has been an increasing interest in behavioral plasticity in the context of personality. In this section, we present one way in which behavioral ecologists study plasticity in a personality context. This approach allows partitioning of variation in phenotypic plasticity between different levels (individuals and genotypes), and can be applied to study variation in plasticity on these levels across environments, contexts, and ages.

Individual differences in behavioral plasticity have been studied from many perspectives by researchers from different fields using different terminologies and methodologies. In addition, several hypotheses have been proposed for explaining the mechanistic or ultimate causes of this phenomenon. In a recent review, Stamps (2016) identified different types of plasticity, discriminating between exogenous (e.g., temperature, conspecifics, predators) and endogenous plasticity (e.g., age, state, circadian cycle, intra-individual variability), contextual (immediate response, e.g., the presence of a novel object) and developmental plasticity (effect of past experience, e.g., learning, maternal effects). All of these types of plasticity can be present within one individual, and time-related plasticity is typically a mixture of several types of plasticity. The behavior of one individual at one moment hence results from the cumulative and interactive effects of its internal state and all of the external stimuli that it has experienced and is experiencing at the moment it is measured. In what follows, most examples and predictions concern contextual plasticity but also age-related plasticity and habituation.

We start this section by outlining an approach commonly used to study plasticity in aspects of personality. In particular, we focus on how ecologists typically use reaction norms to describe between-individual variation in plasticity. We then continue discussing the implication that the presence of between-individual variation in plasticity has for personality, and link these ideas to what empirical studies have found.

The Reaction Norm Approach

Between individual variation in plasticity is studied using a reaction norm approach (Dingemanse et al. 2010; Nussey et al. 2007). Reaction norms are classically used in ecology to represent the diversity of phenotypes that can be produced by one genotype across an environmental gradient. A reaction norm in the context of this chapter is a function specifying, for every individual (or genotype), its expected behavior across what we here term a contextual gradient. This contextual gradient may reflect an extrinsic environment gradient (e.g., temperature), or an ordinal gradient (e.g., age, time, repeated measures) or (two) different contexts (e.g., low and high predation risk), or ontogenetic stages (e.g., juvenile and adult). Reaction norms allow breaking down the population-level plasticity into how we expect individuals/genotypes to respond to this gradient (see Fig. 5.2). As a further consequence, the study of reaction norms provides us with a description of how the variance between individuals (or between genotypes) changes over the contextual gradient, as well as the covariance between environmental contexts (in terms of the ranking of the "lines") (see Fig. 5.2). Reaction norms are not necessarily linear but are here represented as linear slopes for simplicity.

Every reaction norm can be characterized by its elevation and its slope, and individuals may or may not vary in either parameter (Nussey et al. 2007). In the context of personality, the elevation represents each individual's (or genotype's)

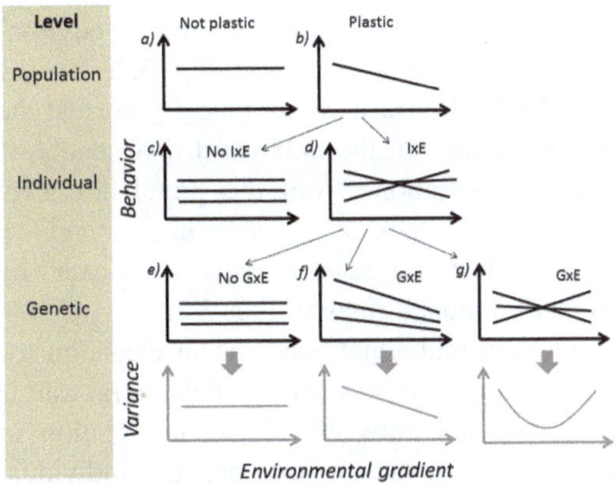

Fig. 5.2 Schematic illustration showing behavioral plasticity on different levels. For simplicity, only linear plasticity is drawn here, but the same hierarchical structure applies to nonlinear relationships. On the population level, **a** the environment-specific mean behavior may be invariant across the environmental gradient, but **b** may also vary. On the individual and genetic levels, deviations from these environment-specific means are considered. **c** All individuals show the same deviation from the average behavior at every value of the environmental gradient, and there is no between-individual variation in plasticity (no IxE). Alternatively, **d** individuals differ in their environment-specific deviation from the environment-specific means, showing variation in plasticity (IxE). Despite the presence of IxE, **e** Genotype—environment interaction (GxE) may be absent, or **f** GxE occurs without the ranking of genotypes changing across environment (reaction norms not crossing within the environmental gradient), or **g** GxE where the ranking of genotypes changes (reaction norms cross). As a result of GxA, additive-genetic variance can change across the environmental gradient. For instance, it can decrease when lines are "fanning in" (**f**) or it can show a curvilinear pattern (**g**)

expected value of personality (at the average of the contextual gradient). Thus, the existence of personality implies that there is between-individual variation in the elevation of the reaction norms. Each reaction norm slope represents the individual's (or genotype's) plastic response of personality across the contextual gradient. An important aspect is that plasticity may occur at the population, individual and genetic level, but that individuals/genotypes need not show variation in their plasticity. When there is no variation in plasticity on the individual level (no individual by environment interactions, IxE), the behavior of each individual changes identically across the contextual gradient. Similarly, the presence of plasticity and even of IxE need not imply the presence of genotype by environment interactions (GxE) (see Fig. 5.2; Nussey et al. 2007). Because we graphically present the reaction norms on the individual and genotypic levels as deviations from the population-level reaction norm, lack of variation in plasticity on the individual and genetic levels is indicated by parallel horizontal lines in Fig. 5.2.

Between-individual variation in elevation (or the intercept) indicates that the behavioral trait is repeatable and between-individual variation in slopes indicates that there is between-individual variation in plasticity or individual–environment interaction (IxE or IxA for individual-age interaction) (see Fig. 5.2d). Whether this between-individual variation in plasticity is due to plasticity itself being heritable (GxE) can be investigated by testing whether the same pattern is found on the genetic level (see Fig. 5.2f, g). The main statistical approach to estimate IxE and GxE is to use random regression phenotypic or random regression animal models (RRPM and RRAM respectively, Nussey et al. 2007). These are linear mixed models where the variance in random effects is allowed to vary according to an environmental variable. The environmental variable typically is continuous, but the models can be applied to ordinal (e.g., age) or different contexts (e.g., low and high predation risk). The RRPM is an extension of Eq. (5.1), where the behavioral response *beh* of individual *i* at trial *t* is modeled as:

$$beh_{i,t} = \mu_{beh} + \mu_F + \mathrm{Env}_t + \mathrm{f}_{ind}(x, E) + \varepsilon_{i(,t)}, \tag{5.4}$$

where Env_t is a generic notation used to describe the effect of the environmental value *E* at trial *t* on the expected behavior *beh*. Statistically, Env_t is a fixed effect denoting, for example, the linear effect of *E* on *beh*, or, in case *E* is a factor (e.g., two contexts), the contrast between the levels of this factor. In any case, Env_t describes the general population-wide response in *beh* for all individuals to variation in the environment. The random regression function $\mathrm{f}_{ind}(x, E)$ describes an orthogonal polynomial of order *x* on the individual level and captures the deviations in behavior *beh* as a function of *E* relative to the overall effect as described by Env (Henderson 1982). For instance, the random regression function $\mathrm{f}_{ind}(0, E)$ assumes behavior *beh* has an individual-specific intercept $ind_{0,\mathrm{I}}$ but individuals respond similarly to the environmental gradient. Their slope is hence the population-level response described by Env (fixed effect). The random regression function $\mathrm{f}_{ind}(1, E)$ assumes that individuals differ in their intercept ($ind_{0,i}$) and linear slope ($ind_{1,i} \times E$,) and so on for higher order polynomials. The random regression parameters (ind_0, ind_1, etc.) constitute a random effect, assumed to be (multivariate) normally distributed with a mean of zero and (co)variances to be estimated. Thus, for order $x > 0$, a covariance between all random regression parameters is estimated. For instance, when $x = 1$, the covariance between individuals' intercepts and slopes is estimated. Finally, $\varepsilon_{i,t}$ is the residual for individual *i* at the trial *t* it is measured. Residual errors can be environment-specific (heterogeneous) or correlated across the environments (homogeneous, noted $\varepsilon_{i,}$). Statistical tests can be conducted to evaluate the order of the random regression and whether residuals are homogeneous or heterogeneous (cf. Brommer et al. 2010). If a significant IxE is found, the RRPM can be expanded to a random regression animal model (RRAM), where the function $f_{ind}(x, E)$ is partitioned into $f_a(x, E)$ and $f_{pe}(x, E)$ describing orthogonal polynomials of order x on the level of additive-genetic and permanent environment effects respectively. As for IxE, the presence of GxE and the order x of the random regression can be statistically tested for (Nussey et al. 2007).

As mentioned previously, one important feature of reaction norms is the covariation between slope and elevation. This covariance can generate changes in between-individual or genetic variance over the contextual gradient (Brommer 2013a). For instance, a negative covariance between elevation and slope may cause expected behaviors expressed by individuals to be "fanning in", and hence the variance between individuals or genotypes will decrease over the contextual gradient (see, e.g., Fig. 5.2f). Other patterns (crossing of reaction norms) may generate an initial decrease in variance, which can be followed by an increase when lines are crossing within the range of the contextual gradient (see Fig. 5.2g). Again, changes in between-individual variance observed to occur across the contextual gradient may or may not be caused by changes in additive-genetic variance (see Fig. 5.2).

Equations (5.1) and (5.3) are sub-models of the more general RRPM and RRAM, where the order x of the random regression functions is assumed to be zero. Conceptually, random regression models allow one to distinguish between the hierarchical levels of plasticity depicted in Fig. 5.2. Thus, random regression models provide a way to determine whether the behavior *beh* is plastic (there is an effect of E on *beh*), whether there is between-individual variation in plasticity of *beh* (IxE), and whether part of this between-individual variation in plasticity of *beh* is heritable (GxE).

Empirical studies have used random regression to test for IxE in personality traits. Indeed, there is evidence for between-individual variation in behavioral plasticity as a function of food availability (Kontiainen et al. 2009), predation risk (Mathot et al. 2011), temperature (Betini and Norris 2012), time of the year (Dingemanse et al. 2012), number of trials (Ensminger and Westneat 2012), or reproductive status (Favreau et al. 2014). A few studies also used random regression to test for between-individual variation in behavioral plasticity over different ages (individual by age interactions, IxA) (Class and Brommer 2016; Fisher et al. 2015; Polverino et al. 2016). On the genetic level, evidence for heritable behavioral plasticity mainly comes from selection experiments reported in the coping-style literature. There is abundant evidence that animals from different lines selected for high and low personality scores differ in their plasticity (Carere et al. 2005; Koolhaas et al. 1999; Øverli et al. 2005). However, evidence for GxE in personality traits is scarce (Dingemanse et al. 2009; Dingemanse et al. 2012), presumably due to a lack of power (Brommer 2013b).

The Character-State Approach

Random regression models are an example of the function-valued trait approach, where trait values are assumed to follow a specific function of some covariate (reviewed by Stinchcombe and Kirkpatrick 2012). Other approaches also exist, but they are not often used to model plasticity in behavior. All of these approaches are a simplification of the so-called character-state approach (Lynch and Walsh 1998), where behavior expressed in each context (e.g., environment, ontogenetic stage) is considered a separate trait and the objective is to estimate the variances at each character state as well as the covariances between these character states. When the environmental gradient is continuous or ordinal (age) it typically covers a considerable number of character states. Function-valued traits are then typically needed to reduce the number of parameters. These approaches at their best will provide a reasonable description of the underlying (co)variances between all character-states, but, because they are based on simplifying assumptions, they may also seriously underrepresent the complexity present across all character-states (Brommer 2013b).

Variation in Plasticity, Repeatability, and Correlations Across Ontogeny

The approaches outlined above present one statistical framework to analyze how aspects of personality change across contextual gradients. In this section, we outline some of the expectations of what the study of variation in plasticity would entail and link this to what has been found. An aspect of personality may be plastic, but when individuals do not vary in their plasticity (no IxE, see Fig. 5.2c), the repeatability is still constant across the contextual gradient. When between-individual variation in plasticity is present (IxE), it can cause (i) changes in between-individual variance, and (ii) changes in individual rankings across the contextual gradient. Thus, between-individual variation in plasticity is expected to impact behavioral repeatability within each context and the correlation of behaviors across contexts. Whereas random regression allows assessing whether IxE is present, it does not directly inform about changes in between-individual variation or changes in ranking of individuals across contexts. For example, a low correlation between elevation and slope may lead to a reduction in between-individual variance (see Fig. 5.2f), or to a humped distribution (see Fig. 5.2g), depending on the exact values of the covariances in elevation and slope (Brommer 2013). In addition, the between-individual variance in behaviors can be similar in two different contexts despite changes in ranking. For example, if the cross-context correlation is −1, the rankings of behavior will reverse across contexts but their variance will be the same. Therefore, the character-state approach and the random regression approach need to be considered jointly to get insights into the consequences of behavioral plasticity for changes in repeatability and changes in ranking across contexts (Brommer 2013).

We here consider in some detail three studies that have attempted to compile estimates from the literature of correlations in behaviors across a contextual gradient. First, Brommer (2013b) used published random regression (Eq. 5.3) estimates of nine studies to calculate the cross-context correlation. Killen et al. (2016) compiled estimates of cross-context correlations of behavioral studies on fish. Finally, Brommer and Class (2015) compiled published correlations of behavior expressed across ontogenetic classes (typically age classes). These three studies all found that the correlations across the contextual gradient generally are moderate with a mean estimate of at most 0.5 (see Fig. 5.3). Thus, only 25% or less of the variation in an aspect of personality in one context or age class is explained by the variation in the other context or age class. Although there clearly are studies reporting a high correlation (close to +1) across a contextual gradient, there also is a sizeable fraction of very low correlations (0 or lower) across contexts or ages (see Fig. 5.3). Furthermore, the repeatability of behavior typically differs between contexts and developmental stages.

A particular form of plasticity in an aspect of personality is when personality changes as individuals age. Repeated records collected on individuals at varying ages can be used to establish that such a pattern, indeed, is attributable to within-individual age-related changes, as opposed to selection removing individuals with either extreme low or high values (Class and Brommer 2016). Within-individual age-related changes in personality are relatively well studied in humans and other primates (e.g., Weiss and King 2015), but remain poorly explored in behavioral ecology. As discussed above, theory based on asset protection predicts that individuals will take more risks as they age, although most studies investigate this issue using cross-sectional analyses. Another process that may cause within-individual age-related changes in personality is senescence. Interestingly, evolutionary theories of senescence predict that any fitness-related trait undergoing senescence should show a genotype-age interaction (GxA),

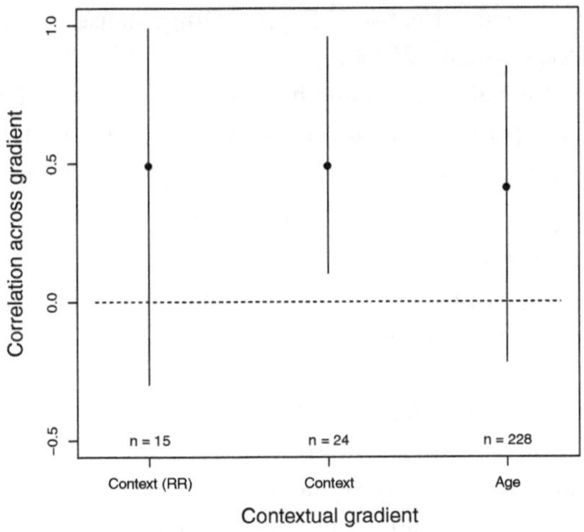

Fig. 5.3 Summary of three studies that have compiled estimates of correlations across a contextual gradient. Plotted are the mean (*filled circle*) and 95% interval (*lines*) based on cross-context correlations for (1) "Context (RR)" by Brommer (2013) who derived cross-context correlations from published random regression models across various context gradients, (2) "Context" by Killen et al. (2016) who compiled correlations across various contexts reported by studies on fish, and "Age" by Brommer and Class (2015) who compiled correlations across ontogenetic classes, for which we here plot the statistics based on correlations across the first two age classes. The number of estimates used is specified in the plot. Some studies have contributed more correlations than others and no correction is made for this non-independence. Studies also differed substantially in their contextual gradient. No published estimate was used in more than one study. The information plotted here should thus be considered indicative

which has been found in several wild populations for life-history traits (Charmantier et al. 2014). Recently, Class and Brommer (2016) found partial support for this prediction for an aspect of personality that showed senescence. In general, however, the hypothesis of evolutionary senescence of personality as an explanation for age-related changes in behavior has received little attention thus far.

Taken together, there is considerable evidence that the repeatability of behaviors is context-dependent and of low correlations of aspects of personality across contexts and across ontogenetic stages. Thus, between-individual variation in plasticity of aspects of personality across ontogenetic or environmental contexts appears to be the norm in nature. As a consequence, the consistency of behavior is arguably not as great as originally believed. This insight is directly derived from explicitly including plasticity into the analyses of aspects of personality, thereby linking the expression of behaviors of individuals across environments, contexts, and ages.

Behavioral Syndromes

The hierarchical (genes to phenotype) view of behavior (see Fig. 5.1) provides an intuitive explanation for why personality exists. Furthermore, this view leads to the expectation that behaviors may be correlated, because the same underlying (genetically coded) processes are likely to affect multiple behaviors (Réale et al. 2007). Behavioral ecologists refer to correlated behaviors as "behavioral syndrome," and they are commonly observed. Behavioral ecologists' interest in behavioral syndromes is twofold. Firstly, the covariance of several aspects of personality suggests a common driver, related, for example, to aspects of physiology (Koolhaas et al. 1999). Second, from an evolutionary perspective, behavioral syndromes imply evolutionary constraints, because multiple correlated traits cannot respond independently to selection (Lynch and Walsh 1998). The evolutionary constraints acting on behavioral syndromes appear to be substantial (Dochtermann and Dingemanse 2013). Correlated behaviors may therefore aid in understanding proximate and ultimate factors underlying personality.

As a construct, the multidimensionality of a behavioral syndrome approaches the psychological concept of personality better than the repeatable single behavior discussed thus far, although it is not identical (Koski 2011). There is some evidence for a genetic underpinning of behavioral syndromes (Dochtermann 2011;

Laine and van Oers, this volume). The bulk of studies documenting behavioral syndromes are, however, based on correlations between multiple behaviors at the phenotypic level. There is a conceptual conflict when defining personality as repeatable behavior (i.e., at the level of between-individual variation), but behavioral syndrome on the phenotypic level (i.e., covariance between the behaviors measured). To see this, consider expanding Eq. (1) to consider two behaviors

$$\begin{aligned} beh1_{it} &= \mu^{beh1} + \mu_F^{beh1} + ind_i^{beh1} + \varepsilon_{it}^{beh1} \\ beh2_{it} &= \mu^{beh2} + \mu_F^{beh2} + ind_i^{beh2} + \varepsilon_{it}^{beh2}, \end{aligned} \tag{5.5}$$

where the symbols are as above except now specific to respective behavior. The deviations from the fixed-effect mean in Eq. (4) are specified by

$$\begin{aligned} \begin{bmatrix} ind_i^{beh1} \\ ind_i^{beh2} \end{bmatrix} &\sim BVN(0, \Sigma_{ind}), \Sigma_{ind} \\ &= \begin{bmatrix} \sigma_{ind}^{2beh1} & \sigma_{ind}^{beh1-beh2} \\ \sigma_{ind}^{beh1-beh2} & \sigma_{ind}^{2beh2} \end{bmatrix} \\ \begin{bmatrix} \varepsilon_{it}^{beh1} \\ \varepsilon_{it}^{beh2} \end{bmatrix} &\sim BVN(0, \Sigma_{res}), \Sigma_{res} \\ &= \begin{bmatrix} \sigma_{res}^{2beh1} & \sigma_{res}^{beh1-beh2} \\ \sigma_{res}^{beh1-beh2} & \sigma_{res}^{2beh2} \end{bmatrix} \\ \Sigma_{pheno} &= \Sigma_{ind} + \Sigma_{res} \\ &= \begin{bmatrix} \sigma_{pheno}^{2beh1} & \sigma_{pheno}^{beh1-beh2} \\ \sigma_{pheno}^{beh1-beh2} & \sigma_{pheno}^{2beh2} \end{bmatrix}, \end{aligned} \tag{5.6}$$

where BVN is the bivariate normal distribution, $\sum$ its variance-covariance matrix and σ the covariance between behavior *beh*1 and *beh*2 on the level of the individual (*ind*) and the residual (*res*). Statistically, Eqs. (5.5) and (5.6) describe *ind* and *res* as stemming from independent and identical distributions (Searle 1961). Hence, the phenotypic (co)variance matrix (conditional on fixed effects) $\sum_{\text{pheno}}$ is the sum of $\sum_{\text{ind}}$ and $\sum_{\text{res}}$. For details on how such models can be solved we refer the reader to a hands-on statistical treatise (e.g., Dingemanse and Dochtermann 2013).

Repeatability can be calculated from the appropriate elements of the variance-covariance matrices (Eq. 6) as the ratios $\sigma_{ind}^{2beh1}/\sigma_{pheno}^{2beh1}$ and $\sigma_{ind}^{2beh2}/\sigma_{pheno}^{2beh2}$ for behaviors *beh*1 and *beh*2 respectively.

Equation (5.6) clarifies that the correlation between the two behaviors can be considered on multiple levels. Importantly, there are clear expectations that the correlations on these various levels will not align. This is because residual variation typically is understood as random noise (barring measurement error), and random noise affecting behavior *beh*1 is not expected to covary with random noise affecting behavior *beh*2. Hence, the residual covariance $\sigma_{res}^{beh1-beh2}$—under the strict interpretation of Eqs. (5.5) and (5.6) is expected to be zero and thus the residual correlation of behaviors *beh*1 and *beh*2 is expected to be zero. Covariances on the individual level can clearly be nonzero. However, there are many possibilities for the residual correlation to also become nonzero. This is because the residual of a behavior will contain, apart from noise, variation caused by any non-modeled factor. Two behaviors may covary because of such unmodeled factors. Especially in behavioral ecology studies carried out in wild populations, there will be potentially many aspects of the environment that are not controlled for and which may lead to (co)variances. One example is spatial variation in the food resources in the local environment (e.g., territory) of individuals.

In principle, it is thus possible that two behaviors, say the rate of attacking a simulated territorial intruder and the rate of feeding offspring, covary positively on the individual level (an individual with a high *ind* value for attack rate towards a territorial intruder tends to have a high *ind* value for feeding its offspring), but not at all on the residual level, or vice versa. Alternatively, the magnitude of these correlations may differ, or, in extreme cases, they could differ in sign. The phenotypic correlation between behaviors *beh*1 and *beh*2 is an average of the correlations on all the hierarchically lower levels, where each correlation's impact on the phenotypic correlation is weighted by the proportion of phenotypic variance explained by that level. That is (Brommer 2014),

$$r_{pheno}^{beh1-beh2} = \sqrt{(R^{beh1}R^{beh2})} \cdot r_{ind}^{beh1-beh2} + \sqrt{(1-R^{beh1})(1-R^{beh2})} \cdot r_{res}^{beh1-beh2}, \quad (5.7)$$

where (dropping sub and super-scripts), r is the correlation between the two behaviors on the phenotypic, between-individual and residual levels respectively, and R is the repeatability.

Equation (5.7) clarifies that the phenotypic correlation between two behaviors will necessarily reflect a mix of the correlations at the various levels. Because the average repeatability of a behavior measured under ecologically relevant conditions is approximately 37% (Bell et al. 2009), the individual-level correlation will necessarily determine only a relatively minor part of the phenotypic correlation. Or, arguably more importantly, a correlation between measured (i.e., phenotypic) behaviors need not be representative of the individual-level correlation between these traits. For example, behaviors *beh*1 and *beh*2 show a positive phenotypic correlation (see Fig. 5.4a, b), but the individual-level correlation may be of different magnitude (see Fig. 5.4c, d). In particular, uncorrelated residuals have the potential to mask even strong between-individual correlations (see Fig. 5.4, right-hand side panels).

What, then, constitutes a behavioral syndrome? Is it sufficient to demonstrate that the phenotypic measures of multiple behaviors covary? At present, the bulk of literature on behavioral syndromes considers only such phenotypic correlations. Some authors have argued that, given that behavioral ecologists consider the presence of *between-individual* variance (i.e., repeatability) as the hallmark of personality, then —analogously—*between-individual* covariance should mark behavioral syndromes (Dingemanse et al. 2012). Inferences on the presence of behavioral syndromes based on phenotypic correlations alone should, at the very least, be considered as taking the "individual gambit" (Brommer 2013c). That is, phenotypic correlations in behaviors may or may not be representative of underlying correlations in individual-specific values for these behaviors. Use of the phenotypic correlation to describe behavioral syndromes may be overly conservative and underappreciate the strength of the associations of aspects of personality. This is because (part of) the residuals of behaviors are likely to be uncorrelated noise. The between-individual correlation is thus anticipated to be higher than the phenotypic correlation (Brommer 2013c). However, in contrast to this expectation, a compilation of published estimates showed that the between-individual correlation is described reasonably well by the phenotypic correlation (see Fig. 5.5, Brommer and Class 2017). At the same time, however, there clearly are studies for which the phenotypic correlation is a very poor proxy for the between-individual correlation underlying the behavioral syndrome (see Fig. 5.5). Thus, while the phenotypic correlations in aspects of personality may *on average* perform reasonably well, estimation of the between-individual correlation for any particular system is likely to be highly valuable.

Assuming that behavioral variation is heritable, behavioral syndromes can be explained by evolutionary quantitative genetics by invoking correlational selection. Correlational selection operates whenever certain combinations of traits are favored over other combinations. Correlational selection has been proposed as a mechanism that maintains genetic variation in behavioral traits. When specific combinations of traits have the same high fitness, a "fitness ridge" is created (Dingemanse and Réale 2013; Sinervo and Svensson 2002). For instance, Brodie (1992) found evidence for correlational selection on color and antipredator behavior in the garter snake and Le Galliard et al. (2013) found negative correlational survival selection between

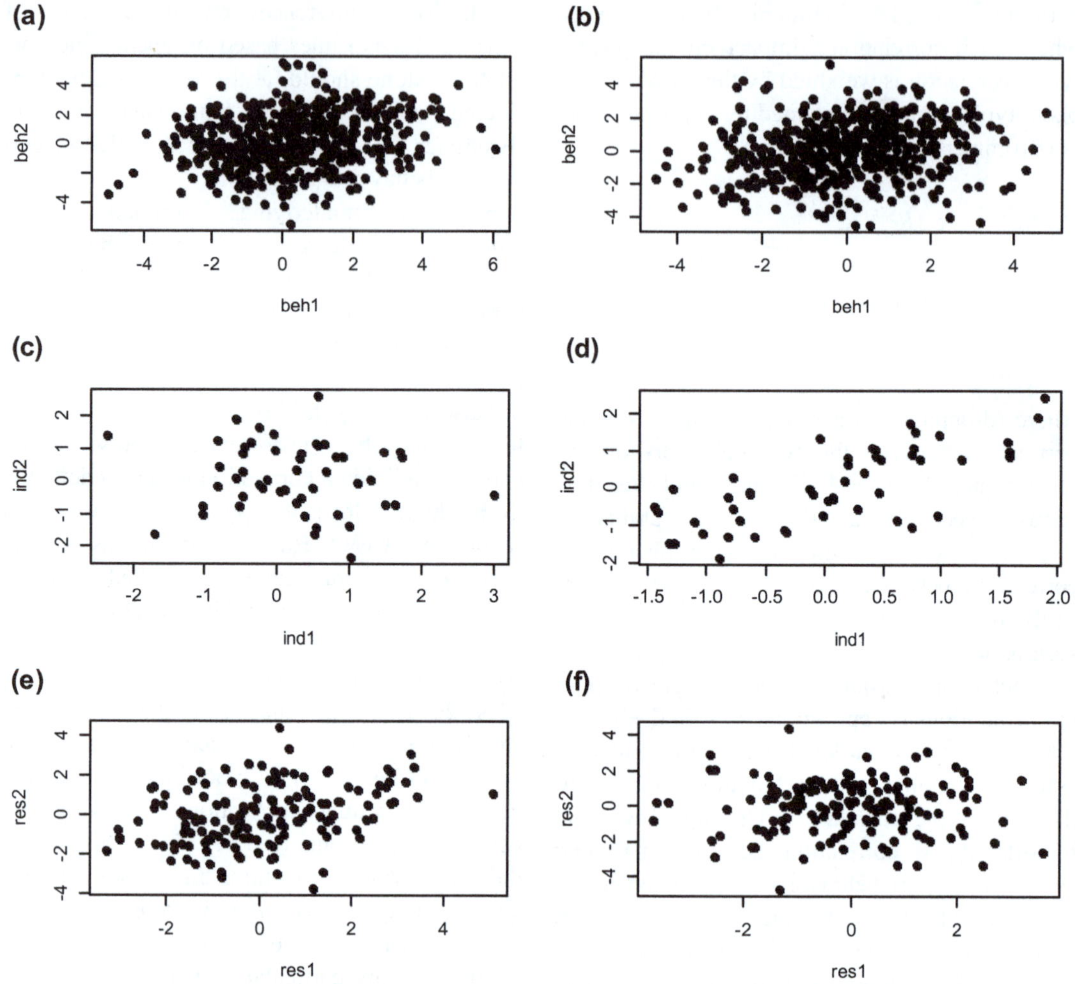

Fig. 5.4 Illustration of how a phenotypic correlation between two behaviors (**a**, **b**) can be partitioned into non-aligning correlations on the between-individual level (**c**, **d**) and residual levels (**e**, **f**). Plotted are random draws for 50 individual measured in 10 trials each for a behavioral syndrome as described in Eqs. (5.5) and (5.6) with an overall mean of 0, and between-individual and residual variances of 1 and 2 for both behaviors respectively (i.e., $R = 33\%$). In one case (**a**, **c**, **e**), the phenotypic correlation (**a**, $r = 0.24$) arose with a between-individual correlation (**c**) of zero and a residual correlation (**e**) of 0.37. In the second case (**b**, **d**, **f**), the same phenotypic correlation (**b**, $r = 0.24$) arose due to a strong between-individual correlation (**d**, $r = 0.75$) which was masked by an absence of correlated residuals (**f**)

exploration and RMR. Correlational selection on aspects of personality is, however, not ubiquitous, for there are studies that failed to detect correlational selection in the wild (Bell and Sih 2007; reviewed in Dingemanse and Réale 2013; Bouwhuis et al. 2014; Réale et al. 2009; Réale and Festa-Bianchet 2003; Sweeney et al. 2013). On the other hand, it is perhaps inherent in the nature of correlational selection to expect different findings. This is because correlational selection can rapidly generate (through linkage disequilibrium), change, or break up genetic correlations between behaviors. Hence, behavioral syndromes can appear rapidly through natural selection (Bell and Sih 2007) or disappear if selection favors opposite combinations of traits (Taylor et al. 2014). As a result, behavioral syndromes can differ between populations facing different environmental conditions (Bell 2005; Karlsson Green et al. 2016).

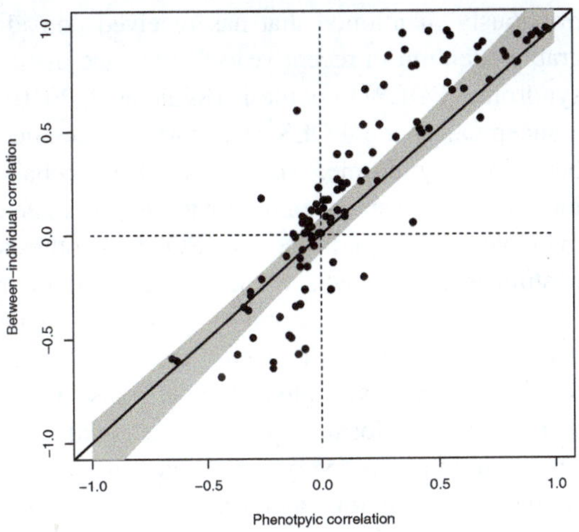

Fig. 5.5 Plot of 109 between-individual correlations against phenotypic correlations of 94 different aspects of personality, based on 39 studies (Brommer and Class 2017). The black abscissa shows a 1:1 correspondence and the polygon the 95% interval of the regression which clearly does not differ significantly from the expected perfect correspondence

Behaviors can also be correlated because they share underlying proximate mechanisms (genetic pleiotropy). Coping styles are a good example of behavioral syndromes caused by genetic pleiotropy: it has been shown using selection lines, that differences in how individuals react to stressful situations (proactive or reactive coping styles) are related to differences in their neuroendocrine profiles (Baugh et al. 2012; Carere et al. 2003; Koolhaas et al. 1999). In the case of pleiotropy, genetic correlations are predicted to be more difficult to break apart by selection than correlations caused by linkage disequilibrium and therefore, behavioral syndromes might be more evolutionarily constrained (Sih et al. 2004). In extreme cases, genetic constraints can generate behavioral syndromes that are maladaptive in some situations (see Arnqvist and Henriksson 1997 for an example involving precopulatory cannibalism in female spiders) but in most cases genetic correlations between behaviors likely constrain their independent evolution to some extent (Dochtermann and Dingemanse 2013).

In some cases, correlations between many measured behaviors could be interpreted as all these measures essentially reflecting one (or several) latent factors, which could be described and distinguished using latent variable modeling (e.g., Dingemanse et al. 2010; Weiss this volume). This approach to describing behavioral syndromes borrows from the psychological approaches where personality is viewed as a latent factor upon which several rated behaviors load (Koski 2011). From this perspective, the correlation between measured behaviors arise because one or more latent factors (which themselves may be correlated) are underlying variation in these measured behaviors. A particularly interesting aspect of this approach is that it allows partitioning of the putative causal pathways into its hierarchical levels, distinguishing the between-individual from the residual level (Araya-Ajoy et al. 2014). Again, behavioral ecologists will be primarily interested in establishing whether presumed latent factors operate on the between-individual level because patterns on the residual level can be caused by numerous less-interesting processes.

Hypotheses for the Existence of Behavioral Syndromes

In the preceding section, we outlined the quantitative genetic (variance partitioning) view with which behavioral ecologists quantify behavioral syndromes. In this section, we briefly discuss how theoretical considerations apply to behavioral syndromes and what kind of empirical work has been conducted from this perspective.

The same phenotypic mechanisms (state-dependent feedback loops and negative frequency-dependent selection) that explain the existence of repeatable behaviors (see above) can explain why individuals vary in a suite of behavioral traits. Using simulations, Luttbeg and Sih (2010) showed that behavioral syndromes can emerge and be maintained by state-dependent feedback loops and the relative strength of state-dependent safety compared to asset protection and starvation avoidance. As for single personality traits, metabolism has been suggested as one possible mechanism that generates behavioral syndromes (through a positive feedback loop). Furthermore, in a simulation study, Wolf and McNamara (2012) showed that, when considering physiology, negative frequency-dependent selection can explain consistent between-individual differences in behavior and coevolution between physiology and behavior. Another simulation study (Mathot and Dall 2013) also showed that individual differences in metabolic rate affect the use of "sampling" and "insurance," both of which are strategies that individuals can adopt to deal with uncertainty in resources and which can generate between-individual variation in suites of behavioral traits, through positive state-dependent feedbacks (Mathot et al. 2012).

Some mechanisms generating personality and behavioral syndromes can be phenotypic or genetic. A good example is non-equilibrium dynamics. In some cases, variation in behavioral types can be maintained by different mechanisms even if the different behavioral strategies existing in the population do not reach equilibrium. Indeed, stochastic environmental variations in space and time that select for different behavioral types have the potential to maintain behavioral variation in the long-term (Wolf and Weissing 2010; Wolf et al. 2013). One good example is the coexistence of dispersers and phylopatric individuals within a population experiencing occasional environmental perturbations. Although dispersers are favored over non-dispersers by perturbations because they can colonize new habitats, their fitness and frequency will decline once a stable population has been established. This is because the dispersal tendency is coupled with other traits facilitating the colonization of new habitats, but which are maladaptive in stable populations. This is an example of nonequilibrium dynamics that is empirically illustrated by studies in western bluebirds (Duckworth and Badyaev 2007), where dispersing individuals are more aggressive and benefit from being more aggressive in colonizing a new habitat. Once established, however, aggressive individuals are outcompeted by less aggressive ones because the latter have higher reproductive output.

One hypothesis at the intersection of the different phenotypic and genetic mechanisms previously mentioned that has received considerable attention in recent years is the pace-of-life syndrome (POLS) hypothesis (Réale et al. 2010). Conceptually, the POLS is a sort of extended behavioral syndrome, where, apart from behaviors, other key aspects of fitness are hypothesized to be part of the syndrome. The POLS hypothesis postulates that aspects of personality and their covariation (i.e., behavioral syndromes) can be explained because personality coevolved with life-history traits and physiological traits. Together all of these form a "pace-of life" syndrome. Thus, individuals vary in their life-history strategies, behavior and physiology, which allow them to either live a "fast" or a "slow" life.

The POLS framework was first used to describe between-species differences in life-history and physiological traits, but was recently extended to within-species differences and personality variation. This is because individuals within species also vary in their life-history strategies and physiology, and because personality is often heritable (Dochtermann et al. 2015) and related to individuals' survival and reproduction (Smith and Blumstein 2008). After Réale et al. (2010) introduced the idea of personality being part of POLS, work on this topic has been substantial (see Table 5.4). Nevertheless, also before this seminal paper, other authors explored the notion of coevolution of personality with other organismal traits. For example, personality has been suggested to have coevolved with growth (Stamps 2007), metabolism (Careau et al. 2008), and life-history strategies (Biro and Stamp, 2008; Wolf et al. 2007). Specific predictions that studies have taken up from the POLS hypothesis as detailed by Réale et al. (2010) are that individuals that are bolder, or faster explorers, mature and reproduce earlier, have a poorer body condition and weaker immune system, but a higher metabolism and shorter lifespan. Another prediction is that these individuals provide less parental care.

To provide an overview of empirical studies on this topic, we compiled phenotypic or between-individual correlations between personality traits (raw or PCA scores) and life-history traits (growth, longevity, age at first

Table 5.4 Table of empirical studies testing for a Pace-of-Life Syndrome (POLS)

Type	Support	Partial support	No support	Opposite	*N*
Physiology	24	8	50	8	90
Life-history trait	10	9	25	8	52
Parental care	0	5	1	9	15

All studies citing Réale et al. (2010), refined by searching for "personality" were included as well as studies cited in these studies. All studies considered were published from 2010 onwards. We included information on 157 aspects of personality in 65 studies, and tallied for each aspects of personality whether it covaried with an aspect of physiology, life-history or parental care in the direction predicted by POLS ((partial) support), did not covary (no support) or covaried in the direction opposite to the prediction (opposite)

reproduction, age at optimal reproduction), physiological characteristics (stress hormones corticosterone or cortisol, metabolic rate, immune system, body condition), and parental care that have been published since 2010 (see Table 5.4). What appears from these results is that predictions from the POLS hypothesis are not always supported. Approximately 36% (32/90) of the reported correlations support predictions from POLS for physiology traits, but more than half (64%, 58/90) of the estimates indicated no evidence or the opposite of such an association (see Table 5.4). Similarly, in terms of the association of aspects of personality with life-history traits (i.e., traits related to reproduction and/or survival), POLS was (partially) supported by 37% (19/52) of estimates, but POLS was not supported or the opposite was found in 63% (32/52) of estimates (see Table 5.4). In terms of parental care, again, 33% (5/15) of the estimates were in partial support of POLS, but 67% (10/15) were in opposite direction (see Table 5.4). Thus, there is a strikingly similar degree of low partial or full support in about one in three estimates for different aspects of POLS (see Table 5.4).

This literature overview presented in Table 5.4 is not exhaustive. It does, however, indicate that evidence for POLS remains equivocal. Concerning parental care, predictions from the asset protection principle, which are opposite to the prediction from the POLS framework, appear to receive better support (see Table 5.2). As already pointed out by Réale et al. (2010), contradicting results suggest that the predicted relationships between personality and life-history traits and physiology might not be universal and may evolve only under certain conditions. In addition to this reservation, we note that the majority of studies considered in our overview have not formally shown that the physiological or life-history trait considered is related to the pace of life in their system. For example, a start of reproduction early in life will only translate to a faster pace of life if it also entails a shorter lifespan. Although such an early versus late life trade-off may be expected in general (Stearns 1992), it may be presumptuous to apply it to any given system, and it may especially be problematic to a priori assume these patterns apply to the individual level.

The POLS hypothesis' merit is that it succinctly places personality into the (much wider) perspective of covarying fitness-related traits. Behavioral ecologists have been enthused by this concept, as evidenced by the number of studies accumulating in a relatively short time (see Table 5.4). At the same time, support for this hypothesis is not overwhelming and more work is needed to carve out and specify testable predictions of the POLS hypothesis.

Plasticity and Behavioral Syndromes

As we outlined above, between-individual variation in plasticity is likely to be common for many aspects of personality. When considering multiple behaviors, between-individual variation in plasticity across contextual environments (IxE) or across ages (IxA) will have consequences on the stability of behavioral syndromes across environments or ontogeny. Consider, for example, a behavioral syndrome where IxE affects one behavior. Because of IxE in this one behavior, the correlation with the other behavior will change, also when this second behavior does not show IxE (see Fig. 5.6). The correlation between these two aspects of personality may, hence, disappear or change sign across environments (see Fig. 5.6).

A review of published age-specific correlations between aspects of personality (Brommer and Class 2015) indeed shows that behavioral syndromes are considered (by the authors of the original studies) to appear or disappear over age classes. A plot of how the correlation between aspects of personality changes with age in published estimates (see Fig. 5.7) also suggests a pattern consistent with one or more aspects of personality showing IxA. That is, some behavioral syndromes get stronger with age, whereas others get weaker (cf. Bell and Stamps 2004). Similarly, comparison of the correlations underlying behavioral syndromes across environmental contexts (Bell and Sih 2007) indicates that behavioral syndromes might change as a result of between-individual differences in plasticity.

An alternative explanation for why the correlation underlying a behavioral syndrome changes across age classes is that correlational selection (see above) "picks out" individuals whose behaviors follow a specific correlation. For example, correlational selection acting between the juvenile and adult age classes on a

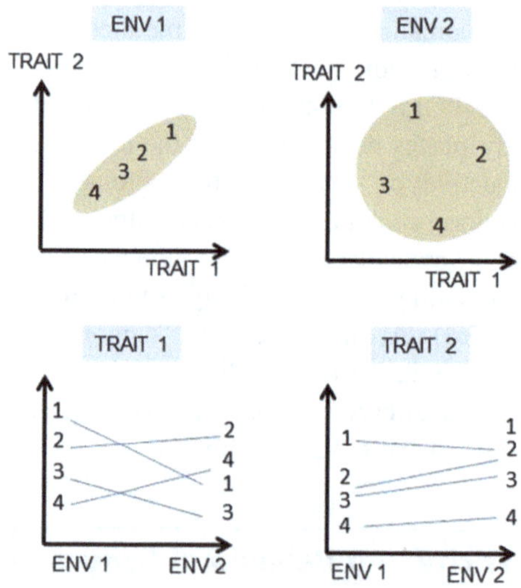

Fig. 5.6 Schematic illustration of the consequences of IxE on the stability of a behavioral syndrome across two discrete environments (*ENV1* and *ENV2*). At *ENV1*, both aspects of personality are tightly correlated on the between-individual level, thus forming a behavioral syndrome. However, IxE in the first aspect of personality causes rank-order changes between individuals across the two environmental context, whereas the rank orders in the second aspect of personality remains the same across environments. As a consequence, the behavioral syndrome in the *ENV2* disappears

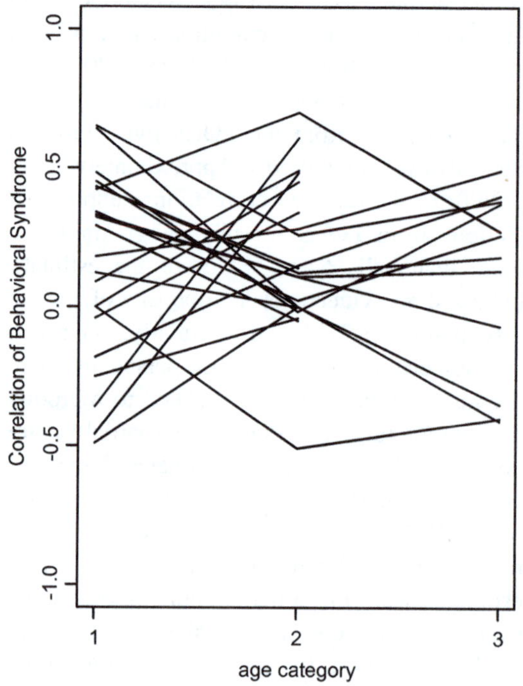

Fig. 5.7 Illustration of the changes in the phenotypic correlation of behavioral syndromes when this correlation is estimated at different ages. Lines connect correlations between two aspects of personality estimated at different age classes. Data plotted is from a compilation by Brommer and Class (2015), and consists of nine studies with 24 pairwise correlations of aspects of personality estimated at two or more age classes. These correlations are plotted here as an illustration of the variability of estimates of behavioral syndrome correlations across ages. The crossing of the lines implies that some behavioral syndromes become stronger with age whereas others become weaker. Note, however, that the estimates do not allow a formal evaluation of whether this "crossing" exceeds what would be expected on the basis of stochasticity alone and the pattern is hence indicative

behavioral syndrome may cause a syndrome to appear or disappear at various ages. Clearly, this explanation need not apply to all systems as survival between age classes may be very high. One study (Class and Brommer 2015) has demonstrated how genotype-age interactions, and not selection, breaks down a strong behavioral syndrome in juveniles, leading it to be absent in adults. In general, however, we know little about how and why relationships between multiple aspects of personality change as organisms develop. In particular, the connection between theory on age-related behavioral plasticity and its predictions regarding behavioral syndromes has received little attention thus far.

The idea of plasticity as an inherent part of personality has been receiving increasing interest in the last decade. This notion views plasticity as a latent trait (a trait that cannot be measured directly on an individual). As a corollary, it implies that plasticity should be repeatable and heritable (Araya-Ajoy et al. 2015). Individual variation in plasticity has already been integrated into the coping-style framework, where proactive individuals are predicted to be less flexible (reactive to changes in familiar situations) and more prone to form routines than reactive individuals, which are more sensitive to environmental cues (Koolhaas et al. 1999). It has thus been proposed that proactive and reactive animals differ in neurobiological properties underlying their response to environmental cues or stressful events (Coppens et al. 2010; Cockrem 2007). Hence, "environmental sensitivity" (the ability to detect changes in the environment) has been hypothesized to be important in animal personalities (Sih and Bell 2008). Such sensitivity, for example, negatively covaries with exploratory behavior (Minderman et al. 2009; Verbeek et al. 1994). In addition, different forms of developmental plasticity, such as learning, have been found to covary with personality. For instance, Guenther et al. (2014) found that aggression, boldness, and exploration are related to learning speed and reversal learning in cavies.

From a reaction norm point of view, the above implies that a steeper reaction norm slope is expected for shy or reactive individuals compared to bold and proactive ones. There have been empirical tests of this prediction. For example, in wild blue tits, the most flexible individuals were the most neophobic and, surprisingly, the most explorative individuals (Herborn et al. 2014).

Between-individual variation in behavioral plasticity can also be integrated with the POLS syndrome where it is argued to covary with life-history traits, physiological traits, and fitness (Urszán et al. 2015; Betini and Norris 2012). For instance, Urszán et al. (2015) proposed that low within-individual variation would be associated with a fast pace of life and favored under stable environments. They indeed found that low intra-individual variation in risk taking was associated with low growth rate in tadpoles but they also found the opposite relationship for exploration and activity. As previously discussed, looking at correlations between pairs of traits might not be the best approach to test the POLS hypothesis, because such an approach does not address causality, which is why these results should be considered to provide tentative support. In addition, there is no clear prediction of the relationship between behavioral plasticity and POLS. These relationships might vary according to the type of plasticity and the context in which it is expressed. For instance, Verdolin and Harper (2013) found that, in mouse lemurs, shy individuals tended to have a lower intra-individual variation than bold individuals, which contrasts with the predictions of Urszán and colleagues (2015). A possibly fruitful research avenue would be to investigate the link between POLS and behavioral plasticities that have documented impacts on fitness or mechanistic links with coping styles (i.e., learning).

Betini and Norris (2012) proposed two alternative evolutionary mechanisms based on the assumption that plasticity is heritable and under correlated selection with personality. Under the "individual quality" hypothesis, only high quality individuals can express high levels of a behavior and afford the costs of high plasticity. The other mechanism, the "compensatory hypothesis," states that individuals scoring high on personality tests do not need to be plastic to optimize their fitness, while individuals scoring low can compensate for their low score by adjusting their behavior to environmental conditions. Whereas Betini and Norris (2012) found support for the compensatory hypothesis, the quality hypothesis was supported by Kontinainen et al. (2009). We can argue here that the relationship between personality and plasticity would actually depend on the personality trait and the type of plasticity and its associated costs.

Another evolutionary explanation for the existence of between-individual differences in personality-related plasticity has been proposed by Mathot et al. (2012). In a population, different tactics ("sampling," "variance sensitivity," and "insurance") can be used by individuals to deal with environmental uncertainty (in the context of foraging but also other behaviors) and generate between-individual differences in behavior and in behavioral plasticity. Between-individual (co)-variation (more insurance favors higher investment in sampling and more variance-averse behavior) in these different tactics linked with personality can be generated by even small initial state differences and can be maintained by state-dependent feedback loops. Alternatively, between-individual (co)variation in these tactics might reflect different strategies that are heritable and evolutionary stable, maintained, for instance, by negative frequency-dependent or fluctuating selection. According to Stamps (2016), this argumentation applies primarily to learning or short-term developmental plasticity.

Conclusions

Behavioral ecologists' interest in personality started with the observations that individuals do not facultatively adjust their behavior to any given situation. Animals show consistent between-individual differences. Furthermore, between-individual differences in one behavior tend to correlate with between-individual differences in other behaviors, forming behavioral syndromes. Behavioral syndromes perhaps come closest to the psychological approach of studying personality, especially when latent factors underlying syndromes are recognized (Araja-Ajoy and Dingemanse 2014). At the same time, a second objective for behavioral ecologists has been to distinguish between hierarchical levels of (co)variation underlying personality and behavioral syndromes, mainly contrasting the individual and residual (or within-individual) levels. As a corollary, personality research in behavioral ecology during the last decade has embraced plasticity (sensu Nussey et al. 2007) as studying it provides further insights in the generation of within-individual variation. As we have emphasized at several instances in this chapter, integration of the notion that individuals may vary in their behavioral plasticity paves the way to a realization that personality and behavioral syndromes are not as consistent as originally perceived. We see this realization as a challenge rather than a problem. Behavioral plasticity, and the variation in personality and behavioral syndromes it creates when investigating these phenomena over contextual environments and ages, appears to be an important and profound part of personality research, and the key challenge is to integrate its causes and consequences for personality and behavioral syndromes. It is perhaps on this front that behavioral ecology has much to add to personality research.

Acknowledgements We thank the Academy of Finland (Grant 289456 to JEB) and the graduate school in Biology, Geography and Geology (to BC) for financial support.

References

Ahlgren, J., Chapman, B. B., Nilsson, P. A., & Brönmark, C. (2015). Individual boldness is linked to protective shell shape in aquatic snails. *Biology Letters, 11*, 20150029.

Andersson, M. Å., & Höglund, E. (2012). Linking personality to larval energy reserves in rainbow trout (*Oncorhynchus mykiss*). *PLoS ONE, 7*, e49247.

Araya-Ajoy, Y. G., & Dingemanse, N. J. (2014). Characterizing behavioural "characters": An evolutionary framework. *Proceedings of the Royal Society of London. Series B, Biological Sciences, 281*, 20132645.

Araya-Ajoy, Y. G., Mathot, K. J., & Dingemanse, N. J. (2015). An approach to estimate short-term, long-term and reaction norm repeatability. *Methods in Ecology and Evolution, 6,* 1462–1473.

Ariyomo, T. O., & Watt, P. J. (2013). Disassortative mating for boldness decreases reproductive success in the guppy. *Behavioral Ecology, 24,* 1320–1326.

Arnqvist, G., & Henriksson, S. (1997). Sexual cannibalism in the fishing spider and a model for the evolution of sexual cannibalism based on genetic constraints. *Evolutionary Ecology, 11,* 255–273.

Baugh, A. T., Schaper, S. V., Hau, M., Cockrem, J. F., de Goede, P., & van Oers, K. (2012). Corticosterone responses differ between lines of great tits (*Parus major*) selected for divergent personalities. *General and Comparative Endocrinology, 175,* 488–494.

Bell, A. M. (2005). Behavioural differences between individuals and two populations of stickleback (*Gasterosteus aculeatus*). *Journal of Evolutionary Biology, 18,* 464–473.

Bell, A. M., Hankison, S. J., & Laskowski, K. L. (2009). The repeatability of behaviour: A meta-analysis. *Animal Behaviour, 77,* 771–783.

Bell, A. M., & Stamps, J. A. (2004). Development of behavioural differences between individuals and populations of sticklebacks, *Gasterosteus aculeatus*. *Animal Behaviour, 68,* 1339–1348.

Bell, A. M., & Sih, A. (2007). Exposure to predation generates personality in threespined sticklebacks (*Gasterosteus aculeatus*). *Ecology Letters, 10,* 828–834.

Bergeron, P., Montiglio, P. O., Réale, D., Humphries, M. M., Gimenez, O., & Garant, D. (2013). Disruptive viability selection on adult exploratory behaviour in eastern chipmunks. *Journal of Evolutionary Biology, 26,* 766–774.

Betini, G. S., & Norris, D. R. (2012). The relationship between personality and plasticity in tree swallow aggression and the consequences for reproductive success. *Animal Behaviour, 83,* 137–143.

Bijleveld, A. I., Massourakis, G., van der Marel, A., Dekinga, A., Spaans, B., van Gils, J. A., et al. (2014). Personality drives physiological adjustments and is not related to survival. *Proceedings of the Royal Society of London. Series B, Biological Sciences, 281,* 20133135.

Biro, P. A., & Stamps, J. A. (2010). Do consistent individual differences in metabolic rate promote consistent individual differences in behavior? *Trends in Ecology & Evolution, 25,* 653–659.

Biro, P. A., & Stamps, J. A. (2008). Are animal personality traits linked to life-history productivity? *Trends in Ecology & Evolution, 23,* 361–368.

Both, C., Dingemanse, N. J., Drent, P. J., & Tinbergen, J. M. (2005). Pairs of extreme avian personalities have highest reproductive success. *Journal of Animal Ecology, 74,* 667–674.

Bouwhuis, S., Quinn, J. L., Sheldon, B. C., & Verhulst, S. (2014). Personality and basal metabolic rate in a wild bird population. *Oikos, 123,* 56–62.

Briffa, M., & Twyman, C. (2011). Do I stand out or blend in? Conspicuousness awareness and consistent behavioural differences in hermit crabs. *Biology Letters, 7,* 330–332.

Brodie, E. D. (1992). Correlational selection for color pattern and antipredator behavior in the garter snake *Thamnophis ordinoides*. *Evolution, 46,* 1284–1298.

Brommer, J. E. (2000). The evolution of fitness in life-history theory. *Biological Reviews of the Cambridge Philosophical Society, 75,* 377–404.

Brommer, J. E., Rattiste, K., & Wilson, A. (2010). The rate of ageing in a long-lived bird is not heritable. *Heredity, 104,* 363–370.

Brommer, J. E. (2013a). Phenotypic plasticity of labile traits in the wild. *Current Zoology, 59,* 485–505.

Brommer, J. E. (2013b). Variation in plasticity of personality traits implies that the ranking of personality measures changes between environmental contexts: Calculating the cross-environmental correlation. *Behavioral Ecology and Sociobiology, 67,* 1709–1718.

Brommer, J. E. (2013c). On between-individual and residual (co)variances in the study of animal personality: Are you willing to take the "individual gambit"? *Behavioral Ecology and Sociobiology, 67,* 1027–1032.

Brommer, J. E. (2014). Using average autonomy to test whether behavioral syndromes constrain evolution. *Behavioral Ecology and Sociobiology, 68,* 691–700.

Brommer, J. E., & Class, B. (2015). The importance of genotype-by-age interactions for the development of repeatable behavior and correlated behaviors over lifetime. *Frontiers in Zoology, 12*(Suppl 1), S2.

Brommer, J. E., & Class, B. (2017). Phenotypic correlations capture between-individual correlations underlying behavioral syndromes. *Behavioral Ecology and Sociobiology, 71*(3).

Careau, V., Humphries, T. D., & Réale, D. (2008). Energy metabolism and animal personality. *Oikos, 117,* 641–653.

Careau, V., Montiglio, P. O., Garant, D., Pelletier, F., Speakman, J. R., Humphries, M. M., et al. (2015). Energy expenditure and personality in wild chipmunks. *Behavioral Ecology and Sociobiology, 69,* 653–661.

Carere, C., Groothuis, T. G., Möstl, E., Daan, S., & Koolhaas, J. (2003). Fecal corticosteroids in a territorial bird selected for different personalities: Daily rhythm and the response to social stress. *Hormones and Behavior, 43,* 540–548.

Carere, C., Drent, P. J., Privitera, L., Koolhaas, J. M., & Groothuis, T. G. G. (2005). Personalities in great tits, Parus major: Stability and consistency. *Animal Behaviour, 70,* 795–805.

Carere, C., & Maestripieri, D. (Eds.). (2013). *Animal personalities: Behavior, physiology, and evolution.* Chicago: The University of Chicago Press.

Carter, A. J., English, S., & Clutton-Brock, T. H. (2014). Cooperative personalities and social niche specialization in female meerkats. *Journal of Evolutionary Biology, 27,* 815–825.

Carter, A. J., Feeney, W. E., Marshall, H. H., Cowlishaw, G., & Heinsohn, R. (2013). Animal personality: What are behavioural ecologists measuring? *Biological Reviews of the Cambridge Philosophical Society, 88,* 465–475.

Charmantier, A., Brommer, J. E., & Nussey, D. H. (2014). The quantitative genetics of ageing. In A. Charmantier, D. H. Garant, & L. E. B. Kruuk (Eds.), *Quantitative genetics in the wild* (pp. 68–83). Oxford: Oxford University Press.

Clark, C. W. (1994). Antipredator behavior and the asset-protection principle. *Behavioral Ecology, 5,* 159–170.

Class, B., & Brommer, J. E. (2015). A strong genetic correlation underlying a behavioural syndrome disappears during development because of genotype-age interactions. *Proceedings of the Royal Society of London. Series B, Biological Sciences, 282,* 20142777.

Class, B., & Brommer, J. E. (2016). Senescence of personality in a wild bird. *Behavioral Ecology and Sociobiology, 70,* 733–744.

Class, B., Kluen, E., & Brommer, J. E. (2014). Evolutionary quantitative genetics of behavioral responses to handling in a wild passerine. *Ecology and Evolution, 4,* 427–440.

Cleasby, I. R., Nakagawa, S., & Schielzeth, H. (2015). Quantifying the predictability of behaviour: Statistical approaches for the study of between-individual variation in the within-individual variance. *Methods in Ecology and Evolution, 6,* 27–37.

Cockrem, J. F. (2007). Stress, corticosterone responses and avian personalities. *Journal of Ornithology, 148,* 169–178.

Coppens, C. M., de Boer, S. F., & Koolhaas, J. M. (2010). Coping styles and behavioural flexibility: Towards underlying mechanisms. *Philosophical Transactions of the Royal Society of London. Series B, Biological Sciences, 365,* 4021–4028.

Cote, J., Dreiss, A., & Clobert, J. (2008). Social personality trait and fitness. *Proceedings of the Royal Society of London. Series B, Biological sciences, 275,* 2851–2858.

Daan, S., & Tinbergen, J. M. (1997). Adaptations of life histories. In J. R. Krebs & N. B. Davies (Eds.), *Behavioural ecology. An evolutionary approach* (pp. 311–333). Oxford: Blackwell.

Dall, S. R. X., Houston, A. I., & McNamara, J. M. (2004). The behavioural ecology of personality: Consistent individual differences from an adaptive perspective. *Ecology Letters, 7,* 734–739.

Dammhahn, M., & Almeling, L. (2012). Is risk taking during foraging a personality trait? A field test for cross-context consistency in boldness. *Animal Behaviour, 84,* 131–1139.

David, M., Auclair, Y., Giraldeau, L. A., & Cézilly, F. (2012). Personality and body condition have additive effects on motivation to feed in Zebra Finches *Taeniopygia guttata. Ibis, 154,* 372–378.

David, M., & Dall, S. R. X. (2016). Unravelling the philosophies underlying "animal personality" studies: A brief re-appraisal of the field. *Ethology, 122,* 1–9.

David, M., Pinxten, R., Martens, T., & Eens, M. (2015). Exploration behavior and parental effort in wild great tits: Partners matter. *Behavioral Ecology and Sociobiology, 69,* 1085–1095.

De Winter, G., Martins, H. R., Trovo, R. A., & Chapman, B. B. (2016). Knights in shining armour are not necessarily bold: Defensive morphology correlates negatively with boldness, but positively with activity, in wild threespine stickleback, *Gasterosteus aculeatus. Evolutionary Ecology Research, 17,* 279–290.

Dingemanse, N. J., Barber, I., Wright, J., & Brommer, J. E. (2012a). Quantitative genetics of behavioural reaction norms: Genetic correlations between personality and behavioural plasticity vary across stickleback populations. *Journal of Evolutionary Biology, 25,* 485–496.

Dingemanse, N. J., Dochtermann, N. A., & Nakagawa, S. (2012b). Defining behavioural syndromes and the role of "syndrome deviation" in understanding their evolution. *Behavioral Ecology and Sociobiology, 66,* 1543–1548.

Dingemanse, N. J., Dochtermann, N., & Wright, J. (2010a). A method for exploring the structure of behavioural syndromes to allow formal comparison within and between data sets. *Animal Behaviour, 79,* 439–450.

Dingemanse, N. J., Kazem, A. J. N., Réale, D., & Wright, J. (2010b). Behavioural reaction norms: Animal personality meets individual plasticity. *Trends in Ecology & Evolution, 25,* 81–89.

Dingemanse, N. J., Both, C., Drent, P. J., & Tinbergen, J. M. (2004). Fitness consequences of avian personalities in a fluctuating environment fitness consequences of avian personalities in a fluctuating environment. *Proceedings of the Royal Society of London. Series B, Biological Sciences, 271,* 847–852.

Dingemanse, N. J., Bouwman, K. M., van de Pol, M., van Overveld, T., Patrick, S. C., Matthysen, E., et al. (2012c). Variation in personality and behavioural plasticity across four populations of the great tit *Parus major. Journal of Animal Ecology, 81,* 116–126.

Dingemanse, N. J., & Wolf, M. (2010). Recent models for adaptive personality differences: A review. *Philosophical Transactions of the Royal Society of London. Series B, Biological Sciences, 365,* 3947–3958.

Dingemanse, N. J., Van der Plas, F., Wright, J., Réale, D., Schrama, M., Roff, D. A., et al. (2009). Individual experience and evolutionary history of predation affect expression of heritable variation in fish personality and morphology. *Proceedings of the Royal Society of London. Series B, Biological sciences, 276,* 1285–1293.

Dingemanse, N. J., & Réale, D. (2013). What is the evidence that natural selection maintains variation in animal personalities? In C. Carere & D. Maetripieri (Eds.), *Animal personalities: Behavior, physiology, and evolution* (pp. 201–220). Chicago: The University of Chicago Press.

Dochtermann, N. A., & Dingemanse, N. J. (2013). Behavioral syndromes as evolutionary constraints. *Behavioral Ecology, 24,* 806–811.

Dochtermann, N. A. (2011). Testing Cheverud's conjecture for behavioral correlations and behavioral syndromes. *Evolution, 65,* 1814–1820.

Dochtermann, N. A., & Roff, D. A. (2010). Applying a quantitative genetics framework to behavioural syndrome research. *Philosophical Transactions of the Royal Society of London. Series B, Biological Sciences, 365,* 4013–4020.

Dochtermann, N. A., Schwab, T., & Sih, A. (2015). The contribution of additive genetic variation to personality variation: Heritability of personality. *Proceedings of the Royal Society of London. Series B, Biological Sciences, 282,* 20142201.

Dosmann, A. J., Brooks, K. C., & Mateo, J. M. (2015). Within-individual correlations reveal link between a behavioral syndrome, condition, and cortisol in free-ranging Belding's ground squirrels. *Ethology, 121,* 125–134.

Duckworth, R. A., & Badyaev, A. V. (2007). Coupling of dispersal and aggression facilitates the rapid range expansion of a passerine bird. *Proceedings of the National Academy of Sciences of the United States of America, 104,* 15017–15022.

Ensminger, A. L., & Westneat, D. F. (2012). Individual and Sex differences in habituation and neophobia in house sparrows (*Passer domesticus*). *Ethology, 118,* 1085–1095.

Falconer, D. S., & MacKay, T. F. C. (1996). *Introduction to quantitative genetics* (4th ed.). Harlow: Longman.

Favreau, F., Goldizen, A. W., Fritz, H., Blomberg, S. P., Best, E. C., & Pays, O. (2014). Within-population differences in personality and plasticity in the trade-off between vigilance and foraging in kangaroos. *Animal Behaviour, 92,* 175–184.

Fisher, D. N., David, M., Tregenza, T., & Rodríguez--Muñoz, R. (2015). Dynamics of among-individual behavioral variation over adult lifespan in a wild insect. *Behavioral Ecology, 26,* 975–985.

Fisher, R. (1958). Polymorphism and natural selection. *Journal of Ecology, 46,* 289–293.

Fitzpatrick, M. J., Feder, E., Rowe, L., & Sokolowski, M. B. (2007). Maintaining a behaviour polymorphism by frequency-dependent selection on a single gene. *Nature, 447,* 210–212.

Gabriel, P. O., & Black, J. M. (2012). Behavioural syndromes, partner compatibility and reproductive performance in Steller's jays. *Ethology, 118,* 76–86.

Guenther, A., Brust, V., Dersen, M., & Trillmich, F. (2014). Learning and personality types are related in cavies (*Cavia aperea*). *Journal of Comparative Psychology, 128,* 74–81.

Hall, M. L., van Asten, T., Katsis, A. C., Dingemanse, N. J., Magrath, M. J. L., & Mulder, R. A. (2015). Animal personality and pace-of-life syndromes: Do fast-exploring fairy-wrens die young? *Frontiers in Ecology and Evolution, 3,* 28.

Harris, M. R., & Siefferman, L. (2014). Interspecific competition influences fitness benefits of assortative mating for territorial aggression in Eastern Bluebirds (*Sialia sialis*). *PLoS ONE, 9*(2), 1–7.

Henderson, C. R. (1982). Analysis of covariance in the mixed model: Higher-level, nonhomogeneous, and random regressions. *Biometrics, 38*(3), 623.

Herborn, K. A., Heidinger, B. J., Alexander, L., & Arnold, K. E. (2014). Personality predicts behavioral flexibility in a fluctuating, natural environment. *Behavioral Ecology, 25,* 1374–1379.

Houston, A. I., & McNamara, J. M. (1999). *Models of adaptive behaviour: An approach based on state.* Cambridge: Cambridge University Press.

Hulthén, K., Chapman, B. B., Nilsson, P. A., Hollander, J., & Brönmark, C. (2013). Express yourself: Bold individuals induce enhanced morphological defences. *Proceedings of the Royal Society of London. Series B, Biological Sciences, 281,* 1–8.

Huntingford, F. A., Andrew, G., Mackenzie, S., Morera, D., Coyle, S. M., Pilarczyk, M., et al. (2010). Coping strategies in a strongly schooling fish, the common carp *Cyprinus carpio. Journal of Fish Biology, 76,* 1576–1591.

Japyassú, H. F., & Malange, J. (2014). Plasticity, stereotypy, intra-individual variability and personality: Handle with care. *Behavioural Processes, 109,* 40–47.

Johnstone, R. A. (2001). Eavesdropping and animal conflict. *Proceedings of the National Academy of Sciences of the United States of America, 98,* 9177–9180.

Johnstone, R. A., & Manica, A. (2011). Evolution of personality differences in leadership. *Proceedings of the National Academy of Sciences of the United States of America, 108,* 8373–8378.

Karlsson Green, K., Eroukhmanoff, F., Harris, S., Pettersson, L. B., & Svensson, E. I. (2016). Rapid changes in genetic architecture of behavioural syndromes following colonization of a novel environment. *Journal of Evolutionary Biology, 29,* 144–152.

Killen, S. S., Adriaenssens, B., Marras, S., Claireaux, G., & Cooke, S. J. (2016). Conservation physiology of marine fishes context dependency of trait repeatability and its relevance for management and conservation of fish populations. *Conservation Physiology, 4,* 1–19.

Kontiainen, P., Pietiainen, H., Huttunen, K., Karell, P., Kolunen, H., & Brommer, J. E. (2009). Aggressive Ural owl mothers recruit more offspring. *Behavioral Ecology, 20,* 789–796.

Koolhaas, J. M., Korte, S. M., De Boer, S. F., Van Der Vegt, B. J., Van Reenen, C. G., Hopster, H., et al. (1999). Coping styles in animals: Current status in behavior and stress-physiology. *Neuroscience and Biobehavioral Reviews, 23,* 925–935.

Koolhaas, J. M., De Boer, S. F., Buwalda, B., & Van Reenen, K. (2007). Individual variation in coping with stress: A multidimensional approach of ultimate and proximate mechanisms. *Brain, Behavior and Evolution, 70,* 218–226.

Koski, S. E. (2011a). How to measure animal personality and why does it matter? Integrating the psychological and biological approaches to animal personality. In M. Inoue-Murayama, S. Kawamura, & A. Weiss (Eds.), *From genes to animal behavior: Social structures, personalities, communication by color* (pp. 115–136). Tokyo: Springer.

Kralj-Fišer, S., & Schneider, J. M. (2012). Individual behavioural consistency and plasticity in an urban spider. *Animal Behaviour, 84,* 197–204.

Kralj-Fišer, S., Sanguino Mostajo, G. A., Preik, O., Pekár, S., & Schneider, J. M. (2013). Assortative mating by aggressiveness type in orb weaving spiders. *Behavioral Ecology, 24,* 824–831.

Krebs, J. R., & Davies, N. B. (Eds.). (2009). *Behavioural ecology: An evolutionary approach* (4th ed.). Hoboken: Wiley.

Kuo, C. Y., Irschick, D. J., & Lailvaux, S. P. (2015). Trait compensation between boldness and the propensity for tail autotomy under different food availabilities in similarly aged brown anole lizards. *Functional Ecology, 29,* 385–392.

Lande, R. (1976). The maintenance of genetic variability by mutation in a polygenic character with linked loci. *Genetical Research, 26,* 221–235.

Laskowski, K. L., & Bell, A. M. (2014). Strong personalities, not social niches, drive individual differences in social behaviours in sticklebacks. *Animal Behaviour, 90,* 287–295.

Laskowski, K. L., & Pruitt, J. N. (2014). Evidence of social niche construction: Persistent and repeated social interactions generate stronger personalities in a social spider. *Proceedings of the Royal Society of London. Series B, Biological Sciences, 281,* 20133166.

Le Cœur, C., Thibault, M., Pisanu, B., Thibault, S., Chapuis, J. L., & Baudry, E. (2015). Temporally fluctuating selection on a personality trait in a wild rodent population. *Behavioral Ecology, 26,* 1285–1291.

Le Galliard, J. F., Paquet, M., & Mugabo, M. (2015). An experimental test of density-dependent selection on temperament traits of activity, boldness and sociability. *Journal of Evolutionary Biology, 28,* 1144–1155.

Le Galliard, J. F., Paquet, M., Cisel, M., & Montes-Poloni, L. (2013). Personality and the pace-of-life syndrome: Variation and selection on exploration, metabolism and locomotor performances. *Functional Ecology, 27,* 136–144.

Luttbeg, B., & Sih, A. (2010). Risk, resources and state-dependent adaptive behavioural syndromes. *Philosophical Transactions of the Royal Society of London. Series B, Biological Sciences, 365,* 3977–3990.

Lynch, M., & Walsh, B. (1998). *Genetics and analysis of quantitative traits.* Sunderland, MA: Sinauer Associates.

Mathot, K. J., & Dingemanse, N. J. (2015). Energetics and behavior: Unrequited needs and new directions. *Trends in Ecology & Evolution, 30,* 199–206.

Mathot, K. J., Nicolaus, M., Araya-Ajoy, Y. G., Dingemanse, N. J., & Kempenaers, B. (2015). Does metabolic rate predict risk-taking behaviour? A field experiment in a wild passerine bird. *Functional Ecology, 29,* 239–249.

Mathot, K. J., ven den Hout, P., Piersma, T., Kampenaers, B., Réale, D., & Dingemanse, N. J. (2011). Disentangling the roles of frequency-vs. state-dependence in generating individual differences in behavioural plasticity. *Ecology Letters, 14,* 1254–1262.

Mathot, K. J., & Dall, S. R. X. (2013). Metabolic rates can drive individual differences in information and insurance use under the risk of starvation. *The American Naturalist, 182,* 611–620.

Mathot, K. J., Wright, J., Kempenaers, B., & Dingemanse, N. J. (2012). Adaptive strategies for managing uncertainty may explain personality-related differences in behavioural plasticity. *Oikos, 121,* 1009–1020.

Maynard Smith, J. (1982). *Evolution and the theory of games*. Cambridge: Cambridge University Press.

McCrae, R. R., & Costa, P. T., Jr. (1999). A five-factor theory of personality. *Handbook of personality: Theory and Research, 2,* 139–153.

McElreath, R., Luttbeg, B., Fogarty, S. P., Brodin, T., & Sih, A. (2007). Evolution of animal personalities. *Nature, 450,* 5–6.

McKenzie, D. J., Belao, T. C., Killen, S. S., & Rantin, F. T. (2015). To boldly gulp: Standard metabolic rate and boldness have context-dependent influences on risk-taking to breathe air in a catfish. *Journal of Experimental Biology, 218,* 3762–3770.

McNamara, J. M., Barta, Z., Fromhage, L., & Houston, A. I. (2008). The coevolution of choosiness and cooperation. *Nature, 451,* 189–192.

Merilä, J., Kruuk, L. E., & Sheldon, B. C. (2001). Cryptic evolution in a wild bird population. *Nature, 412*(July), 76–79.

Minderman, J., Reid, J. M., Evans, P. G. H., & Whittingham, M. J. (2009). Personality traits in wild starlings: Exploration behavior and environmental sensitivity. *Behavioral Ecology, 20,* 830–837.

Mishra, S., Logue, D. M., Abiola, I. O., & Cade, W. H. (2011). Developmental environment affects risk-acceptance in the hissing cockroach, *Gromphadorhina portentosa. Journal of Comparative Psychology, 125,* 40–47.

Modlmeier, A. P., Laskowski, K. L., DeMarco, A. E., Coleman, A., Zhao, K., Brittingham, H. A., et al. (2014). Persistent social interactions beget more pronounced personalities in a desert-dwelling social spider. *Biology Letters, 10,* 20140419.

Monestier, C., Morelle, N., Gaillard, J. M., Cargnelutti, B., Vanpé, C., & Hewison, A. J. M. (2015). Is a proactive mum a good mum? A mother's coping style influences early fawn survival in roe deer. *Behavioral Ecology, 26*(5), 1395–1403.

Montiglio, P.-O., Garant, D., Bergeron, P., Dubuc Messier, G., & Réale, D. (2014). Pulsed resources and the coupling between life-history strategies and exploration patterns in eastern chipmunks (*Tamias striatus*). *Journal of Animal Ecology, 83,* 720–728.

Nakagawa, S., & Schielzeth, H. (2010). Repeatability for Gaussian and non-Gaussian data: A practical guide for biologists. *Biological Reviews of the Cambridge Philosophical Society, 85,* 935–956.

Nettle, D. (2006). The evolution of personality variation in humans and other animals. *American Psychologist, 61,* 622–631.

Nicolaus, M., Tinbergen, J. M., Bouwman, K. M., Michler, S. P. M., Ubels, R., Both, C., et al. (2012). Experimental evidence for adaptive personalities in a wild passerine bird. *Proceedings of the Royal Society of London. Series B, Biological Sciences, 279,* 4885–4892.

Nicolaus, M., Tinbergen, J. M., Ubels, R., Both, C., & Dingemanse, N. J. (2016). Density fluctuations represent a key process maintaining personality variation in a wild passerine bird. *Ecology Letters, 19,* 478–486.

Niemelä, P. T., Dingemanse, N. J., Alioravainen, N., Vainikka, A., & Kortet, R. (2013). Personality pace-of-life hypothesis: Testing genetic associations among personality and life history. *Behavioral Ecology, 24,* 935–941.

Nussey, D. H., Wilson, A., & Brommer, J. (2007). The evolutionary ecology of individual phenotypic plasticity in wild populations. *Journal of Evolutionary Biology, 20,* 831–844.

Øverli, Ø., Winberg, S., & Pottinger, T. G. (2005). Behavioral and neuroendocrine correlates of selection for stress responsiveness in rainbow trout: A review. *Integrative and Comparative Biology, 45*(3), 463–474.

Patrick, S. C., Chapman, J. R., Dugdale, H. L., Quinn, J. L., & Sheldon, B. C. (2012). Promiscuity, paternity and personality in the great tit. *Proceedings of the Royal Society of London. Series B, Biological Sciences, 279,* 1724–1730.

Penke, L., Denissen, J. J., & Miller, G. F. (2007). The evolutionary genetics of personality. *European Journal of Personality, 21,* 549–587.

Polverino, G., Cigliano, C., Nakayama, S., & Mehner, T. (2016). Emergence and development of personality over the ontogeny of fish in absence of environmental stress factors. *Behavioral Ecology and Sociobiology, 70,* 2027–2037.

Pruitt, J. N., & Riechert, S. E. (2009a). Frequency-dependent success of cheaters during foraging bouts might limit their spread within colonies of a socially polymorphic spider. *Evolution, 63,* 2966–2973.

Pruitt, J. N., & Riechert, S. E. (2009b). Sex matters: Sexually dimorphic fitness consequences of a behavioural syndrome. *Animal Behaviour, 78,* 175–181.

Quinn, J. L., Patrick, S. C., Bouwhuis, S., Wilkin, T. A., & Sheldon, B. C. (2009). Heterogeneous selection on a heritable temperament trait in a variable environment. *The Journal of Animal Ecology, 78,* 1203–1215.

Réale, D., Martin, J., Coltman, D. W., Poissant, J., & Festa-Bianchet, M. (2009). Male personality, life-history strategies and reproductive success in a promiscuous mammal. *Journal of Evolutionary Biology, 22,* 1599–1607.

Réale, D., Dingemanse, N. J., Kazem, A. J. N., & Wright, J. (2010a). Evolutionary and ecological approaches to the study of personality. *Philosophical Transactions of the Royal Society of London. Series B, Biological Sciences, 365,* 3937–3946.

Réale, D., & Festa-Bianchet, M. (2003). Predator-induced natural selection on temperament in bighorn ewes. *Animal Behaviour, 65,* 463–470.

Réale, D., Garant, D., Humphries, M. M., Bergeron, P., Careau, V., & Montiglio, P. (2010b). Personality and the emergence of the pace-of-life syndrome concept at the population level. *Philosophical Transactions of the Royal Society of London. Series B, Biological Sciences, 365,* 4051–4063.

Réale, D., Reader, S. M., Sol, D., McDougall, P. T., & Dingemanse, N. J. (2007). Integrating animal temperament within ecology and evolution. *Biological Reviews of the Cambridge Philosophical Society, 82,* 291–318.

Rice, W. R., & Chippindale, A. K. (2001). Intersexual ontogenetic conflict. *Journal of Evolutionary Biology, 14,* 685–693.

Rosenzweig, M. R., & Bennett, E. L. (1996). Psychobiology of plasticity: Effects of training and experience on brain and behavior. *Behavioural Brain Research, 78,* 57–65.

Royauté, R., Greenlee, K., Baldwin, M., & Dochtermann, N. A. (2015). Behaviour, metabolism and size: Phenotypic modularity or integration in *Acheta domesticus*? *Animal Behaviour, 110,* 163–169.

Schuett, W., Dall, S. R. X., & Royle, N. J. (2011). Pairs of zebra finches with similar "personalities" make better parents. *Animal Behaviour, 81,* 609–618.

Searle, S. (1961). Phenotypic, Genetic and Environmental Correlations. *Biometrics, 17,* 474–480.

Sih, A., & Bell, A. M. (2008). Chapter 5 insights for behavioral ecology from behavioral syndromes. In *Advances in the study of behavior* (Vol. 38, pp. 227–281).

Sih, A., Bell, A., & Johnson, J. C. (2004). Behavioral syndromes: An ecological and evolutionary overview. *Trends in Ecology & Evolution, 19,* 372–378.

Sih, A., Mathot, K. J., Moirón, M., Montiglio, P., Wolf, M., & Dingemanse, N. J. (2015). Animal personality and state—Behaviour feedbacks: A review and guide for empiricists. *Trends in Ecology & Evolution, 30,* 50–60.

Sinervo, B., & Lively, C. M. (1996). The rock–paper–scissors game and the evolution of alternative male strategies. *Nature, 380*(March), 240–243.

Sinervo, B., & Svensson, E. (2002). Correlational selection and the evolution of genomic architecture. *Heredity, 89,* 329–338.

Sinn, D. L., Apiolaza, L. A., & Moltschaniwskyj, N. A. (2006). Heritability and fitness-related consequences of squid personality traits. *Journal of Evolutionary Biology, 19,* 1437–1447.

Smith, B. R., & Blumstein, D. T. (2008). Fitness consequences of personality: A meta-analysis. *Behavioral Ecology, 19,* 448–455.

Spoon, T. R., Millam, J. R., & Owings, D. H. (2006). The importance of mate behavioural compatibility in parenting and reproductive success by cockatiels, *Nymphicus hollandicus*. *Animal Behaviour, 71,* 315–326.

Stamps, J. A. (2007). Growth-mortality tradeoffs and 'personality traits' in animals. *Ecology Letters, 10*(5), 355–363.

Stamps, J. A. (2016). Individual differences in behavioural plasticities. *Biological Reviews of the Cambridge Philosophical Society, 91,* 534–567.

Stearns, S. C. (1992). *The evolution of life histories*. Oxford: Oxford University Press.

Stinchcombe, J. R., & Kirkpatrick, M. (2012). Genetics and evolution of function-valued traits: Understanding environmentally responsive phenotypes. *Trends in Ecology & Evolution, 27,* 637–647.

Sweeney, K., Gadd, R. D. H., Hess, Z. L., McDermott, D. R., MacDonald, L., Cotter, P., et al. (2013). Assessing the effects of rearing environment, natural selection, and developmental stage on the emergence of a behavioral syndrome. *Ethology, 119,* 436–447.

Taylor, R. W., Boutin, S., Humphries, M. M., & Mcadam, A. G. (2014). Selection on female behaviour fluctuates with offspring environment. *Journal of Evolutionary Biology, 27,* 2308–2321.

Tinbergen, N. (1963). On aims and methods of Ethology. *Zeitschrift Für Tierpsychologie, 20,* 410–433.

Tinker, M. T., Mangel, M., & Estes, J. A. (2009). Learning to be different: Acquired skills, social learning, frequency dependence, and environmental variation can cause behaviourally mediated foraging specializations. *Evolutionary Ecology Research, 11,* 841–869.

Urszán, T. J., Török, J., Hettyey, A., Garamszegi, L. Z., & Herczeg, G. (2015). Behavioural consistency and life history of *Rana dalmatina* tadpoles. *Oecologia, 178,* 129–140.

Verbeek, M. E. M., Drent, P. J., & Wiepkema, P. R. (1994). Consistent individual differences in early exploratory behaviour of male great tits. *Animal Behaviour, 48,* 113–1121.

Verdolin, J. L., & Harper, J. (2013). Are shy individuals less behaviorally variable? Insights from a captive population of mouse lemurs. *Primates, 54,* 309–314.

Verweij, K. J. H., Yang, J., Lahti, J., Veijola, J., Hintsanen, M., Pulkki-Råback, L., et al. (2012). Maintenance of genetic variation in human personality: Testing evolutionary models by estimating heritability due to common causal variants and investigating the effect of distant inbreeding. *Evolution, 66,* 3238–3251.

Vetter, S. G., Brandstätter, C., Macheiner, M., Suchentrunk, F., Gerritsmann, H., & Bieber, C. (2016). Shy is sometimes better: Personality and juvenile body mass affect adult reproductive success in wild boars, *Sus scrofa*. *Animal Behaviour, 115,* 193–205.

Weiss, A., & King, J. E. (2015). Great ape origins of personality maturation and sex differences: A study of orangutans and chimpanzees. *Journal of Personality and Social Psychology, 108,* 648–664.

Weissing, F. J. (1996). Genetic versus phenotypic models of selection: Can genetics be neglected in a long-term perspective? *Journal of Mathematical Biology, 34,* 533–555.

Wilson, A. J., Réale, D., Clements, M. N., Morrissey, M. M., Postma, E., Walling, C. A., et al. (2010a). An ecologist' s guide to the animal model. *Journal of Animal Ecology, 79,* 13–26.

Wilson, A. D. M., Whattam, E. M., Bennett, R., Visanuvimol, L., Lauzon, C., & Bertram, S. M. (2010b). Behavioral correlations across activity, mating, exploration, aggression, and antipredator contexts in the European house cricket, *Acheta domesticus*. *Behavioral Ecology and Sociobiology, 64,* 703–715.

Wolf, M., van Doorn, G. S., & Weissing, F. J. (2011). On the coevolution of social responsiveness and behavioural consistency. *Philosophical Transactions of the Royal Society of London. Series B, Biological Sciences, 278,* 440–448.

Wolf, M., van Doorn, G. S., & Weissing, F. J. (2008). Evolutionary emergence of responsive and unresponsive personalities. *Proceedings of the National Academy of Sciences of the United States of America, 105,* 15825–15830.

Wolf, M., & McNamara, J. M. (2012). On the evolution of personalities via frequency-dependent selection. *The American Naturalist, 179,* 679–692.

Wolf, M., van Doorn, G. S., Leimar, O., & Weissing, F. J. (2007). Life-history trade-offs favour the evolution of animal personalities. *Nature, 447,* 581–584.

Wolf, M., & Weissing, F. J. (2010). An explanatory framework for adaptive personality differences. *Philosophical Transactions of the Royal Society of London. Series B, Biological Sciences, 365,* 3959–3968.

Wolf, M., van Doorn, G. S., Leimar, O., & Weissing, F. J. (2013). The evolution of animal personalities. In C. Carere & D. Maetripieri (Eds.), *Animal personalities: Behavior, physiology, and evolution* (pp. 252–275). Chicago: The University of Chicago Press.

Worth, A. R., Thompson, R. C. A., & Lymbery, A. J. (2014). Reevaluating the evidence for toxoplasma gondii-induced behavioural changes in rodents. *Advances in Parasitology, 85,* 109–142.

Zhang, X. S., & Hill, W. G. (2005). Genetic variability under mutation selection balance. *Trends in Ecology & Evolution, 20,* 468–470.

Articles cited in the table

Alcalay, Y., Ovadia, O., & Scharf, I. (2014). Behavioral repeatability and personality in pit-building antlion larvae under differing environmental contexts. *Behavioral Ecology and Sociobiology, 68,* 1985–1993.

Boulton, K., Grimmer, A. J., Rosenthal, G. G., Walling, C. A., & Wilson, A. J. (2014). How stable are personalities? A multivariate view of behavioural variation over long and short timescales in the sheepshead swordtail, *Xiphophorus birchmanni. Behavioral Ecology and Sociobiology, 68,* 791–803.

Cordes, N., Schmoll, T., & Reinhold, K. (2013). Risk-taking behavior in the lesser wax moth: Disentangling within- and between-individual variation. *Behavioral Ecology and Sociobiology, 67,* 257–264.

DiRienzo, N., Niemelä, P. T., Hedrick, A. V., & Kortet, R. (2016). Adult bacterial exposure increases behavioral variation and drives higher repeatability in field crickets. *Behavioral Ecology and Sociobiology, 70,* 1941–1947.

Fitzsimmons, L. P., & Bertram, S. M. (2013). No relationship between long-distance acoustic mate attraction signals and male fertility or female preference in spring field crickets. *Behavioral Ecology and Sociobiology, 67*(6), 885–893.

Grim, T., Samaš, P., & Hauber, M. E. (2014). The repeatability of avian egg ejection behaviors across different temporal scales, breeding stages, female ages and experiences. *Behavioral Ecology and Sociobiology, 68,* 749–759.

Kluen, E., Siitari, H., & Brommer, J. E. (2014). Testing for between individual correlations of personality and physiological traits in a wild bird. *Behavioral Ecology and Sociobiology, 68,* 205–213.

Koski, S. E. (2011b). Social personality traits in chimpanzees: Temporal stability and structure of behaviourally assessed personality traits in three captive populations. *Behavioral Ecology and Sociobiology, 65,* 2161–2174.

Lapierre, J. M., Mennill, D. J., & MacDougall-Shackleton, E. A. (2011). Spatial and age-related variation in use of locally common song elements in dawn singing of song sparrows *Melospiza melodia*: Old males sing the hits. *Behavioral Ecology and Sociobiology, 65,* 2149–2160.

Lichtenstein, J. L. L., DiRienzo, N., Knutson, K., Kuo, C., Zhao, K. C., Brittingham, H. A., et al. (2016). Prolonged food restriction decreases body condition and reduces repeatability in personality traits in web-building spiders. *Behavioral Ecology and Sociobiology, 70,* 1793–1803.

Mell, H., Josserand, R., Decencière, B., Artacho, P., Meylan, S., & Le Galliard, J. F. (2016). Do personalities co-vary with metabolic expenditure and glucocorticoid stress response in adult lizards? *Behavioral Ecology and Sociobiology, 70,* 951–961.

Michelangeli, M., Chapple, D. G., & Wong, B. B. M. (2016). Are behavioural syndromes sex specific? Personality in a widespread lizard species. *Behavioral Ecology and Sociobiology, 70,* 1911–1919.

Pruitt, J. N., DiRienzo, N., Kralj-Fišer, S., Johnson, J. C., & Sih, A. (2011). Individual- and condition-dependent effects on habitat choice and choosiness. *Behavioral Ecology and Sociobiology, 65,* 1987–1995.

Rödel, H. G., Zapka, M., Talke, S., Kornatz, T., Bruchner, B., & Hedler, C. (2014). Survival costs of fast exploration during juvenile life in a small mammal. *Behavioral Ecology and Sociobiology, 69,* 205–217.

Stein, L. R., & Bell, A. M. (2015). Consistent individual differences in paternal behavior: A field study of three-spined stickleback. *Behavioral Ecology and Sociobiology, 69,* 227–236.

Stoffer, B., & Uetz, G. W. (2015). The effects of social experience with varying male availability on female mate preferences in a wolf spider. *Behavioral Ecology and Sociobiology, 69,* 927–937.

Toscano, B. J., Gatto, J., & Griffen, B. D. (2014). Effect of predation threat on repeatability of individual crab behavior revealed by mark-recapture. *Behavioral Ecology and Sociobiology, 68,* 519–527.

Toscano, B. J., & Monaco, C. J. (2015). Testing for relationships between individual crab behavior and metabolic rate across ecological contexts. *Behavioral Ecology and Sociobiology, 69,* 1343–1351.

Wexler, Y., Subach, A., Pruitt, J. N., & Scharf, I. (2016). Behavioral repeatability of flour beetles before and after metamorphosis and throughout aging. *Behavioral Ecology and Sociobiology, 70,* 745–753.

Wilson, A. J., Grimmer, A., & Rosenthal, G. G. (2013). Causes and consequences of contest outcome: Aggressiveness, dominance and growth in the sheepshead swordtail, Xiphophorus birchmanni. *Behavioral Ecology and Sociobiology, 67*(7), 1151–1161.

Yuen, C. H., Pillay, N., Heinrichs, M., Schoepf, I., & Schradin, C. (2016). Personality traits are consistent when measured in the field and in the laboratory in African striped mice (*Rhabdomys pumilio*). *Behavioral Ecology and Sociobiology, 70,* 1235–1246.

Part III

Personality in Various Taxonomic Groups

Feeling Fishy: Trait Differences in Zebrafish (*Danio Rerio*)

6

Kanza M. Khan and David J. Echevarria

Abstract

Personality in humans refers to the behavioral, affective, and cognitive traits that persist through time and across context. However, specific definitions and methods of quantifying personality vary depending on the specific approach that researchers emphasize. The trait approach allows clinicians to make informed diagnoses about their patients, and demonstrates predictive validity regarding health (e.g., longevity) and personal (e.g., occupational success) outcomes. The trait approach has also been applied to study personality development and personality disorders in nonhuman species. The zebrafish (*Danio rerio*) is a widely used neurobehavioral model organism that demonstrates tremendous translational potential with humans. Zebrafish exhibit several traits that remain consistent with time and across situations and, thus, have some personality traits like those of humans. Many behavioral and genetic differences have been observed between laboratory bred and wild-type zebrafish, which are largely attributable to a decrease in selection pressures in the laboratory setting. Selective breeding of zebrafish allows for the study of particular phenotypes (e.g., anxiety) to gain a deeper understanding of behavioral phenotypes, and provides a model for testing novel drug treatments. Here, we discuss the five major traits exhibited by zebrafish (boldness, exploration, activity, aggression, and sociability), and population (strain) differences in these traits. The use of zebrafish as neurobehavioral models of personality, and potential for the development of drug therapies for personality disorders is discussed.

The study of personality in human and nonhuman animals focuses on patterns of behavioral, cognitive, and affective responses to changes in an individual's environment (Croston et al. 2015; McAdams 2015; Stamps and Groothuis 2010). Broadly speaking, personality refers to an individual's traits, temperament, dispositions, goals, attitudes, and abilities (Gosling 2001; McAdams

K.M. Khan · D.J. Echevarria (✉)
Department of Psychology, The University of Southern Mississippi, 118 College Drive, Box 5025, Hattiesburg, MS 39406, USA
e-mail: David.Echevarria@usm.edu

K.M. Khan
e-mail: Kanza.Khan@usm.edu

J. Vonk et al. (eds.), *Personality in Nonhuman Animals*,
DOI 10.1007/978-3-319-59300-5_6

2015). However, personality psychologists vary in their approach to the study of personality.

McAdams and Pals (2006) developed a model of personality that is congruent with the emotional and cognitive development of the individual. In this view, personality development is construed as three parallel and interconnected lines of psychological continuity over the course of an individual's lifetime: to be a social actor, a motivated agent, and an autobiographical author (McAdams 2015). As a *social actor,* the infant observes individuals in the periphery and begins to make reputation judgments (e.g., "he is nice", "she is aggressive"). These observations lay the groundwork for the dispositional traits that we attribute to others. Repeated observations and judgements become reputations. According to this view, the development of personality begins as the social actor gains different reputations from interacting with others in their periphery (McAdams 2015). As the young actor matures, s/he refines his/her ability to detect personality traits in others and thereby better understands the way in which s/he is judged by others. The second line of development begins at roughly 5–7 years, when the young actor forms directed goals. As a *motivated agent*, the young child must be able to articulate unique motivations and formulate plans to achieve said goals. Though it can be said that humans from a very young age will form and pursue goals (e.g., an infant turning his/her head to nurse), the full manifestation of a motivated agency does not fully develop until several years later. It has been hypothesized that the development of a theory of mind significantly aids in the development of the motivated agent (Apperley 2012). An individual is said to have a theory of mind if he is able to attribute mental states to himself and others (Meltzoff 1999; Wimmer and Perner 1983). With the full manifestation of theory of mind (occurring at 5–7 years), the motivated agent may begin to pursue their goals with what they believe to be true about the world (McAdams 2015). The third line of personality development sees the young child becoming an *autobiographical author*. This begins at roughly 3–4 years, when the young child recounts episodic memories. As the individual matures, s/he begins to form a narrative that explains who s/he is and how s/he came to be the person s/he is today. The narrative that s/he carries becomes an integral part of his/her personality (McAdams 2015).

The psychobiological approach incorporates biological, environmental, and developmental influences on personality development. Several personality systems have been developed based on this approach, including the Temperament and Character Inventory (TCI; Cloninger et al. 1993), Eysenck's three factor (PEN) model (Eysenck 1967), and Gray's reinforcement sensitivity theory (Gray 1970). These models differ on several levels, including the number of latent traits that they operationalize, and the proposed biological mechanisms underlying particular behaviors (Larstone et al. 2002; Zuckerman and Cloninger 1996). Traits refer to the long-standing characteristics that describe the way that individuals feel, think, or behave (Allport 1931). These may include such characteristics as the tendency to be sociable, emotionally stable, or open to new experiences, among others. The PEN model, FFM model and others, are often used in clinical settings to inform the clinician of the individual's general dispositions and to aid in the diagnosis of personality disorders (Dawis 1992; Harkness and Lilienfield 1997; Lubinski 2000; Wiggins 2003).

One of the most widely known and used models of personality is the Five-Factor Model (FFM; also referred to as the Big Five; Goldberg 1990; McCrae and John 1992). Developed through factor analysis, the FFM identifies five factors: (I) extraversion, (II) agreeableness, (III) conscientiousness (or dependability), (IV) neuroticism (versus emotional stability), and (V) openness to experiences (Goldberg 1990). Instruments measuring personality traits such as the Big Five are highly reliable, even across the lifespan (Fleeson and Gallagher 2009; McCrae and Costa Jr. 1987). Trait standing has been widely demonstrated to be associated with health, occupational, and other life outcomes (Deary et al. 2010; Fleeson and Gallagher 2009; Ozer and Benet-Martinez 2006; Roberts et al.

2007). For example, after controlling for extraneous variables (e.g., SES, gender), individuals that score high in positive emotionality (Danner et al. 2001), extraversion (Danner et al. 2001), and conscientiousness (Friedman et al. 1995) tend to live longer. The FFM is widely applied to the study of normal personality, but recently researchers have begun to examine the relationship of the FFM with personality disorders.

The fifth edition of the Diagnostic and Statistical Manual of Mental Disorders (*DSM-V*; American Psychiatric Association 2013) defines personality disorders as an enduring set of thought and behavior patterns that are markedly different from the expectations of one's culture (p. 645). The DSM-V currently acknowledges 11 personality disorders, which fall into one of three clusters. The clusters of personality disorders are organized based on descriptive similarities: Cluster A includes paranoid, schizoid, and schizotypal personality disorders; Cluster B includes antisocial, borderline, histrionic, and narcissistic personality disorders; and Cluster C includes avoidant, dependent, and obsessive compulsive personality disorders (American Psychiatric Association 2013). However, there are several problems with the diagnostic criteria outlined in the DSM-V (and previous versions of the DSM). First, there is a high degree of comorbidity between personality disorders (Dolan-Sewell et al. 2001), which leads the clinician to question which disorder should be the focus of clinical intervention when the patient presents with two or more personality disorders (Krueger and Eaton 2010). Second, the criteria used in determining whether an individual has a personality disorder is arbitrary (Kamphuis and Noordhof 2009). For instance, to diagnose an individual with borderline personality disorder s/he must present with at least five of the nine diagnostic criteria, however, there is no empirical evidence that supports this threshold (Krueger and Eaton 2010). As such, there has been a recent push in reconceptualizing personality disorders in terms of personality dimensions (Clark 2007; Krueger and Eaton 2010; Widiger and Samuel 2005). The development of a pathological five model (PFM), conceptually similar to the FFM has been suggested; this would include domains such as (I) antagonism, opposite of agreeableness on the FFM, (II) disinhibition, opposite of conscientiousness on the FFM, (III) negative emotionality, encompasses extreme neuroticism on the FFM, (IV) introversion, opposite of extraversion on the FFM and (V) peculiarity (Krueger and Eaton 2010). The first four dimensions of the PFM relate to the domains in the FFM; antagonism and disinhibition are the opposites of agreeableness and conscientiousness respectively (Krueger and Eaton 2010). Negative emotionality is encompassed by extreme neuroticism, whereas introversion is the opposite of extraversion on the FFM (Krueger and Eaton 2010). The last dimension, peculiarity, is narrower than the rest and has a complex relationship to the FFM. Along with openness to new experiences, the dimension of peculiarity is thought to be part of a hierarchical trait domain (Krueger and Eaton 2010). This relationship, however, requires further study. The development of the PFM model would allow the clinician to obtain a multidimensional personality trait model for each patient, which would aid in their diagnosis and treatment (Krueger and Eaton 2010).

Personality traits, adaptive or maladaptive, are heritable, and there is a great emphasis in personality research to identify the biological determinants of behavior (Bouchard and Loehlin 2001; Ebstein et al. 1996; Fullerton et al. 2002; see also Brommer et al. this volume). The evaluation of traits across pairs of differently related individuals (e.g., monozygotic and dizygotic twins, full and half siblings, or parents and offspring) can help to elucidate four sources of variance: additive genetic effects, common environmental effects, nonshared environmental effects, and the interaction of genotype with the environment (Bouchard and Loehlin 2001). The broad-sense heritability (i.e., the proportion of individual differences attributable to additive and nonadditive genetic influences) of personality traits ranges from 40 to 80%, allowing minimal room for environmental influences (Bouchard and Loehlin 2001). Aside from genetic factors, there are several sources of variance influencing personality traits, including but not limited to the sex of the individual (Viken et al. 1994) or the number of genetic loci for the focal trait (Bouchard and Loehlin 2001). Genetic differences among individuals account for variation in allele length (Benjamin et al. 1996) and the availability of post-synaptic receptors (Caravaggio et al. 2016). For instance, high trait standing for novelty seeking (measured as high extraversion and low conscientiousness on the FFM) is associated with longer allele length for the D4 dopamine receptor (*DRD4*) gene (Benjamin et al. 1996). Low dopamine D2/3 receptor levels are associated with increased impulsivity and monotony avoidance (Caravaggio et al. 2016). Although individuals exhibit differences in receptor availability, the related behaviors and pattern of thinking remain within the normal range (Benjamin et al. 1996; Caravaggio et al. 2016). Once the patterns of thinking become disruptive and inhibitory to daily life, problems arise. Molecular genetics allows researchers to identify genetic loci that are associated with particular traits, or the identification of polymorphic genes that play a role in trait development (Ebstein et al. 1996). However, researchers should be careful to not reduce all traits to a single gene or a set of genes as there are many additional factors (e.g., environment, experiences) that influence personality trait development (Claridge and Davis 2013; Fullerton et al. 2002). To gain a better understanding of the influences of environment, and its interaction with genetic predispositions, researchers may choose to study nonhuman models.

Nonhuman species have long been used as a means to better understand the development and progression of human disease states and as models for drug development (Mehta and Gosling 2008). Nonhuman animals, similar to humans, demonstrate individual differences in behavior that remain relatively constant through time and across contexts (Gosling 2001). These behavioral characteristics and individual differences have been termed the animal's temperament (Reale et al. 2007), behavioral syndromes (Moretz et al. 2007a; Sih et al. 2004), or coping styles (Koolhaas et al. 1999). The rodent model has provided great insights into the extent of genetic influence on individual differences (Fullerton et al. 2002). Selective breeding allows for the development of different strains that are predisposed to various states and disorders, such as anxiety (Driscoll et al. 2009) or attention deficit hyperactivity disorder (ADHD; Sagvolden 2000). There are numerous rodent strains that are utilized in neurobehavioral research. Here we discuss a few examples.

Roman high-avoidance (RHA) and Roman low-avoidance (RLA) rats were originally bred for rapid acquisition and non-acquisition of an avoidance task (Driscoll et al. 2009). However, performance in the avoidance task was the result of emotional factors (i.e., stress-induced freezing and grooming), rather than their capacity to learn (Driscoll and Battig 1982). The RLA strain demonstrates greater anxiety-like responses (relative to the RHA strain), as evidenced by an increased frequency of freezing and grooming behaviors, acute increases of plasma ACTH, prolactin, renin, and aldosterone levels (Driscoll et al. 2009). As such, RHA and RLA animals are used in anxiety and fear research.

Genetic models are also widely used in neurobehavioral research, and are commonly used to model depressive states (Porsolt 2000). Depression in rodents is characterized by increased immobility, increases appetite, weight loss, anhedonia, and impaired immune response (Seligman et al. 1975). Presently, knockout models represent the majority of mutant rodents that have been phenotyped for depressive states (Krishnan and Nestler 2011). The use of transgenic rodent models allows researchers to better understand the complexity of the disease (i.e., the types of genes involved, and the wide range of their functions; Barkus 2013). The majority of the transgenic lines involve knocking out a candidate gene for the development of depression, thereby resulting in antidepressive-like phenotypes (Barkus 2013) in tests that are sensitive measures of depressive behavior (for a review of tests that measure depressive behaviors in rodents, see Krishnan and Nestler 2011). Examples include the knockout of genes coding for plasma membrane transporters for dopamine (Perona et al.

2008), serotonin (Barkus 2013; Perona et al. 2008), norepinephrine (Sallinen et al. 1998), and BDNF (Barkus 2013). Several rodent lines have also been developed that increase depressive-like symptoms. Serotonin overexpressing mice demonstrate reduced anxiety—the opposite effect observed in the knockout mice (Barkus 2013), while the knockout of norepinephrine 2A adrenoceptors results in increased immobility (indicative of helplessness; Barkus 2013; Sallinen et al. 1998). Genetic and selectively bred models may serve as the first line of drug screening for antidepressant drugs. Efficacy is tested by introducing the animal to a series of tasks that measure helplessness in rodents; these include the tail suspension test and forced swim test, among others (Krishnan and Nestler 2011).

Although rodent models provide indispensable insight into the neurobiological changes associated with depressive states and related behaviors, researchers must be cautious when developing antidepressant drugs. Rodent knockdown models respond to acute administration of antidepressant drugs, as observed by an increase in the number of escape postures and an increased amount of time searching for an escape in acute stress tests (Krishnan and Nestler 2011; Porsolt 2000). Clinical trials of the efficacy of antidepressant drugs demonstrate high efficacy of the antidepressant effects of novel drug treatments (Slattery et al. 2011). Unfortunately, the effectiveness of antidepressant drugs in humans results are not as substantial, with 30–40% of patients never responding to the therapy (Krishnan and Nestler 2011; Slattery and Cryan 2014). The low response rate is likely attributable to the lack of objective diagnostic criteria (Frazer and Morilak 2005) and the translatability of animal models with humans (Belzung 2014). One of the issues with the use of animal models as models of diseases, for instance, Parkinson's, is the way in which the model is created; i.e., animal models may have a genetic knockout mutation that accounts for a portion of the disorder's symptoms (Garner 2014; Frazer and Morilak 2005). Further, the conditions that animals are exposed to (e.g., single housing) during rearing and testing may account for some of the variation in the development of the disease (Garner 2014). This is especially evident in the development of tumors in mouse models (Hermes et al. 2009) and in stress responses (Sorge et al. 2014). In light of such evidence, Joseph Garner (2014) has suggested that it would behoove researchers to treat animal experiments as if they were human trials.

However, the failure to respond to drugs in clinical trials does not mean that there is little translational potential between humans and rodent species. The deactivation of Brodman Area 24 in humans by deep brain stimulation is effective in alleviating depressive symptoms in treatment resistant patients (Mayberg et al. 2005). The inactivation of the infralimbic cortex in rodents produces a similar response (Slattery et al. 2011). Thus, there is high translational potential between humans and rodent species, and the discrepancies observed in the efficacy rates during drug trials may not be the result of incorrect targeting of specific pathways. Focusing on the identification of specific sites of action may therefore provide a better understanding of drug effects and hold greater therapeutic potential (Krishnan and Nestler 2011).

In this endeavor, the use of a simpler neurobehavioral model is key. The zebrafish is a small teleost fish (roughly 3–4 cm in length), that is native to shallow waters in India, Pakistan and Bangladesh (Bhat 2004). The use of zebrafish (*Danio rerio*) in neurobehavioral research has steadily increased over the past two decades (Collier and Echevarria 2013; Gerlai 2003; Gerlai et al. 2000; Stewart et al. 2014; Tran and Gerlai 2013). Despite its utility in neurobehavioral research, the rodent model is costly and moves at a relatively slow pace (Gauthier et al. 2010; McFarlane et al. 2008; McTighe et al. 2013). The zebrafish model presents a combination of neural simplicity and behavioral complexity that allows for the translation of behaviors and results to rodents and humans. The entire zebrafish genome has been sequenced, revealing homology between the zebrafish and humans in neurosignaling molecules and pathways (Howe et al. 2013). Although the nervous system of the zebrafish is simple relative to the rodent model, functional homology of neural regions are conserved (e.g., in the absence of a suprachiasmatic nucleus, zebrafish possess a parapineal organ, which, with the retina, function as the main circadian oscillator; Cahill 1996). As such, zebrafish exhibit several comparable behaviors to mammals and are widely utilized as models of anxiety (Egan et al. 2009), addiction (Collier and Echevarria 2013), and learning (Best et al. 2008), among others. Zebrafish produce a wide range of behaviors, making them a valuable model in studies of, for example, anxiety, addiction, and learning (Best et al. 2008; Collier and Echevarria 2013; Egan et al. 2009). The short reproduction cycle, and relative ease of developing mutant lines, makes the zebrafish a viable model for transgenic studies as well. In recent years, the study of zebrafish personality has gained momentum.

We begin the next section with an overview of trait differences in zebrafish and a review of the common measures used in classifying behavioral traits in the animal. From there, we discuss different zebrafish strains and genetic models that have been utilized in studying mood disorders and discuss some of the limitations of the zebrafish model.

Studying Personality in Zebrafish; Measurements, Fitness, and Population Differences

The ontogeny of personality traits in zebrafish, and other vertebrate animals, is the product of natural selection via reproductive success and predation (Dingemanse et al. 2009; Dingemanse and Reale 2005; Stamps and Groothuis 2010). Presently, there are five main behavior categories in the zebrafish: boldness, exploratory behavior, aggression, sociability, and locomotor activity (Ariyomo et al. 2013; Moretz et al. 2007a; Sailer et al. 2012; Tran and Gerlai 2013; Wright et al. 2003). In the following sections, we discuss the methods of measuring trait belongingness in zebrafish and discuss population level differences between strains, where applicable. Because personality research in zebrafish is relatively new, the population differences between strains is not fully documented. Commonly used strains in behavioral research include the AB, TU (Tubingen), and *Casper*. The AB strain was developed in the 1970s and was the product of mixing two separate strains, A and B (Howe et al. 2013). The AB and the TU lines are laboratory standards, and are roughly 50 generations removed from the wild-type stock (Norton and Bally-Cuif 2010). *Casper* is a mutant line that is transparent and is largely used for optogenetics and in vivo imaging (White et al. 2008). As will be discussed, the domestication and rearing of animals in nonnatural settings involves the removal of some selection pressures (e.g., predation and foraging; Wright et al. 2006). The removal of various selection pressures may result in behavioral (e.g., reduced aggression; Ariyomo et al. 2012; Wright et al. 2003) and morphological (e.g., fin size; Plaut 2000) changes, as well a loss in genetic diversity. As such, behavioral research laboratories must exercise caution when generalizing results between strains.

Shyness–Boldness

Boldness in zebrafish is a suite of characteristics in which the animal demonstrates reduced anxiety or fear in approaching a novel object (Kalueff et al. 2013). It is characterized by the tendency to take risks and seek novel environments, the reduced tendency to flee from a predator, and an increase in biting behaviors (Ariyomo et al. 2013; Conrad et al. 2011; Miklosi and Andrew 2006; Sailer et al. 2012; Shaklee 1963). Bold animals will approach a predator (not flee) in order to gain more information about the predator's identity, hunger state, and the likelihood of an attack (Dugatkin and Godin 1992; Magurran 1986; Moretz et al. 2007a; Sailer et al. 2012).

Boldness is most often measured via the open field test, whereby the animal is introduced to a novel environment and allowed to explore for a set amount of time (Ariyomo et al. 2013; Ariyomo and Watt 2012). In this task, the circular (Cachat et al. 2013), rectangular (Ariyomo et al. 2013) or square (Sison and Gerlai 2011) tank is

Table 6.1 Tasks used to quantify traits in adult zebrafish

Trait	Test/task	Behavioral response	Reference
Aggression	Mirror image test	Short bouts of fast swimming towards the mirror	(Ariyomo et al. 2013; Gerlai 2003)
	Dyadic fights	Striking, biting, circling, or chasing the opponent. When in close proximity, the fish may erect its dorsal and anal fins, and flare its body toward the opponent	(Oliveira et al. 2011)
	Pigment response	Aggressive animals display darker and more vivid colors	(Gerlai et al. 2000)
Boldness	Open field test	Tendency to cover a greater swim distance, and swim towards the middle of the tank. Indicative of risk taking ability	(Ariyomo et al. 2013)
	Feeding	Low latency to eat following a stressor	(Moretz et al. 2007a)
Sociability	Social interaction (stranger)	Focal zebrafish spend a greater amount of time interacting with a 'stranger' fish in a separate compartment, compared to an empty compartment	(Barba-Escobedo and Gould 2012; Riehl et al. 2011)
	Social novelty	Preferring to interact with an unfamiliar non-kin, as compared to a familiar non-kin	(Barba-Escobedo and Gould 2012)
	Shoal cohesion	Maintain a smaller interfish distance in groups of 10 fish	(Buske and Gerlai 2012)
	Mirror test	Biting at the mirror (to remove a barrier and interact with a social partner); low responses may be used to quantify ASD	(Moretz et al. 2007b; Stewart et al. 2013)
Activity	Observation cylinder test	Latency to top, time spent in the top, entries to top, duration and number of freezing bouts	(Grossman et al. 2010)
	Open field	Total distance swum	(Tran and Gerlai 2013)
Exploration	Serial open field task	Latency to explore occluded regions of a tank	(Sailer et al. 2012; Wisenden et al. 2011)
	Light dark box	Latency to enter the white region; fewer entries to the white region	(Grossman et al. 2011)
	T-maze exploration	Latency to leave the start box	(Vignet et al. 2013)
	Novel tank test	Latency to enter the top half of the tank	(Grossman et al. 2011)

divided into several smaller sections. Animals that cross more lines, and swim in the inner regions and outer regions of the tank are deemed bold, whereas those that cross the fewest lines are deemed shy (Ariyomo et al. 2013). Another common task is the predator-inspection test, in which the animal is exposed to a natural or sympatric predator. Bold behavior may also be evaluated through the introduction of a novel object into the environment. In both tasks, boldness can be measured as a function of the amount of time that animals remain within a short distance (e.g., 1.5 body lengths) of the novel object (Wright et al. 2003). See Table 6.1 for a list of currently used tasks that measure behavioral traits in zebrafish.

The behavioral tasks used to measured boldness in zebrafish allow the researcher to rank the zebrafish on a scale that ranges from shy to middling to bold (Ariyomo and Watt 2012). Typically, male zebrafish are bolder than their female counterparts (Ariyomo et al. 2013). Boldness is also associated with social rank in newly formed zebrafish dyads (Dadda et al. 2010). Social rank may be directly related to the ability to garner mates and resources. As such, bold male zebrafish, relative to middling and shy male zebrafish, fertilize a greater number of the female's eggs (Ariyomo and Watt 2012), and produce offspring that are more likely to be bold, themselves (Ariyomo et al. 2013). This is highly advantageous from an evolutionary standpoint, as bold animals may have a reduced lifespan (Sih et al. 2004; see also Watters et al., this volume).

Mathematical techniques have been used to identify section(s) of DNA that correlate with various functional or morphological phenotypes. A quantitative trait locus (QTL) analysis is one such technique and has been used to localize the trait of boldness to chromosomes 9, 16 and 21 in zebrafish (Wright et al. 2006). Genomic differences between wild-derived (i.e., first or second generation offspring of wild-type animals) have been detected between the AB and Santal (a founder population sampled from a small town in Bangladesh) strains (Wright et al. 2006). This is consistent with behavioral differences in a novel object approach task (Wright et al. 2006, 2003). Relative to the AB strain, the Santal strain (and other wild-derived strains, named after the sample sites: Canal, Tangail, and Nepal) spent a greater amount of time within 1.5 body lengths of the novel object and approached the novel object a greater number of times (Wright et al. 2006, 2003). Differences in boldness were also observed within each of the four wild-derived populations, indicative of genetically based behavioral differences (Wright et al. 2003).

Exploration-Avoidance

Exploratory behavior is characterized as the tendency for the animal to explore its environment and may be assayed through several tasks. For example, in tasks measuring exploratory behavior, the zebrafish is introduced to a novel arena, and allowed to explore the environment. In the novel tank test, the zebrafish is allowed to explore a relatively narrow tank that largely affords only vertical swimming (Egan et al. 2009). On introduction to the novel tank, the zebrafish typically dive to the bottom of the tank (Egan et al. 2009; Stewart et al. 2012). The latency to explore the top region of the tank, and the total time spent exploring the top region are measured as indicators of exploratory behavior (Vignet et al. 2013). In a modified version of the T-maze task, the animal is placed in the start box and allowed to explore the environment for a 5-minute period (Vignet et al. 2013). The latency to leave the start box, time spent in each region of the T-maze and the distance traveled per minute inform the researcher of where the zebrafish falls on the exploratory continuum. Exploratory animals will explore each region, have a lower latency to leave the start box, and will tend to swim larger distances per minute (Vignet et al. 2013).

Another task used to measure exploratory behavior is the serial open field task. In this task, a rectangular tank is divided into five sections and PVC pipes allow passage between sections (Sailer et al. 2012; Wisenden et al. 2011). The zebrafish is initially introduced to one of the sections, and allowed to acclimate to the new environment. After seven days, animals that travel through the different sections and arrive at the final section of the serial open field task are deemed exploratory (Sailer et al. 2012). Animals that remain in the first (or first few) section are deemed non-exploratory (Sailer et al. 2012).

The latency to enter and explore new environments varies across strains (Vignet et al. 2013). For instance, adult AB zebrafish tend to leave the start box in a T-maze exploration task sooner than TU adults (Vignet et al. 2013). The tendency to explore can be influenced by the administration of stimulants and depressants. Stimulants, such as alarm pheromone or caffeine, reduce the total time spent exploring the environment (Wong et al. 2010). The chronic administration of fluoxetine (a selective serotonin reuptake inhibitor; SSRI), and acute administration of low doses of alcohol reduce the latency of the animal to explore the top regions of the environment in the novel tank test (Wong et al. 2010).

Aggression

Aggression is defined as the propensity of the animal to be confrontational (Olivier and Young 2002). In the zebrafish, aggression is characterized by nipping at or biting an opponent, and charging (short bouts of fast swimming) at the opponent (Ariyomo and Watt 2012; Gerlai et al. 2000; Moretz et al. 2007a). Aggression is most often measured using a mirror aggression test. In this task, the animal is placed in a small rectangular tank that has a mirror placed at an incline on the back wall. The zebrafish may perceive its reflection to be an image of a nearby conspecific and any schooling or aggressive behavior are recorded (Gerlai et al. 2000). In this task, the amount of time the zebrafish spends close to and far away from the mirror are recorded, as well as the frequency of attack behaviors (Gerlai et al. 2000). Animals that spend more time in the region farther away from the mirror are classified as avoidant, whereas those that approach the mirror, and display attack behavior are classified as aggressive (Gerlai et al. 2000). Aggressive zebrafish are generally also darker and assume a stereotypic posture: erection of the dorsal, caudal, pectoral, and anal fins (Gerlai et al. 2000).

Much like boldness, being aggressive confers certain evolutionary advantages. For example, zebrafish demonstrate aggressive behaviors towards conspecifics to gain or maintain a relatively higher dominance rank, which enables them to garner resources and mates (Ariyomo et al. 2013; Spence et al. 2008). Selective breeding of zebrafish also produced differences between strains of zebrafish. The TM1 (roughly 30 generations removed from the wild-type population) strain is more aggressive than the Nadia strain at five generations removed from wild-type populations (Moretz et al. 2007b). In a study that evaluated aggression, sociability, and boldness, Moretz et al. (2007b) mixed two strains of zebrafish, and Nadia. When in a mixed shoal, animals from either strain exhibited increases in aggression as characterized by an increase in biting behavior (Moretz et al. 2007b). The increase in aggression is likely the result of the Nadia strain reacting to the high baseline aggression of the TM1 strain. The relatively lower genetic basis for aggression could also allow for the influence of external factors, including age of the animal, size, motivational state, and residency status (Moretz et al. 2007b).

Sociability

Sociability is the tendency of the animal to shoal (a group of 2–10 zebrafish; Ruhl and McRobert 2005; Saverino and Gerlai 2008). As social animals, zebrafish prefer to remain in close proximity to conspecifics (Pitcher 1983). Remaining within a shoal poses evolutionary advantages, as the animal is better able to acquire resources and is protected from predators (Wright et al. 2006). Within a shoal, sociability is assessed by measuring the average interfish distance between conspecifics (Pham et al. 2012). Zebrafish that remain within four average body lengths of a neighboring zebrafish are considered to be part of the shoal (Pham et al. 2012). The average interfish distance maintained between neighboring fish is dependent on the size of the group, and the strain tested. For instance, when group sizes are small, zebrafish of the AB strain tend to less cohesive and remain interested in the heterogeneities of the environment (Seguret et al. 2016). This effect is diminished with increasing group sizes, i.e., group cohesion increases (Seguret et al. 2016).

Other measures of sociability include social preference tests and the mirror biting test (Pham et al. 2012). For example, in the two-chamber test, a live conspecific is placed in an exposure compartment (separated from the rest of the arena by transparent divider). The amount of time the focal zebrafish spends near the conspecific is measured (Pham et al. 2012). This task may also be conducted using animated images (Saverino and Gerlai 2008). A greater proportion of time spent near the conspecific, real or animated, is indicative of social behavior (Saverino and Gerlai 2008). The mirror biting test, which as noted above, is typically used to measure aggression, may also be used to measure of sociability (Pham et al. 2012). In the mirror biting task, the zebrafish may regard its reflection as a conspecific and bite at the mirror as a means to remove a barrier, which would allow the focal zebrafish to interact more closely with the perceived conspecific (Moretz et al. 2007b). In the mirror biting task, the zebrafish is introduced into a small tank that either (a) already has a mirror, or (b) has a mirror introduced after the zebrafish acclimates to the novel environment (Pham et al. 2012). In both mirror biting tasks, the number of mirror approaches, mirror contacts (i.e., biting), and latency to approach the mirror are measured. A greater number of approaches and contacts are indicative of sociability in the zebrafish (Pham et al. 2012).

Activity

In comparison with the behavioral traits already mentioned, activity level in the zebrafish has received the least amount of attention (Tran and Gerlai 2013). The high degree of within-group variance for locomotor activity could be the result of several factors, including observer error and random variance. More recently, genetic predispositions have been hypothesized to be one of the determining factors in activity level (Tran and Gerlai 2013).

Several parameters are measured to assess locomotor activity: total distance swum, average swim speed, and proportion of time spent swimming (Lange et al. 2013). Locomotor ability undergoes drastic changes at each stage of development (i.e., larval, juvenile, and adult stages; Lange et al. 2013). Differences have also been observed between laboratory strains (EK, WIK, TU, AB) at each developmental stage, suggesting that genetic differences underlie variation in this trait (Lange et al. 2013). A recent study by Tran and Gerlai (2013) found that activity levels of three groups of zebrafish (high, medium, and low activity) remain consistent over several days' observation. For seven consecutive days, individual zebrafish were placed in an open field task and their activity levels were measured over a 10-minute period (Tran and Gerlai 2013). Animals were rank ordered as high, medium, or low activity, based on their activity during behavioral testing on day one (zebrafish were allowed to swim freely in a 37L tank; Tran and Gerlai 2013). For six days following the first observation, the rank order of activity remained constant (although actual activity levels did exhibit a gradual decline; Tran and Gerlai 2013). On the 8th day of observation, animals were placed in a novel environment; those that were originally classified as highly active exhibited the greatest amount of locomotor activity (Tran and Gerlai 2013). An identical effect was observed with medium and low activity zebrafish, supporting the claims that genetic predispositions affect activity levels in the zebrafish, and that locomotion presents with limited behavioral plasticity across contexts.

Current Perspectives and Future Directions in Zebrafish Models of Human Personality and Personality Disorders

The relatively high genetic homology of the zebrafish with humans, and the ability to develop mutant lines via larval exposure to mutagens allow for the development of models of human diseases such as depression and anxiety disorders (Kalueff et al. 2012). In the following sections, we discuss current research with zebrafish mutant models and the effects of drug administration, as they relate to these disorders.

Depressive Disorders

In humans, the stress response is mediated by the hypothalamus–pituitary–adrenal (HPA) axis. Corticotropin-releasing hormone (CRH) is released from the hypothalamus in response to stress, physical illness, and activity (Smith and Vale 2006). An increased production of CRH promotes the release of adrenocorticotropic hormone from the pituitary, which stimulates the production of glucocorticoids in the adrenal gland (Griffiths et al. 2012). The main glucocorticoid involved in the stress response in humans is cortisol (Smith and Vale 2006). The release of cortisol serves a homeostatic function in the acute stress response by providing negative feedback to stress circuits (Griffiths et al. 2012). Activated glucocorticoid receptors (GR) bind to DNA and regulate the transcription of several genes including those relating to the HPA axis (Griffiths et al. 2012). However, repeated activation of the HPA axis may result in long term neural adaptations, including the down regulation of GR expression (Howell et al. 2011). This has been linked to the onset of depressive symptoms in rodent models (Zhou et al. 2011). Zebrafish have a comparable stress axis to the HPA, termed the hypothalamus–pituitary–interrenal (HPI) axis (Alsop and Vijayan 2008). Several components of the stress response in zebrafish and humans are identical (CRH, cortisol, ACTH; Alsop and Vijayan 2008). By 4 days post fertilization (dpf), larval zebrafish also exhibit GR-mediated negative feedback (Griffiths et al. 2012; Alsop and Vijayan 2009). Recently, researchers discovered a single nucleotide substitution in the zebrafish genome in a region that is essential for DNA binding (Ziv et al. 2012). As a result, cortisol may still bind to the mutant GR, gr^{s357}, and translocate to the nucleus, but it does not bind to the DNA, and there is no transcriptional activity (Ziv et al. 2012).

In the novel tank test, the gr^{s357} mutants are more immobile and exhibit reduced exploration (Ziv et al. 2012). Whole body cortisol is also markedly higher in the gr^{s357} mutant, relative to wild-type controls (Griffiths et al. 2012). Administration of fluoxetine attenuates the behavioral (Ziv et al. 2012) and physiological stress response (Griffiths et al. 2012). The administration of betamethasone (synthetic glucocorticoid) suppresses the transcription of *pomc* in wild-type animals, but does not have any effect on the gr^{s357} mutants (Griffiths et al. 2012), further supporting the claim that GR knockout in zebrafish eliminates GR binding to the DNA.

Impulsive-Disinhibited Personality

Impulsive-disinhibited personality (IDP) is a behavioral trait that is chiefly characterized by seeking immediate gratification (Laplana et al. 2014). It is related to novelty seeking, and as such, the two traits have been historically combined into a single construct (Zuckerman 1996). Recently, there has been a push to separate the two traits as unique, as they may predict different types of behaviors (Magid et al. 2007). For instance, in an undergraduate population, the trait of impulsivity and sensation, measured via the Impulsiveness and Monotony Avoidance subscales from the Impulsiveness Monotony Avoidance (IMA) scale were differentially related to alcohol use and alcohol problems (Magid et al. 2007). In the undergraduate population, high trait belongingness for sensation seeking was more closely related to alcohol use and impulsivity more closely predicted alcohol-related problems (Magid et al. 2007). Impulsivity may be evaluated via instruments such as the Eysenck Personality Inventory (EPI; Acton 2003), the Eysenck Personality Questionnaire (EPQ; Acton 2003), and the NEO-PI-R (a self-report inventory designed to measure the basic components of personality identified on the FFM; Costa and McCrae 1992). The behavioral trait of impulsivity is a facet of the global trait of extraversion in the EPI, and psychoticism in the EPQ (Acton 2003). Impulsivity, in either instrument, is assessed via self-reports for survey items (Acton 2003). On the NEO-PI-R, impulsivity is related to four facets that load onto several traits on the FFM: sensation seeking (extraversion), lack of premeditation (conscientiousness), lack of perseverance (conscientiousness) and urgency (neuroticism; Whiteside and Lynam 2001).

Impulsive-disinhibited personality is associated with the reduced ability to regulate the tendency to respond to stimuli and is associated with various Cluster B personality disorders including antisocial and histrionic personality disorders (Laplana et al. 2014). To better understand IDP, researchers have sought to understand the heritability of this personality trait, often looking to polymorphic regions in the human genome (Laplana et al. 2014). Copy-number variants (CNVs) are polymorphic regions and are associated with several behavioral traits (Cook and Scherer 2008). Genome wide analysis reveal CNV affecting a 30.1 Kb region in SIRPB1 intron 1 that is correlated with impulsivity and sensation seeking in a dose-dependent manner (Laplana et al. 2014). The *SIRPB1* candidate gene for IDP has been evaluated in the zebrafish as well. Functional assays in larval zebrafish highlight two regions that are associated with the expression of SIRPB1 (Laplana et al. 2014), opening up a new avenue of neurobehavioral research in the zebrafish.

Anxiety Disorders

Stress and anxiety have been widely studied in the zebrafish (Egan et al. 2009; Griffiths et al. 2012; Stewart et al. 2012). The novel tank test is one of the more widely used behavioral measures of anxiety in the zebrafish (Stewart et al. 2012). As previously mentioned, in this task, a zebrafish is introduced to a novel environment that largely permits only vertical swimming. In addition to the behavioral parameters that are routinely collected in the novel tank test (i.e., placement within the tank), two additional behaviors are measured in the study of anxious states, the increased tendency to engage in erratic swimming and an increase in the time spent in an immobile state (freezing; Stewart et al. 2012). Erratic swimming is characterized by rapid direction changes that are accompanied by an increase in swim speed (Kalueff et al. 2013). The stress response in the zebrafish is characterized by an increased frequency and length of freezing bouts, increased bottom dwelling, and greater whole body cortisol levels (Egan et al. 2009). Outside of the novel tank test, the activation of the stress responses induces increased shoal cohesion and decreased aggression (Egan et al. 2009). The anxiety response is mediated by genetic factors and is susceptible to pharmacological intervention. Commercially available mutant strains (leopard, long fin, and albino) exhibit greater latency to explore the top regions of their environment and have fewer transitions to the top regions of the environment (Egan et al. 2009).

Administration of a single dose of caffeine results in an increased latency to explore the top region of the novel tank (Egan et al. 2009), suggestive of an anxiety-like response. Chronic exposure to fluoxetine and low doses of ethanol (0.3% v/v) reduce behavioral and physiological markers of anxiety. Chronic exposure to fluoxetine and low doses of alcohol markedly increases total exploration of the top region of the tank and significantly reduces cortisol expression, relative to wild-type controls (Egan et al. 2009). The acute and chronic administration of monoamine oxidase inhibitor agent, tranylcypromine (TCP; Stewart et al. 2011), and, following acute and chronic treatments with lysergic acid diethlamide (LSD; Stewart et al. 2011; Grossman et al. 2010) modulate serotonin neurotransmission (Backstrom et al. 1999) and reduces anxiety in the zebrafish (shorter latency to top, increased transitions to top, reduced freezing and erratic movements).

Discussion

In the interest of furthering our understanding of the neural and biological underpinnings of psychological and personality disorders, the use of animal models is invaluable. Work with rodent models has been indispensable to the discovery of genes relating to certain traits (e.g., neuroticism), and has played a crucial role in the development and improvement of drugs used to treat symptoms of depression and other psychiatric disorders. However, the complexity of the rodent's nervous system can result in inconsistencies in the translation of drug responses to humans (Krishnan and Nestler 2011; Mayberg et al. 2005).

The zebrafish has been widely used as a neurobehavioral model, and has proven to be a valuable model in the discovery of candidate genes for personality traits, and in the screening of novel drug treatments. This animal holds tremendous translational potential as the necessary neurotransmission pathways for emotion regulation, homologous brain regions for behavior control, and genetic orthologues are highly conserved with humans (Howe et al. 2013). The use of inbred strains allows researchers to control for genetic variance, and rearing animals under constant conditions works to limit environmental variation during early life. However, as has been noted, inbreeding and selective breeding produces strains with specific behavioral phenotypes (e.g., higher baseline for activity or anxiety). Thus, researchers should be cautious when studying the effects of, for instance, anxiogenic or anxiolytic substances, paying close attention to the strain tested and their baseline behaviors. In addition to selective breeding, the zebrafish model allows for easy genetic manipulation. Genetic knockdown and knockout models allow researchers to use study conditions and traits such as IPD (Laplana et al. 2014), and depression (Ziv et al. 2012). A greater understanding of the neural targets for drugs that alleviate symptoms of such disease would inform researchers of possible new drug targets, paving the way for improving current treatments and work to ultimately improve the human condition.

References

Acton, G. S. (2003). Measurement of impulsivity in a hierarchical model of personality traits: Implications for substance use. *Substance Use and Misuse, 38,* 67–83.

Allport, G. W. (1931). What is a trait of personality? *The Journal of Abnormal and Social Psychology, 25,* 368.

Alsop, D., & Vijayan, M. M. (2008). Development of the corticosteroid stress axis and receptor expression in zebrafish. *American Journal of Physiology-Regulatory, Integrative and Comparative Physiology, 294,* R711–R719.

Alsop, D., & Vijayan, M. M. (2009). The zebrafish stress axis: Molecular fallout from the teleost-specific genome duplication event. *International Fish Endocrine Symposium Special Issue, 161,* 62–66.

American Psychiatric Association. (2013). *Diagnostic and statistical manual of mental disorders* (5th ed.). Arlington, VA: American Psychiatric Publishing.

Apperley, L. A. (2012). What is "theory of mind"? Concepts, cognitive processes, and individual differences. *The Quarterly Journal of Experimental Psychology, 65,* 825–839.

Ariyomo, T. O., Carter, M., & Watt, P. J. (2013). Heritability of boldness and aggressiveness in the zebrafish. *Behavior Genetics, 43,* 161–167.

Ariyomo, T. O., & Watt, P. J. (2012). The effect of variation in boldness and aggressiveness on the reproductive success of zebrafish, *Danio rerio*. *Animal Behaviour, 83,* 41–46.

Backstrom, J. R., Chang, M. S., Chu, H., Niswender, C. M., & Sanders-Bush, E. (1999). Agonist-directed signaling of serotonin 5-HT2C receptors: Differences between serotonin and lysergic acid diethylamide (LSD). *Neuropsychopharmacology, 21*(2 Suppl), 77S–81S.

Barba-Escobedo, P. A., & Gould, G. G. (2012). Visual social preferences of lone zebrafish in a novel environment: Strain and anxiolytic effects. *Genes, Brain and Behavior, 11,* 366–373.

Barkus, C. (2013). Genetic mouse models of depression. *Current Topics in Behavioral Neuroscience, 14,* 55–78.

Belzung, C. (2014). Innovative drugs to treat depression: Did animal models fail to be predictive or did clinical trials fail to detect effects? *Neuropsychopharmacology, 39,* 1041–1051.

Benjamin, J., Li, L., Patterson, C., Greenberg, B. D., Murphy, D. L., & Hamer, D. H. (1996). Population and familial association between the D4 dopamine receptor gene and measures of novelty seeking. *Nature Genetics, 12,* 81–84.

Best, J. D., Berghmans, S., Hunt, J. J., Clarke, S. C., Fleming, A., Goldsmith, P., et al. (2008). Non-associative learning in larval zebrafish. *Neuropsychopharmacology, 33,* 1206–1215.

Bhat, A. (2004). Patterns in the distribution of freshwater fishes in rivers of Central Western Ghats, India and their associations with environmental gradients. *Hydrobiologia, 529,* 83–97.

Bouchard, T., Jr., & Loehlin, J. C. (2001). Genes, evolution, and personality. *Behavior Genetics, 31,* 243–273.

Buske, C., & Gerlai, R. (2012). Maturation of shoaling behavior is accompanied by change in the dopaminergic and serotoninergic systems in zebrafish. *Developmental Psychobiology, 54,* 28–35.

Cachat, J. M., Kyzar, E., Collins, C., Gaikwad, S., Green, J., Roth, A., et al. (2013). Unique and potent effects of acute ibogaine on zebrafish: The developing utility of novel aquatic models for hallucinogenic drug research. *Behavioural Brain Research, 236,* 258–269.

Cahill, G. M. (1996). Circadian regulation of melatonin production in cultured zebrafish pineal and retina. *Brain Research, 708,* 177–181.
Caravaggio, F., Fervaha, G., Chung, J. K., Gerretsen, P., Nakajima, S., Plitman, E., et al. (2016). Exploring personality traits related to dopamine D2/3 receptor availability in striatal subregions of humans. *European Neuropsychopharmacology, 26,* 644–652.
Claridge, G., & Davis, C. (2013). Personality dimensions: description and biology. In *Personality and psychological disorders* (pp. 31–59). London: Routledge.
Clark, L. A. (2007). Assessment and diagnosis of personality disorder: Perennial issues and an emerging reconceptualization. *Annual Review of Psychology, 58,* 227–257.
Cloninger, C. R., Svarakic, R., Dragan, M., & Przybeck, T. R. (1993). A psychobiological model of temperament and character. *Archives of General Psychiatry, 50,* 975–990.
Collier, A. D., & Echevarria, D. J. (2013). The utility of the zebrafish model in conditioned place preference to assess the rewarding effects of drugs. *Behavioural Pharmacology, 24,* 375–383.
Conrad, J. L., Weinersmith, K. L., Brodin, T., Saltz, J. B., & Sih, A. (2011). Behavioural syndromes in fishes: A review with implications for ecology and fisheries management. *Joural of Fish Biology, 78,* 395–435.
Cook, E. H. E., & Scherer, S. W. S. (2008). Copy-number variations associated with neuropsychiatric conditions. *Nature, 455,* 919–923.
Costa, P. T., Jr., & McCrae, R. R. (1992). NEO-PI-R: NEO personality inventory-revised. In G. J. Boyle, G. Matthews, & D. H. Saklofske (Eds.), *The SAGE handbook of personality theory and assessment: personality measurement and testing*, volume 2 (pp. 179–198). Great Britian: The Cromwell Press Ltd.
Croston, R., Branch, C. L., Kozlovsky, D. Y., Dukas, R., & Pravosudov, V. V. (2015). Heritability and the evolution of cognitive traits. *Behavioral Ecology, 00* (00), 1–13.
Dadda, M., Domenichini, A., Piffer, L., Argenton, F., & Bisazza, A. (2010). Early differences in epithalamic left-right asymmetry influence lateralization and personality of adult zebrafish. *Behavioral Brain Research, 206,* 208–215.
Danner, D. D., Snowdon, D. A., & Friesen, W. V. (2001). Positive emotions in early life and longevity: Findings from the nun study. *Personality Processes and Individual Differences, 85,* 804–813.
Dawis, R. V. (1992). The individual differences tradition in counseling psychology. *Journal of Counseling Psychology, 39,* 7–19.
Deary, I. J., Weiss, A., & Batty, G. D. (2010). Intelligence and personality as predictors of illness and death: How researchers in differential psychology and chronic disease epidemiology are collaborating to understand and address health inequalities. *Psychological Science in the Public Interest, 11,* 53–79. doi:10.1098/rspb.2008.1555.
Dingemanse, N. J., der Plas, F. V., Wright, J., Reale, D., Schrama, M., Roff, D. A., ... Barber, I. (2009). Individual experience and evolutionary history of predation affect expression of heritable variation in fish personality and morphology. *Proceedings of the Royal Society of London B: Biological Sciences, rspb-2008, 276,* 1285–1293.
Dingemanse, N. J., & Reale, D. (2005). Natural selection and animal personality. *Behavior, 142,* 1159–1184.
Dolan-Sewell, R. T., Krueger, R. F., & Shea, M. T. (2001). Co-occurrence with syndrome disorders. In W. J. Livesley (Ed.), *Handbook of personality disorders.* New York: Guilford Press.
Driscoll, P., & Bättig, K. (1982). Behavioral, emotional and neurochemical profiles of rats selected for extreme differences in active, two-way avoidance performance. In I. Lieblich (Ed.), *Genetics of the brain* (pp. 96–123). Amsterdam: Elevier.
Driscoll, P., Fernández-Teruel, A., Corda, M. G., Giorgi, O., & Steimer, T. (2009). Some guidelines for defining personality differences in rats. In Y. K. Kim (Ed.), *Handbook of behavior genetics* (pp. 281–300). New York: Springer.
Dugatkin, L. A., & Godin, J.-G. (1992). Prey approaching predators: A cost-benefit perspective. *Annales Zoologici Fennici, 29,* 233–252.
Ebstein, R. P., Novick, O., Umansky, R., Priel, B., Osher, Y., Blaine, D., et al. (1996). Dopamine D4 receptor (D4DR) exon III polymorphism associated with the human personality trait of Novelty Seeking. *Nature Genetics, 12,* 78–80.
Egan, R. J., Bergner, C. L., Hart, P. C., Cachat, J. M., Canavello, P. R., Elegante, M. F., et al. (2009). Understanding behavioral and physiological phenotypes of stress and anxiety in zebrafish. *Behavioural Brain Research, 205,* 38–44.
Eysenck, H. J. (1967). *The biological basis of personality.* Springfield, IL: Thomas.
Fleeson, W., & Gallagher, M. P. (2009). The implications of the big-five standing for the distribution of trait manifestation in behavior: Fifteen experience-sampling studies and a meta-analysis. *Journal of Personal and Social Psychology, 97,* 1097–1114.
Frazer, A., & Morilak, D. A. (2005). What should animal models of depression model? *Neuroscience & Biobehaivoral Reviews, 29,* 515–523.
Friedman, H. S., Tucker, J. S., Schwartz, J. E., Tomlinson-Keasey, C., Martin, L. R., Wingard, D. L., et al. (1995). Psychosocial and behavioral predictors of longevity: The aging and death of the "Termites". *American Psychologist, 50,* 69–78.
Fullerton, J., Cubin, M., Tiwari, H., Wang, C., Bomhra, A., Davidson, S., et al. (2002). Linkage analysis of extremely discordant and concordant sibling pairs identifies quantitative-trait loci that influence variation in the human personality trait neuroticism. *American Journal of Human Genetics, 72,* 879–890.
Garner, J. P. (2014). The significance of meaning: Why do over 90% of behavioral neuroscience results fail to

translate to humans, and what can we do to fix it? *ILAR Journal, 55,* 438–456.

Gauthier, J., Champagne, N., Lafreniere, R. G., Xiong, L., Spiegelman, D., Brustein, E., et al. (2010). De novo mutations in the gene encoding the synaptic scaffolding protein SHANK3 in patients ascertained for schizophrenia. *Proceedings of the National Academy of Sciences, 107,* 7863–7868.

Gerlai, R. (2003). Zebra fish: An uncharted behavior genetic model. *Behavior Genetics, 33,* 462–468.

Gerlai, R., Lahav, M., Guo, S., & Rosenthal, A. (2000). Drinks like a fish: Zebra fish (*Danio rerio*) as a behavior genetic model to study alcohol effects. *Pharmacology, Biochemistry and Behavior, 67,* 773–782.

Goldberg, L. R. (1990). An alternative "description of personality": The big-five factor structure. *Journal of Personality and Social Psychology, 59,* 1216.

Gosling, S. D. (2001). From mice to men: What can we learn about personality from animal research? *Psychological Bulletin, 127,* 45–86.

Gray, J. A. (1970). The psychophysiological basis of introversion–extraversion. *Behaviour Research and Therapy, 8,* 249–266.

Griffiths, B. B., Schoonheim, P. J., Ziv, L., Voelker, L., Baier, H., & Gahtan, E. (2012). A zebrafish model of glucocorticoid resistance shows serotonergic modulation of the stress response. *Frontiers in Behavioral Neuroscience, 6*(68), 1–10.

Grossman, L., Stewart, A., Gaikwad, S., Utterback, E., Wu, N., DiLeo, J., et al. (2011). Effects of piracetam on behavior and memory in adult zebrafish. *Brain Research Bulletin, 85,* 58–63.

Grossman, L., Utterback, E., Stewart, A., Gaikwad, S., Chung, K. M., Suciu, C., et al. (2010). Characterization of behavioral and endocrine effects of LSD on zebrafish. *Behavioural Brain Research, 214,* 277–284.

Harkness, A. R., & Lilienfield, S. O. (1997). Individual differences science for treatment planning: Personality traits. *Psychological Assessments, 9,* 349–360.

Hermes, G. L., Delgado, B., Tretiakova, M., Cavigelli, S. A., Krausz, T., Conzen, S. D., et al. (2009). Social isolation dysregulates endocrine and behavioral stress while increasing malignant burden of spontaneous mammary tumors. *Proceedings of the National Academy of Sciences, 106,* 22393–22398.

Howe, D., Bradford, Y. M., Conlin, T., Eagle, A. E., Fashena, D., Frazer, K., et al. (2013a). ZFIN, the zebrafish model organism database: Increased support for mutants and transgenics. *Nucleic Acids Research, 41,* 854–860.

Howe, K., Clark, M. D., Torroja, C. F., Torrance, J., Berthelot, C., Muffato, M., et al. (2013b). The zebrafish reference genome sequence and its relation to the human genome. *Nature, 496,* 498–503. doi:10.1038/nature12111.

Howell, K. R., Kutiyanawalla, A., & Pillai, A. (2011). Long-term continuous corticosterone treatment decreases VEGF receptor-2 expression in frontal cortex. *PLoS ONE, 6*(5), e20198.

Kalueff, A. V., Gebhardt, M., Stewart, A., Cachat, J. M., Brimmer, M., Chawla, J. S., et al. (2013). Towards a comprehensive catalog of zebrafish behavior 1.0 and beyond. *Zebrafish, 10,* 70–86.

Kalueff, A. V., Stewart, A., Kyzar, E., Cachat, J. M., Gebhardt, M., Landsman, S., et al. (2012). Time to recognize zebrafish 'affective' behavior. *Behaviour, 149,* 1019–1036.

Kamphuis, J. H., & Noordhof, A. (2009). On categorical diagnoses in DSM–V: Cutting dimensions at useful points? *Psychological Assessment, 21,* 294–301.

Koolhaas, J. M., Korte, S. M., De Boer, S. F., Van Der Vegt, B. J., Van Reenen, C. G., Hopster, H., et al. (1999). Coping styles in animals: Current status in behavior and stress-physiology. *Neuroscience and Biobehavioral Reviews, 23,* 925–935.

Krishnan, V., & Nestler, E. J. (2011). Animal models of depression: Molecular perspectives. *Current Topics in Behavioral Neuroscience, 7,* 121–147.

Krueger, R. F., & Eaton, N. R. (2010). Personality traits and the classification of mental disorders: Toward a more complete integration in dsm-5 and an empirical model of psychopathology. *Personality Disorders: Theory, Research, and Treatment, 1,* 97–118.

Lange, M., Neuzeret, F., Fabreges, B., Froc, C., Bedu, S., Bally-Cuif, L., et al. (2013). Inter-individual and inter-strain variations in zebrafish locomotor ontogeny. *PLoS ONE, 8,* e70172.

Laplana, M., Royo, J. L., Garcia, L. F., Aluja, A., Gomez-SKarmeta, J. L., & Fibla, J. (2014). SIRPB1 copy-number polymorphism as candidate quantitative trait locus for impulsive-disinhibited personality. *Genes, Brain and Behavior, 13,* 653–662.

Larstone, R. M., Jang, K. L., Livesley, J., Vernon, P. A., & Wolf, H. (2002). The relationship between Eysenck's P-E-N model of personality, the five-factor model of personality, and traits delineating personality dysfunction. *Personality and Individual Differences, 33,* 25–37.

Lubinski, D. (2000). Scientific and social significance of assessing individual differences: "Sinking shafts at a few critical points". *Annual Reviews in Psychology, 51,* 405–444.

Magid, V., MacLean, M. G., & Colder, C. R. (2007). Differentiating between sensation seeking and impulsivity through their mediated relations with alcohol use and problems. *Addictive Behaviors, 21,* 2046–2061.

Magurran, A. E. (1986). Predator inspection behaviour in minnow shoals: Differences between populatios and individuals. *Behavioral Ecology and Sociobiology, 19,* 267–273.

Mayberg, H. S., Lozano, A. M., Voon, V., McNeely, H. E., Seminowicz, D., Hamani, C., et al. (2005). Deep brain stimulation for treatment-resistant depression. *Neuron, 45,* 651–660.

McAdams, D. P. (2015). Three lines of personality development: A conceptual itinerary. *European Psychologist, 20,* 252–264.

McAdams, D. P., & Pals, J. L. (2006). A new big five: fundamental principles for an integrative science of personality. *American Psychologist, 61*(3), 204–217.
McCrae, R. R., & Costa, P. T., Jr. (1987). Validation of the five-factor model of personality across instruments and observers. *Journal of Personality and Social Psychology, 52,* 81–90.
McCrae, R. R., & John, O. P. (1992). An introduction to the five-factor model and its applications. *Journal of Personality, 60,* 175–215.
McFarlane, H. G., Kusek, G. K., Yang, M., Phoenix, J. L., Bolivar, V. J., & Crawley, J. N. (2008). Autism-like behavioral phenotypes in BTBR T + tf/J mice. *Genes, Brain and Behavior, 7,* 152–163.
McTighe, S. M., Neal, S. J., Lin, Q., Hughes, Z. A., & Smith, D. G. (2013). The btbr mouse model of autism spectrum disorders has learning and attentional impairments and alterations in acetylcholine and kynurenic acid in prefrontal cortex. *PLoS ONE, 8,* 62189.
Mehta, P. H., & Gosling, S. D. (2008). Bridging human and animal research: a comparative approach to studies of personality and health. *Brain Behavior and Immunity, 22,* 651–661.
Meltzoff, A. N. (1999). Origins of the theory of mind, cognition, and communication. *Journal of Communication Disorders, 32,* 251–269.
Miklosi, A., & Andrew, R. J. (2006). The zebrafish as a model for behavioral studies. *Zebrafish, 3*(2), 227–234.
Moretz, J. A., Martins, E. P., & Robinson, B. D. (2007a). Behavioral syndromes and the evolution of correlated behavior in zebrafish. *Behavioral Ecology, 18,* 556–562.
Moretz, J. A., Martins, E. P., & Robinson, B. D. (2007b). The effects of early and adult social environment on zebrafish (*Danio rerio*) behavior. *Environmental Biology of Fishes, 80,* 91–101.
Norton, W., & Bally-Cuif, L. (2010). Adult zebrafish as a model organism for behavioural genetics. *BMC Neuroscience, 11*(1), 1–11.
Oliveira, R. F., Silva, R. H., & Simoes, J. M. (2011). Fighting zebrafish: Characterization of aggressive behavior and winner-loser effects. *Zebrafish, 8,* 74–81.
Olivier, B., & Young, L. J. (2002). Animal models of aggression. In K. L. Davis, D. Charney, J. T. Coyle, & C. Nemeroff (Eds.), *Neuropsychopharmacology: The fifth generation of progress* (pp. 1699–1708). Philadelphia: Lippincott, Williams and Wilkins.
Ozer, D. J., & Benet-Martinez, V. (2006). Personality and the prediction of consequential outcomes. *Annual Review of Psychology, 57,* 401–421.
Perona, M. T. G., Waters, S., Hall, F. S., Sora, I., Lesch, K.-P., Murphy, D. L., et al. (2008). Animal models of depression in dopamine, serotonin, and norepinepherine transporter knockout mice: Prominent effects of dopamine transporter deletions. *Behavioural Pharmacology, 19,* 566–574.
Pham, M., Raymond, J., Hester, J., Kyzar, E., Gaikwad, S., Bruce, I., … & Zapolsky, I. (2012). Assessing social behavior phenotypes in adult zebrafish: shoaling, social preference, and mirror biting tests. In A. V. Kalueff & A. M. Stewart (Eds.) *Zebrafish protocols for neurobehavioral research* (pp. 231–246). New York, NY: Humana Press.
Pitcher, T. J. (1983). Heuristic definitions of fish shoaling behavior. *Animal Behaviour, 31,* 611–612.
Plaut, I. (2000). Effects of fin size on swimming performace, swimming behaviour and routine activity of zebrafish danio rerio. *The Journal of Experimental Biology, 203,* 813–820.
Porsolt, R. D. (2000). Animal models of depression: Utility for transgenic research. *Reviews in the Neurosciences, 11,* 53–58.
Reale, D., Reader, S. M., Sol, D., McDougall, P. T., & Dingemanse, N. J. (2007). Integrating animal temperament within ecology and evolution. *Biological Reviews, 82,* 291–318.
Riehl, R., Kyzar, E., Allain, A., Green, J., Hook, M., Monnig, L., et al. (2011). Behavioral and physiological effects of acute ketamine exposure in adult zebrafish. *Neurotoxicology and Teratology, 33,* 658–667.
Roberts, B. W., Kuncel, N. R., Shiner, R. L., Caspi, A., & Goldberg, L. R. (2007). The power of personality: The comparative validity of personality traits, socioeconomic status, and cognitive ability for predicting important life outcomes. *Perspectives on Psychological Science, 2,* 313–345.
Ruhl, N., & McRobert, S. P. (2005). The effect of sex and shoal size on shoaling behaviour in *Danio rerio. Journal of Fish Biology, 67,* 1318–1326.
Sagvolden, T. (2000). Behavioral validation of the spontaneously hypertensive rat (SHR) as an animal model of attention-deficit/hyperactivity disorder (AD/HD). *Neuroscience & Biobehavioral Reviews, 24*(1), 31–39. doi:10.1016/S0149-7634(99)00058-5.
Sailer, C. D., Radenic, S. J., & Wisenden, B. D. (2012). A method for sorting zebrafish on the exploratory-boldness behavioral axis. In A. V. Kalueff & A. M. Stewart (Eds.) *Zebrafish protocols for neurobehavioral research* (pp. 145–151). New York, NY: Humana Press.
Sallinen, J., Haapalinna, A., Viitamaa, T., Kobilka, B. K., & Scheinin, M. (1998). Adrenergic alpha2-receptors modulate the acoustic startle reflex, prepulse inhibition, and aggression in mice. *Journal of Neuroscience, 18,* 3035–3042.
Saverino, C., & Gerlai, R. (2008). The social zebrafish: Behavioral responses to conspecific, heterospecific, and computer animated fish. *Behavioral Brain Research, 191,* 77–87.
Seguret, A., Collignon, B., & Halloy, J. (2016). Strain differences in the collective behaviour of zebrafish (*Danio rerio*) in heterogenous environment. arXiv preprint arXiv, 1602.00024.
Seligman, M. E., Rosellini, R. A., & Kozak, M. J. (1975). Learned helplessness in the rat: time course, immunization, and reversibility. *Journal of Comparative and Physiological Psychology, 88,* 542–547.

Shaklee, A. B. (1963). Comparative studies of temperament: Fear responses in different species of fish. *The Journal of Genetic Psychology: Research and Theory on Human Development, 102,* 295–301.

Sih, A., Bell, A., & Johnson, J. C. (2004). Behavioural syndromes: An ecological and evolutionary overview. *Trends in Ecology & Evolution, 19,* 372–378.

Sison, M., & Gerlai, R. (2011). Behavioral performance altering effects of MK-801 in zebrafish (*Danio rerio*). *Behavioural Brain Research, 220,* 331–337.

Slattery, D. A., & Cryan, J. F. (2014). The ups and downs of modelling mood disorders in rodents. *ILAR Journal, 55,* 297–309.

Slattery, D. A., Neumann, I. D., & Cryan, J. F. (2011). Transient inactivation of the infralimbic cortex induces antidepressant-like effects in the rat. *Journal of Psychopharmacology, 25,* 1295–1303.

Smith, S. M., & Vale, W. W. (2006). The role of the hypothalamic-pituitary-adrenal axis in neuroendocrine responses to stress. *Dialogues in Clinical Neuroscience, 8,* 383–395.

Sorge, R. E., Martin, L. J., Isbester, K. A., Sotocinal, S. G., Rosen, S., Tuttle, A. H., et al. (2014). Olfactory exposure to males, including men, causes stress and related analgesia in rodents. *Nature Methods, 11,* 629–632.

Spence, R., Gerlach, G., Lawrence, C., & Smith, C. T. (2008). The behaviour and ecology of the zebrafish *Danio rerio*. *Biological Reviews, 83,* 13–34.

Stamps, J., & Groothuis, T. G. G. (2010). The development of animal personality: Relevance, concepts and perspectives. *Biological Reviews, 85,* 301–325.

Stewart, A. M., Braubach, O., Spitsbergen, J., Gerlai, R., & Kalueff, A. V. (2014). Zebrafish models for translational neuroscience research: From tank to bedside. *Trends in Neurosciences, 37,* 264–278.

Stewart, A. M., Gaikwad, S., Kyzar, E., Green, J., Roth, A., & Kalueff, A. V. (2012). Modeling anxiety using adult zebrafish: A conceptual review. *Neuropharmacology, 62,* 135–143.

Stewart, A. M., Nguyen, M., Wong, K., Poudel, M. J., & Kalueff, A. V. (2013). Developing zebrafish models of autism spectrum disorder (ASD). *Progress in Neuro-Psychopharmacology and Biological Psychiatry, 50,* 27–36.

Stewart, A. M., Wong, K., Cachat, J. M., Gaikwad, S., Kyzar, E., Wu, N., et al. (2011). Zebrafi sh models to study drug abuse-related phenotypes. *Reviews in the Neurosciences, 22,* 95–105.

Tran, S., & Gerlai, R. (2013). Individual differences in activity levels in zebrafish (*Danio rerio*). *Behavioral Brain Research, 257,* 224–229.

Vignet, C., Begout, M.-L., Pean, S., Lyphout, L., Leguay, D., & Cousin, X. (2013). Systematic screening of behavioral responses in two zebrafish strains. *Zebrafish, 10,* 365–375.

Viken, R. J., Rose, R. J., Kaprio, J., & Koskenvuo, M. (1994). A developmental genetic analysis of adult personality: Extraversion and neuroticism from 18 to 59 years of age. *Journal of Personal and Social Psychology, 66,* 722–730.

White, R. M., Sessa, A., Burke, C., Bowman, T., & LeBlanc, J. (2008). Transparent adult zebrafish as a tool for in vivo transplantation analysis. *Cell Stem Cell, 2,* 183–189.

Whiteside, S. P., & Lynam, D. R. (2001). The five factor model and impulsivity: Using a structural model of personality to understand impulsivity. *Personality and Individual Differences, 30,* 669–689.

Widiger, T. A., & Samuel, D. B. (2005). Diagnostic categories or dimensions? A question for the diagnostic and statistical manual of mental disorders—Fifth edition. *Journal of Abnormal Psychology, 114,* 494–504.

Wiggins, J. S. (2003). *Paradigms of personality assessment*. New York: Guilford Press.

Wimmer, H., & Perner, J. (1983). Beliefs about beliefs: Representation and constraining function of wrong beliefs i young children's understanding of deception. *Cognition, 13,* 103–128.

Wisenden, B. D., Sailer, C. D., Radenic, S. J., & Sutrisno, R. (2011). Maternal inheritance and exploratory-boldness behaviourl syndrome in zebrafish. *Behaviour, 148,* 1443–1456.

Wong, K., Elegante, M. F., Bartels, B. K., Elkhayat, S. I., Tien, D., Roy, S., et al. (2010). Analyzing habituation responses to novelty in zebrafish (*Danio rerio*). *Behavioural Brain Research, 208,* 450–457.

Wright, D., Nakamichi, R., Krause, J., & Butlin, R. K. (2006). QTL analysis of behavioral and morphological differentiation between wild and laboratory zebrafish (*Danio rerio*). *Behavior Genetics, 36,* 271–284.

Wright, D., Rimmer, L. B., Pritchard, V. L., Krause, J., & Butlin, R. K. (2003). Inter and intra-population variation in shoaling and boldness in the zebrafish (*Danio rerio*). *The Science of Nature, 90,* 374–377.

Zhou, Q. G., Zhu, L. J., Wu, H. Y., Luo, C. X., Chang, L., & Zhu, D. Y. (2011). Hippocampal neuronal nitric oxide synthase mediates the stress-related depressive behaviors of glucocorticoids by downregulating glucocorticoid receptor. *Journal of Neuroscience, 31,* 7579–7590.

Ziv, L., Muto, A., Schoonheim, P. J., Meijsing, S. H., Strasser, D., Ingraham, H. A., et al. (2012). An affective disorder in zebrafish with mutation of the glucocorticoid receptor. *Molecular Psychiatry, 18,* 1–11.

Zuckerman, M. (1996). The psychological model for impulsive unsocialized sensation seeking: A comparative approach. *Neuropsychobiolgy, 34,* 125–129.

Zuckerman, M., & Cloninger, C. R. (1996). Relationships between Cloninger's, Zuckerman's, and Eysenck's dimensions of personality. *Personality and Individual Differences, 21,* 283–285.

Personality in Elasmobranchs with a Focus on Sharks: Early Evidence, Challenges, and Future Directions

7

J.S. Finger, F. Dhellemmes and T.L. Guttridge

Abstract

Because of its impact on our understanding of evolution and ecology, animal personality has become an important area of research within behavioral ecology. Indeed, individual variation is no longer considered random noise but as a consistent phenomenon that impacts animal biology. However, research on animal personality and individual differences has largely focused on small-bodied species, which means that sharks and other elasmobranchs are dramatically underrepresented. The aim of this chapter is to illustrate, using existing studies, the opportunities and challenges involved in studying Elasmobranchs in captivity and in the field. While doing so, we discuss how this work informs the fields of animal personality and elasmobranch conservation. Although the lack of data necessitates a focus on sharks with only a mention of rays, we hope that this chapter will stimulate further research on personality in this underrepresented group.

Introduction

Personality in animals describes consistent behavioral differences across time and contexts between individuals of the same population (Stamps and Groothuis 2010). Animal personality can be approached from different perspectives and levels. For instance, individuals can consistently differ in certain behaviors, referred to

J.S. Finger (✉) · F. Dhellemmes
Faculty of Life Sciences, Humboldt Universität zu Berlin, Berlin, Germany
e-mail: js.finger@yahoo.fr

J.S. Finger · F. Dhellemmes
Leibniz-Institute of Freshwater Ecology and Inland Fisheries, Berlin, Germany

F. Dhellemmes · T.L. Guttridge
Bimini Biological Field Station Foundation, Bimini, Bahamas

J. Vonk et al. (eds.), *Personality in Nonhuman Animals*,
DOI 10.1007/978-3-319-59300-5_7

hereafter as behavioral axes. Five axes are commonly used, and provide a framework with which to explore animal personality (Réale et al. 2007). These axes are sociability, exploration, boldness, aggression in social contexts, and activity. These axes can be correlated, forming what are referred to as behavioral syndromes (Garamszegi et al. 2013). The interest triggered by animal personality, coupled with its important evolutionary and ecological consequences (Sih et al. 2012; Wolf and Weissing 2012), has driven this field to become a major topic in behavioral ecology. However, most of the work to date comes from observations of captive animals (see Archard and Braithwaite 2010 for examples and exceptions). This has led to a paucity of data on larger bodied, wild animals. Sharks and other elasmobranchs are no exception to this trend despite presenting interesting evolutionary and ecological characteristics on their own or as models for large aquatic animals.

Sharks are part of a highly diverse group of marine vertebrates, known as the cartilaginous fishes (Chondrichthyes). Chondrichthyans evolved independently of bony fishes (Osteichthyes) at least 400 million years ago (Pough et al. 1999) and, therefore, should not be considered "typical" fish (Sims 2003). The vertebrate class Chondrichthyes includes Holocephali (Chimaeras) and elasmobranchs (sharks, skates, and rays); however, due to the scarcity of behavioral data on most chondrichthyan fishes, the present chapter mainly focuses on sharks. Sharks range from planktivores to apex predators, exhibit diverse modes of reproduction, display ontogenetic shifts in their diet and habitat selection, and show considerable variation in brain size and complexity (Grubbs 2010; Lowry and Motta 2008; Yopak 2012). A large proportion of shark species present slow growth and reproduction rates, and long life spans. As described in this chapter, their diversity, important functional roles in the top-down control of marine ecosystem structure and function (Ferretti et al. 2010; Heithaus and Dill 2002), and life history strategy make them an interesting alternative group to study relative to most aquatic animals commonly investigated in animal personality, e.g., guppies (*Poecilia reticulata*; Burns 2008; Irving and Brown 2013), mosquitofish (*Gambusia* spp.; Burns et al. 2012; Cote et al. 2010), and sticklebacks (*Gasterosteus aculeatus*; Harcourt et al. 2009; Ward et al. 2004). Due to these life history traits and overexploitation in global fisheries, many shark populations are in decline (Dulvy et al. 2014; Worm et al. 2013), which might, in turn, impact entire ecosystems through trophic cascades and interactions (e.g., Heithaus et al. 2008; Stevens et al. 2000). These environmental concerns combined with the recent realization that personality should be included within environmental and fishery management programs (Conrad et al. 2011; Mittelbach et al. 2014) argue for a better understanding of shark behavior at the individual level.

To illustrate the relevance and feasibility of studying personality in sharks, despite the scarcity of research available, this chapter takes a holistic approach and discusses both proofs of personality and individual differences. The first section focuses on investigations of personality in sharks. Here, we describe different research methods, setups (e.g., captivity vs. semi-captivity), and the species used to study personality. Further, we explain why the juvenile lemon shark (*Negaprion brevirostris*) is an interesting model for the study of personality. This section aims to first illustrate how sharks can contribute important insights to contemporary discussions of animal personality and to stimulate the development and identification of other elasmobranch species with such potential. Next, we discuss the challenges of studying shark behavior under controlled conditions, which leads to the second section where the possibility and potential of studying shark personality in the wild is explored. Studies on the Port Jackson (*Heterodontus portusjacksoni*) and juvenile bull shark (*Carcharhinus leucas*) are introduced to demonstrate the practicability of investigating consistent individual differences in their behavior, along with an account of an ongoing project investigating novel object inspection in great white sharks (*Carcharodon carcharias*). Next, we review various field studies that describe interindividual variations in key behaviors of sharks (e.g., refuging and excursions, large-scale movements, feeding specialization, human interactions, and social behavior) illustrating the potential for informative studies of shark personality. These examples are included with arguments supporting the importance of considering input of shark personality on management of their populations and ecosystems. Of equal interest are the differences and similarities that shark biology displays relative to the other classical aquatic species intensively studied. In addition, we describe some of the available tools (e.g., technology, revisiting old data) that we believe could stimulate personality research in sharks. Finally, the chapter concludes by suggesting potential avenues investigators could take to advance our knowledge of elasmobranch personality.

Personality of Sharks in Captivity

To date, only a handful of studies have investigated elasmobranch personality. These have focused on testing for the presence of personality traits in a few shark species. In addition to the lemon shark and Port Jackson shark, personality has been studied in the small-spotted catshark (*Scyliorhinus canicula*; Jacoby et al. 2014). Most studies to date have used behavioral assays in captive or semi-captive environments (Byrnes and Brown 2016; Byrnes et al. 2016b; Finger et al. 2016; Jacoby et al. 2014; Wilson et al. 2015). These studies differed in their methods and ensuing results, with some demonstrating consistent individual differences, but all providing useful guidelines for future investigations.

As is the case for most animals, sharks reared in captivity are more readily accessible for behavioral assays and personality investigations than their wild counterparts. Taking advantage of the ability of juvenile small-spotted catsharks (*S. canicula*), to thrive in captivity, Jacoby et al. (2014) investigated consistent individual differences in their social network positions. Catsharks are a small-bodied benthic species, regularly observed resting on the seabed in social groups (Jacoby et al. 2014; Sims et al. 2001). Such grouping can be defined as socially connected when one shark rests within one body length another. This obvious and easily observable social behavior allows for the construction of social networks (Croft et al. 2011; Krause et al. 2009). Social network analyses provide measures to characterize the social dynamic of groups. Within these measures, some are focused on individuals, and so provide a method for quantifying individual sociality or social personality. Jacoby et al. (2014) repeatedly tested groups of the same individuals across different habitat types, e.g., simple to complex. They found that individual social network strength (sum of an individual's association index with all other individuals) was consistent across habitats. However, when preference for group size was controlled for (average group size each individual was observed in), the consistency in social behavior was less pronounced and no longer significant, suggesting that group size preferences drive social consistency, but only within small groups. In their discussion, the authors suggested rearranging individuals between test groups to determine if this consistency was due to individual personality or the group dynamic (average group size available). This would certainly be an important next step for this species and would help to elucidate the role of personality traits in grouping behavior of a marine predator. Additionally, this study illustrated the feasibility and potential of using captive-reared sharks for personality research. In a similar vein, exploring possibilities of collaborating with public aquariums to obtain data on a large range of elasmobranch species could provide researchers with a practical model to study personality in sharks. The availability of other large animals in captivity has for instance been productive for researchers studying great apes (e.g., King and Figueredo 1997) or dolphins (e.g., Highfill and Kuczaj 2007). One other idea that should be explored in future work is to manipulate the genetic background (for instance selecting lines of particular personality types) and/or manipulating the rearing environment of sharks to understand the respective roles of heredity and rearing in personality development and maintenance (see also Brommer and Class, this volume; Laine and van Oers, this volume). This approach has, for instance, provided interesting insights into personality in species of bony fish (e.g., Brown et al. 2007), reptiles (reviewed in Waters, Bowers and Burghardt, this volume), and birds (e.g., Groothuis and Carere 2005), and is achievable in some elasmobranch species (Griffiths et al. 2011).

In addition to studying personality in captive-reared sharks, personality tests can be performed on wild-caught sharks. As with other populations, demonstrating the presence of personality is a necessary step before showing how personality impacts the natural behavior and ecology of these animals. Personality tests on wild-caught animals were performed in two species: juvenile lemon sharks and Port Jackson sharks. Whereas Finger et al. (2016) and Wilson et al. (2015) used semi-wild captive experiments (outdoor enclosures in the sharks' natural environments; see Guttridge et al. 2009) to investigate personality in juvenile lemon sharks, Byrnes and Brown (2016) used wild-caught juvenile Port Jackson sharks in a laboratory setting.

To test for consistency in social network position, Wilson et al. (2015) observed ten wild-caught juvenile lemon sharks from Eleuthera, Bahamas, in a mangrove enclave that was artificially closed. Individuals equipped with tri-axial accelerometer loggers (see Table 7.2) were observed three times a day over 8 days for their sociability (i.e., being within 1 body length of a conspecific) and leadership (i.e., being at the front of a pair or group), as well as locomotor profile (e.g., time swimming fast, swimming duration, frequency of burst swimming events, number of transitions between activity states, activity period). In contrast to catsharks assessed previously, juvenile lemon sharks in this experiment did not show consistent differences in their network positions. The dynamic of social interactions was related to individuals' locomotor profiles. Furthermore, the authors found that more social individuals were more active than less social individuals. This relationship between sociality and activity level might indicate the presence of a behavioral syndrome similar to that observed in Bony fish (e.g., Cote et al. 2010; Irving and Brown 2013). However, this remains to be demonstrated due to the lack of consistency in social behavior and the fact that the behaviors were not tested independently, thus creating a contextual overlap (Garamszegi et al. 2013).

More recently, Finger et al. (2016) captured wild juvenile lemon sharks and repeatedly subjected them to a novel open field test. They found consistent individual differences in their rate of movement. Subsequent repetitions of the novel open field test were used to test whether individual sharks were reacting to the novelty of the open field (i.e., habituating) or simply displaying their normal swimming behavior (i.e., showing no habituation; but see Finger et al. 2016 for other possible interpretations). While exploring habituation, this investigation revealed considerable variation in habituation rates between individuals, but also a covariance between these rates and exploration during the first open field visit. In other words, Finger et al. (2016) found a relation between personality and plasticity with faster explorers showing more rapid habituation to the novel open field test. Similar results have been obtained in other taxa (e.g., Dingemanse et al. 2012; Rodríguez-Prieto et al. 2011) suggesting that individual differences in plasticity are widespread and should be further investigated to better understand how changes in the environment (naturally- or human-induced) could differentially impact individuals.

To investigate the presence of consistent individual differences in an emergence test and stress responses to handling, Byrnes and Brown (2016) collected wild Port Jackson sharks and housed them temporarily in a laboratory. Sharks were scored based on time taken to move out of cover from a "refuge box" (emergence test) and the increase or decrease in activity relative to their baseline activity (reaction to stress test). The sharks showed consistent individual differences in both tests across repeated testing. Furthermore, the sharks that emerged faster from cover, i.e., bolder individuals, were more active after handling compared to those that were less bold. The authors interpreted this as evidence for the existence of a proactive–reactive coping style in sharks, with reactive individuals being more passive toward their environment than were proactive individuals. In addition, individuals that were more active after a stressful event exhibited greater lateralization strength, i.e., the propensity to choose one turning direction over another compared to individuals that were less active (Byrnes et al. 2016b). Due to the lack of comparative data in elasmobranchs, these relationships are difficult to interpret. Taken together, however, these early findings offer guidelines that could be applied to studies of different shark species, enabling other researchers to gather much-needed data to draw conclusions about the evolution of personality in sharks and other elasmobranchs.

Prior to the investigation described above (Finger et al. 2016), juvenile lemon sharks of Bimini (the Bahamas) had been extensively studied (Fig. 7.1 shows a juvenile lemon shark with its typical color tag for individual identification). In our opinion and experience, this particular system has characteristics that make the juvenile lemon sharks of Bimini a promising model for the study of elasmobranch personality. Because similar systems that have not yet been identified might already be available for studying personality, we wanted to describe what makes the juvenile lemon sharks of Bimini pertinent to the study of personality in sharks and elasmobranchs. This could lead to the development of alternative models that would provide the comparative data necessary for the understanding of proximate and ultimate causes of personality. In Bimini, during their first 3 years of life, juvenile lemon sharks use nursery areas (mangrove-fringed lagoons) as protection from predators (Chapman et al. 2009; Guttridge et al. 2012). They have small home ranges and display high site fidelity (Morrissey and Gruber 1993a, b; Sundström et al. 2001), and, as a result, can be successfully recaptured. Furthermore, annual sampling of the majority of a population's newborns in the main nursery area allows for the long-term study of their ecology (e.g., life

Fig. 7.1 Juvenile lemon sharks used for behavioral assays and identified with the *purple color tag* attached to its first dorsal fin. *Credit* Eugene Kitsios

history: Dibattista et al. 2007; philopatry: Feldheim et al. 2014; survival: Gruber et al. 2001). Acoustic telemetry studies revealed excursions out of the typical home range into deeper water, and away from the safety of the mangroves (Morrissey and Gruber 1993b) and visual observations have shown differences in refuge use (Guttridge et al. 2012). Although none of these investigations explicitly discussed personality, their observations indicate possible individual differences in the inclination of individual sharks to take risks. The fact that individual differences in a novel open field could be related to these existing observations led Finger et al. (2016) to suggest that risk–benefit trade-offs (Stamps 2007; Wolf et al. 2007) could play a major role in explaining the presence of personality in the juvenile lemon sharks of Bimini. This hypothesis is further supported by the finding that large and fast-growing 1-year-old individuals have a higher mortality rate (Dibattista et al. 2007), which supports the idea that higher risks bring greater benefits (i.e., increased growth for higher mortality). Moreover, stable isotope analyses in juvenile lemon sharks from Bimini revealed marked individual differences in their feeding habits (Hussey et al. 2017). Some sharks preferentially feed close to shore (mangroves; refuge habitat) whereas others are more wide ranging (seagrass flats; risky habitats). This relationship between personality differences in sharks and resulting differential risks and benefits is still hypothetical, as these studies have been performed independently; however, they provide a framework for future in-depth studies. Indeed, proximate and ultimate causes of personality are not fully understood, and trade-off hypotheses need to be tested. However, if predation risk plays an important role in personality development and maintenance (Bell and Sih 2007; Brown et al. 2007; Dingemanse et al. 2007; Urszán et al. 2015) then this might not be true for all shark species. As described above, evidence for personality was also found in juvenile Port Jackson sharks, despite Byrnes and Brown (2016) suggesting that there was little to no predation on this population. If such difference in predation rates exists (which depends upon further investigations), then comparative studies between these systems could give us insight into animal personality development and maintenance.

Experimental Challenges and Constraints

Along with demonstrating the feasibility and the potential of studying shark personality in captive settings, it is important to discuss the limitations and difficulties associated with doing so. Many shark species are difficult to handle and are easily stressed during capture (Gallagher et al. 2014; Morgan and Burgess 2007) and transportation (Young et al. 2002). Furthermore, most species are difficult (e.g., Crow and Hewitt 1988; Gruber and Keyes 1981; Stevens 1994), if not impossible, to maintain in captivity as they require unique husbandry methods (Smith et al. 2004), which would likely stress the animal, and probably lead to early mortality. Even if these practical and ethical issues were not limiting factors, performing behavioral tests on unhealthy or stressed subjects would likely result in inaccurate and/or biased findings. Despite these challenges, some public aquariums have managed to keep large sharks (e.g., whale sharks, *Rhincodon typus*, at the Georgia Aquarium, Atlanta, GA, USA); however, doing so for a large enough sample size would be expensive. These potential problems should be carefully considered before further experimental planning.

Obtaining a large enough sample size is a common issue when studying personality. Wilson et al. (2015) illustrate how, even with the substantial effort of capturing the majority of the population, only a small number of individuals could be captured ($N = 10$) leading to potential failure in detecting consistency. Furthermore, despite being able to demonstrate consistent individual differences in personality, Finger et al. (2016) encountered a similar problem with their sample being too small to show covariance between plasticity and personality, using mixed modeling approaches. We thus strongly recommend using power simulations (e.g., Finger et al. 2016) when possible, to avoid type II errors, which could deter future researchers from studying personality in sharks. It should also be borne in mind that, even when a large enough population is located, capturing the necessary sample size is a long and often expensive process, which usually demands a large team of scientists and/or a long period of field work (e.g., Gruber et al. 2001).

Compared to most animals investigated in behavioral studies, many sharks are large-bodied and so, to circumvent this issue, studying juvenile sharks may be more feasible. Still, it is important to understand that the size of juvenile sharks represents an experimental design challenge. For example, although juvenile lemon sharks are typically <1 m in total length, the required size of the semi-captive behavioral pen for observation by Finger et al. (2016) was large (12 × 6 m). Figure 7.2 illustrates the typical pen size required to observe shark behavior in a social and novel open field context (Finger, unpublished data). Further, depending on the species, researchers should consider how to transfer sharks from holding to experimental pens, considering handling stress, time, and effort. To limit potential stress, investigators have used a technique known as "ushering," which involves gently guiding the shark into the desired location, via a series of channels and temporary doors (Finger et al. 2016; Fuss et al. 2013; Guttridge and Brown 2013). Ushering has shown its effectiveness, but, depending on the size of the pen and the mobility of the species, this method can still require considerable time and personnel. Finally, it is important to note that if personality is demonstrated in juvenile stages, it does not lead to automatic stability throughout adulthood (e.g., Castanheira et al. 2016; Petelle et al. 2013) and results should be interpreted accordingly. However, the study of personality in juvenile sharks should not be neglected as successful management programs also depend on survival at this early stage of development (Heupel et al. 2007; Ward-Paige et al. 2015).

Summary

Overall, early investigations of shark personality have developed effective and reliable investigative methods, which will hopefully inspire

Fig. 7.2 Example of a semi-captive setup established in the shallow waters of Bimini, The Bahamas to test for personality in the juvenile lemon shark, *Negaprion brevirostris*

research into the personalities of sharks and other elasmobranchs. Consistent individual differences were demonstrated in two taxonomic orders showing that it is possible to obtain a diverse taxonomic representation using experimental studies. Nevertheless, the limited species and personality traits tested thus far do not enable one to draw general conclusions about whether personality is present in other elasmobranchs. Fortunately, there are numerous species that have been used for other behavioral investigations, including gray bamboo sharks (*Chiloscyllium griseum*, Fuss et al. 2013), coral catsharks (*Atelomycterus marmoratus*, Fuss et al. 2013), ocellate river stingrays (*Potamotrygon motoro*, Schluessel and Bleckmann 2005), and bonnethead sharks (*Sphyrna tiburo*, Myrberg and Gruber 1974) that could potentially be tested for personality. Furthermore, it is likely that other wild populations of sharks might provide good models for studying personality. However, it is important to emphasize that, when planning, conducting, and interpreting such experiments, investigators should take into account the constraints associated with testing sharks.

Personality and Individual Differences in the Field

The field of animal personality continues to benefit from controlled captive experimental approaches (Campbell et al. 2009; Herborn et al. 2010). However, large-bodied marine animals, including many sharks and close to half of chondrichthyan species, undertake migrations as part of their life cycle (Grubbs et al. 2010) and require considerable space to demonstrate natural prey-capture behavior (e.g., breaching behavior in great white sharks: Martin et al. 2005; patch foraging in basking sharks, *Cetorhinus maximus*; Sims et al. 2003), or reside in depths exceeding 200 m (Cotton and Grubbs 2015). As such, experiments of captive sharks are difficult, if not impossible to implement, which could prevent personality testing in many species. Such limitations generate challenges in understanding personality in a natural context and thus hinder our ability to understand the evolution and ecology of personality in sharks. This could, over time, drastically impact our ability to develop appropriate and realistic management programs that incorporate information on personality.

One concern with ignoring individual differences has been illustrated in other taxa. Because some personality types have a greater probability of being captured (Biro and Post 2008; Biro and Sampson 2015; Uusi-Heikkilä et al. 2008), traditional management programs are bound to be less effective as the estimations of fishing pressures on a population are biased (e.g., unappreciated artificial selection against fast-grower, bold individuals by passive fishing gears). Therefore, along with developing experimental models, some effort should be directed toward investigating sharks' personality in the field. However, regardless of species, field work is associated with difficulties, for example, in continuously observing animals and/or not being able to control environmental conditions. In addition, the fact that many sharks live in an environment that is difficult to work in (e.g., deep water in the open ocean), usually avoid human contact, and are highly mobile (e.g., traveling thousands of kilometers), makes this task almost impossible. However, despite these challenges, in the following section we describe studies investigating shark personality in the wild, and highlight studies that identified individual differences in ecologically relevant behaviors.

Evidence for Personality in the Field

Evidence for personality was recently demonstrated in a study that involved capturing and testing Port Jackson sharks in the field (Byrnes et al. 2016a). The researchers used a docility test that measured the shark's response to human handling (e.g., Martin and Réale 2008), and a lateralization test that recorded if and which side a shark rolled onto once it landed on a boat. Individuals consistently differed in docility, varying from sharks that struggled to those that did not move during handling, despite controlling for potential confounding effect of size, sex, and population. Consistency of lateralization was not investigated due to the lack of repeated trials, but individual differences were detected. The authors found no covariance between these two tests. This study thus described a new method of assessing personality that is applicable to numerous species, including bottom-dwelling sharks, such as nurse sharks (*Ginglymostoma cirratum*) or catsharks. Using this study as inspiration could lead to a large comparative data set based on observations in the wild and provide an interesting avenue to test for personality in relation to natural shark behavior. However, the approach used here would have to be modified for larger and/or more mobile animals.

The Juvenile Bull Shark as a Model Species for Personality Research

Investigations of juvenile bull sharks were successfully used to detect consistent individual differences in behavior (Matich et al. 2011), revealing their systems as particularly promising for the study of personality in sharks. Prior to discussing this study, it is interesting to note parallels with the juvenile lemon shark studies described above. Like lemon sharks, which use mangrove-fringed habitats as nursery areas, juvenile bull sharks can also be found using estuarine or freshwater habitats. Some of these habitats are known to be safer (less predation) but less productive (less food) further upstream (further away from the marine environment, Matich et al. 2011), and so are comparable to what is experienced by juvenile lemon sharks in Bimini. These similarities suggest that juvenile bull sharks face similar trade-offs between benefits and risks, which could promote individual differences. Indeed, Matich et al. (2011) found that individual juvenile bull sharks from their study site differed in their diet. Some individuals fed in the riskier marine food web and others in the safer estuarine habitat, which led the authors to propose that individuals differed in their risk–benefit strategies. More recently, in the same system, juvenile bull sharks were found to differ in their movements and in the portion of the estuary they used, with some being detected more often in riskier locations (e.g., downstream near the mouth) than others (Matich et al. 2011). These consistent individual differences were documented for at least 4 months and were independent of age class. Unfortunately, these investigations were performed on different groups of individuals leaving the existence of a potential relationship between individual differences in estuary use and feeding habit unproven.

It is important to note that estuarine systems can be heterogeneous and so, until there is evidence for personality in a more controlled environment, it is challenging to disentangle what could be related to responses to environmental conditions or to individual behavioral differences. For instance, Ortega et al. (2009) investigated movement in juvenile bull sharks in Florida and in another river system. Similarly, they found differences in movement between two groups of juvenile bull sharks, but explained that these could have been related to the locations within the river and the differences in habitat within these areas. Regardless, juvenile bull sharks are a promising model with which to investigate consistent individual differences in behavior. The heterogeneity of predator pressure and resource abundance in the juvenile bull sharks' nursery habitat contributes to the likelihood that personality differences are present in this species. Longer term studies along with a demonstration of a long-lasting relationship between movement differences and feeding location differences (risky marine web food vs. safer estuary) or life history traits (e.g., growth rate and survival) would strongly suggest that these differences reflect (at least partly) consistent individual differences and demonstrate the ecological importance of such differences. In the best scenario, individual differences should be investigated in controlled captive or semi-captive conditions and then in the field (e.g., Herborn et al. 2010; Yuen et al. 2016). This strategy would, for instance, allow researchers to further investigate if habitat use variability is due to differences in boldness, dominance, sociability, or exploration.

Behavioral Assays in Large Free Ranging Sharks

Being usually shy, wide ranging, and living in concealed environments, shark behavior is difficult to study for many species (Nelson 1977). The need to observe the same individuals multiple times, to demonstrate personality and logically further amplify this difficulty. However, as shark personality receives more attention, systems and methods are likely to be identified and developed or instance, white sharks provide a potential model to study personality in a large, upper trophic predator. White sharks can be attracted to research boats using bait and repeatedly observed over months or years, as demonstrated by the development of ecotourism in South Africa (e.g., Laroche et al. 2007). This opens up the possibility to observe the same individual multiple times. Furthermore, these sharks have been observed inspecting novel objects on the water surface (Hammerschlag et al. 2012). These characteristics were taken advantage of by William Hughes, Marlene Stürup, and colleagues in their investigation of individual differences in the behavior of white sharks, and the preliminary results are encouraging (William Hughes, personal communication). If these tests are applicable to personality research, they might provide a method for testing personality in large sharks in the wild. These methods could be used to further investigate the relationships between personality traits (such as novel object inspection) and individual differences in observed behaviors, such as movements, space use, hunting strategy, social interactions, feeding habits, and more, to be described below.

Evidence for Interindividual Variation in Shark Behavior: Reasons to Study Shark Personality

In the following part of this chapter, we describe the growing evidence for interindividual variability in the natural behavior of elasmobranchs across several species with diverse taxonomy, biology, and behavior (see Table 7.1). These descriptions of variation between individuals do not constitute proof of personality, and we do not intend to use these descriptions as such. Indeed, little attention has been paid to testing if individual differences in behavior are consistent (a key concept to personality) and if these differences have fitness consequences (e.g., growth rate, survival, reproduction) in the field. Despite the dearth of studies that test for consistency, we want to illustrate how, if proven to be related to personality, further investigating the variability between individuals could improve management success and our understanding of animal personality. Because such studies could inspire further work in shark personality, we describe which tools and methods were used in these studies in more detail within Tables 7.1 and 7.2. In this section we selected a few studies to illustrate and discuss differences in movements (e.g., localized and large-scale) with a brief discussion about incorporating personality variables in the design of marine protected areas. In addition, we describe the rationale for studying personality in large, low predation-risk predators, and the potential for strong impacts of individual differences on ecosystems and sensitivity to wildlife tourism. Finally, we discuss preliminary evidence for differences in the social behavior of sharks and emphasize the importance of investigating personality in other elasmobranchs.

Understanding elasmobranch movement is considered as a critical step to improve their conservation (Chapman et al. 2015; Papastamatiou and Lowe 2012). Data generated from these studies are crucial for delineating key areas that require protection, for example sites of parturition or aggregation (e.g., Mucientes et al. 2009). Such engagement has led to an explosion of studies on shark movements, development of technologies, and modern data analytic methods (see review: Hussey et al. 2015; Jacoby and Freeman 2016). This effort has unraveled interindividual differences in aspects of localized and large-scale movements.

The use of core areas has been documented in numerous shark species (see examples in Table 7.1). These demonstrations have provided a large amount of data on movements in sharks

Table 7.1 Examples of investigations used in the chapter describing interindividual differences in Elasmobranchs field studies

Species	Interindividual differences	Method	Source
Blacktip reef shark (*Carcharhinus melanopterus*)	Home range, habitat use and excursions High variability in isotopic signature; Some individuals might rely more on pelagic prey Gregariousness Community differences in their mean sociability	Acoustic telemetry (passive) SPOT tags and SIA Direct observations via SCUBA photo-identification	Papastamatiou et al. (2010) Mourier et al. (2012)
Scalloped hammerhead shark (*Sphyrna lewini*)	Core area use and distance of excursions out of these Refuging and night excursions away from seamount Mention of potential Solitary versus group excursions	Acoustic telemetry (active) Acoustic telemetry (active)	Ketchum et al. (2014) Klimley and Nelson (1984)
Gray reef shark (*Carcharhinus amblyrhynchos*)	Time of day attendance patterns Home range behavior. Reef associated sharks displaying more nomadic patterns than lagoon associated sharks. Nomadic individuals showed more agonistic behaviors toward divers	Acoustic telemetry (active) Acoustic telemetry (active and passive)	Field et al. (2010) McKibben and Nelson (1986)
Blue shark (*Prionace glauca*)	Site fidelity Male large-scale movements and destinations	Satellite telemetry (PSAT and SPOT)	Vandeperre et al. (2014)
Bull shark (*Carcharhinus leucas*)	Tendency to migrate Residency versus migratory behavior Niche specialization among specialist individuals	Acoustic telemetry (passive) Acoustic telemetry (passive) SIA	Heupel et al. (2015) Espinoza et al. (2016) Matich et al. (2011)
Tiger sharks (*Galeocerdo cuvier*)	Migration in mature females Inter-island movements Movement and residency Diving behavior and depth use Habitat use Movement: Path structure, patch size and tortuosity	Acoustic telemetry (passive) Satellite telemetry (PSAT and SPOT) Acoustic telemetry (passive) Satellite telemetry (PSAT) Crittercam Acoustic telemetry (passive) Acoustic telemetry (active)	Papastamatiou et al. (2013) Meyer et al. (2010) Vaudo et al. (2014) Afonso and Hazin (2015) Heithaus et al. (2002) Papastamatiou et al. (2011)
Spiny dogfish (*Squalus acanthias*)	Long distance movements and residency	Mark recapture	McFarlane and King (2003)
Great white shark (*Carcharodon carcharias*)	Hunting strategies Foraging strategies Response to provisioning	Acoustic telemetry (active) SIA Acoustic telemetry (passive)	Towner et al. (2016) Kim et al. (2012) Huveneers et al. (2013)
Juvenile lemon sharks (*Negaprion brevirostris*)	Leading groups Relationship between growth rate and mortality	Direct wild observation and conventional tagging Mark recapture	Guttridge et al. (2011) Dibattista et al. (2007)

(continued)

Table 7.1 (continued)

Species	Interindividual differences	Method	Source
Broadnose sevengill (*Notorynchus cepedianus*)	Movement and diet	SIA Acoustic telemetry (passive)	Abrantes and Barnett (2011)
Cownose ray (*Rhinoptera bonasus*)	Residency time and home range	Acoustic telemetry (passive)	Collins et al. (2007)
Manta ray (*Manta alfredi*)	Movement patterns at varying spatial scales	Acoustic telemetry (active)	Papastamatiou et al. (2012)

Table 7.2 Common methods used to investigate natural behavior of large aquatic animals

Method	Technique
Acoustic telemetry	Acoustic tags fitted to animals autonomously transmit positioning data to static receiving stations (passive tracking) that can be retrieved periodically or to mobile "real-time" receiving stations (active tracking) for example on a pursuit vessel
Satellite telemetry	PSAT (Pop-up Archival Satellite Tags) gather data on a variety of measurements (i.e., temperature, pressure, luminosity) while attached to the animal. Observations are sent to land-based receivers via orbiting satellites once tags are detached from the animal SPOT (Smart Position and Temperature tags) data on a variety of measurements (i.e., temperature, pressure, luminosity). Observations are sent to land-based receivers via orbiting satellites each time the antenna is out of the water (i.e., when the sharks' fin is out of the water for tags fitted to dorsal fins)
Crittercam	Crittercams are small animal-borne video cameras that record the behavior of its bearer. They are mostly used for predator–prey encounter studies
Photo-Identification	This technique consists of identifying unique features on an animal that are consistent through time. Pictures from recreational divers or the scientific team can be compared to previous images and individuals can be identified and observed through time. In Elasmobranchs, fin edges, ventral patterns, scarring, or skin patterns have been successfully used to identify individuals, multiple years in a row
Tri-axial accelerometer	Tri-axial accelerometer data loggers have revolutionized studies of animal behavior and are providing unprecedented new insight into the biomechanics of shark swimming. These devices measure accelerations due to gravity in three dimensions, and can be used to reconstruct a high-resolution record of shark body movements, including tail beat frequency and amplitude
External tagging	Another technique used for individual recognition is through external tagging. Each individual can receive a tag that can be easily recognized (specific tag number, color, shape). This can allow the participation of recreational divers into the studies if necessary

and have revealed individual differences in uses of these areas (see Table 7.1). In addition to being important to animal ecology and survival, the fact that some sharks use core areas represents an interesting methodological advantage. Indeed, sharks' regular use of the same area for long periods of time allows researchers to follow individuals for extended periods (e.g., 3.5 years: Papastamatiou et al. 2010) and to then perform finer scale investigations. Understanding individual differences in movements around and in core areas can contribute to the development of accurate and efficient protected areas (i.e., marine protected areas, no-take zones). For instance, while focusing on protecting only those core areas, sedentary individuals will be artificially selected for (Kaplan et al. 2014). If such differences are shown to be consistent and related to life history traits, this artificial selection could have a strong impact on the conservation of the species. Illustrating this point, many studies have found evidence for and/or discussed individual differences: adult blacktip reef shark, *Carcharhinus melanopterus* (Papastamatiou et al. 2010), scalloped hammerhead, *Sphyrna lewini* (Ketchum et al. 2014; Klimley and Nelson 1984) and gray reef shark, *Carcharhinus amblyrhynchos* (Field et al. 2010; McKibben and Nelson 1986). Interestingly, some of these examples mentioned differences in other behaviors (e.g., social and agonistic behavior; see Table 7.1) revealing these systems as potentially good sources of data for future studies on personality.

Similarly, individual differences could be observed in large-scale movements (e.g., across international boundaries or oceans). Individual differences in such movements were, for instance, documented for blue sharks, *Prionace glauca* (Vandeperre et al. 2014), which were migrating throughout a large part of the North Atlantic in summer months. In Australia, bull sharks (Espinoza et al. 2016; Heupel et al. 2015) showed marked individual differences in their migration pattern. Examples of variation in destination, timing, and/or the extent of migration were also documented in spiny dogfish, *Squalus acanthias* (McFarlane and King 2003) throughout the Pacific, bull sharks in Australia (Espinoza et al. 2016), and in tiger sharks (Meyer et al. 2010; Papastamatiou et al. 2013). In addition, a promising phenomenon describing individual variation is partial migration (i.e., when only a portion of the population migrates). Little work has been performed in sharks, but partial migration is expected to have strong ecological and evolutionary impacts and important implications for fisheries management (Chapman et al. 2012, 2015). Interestingly, differences in migration have been related to personality in fish (Chapman et al. 2011). No such association was investigated in sharks; however, in Hawaii, where some tiger sharks were resident and others were transient, returning for short foraging excursions (Meyer et al. 2010), one proposed explanation for the variation was individual differences in cognitive maps due to initial differences in exploration. If this is the case, then it would strongly support personality as one cause for differences in large-scale movements in tiger sharks. And, if proven to be true and some individuals consistently migrated whereas others did not, then their conservation status might differ (e.g., differing legislation across borders) leading to selection favoring certain personality types.

Investigating personality and its effect on the behavior of large upper trophic predators, such as tiger sharks or great white sharks, is critical for a more complete understanding of individual difference effects on these species' ecology but also throughout their ecosystems. Indeed, these large, predatory animals have a strong impact on their ecosystems. For instance, they play a role in coupling otherwise discreet food webs (Heupel et al. 2015; Matich et al. 2011; Rooney et al. 2006) and can impact prey behaviors. Such effects can be expected to occur throughout the food web where these sharks are present because they are known to target and, therefore, influence the behaviors and habitat use of other top predators, including marine mammals, such as bottlenose dolphins, *Tursiops aduncus* (Heithaus and Dill 2002) or Cape fur seals, *Arctocephalus pusillus* (Towner et al. 2016). If individual differences in movements or feeding habits are consistent, then these differences could have large ecological ramifications throughout the migratory range of these large predators. Accordingly, individual differences were observed in tiger shark movements (Afonso and Hazin 2015; Heithaus et al. 2002; Meyer et al. 2010; Vaudo et al. 2014) and in great white shark hunting strategies (Towner et al. 2016). The differences in hunting strategy were consistent over short time periods (over a month: Towner et al. 2016). As explained by the authors, a long-term investigation would be useful to provide evidence for individual specialization and an understanding of the impact that these findings have on prey behavior and foraging success (Towner et al. 2016).

Along with individual differences in behavior, great white sharks vary in their dietary shift with age and in their diet specialization. Researchers classified individuals as either specialists, generalists, or intermediary (Kim et al. 2012). Unfortunately, these studies were conducted in different locations; therefore, we can only speculate whether such differences in feeding habits are related to individual differences in behaviors of great whites as observed in other taxa (Toscano et al. 2016).

Despite our focus on the most well-known species, other large sharks that have received less attention are as important and have similar impacts on their ecosystems. For instance, the broadnose sevengill shark, *Notorynchus cepedianus*, is known to prey on other elasmobranchs and marine mammals (Abrantes and Barnett 2011). Marked individual differences in movement and habitat use were found in this species. In addition, preliminary evidence suggested that there were diet differences between individuals at the study site. However, the authors explained that, due to unknown winter residency locations for the sharks, it was difficult to draw strong conclusions on this later point. In view of these few studies, it seems important to verify how common individual differences are in large sharks and how these differences impact ecosystems. These observations, combined with the possible existence of individual differences in large-scale movements leading to a complex network of linked ecosystems, suggest that underestimating individual differences could hinder our ability to understand and protect marine ecosystems.

Large charismatic sharks such as the tiger or great white shark have become stars of provisioning wildlife tourism and the potential effect of this activity has raised concerns (Gallagher et al. 2015). Again, it is worth bearing in mind that these effects may not hold for all individuals in the population. Indeed, the amplitude of behavioral changes caused by ecotourism varies between great white sharks (Huveneers et al. 2013). Such differences, if consistent, could be an example of individual differences in attraction to novelty or in reaction to what might be considered as a potentially dangerous stimulus (dimensions of personality termed neophilia and boldness, respectively; Réale et al. 2007). If a link between personality and individual differences in reaction to ecotourism is demonstrated, behavioral observations with the participation of commercial boats could lead to the creation of a personality database.

Studying a variety of organisms with different life history traits and ecological conditions is needed to understand the predictors of animal personality (Réale et al. 2010, 2007). Therefore, testing and investigating personality in large upper trophic predator sharks (see above) could also be rewarding for the field of animal personality. Indeed, large predators have a very low predation risk compared to most species that are investigated in animal personality research. Therefore, exploring personality in these species would add important data to the discussion on predation as one of the potential mechanisms for emergence and maintenance of animal personality. Similarly, the diversity of elasmobranchs does not stop at sharks. In fact, along with skates, rays, and chimaeras, sharks represent their own evolutionary lineage, and, as mentioned before, are highly diverse in their behavior and ecology. This is therefore an opportunity to add important comparative data. Unfortunately, less work has been conducted on other elasmobranchs. Still, such variation has been documented in a handful of ray species. For instance, individual differences in activity and residence time were found in the cownose ray, *Rhinoptera bonasus* (Collins et al. 2007) and in the movement pattern of manta rays, *Manta alfredi* (Papastamatiou et al. 2012). Notably, adult cownose rays and other rays have successfully been held in captivity for extended periods of time (Fisher et al. 2011), and so provide an opportunity to link behavioral traits observed in captivity to those observed in the wild. To our knowledge, no personality research has been conducted on captive rays or skates, but we hope to see research move in this direction.

Along with increasing the diversity of species, the behavioral complexity of elasmobranchs provides an opportunity to increase the range of behaviors screened for individual differences. This is the case for social behavior in sharks, which has attracted a lot of attention (Jacoby et al. 2012), including recent findings on individual differences in sociality (see Jacoby et al. 2014). Understanding more about the dynamics of shark social behavior is important as concentrations of sharks can easily be targeted by spatially focused fishing (Jacoby et al. 2012; Mucientes et al. 2009). Considering that some personality traits (e.g., activity and boldness) can lead to disproportionate capture rates (Biro and Post 2008), understanding the presence and extent of social differences seems necessary to improve fishery and ecosystem management. Furthermore, many questions concerning social dynamics in animals and the role of personality differences in groups remain (Farine et al. 2015; Wolf and Krause 2014). Adding taxonomic diversity will contribute to a more global understanding of personality role in social group structures.

In the wild, juvenile lemon sharks differ in their tendency to lead groups (Guttridge et al. 2011). Differences were found to be size-dependent with larger individuals leading groups more frequently. However, there were instances where smaller individuals were observed to lead groups (Guttridge, personal communication). Future work should investigate how much these differences relate to personality and not confounding factors, such as age and size. Further individual differences were observed in the social behavior of blacktip reef sharks. Researchers measured gregariousness as the size of the group in which the individuals were observed (Mourier et al. 2012). They found that individuals differed consistently in their preferred group size. In addition, this investigation revealed the presence of distinct communities with varying social dynamics. More personality-focused investigations controlling for confounding factors (e.g., size, sex, and social environment) using the same method could give valuable information on shark personality in the wild. Over the long term, this method and system could provide a good opportunity to study the relationship between personality and social dynamics in large marine vertebrates. Furthermore, this could be extended to additional elasmobranch species that form predictable aggregations, for example, whitetip reef sharks, *Triaenodon obesus* (Whitney et al. 2012), whale sharks, *Rhincodon typus*, and manta rays (Rohner et al. 2013).

Another aspect of shark social dynamics worth exploring is the observation of differences in social ranking (e.g., dominance, agonistic behavior). Such differences have been shown to be consistent in bony fish (McGhee and Travis 2010) and related to other personality traits, including boldness and aggressiveness (Colléter and Brown 2011) and reaction to stress and aggressiveness (Øverli et al. 2004). Interestingly, agonistic behaviors (e.g., head shakes or cork-screw swimming) were displayed by scalloped hammerhead sharks (Klimley 1985). These sharks performed such behaviors to retain or obtain a central position within the large school they form. Unfortunately, whether these individuals consistently performed such behaviors and were more successful at occupying central position was not tested. The author interpreted the central position within the school as providing a social advantage to individuals, but not as a means by which individuals can avoid predators. This interpretation for occupying a particular position contrasts with the interpretation of this behavior for most schooling aquatic organisms. Indeed, in fish schools that are subject to high predation, positions are associated with varying degrees of risk (e.g., being at the front of the school is riskier than being inside the school; Krause 1994; Ward et al. 2004). Because of this differing schooling function in the scalloped hammerhead shark, it would be interesting to determine whether individual differences in this behavior exist and how these differences are maintained in this species (or other with similar characteristics).

Tools for Studying Shark Personality in the Wild

In this review, we highlighted studies that provide evidence for individual differences in wild shark behavior along with the tools and methods used to study personality in these populations (see Tables 7.1, 7.2). Following this, we now emphasize the availability of former methods or data that, if revisited, could be used to investigate elasmobranch personality in the wild. We then briefly mention other tools that have been used in ecological and behavioral studies on sharks that will likely enhance the study of personality in these animals.

One behavior that could provide interesting data is the reaction of sharks to divers or other intrusive stimuli. Martin (2007) reviewed agonistic postures of sharks and how scuba divers triggered such reactions. Individual differences in agonistic response to divers have, for instance, been mentioned in gray reef sharks (McKibben ad Nelson 1986) suggesting a promising use of this approach. Using these observations as a starting point, remote-controlled craft equipped with cameras could be used to investigate personality traits such as boldness or neophilia, safely, in numerous species.

Existing data can also be used to test questions regarding animal personality. So far, no studies have taken this approach to investigate personality but present examples of re-analyzed data revealed interindividual differences in movement behavior. For example, Papastamatiou et al. (2011) re-analyzed acoustic tracking data collected from tiger sharks, common thresher sharks, *Alopias vulpinus*, and blacktip reef sharks, and found individual differences in the movement patterns of tiger and thresher sharks. This illustrates the potential of the large amount of available data on elasmobranch movements. Once investigated under the animal personality framework, these data could become a useful source of information.

In their investigation, Wilson et al. (2015) measured individual swimming activity using accelerometers (see Table 7.2 and reviews by Shepard et al. 2008; Wilson et al. 2006). Accelerometers have been successfully used on sharks (Bullock et al. 2015; Gleiss et al. 2013; Whitney et al. 2007). These devices generate large data sets for researchers (i.e., continuous recording of tri-axial acceleration to quantify, for instance, locomotor activity and deduce body motion and posture), and these data could be used to test for and investigate individual differences.

Recently, new tools have been developed to investigate social behavior in such a way that direct observations are not required. Guttridge et al. (2010) made use of 'proximity receivers', small ultrasonic acoustic receivers that can be placed on focal animals to record signals sent by acoustic tags deployed on other individuals. These devices can be set to receive signals at close (e.g., <10 m) or far distances (e.g., 100 m), allowing researchers to record the proximity of interactions. This technology could generate high-resolution data and enable researchers to study social interactions in sharks over long periods of time (see also Haulsee et al. 2016). In the same vein, the development of acoustic technology (Hussey et al. 2015) has seen the parallel development of co-occurrence analysis (Jacoby and Freeman 2016). Co-occurrences between individuals (detection of two or more individuals at different locations in a certain time interval) can be obtained using acoustic passive tools (see Table 7.2) and then used as a proxy for social interactions between individuals (Krause et al. 2013). Furthermore, fine scale positioning using acoustic technology permits researchers to almost continuously and simultaneously record the position of different individuals allowing researchers to investigate social behavior and aggregation of sharks. This method has been used to study the social dynamics of spotted wobbegong sharks, *Orectolobus maculatus* (Armansin et al. 2016), but consistent individual differences have not yet been examined.

To study the ecology of leopard sharks (*Triakis semifasciata*) at the population level, Nosal et al. (2012) used aerial balloons. One could also conceivably use drones to follow particular individuals and observe behaviors of interest. Drones and balloons are useful technologies that allow researchers to observe cryptic animals such as sharks without disturbing these animals (Kiszka et al. 2016). Such methods can be applied only to particular environments, namely those in which one can maintain sight of the focal individual. Furthermore, these technologies have to be used in combination with methods for confirming the identity of the focals (e.g., former localization using acoustic tags). When these conditions are met, drones and other new technologies could become powerful tools for gathering detailed natural behavior about individuals.

Summary

We illustrated how studying personality in sharks could contribute to a better understanding of animal personality evolution and of the ecology of sharks and their ecosystems. Unfortunately, most of the examples do not consider personality as a potential explanation and, therefore, neglect to test if individual differences are consistent. This is unfortunate because, if some of the differences described in this section are consistent, they could have strong ecological implications. We mentioned, for instance, how variation in localized movements around home ranges or refuges, in large-scale movements, in social behavior, and in intra-population differences in feeding habits are potentially related to these behavioral differences. Because these differences can impact elasmobranch conservation or conservation of the ecosystem on a large scale, it is important to consider personality within sharks' and other elasmobranchs' biology. Interestingly, such individual variation was also found in large apex predators under low predation risk. As mentioned before, this is of interest because predation risk is a central cause within the animal personality literature and determining whether such large predators exhibit personality could provide insights into the mechanisms that maintain animal personality variation.

Conclusions and Future Directions

The pattern emerging from this chapter is that research on the personality of sharks and other elasmobranchs is just beginning and that considerable effort is needed to reach the amount of data gathered on some other taxa. On the more positive side, the studies that do exist have not only managed to demonstrate the existence of shark personality, but have also demonstrated interesting systems and methods that will inspire future researchers.

There is a large literature on other taxa that can and should be used to design behavioral experiments for measuring personality in sharks while keeping in mind the inherent difficulties. Furthermore, there are examples in the shark literature of behavioral assays that could be used to investigate these behaviors at the individual level. For example, to investigate dominance hierarchies in captive sharks, two studies looked at a behavior called avoidance (Allee and Dickinson 1954: smooth dogfish, *Mustelus canis*) or give-way (Myrberg and Gruber 1974: bonnethead sharks, *S. tiburo*). This behavior occurs when two individuals are swimming toward each other and one individual changes its direction to avoid collision. Both studies found clear differences between individuals; however, size differences seemed to explain most of the variation in both cases and sex differences also contributed to these differences in Myrberg and Gruber's study. Unfortunately, personality was not considered as an additional cause for differences in avoidance. Another well-used method is the binary choice experiment. So far this has been used only to investigate overall sociality in juvenile lemon sharks (Guttridge et al. 2009) despite its successful application for studying personality in fish (e.g., Cote et al. 2012; Harcourt et al. 2009). These methods are reliable, easy to design, and useful for gathering data on shark personality.

Following captive experiments, if possible, it is important to perform further tests in the wild using the same individuals. This is a crucial step to understand the extent and impact of personality on observed individual differences in the wild. As such data are being collected in more manageable species (i.e., testable in captivity and in the wild), this will help to better interpret individual differences observed in species that cannot be tested in captivity or semi-wild environments (e.g., highly mobile sharks). In this chapter, we provided some examples of species awaiting such transitions, namely juvenile lemon sharks and Port Jackson sharks.

One recurrent issue throughout this chapter is the lack of studies examining the association between life history traits, feeding habit differences, and personality, despite their feasibility (e.g., Tinker et al. 2007). This association is fundamental to understand the evolutionary and ecological consequences of personality in the everyday life of sharks, and could lead to an improvement in fisheries and ecosystem management.

One remaining important question, which represents a major challenge in the study of personality in sharks, and other long-lived animals is whether personality traits are stable throughout the lifetime of the individual, including when individuals experience drastic changes in their environment. For instance, after a few years, predation risk might dramatically decrease (Chapman et al. 2009; Guttridge et al. 2012) and this decrease might disturb former rank-order differences between individuals. Addressing this question requires a long-term study that starts when the animals are juveniles and so presents obvious challenges.

Overall, research on personality in sharks is in its infancy, but, as the importance of individual differences is being appreciated more by researchers, new methods, systems and models for studying personality are being developed. We believe that in a few years, with researchers giving sharks the attention that they deserve, more information will become available that will improve our understanding of shark behavior, assist with shark management and conservation, and enable us to gain a greater insight into the evolution of personality.

Acknowledgements We would like to thank Jennifer Vonk and Alexander Weiss for their comments throughout the writing process as well as Jens Krause and Samuel Gruber for their useful comments. We also thank William Hughes for sharing information and helping with part of this chapter.

References

Abrantes, K. G., & Barnett, A. (2011). Intrapopulation variations in diet and habitat use in a marine apex predator, the broadnose sevengill shark *Notorynchus cepedianus*. *Marine Ecology Progress Series, 442,* 133–148

Afonso, A. S., & Hazin, F. H. V. (2015). Vertical movement patterns and ontogenetic niche expansion in the tiger shark, *Galeocerdo cuvier*. *PLoS ONE, 10,* e0116720. doi:10.1371/journal.pone.0116720

Allee, W. C., & Dickinson, J. C. (1954). Dominance and subordination in the smooth dogfish *Mustelus canis* (Mitchill). *Physiological Zoology, 27,* 356–364

Archard, G. A., & Braithwaite, V. A. (2010). The importance of wild populations in studies of animal temperament. *Journal of Zoology, 281,* 149–160. doi:10.1111/j.1469-7998.2010.00714.x

Armansin, N. C., Lee, K. A., Huveneers, C., & Harcourt, R. G. (2016). Integrating social network analysis and fine-scale positioning to characterize the associations of a benthic shark. *Animal Behaviour, 115,* 245–258. doi:10.1016/j.anbehav.2016.02.014

Bell, A. M., & Sih, A. (2007). Exposure to predation generates personality in three spined sticklebacks (*Gasterosteus aculeatus*). *Ecology Letters, 10,* 828–834. doi:10.1111/j.1461-0248.2007.01081.x

Biro, P. A., & Post, J. R. (2008). Rapid depletion of genotypes with fast growth and bold personality traits from harvested fish populations. *Proceedings of the National Academy of Sciences, 105,* 2919–2922. doi:10.1073/pnas.0708159105

Biro, P. A., & Sampson, P. (2015). Fishing directly selects on growth rate via behaviour: Implications of growth-selection that is independent of size. *Proceedings of the Royal Society of London B: Biological Sciences, 282,* 20142283. doi:10.1098/rspb.2014.2283

Brown, C., Burgess, F., & Braithwaite, V. A. (2007). Heritable and experiential effects on boldness in a tropical poeciliid. *Behavioral Ecology and Sociobiology, 62,* 237–243. doi:10.1007/s00265-007-0458-3

Bullock, R. W., Guttridge, T. L., Cowx, I. G., Elliott, M., & Gruber, S. H. (2015). The behaviour and recovery of juvenile lemon sharks *Negaprion brevirostris* in response to external accelerometer tag attachment. *Journal of Fish Biology, 87,* 1342–1354. doi:10.1111/jfb.12808

Burns, J. G. (2008). The validity of three tests of temperament in guppies (*Poecilia reticulata*). *Journal of Comparative Psychology, 122,* 344

Burns, A. L. J., Herbert-Read, J. E., Morrell, L. J., & Ward, A. J. W. (2012). Consistency of leadership in shoals of mosquitofish (*Gambusia holbrooki*) in novel and in familiar environments. *PLoS ONE, 7,* e36567. doi:10.1371/journal.pone.0036567

Byrnes, E. E., & Brown, C. (2016). Individual personality differences in Port Jackson sharks *Heterodontus portusjacksoni*. *Journal of Fish Biology, 89,* 1142–1157. doi:10.1111/jfb.12993

Byrnes, E. E., Pouca, C. V., Chambers, S. L., & Brown, C. (2016a). Into the wild: Developing field tests to examine the link between elasmobranch personality and laterality. *Behaviour, 153,* 1777–1793. doi:10.1163/1568539X-00003373

Byrnes, E. E., Vila Pouca, C., & Brown, C. (2016b). Laterality strength is linked to stress reactivity in Port Jackson sharks (*Heterodontus portusjacksoni*). *Behavioural Brain Research, 305,* 239–246. doi:10.1016/j.bbr.2016.02.033

Campbell, D. L. M., Weiner, S. A., Starks, P. T., & Hauber, M. E. (2009). Context and control: Behavioural ecology experiments in the laboratory. *Annales Zoologici Fennici, 46,* 112–123. doi:10.5735/086.046.0204

Castanheira, M. F., Martínez Páramo, S., Figueiredo, F., Cerqueira, M., Millot, S., Oliveira, C. C. V., et al. (2016). Are coping styles consistent in the teleost fish *Sparus aurata* through sexual maturation and sex reversal? *Fish Physiology and Biochemistry, 42,* 1441–1452. doi:10.1007/s10695-016-0231-5

Chapman, D. D., Babcock, E. A., Gruber, S. H., Dibattista, J. D., Franks, B. R., Kessel, S. A., et al. (2009). Long-term natal site-fidelity by immature lemon sharks (*Negaprion brevirostris*) at a subtropical island. *Molecular Ecology, 18,* 3500–3507. doi:10.1111/j.1365-294X.2009.04289.x

Chapman, D. D., Feldheim, K. A., Papastamatiou, Y. P., & Hueter, R. E. (2015). There and back again: A review of residency and return migrations in sharks, with implications for population structure and management. *Annual Review of Marine Science, 7,* 547–570. doi:10.1146/annurev-marine-010814-015730

Chapman, B. B., Hulthén, K., Blomqvist, D. R., Hansson, L.-A., Nilsson, J.-Å., Brodersen, J., et al. (2011). To boldly go: Individual differences in boldness influence migratory tendency. *Ecology Letters, 14,* 871–876. doi:10.1111/j.1461-0248.2011.01648.x

Chapman, B. B., Hulthén, K., Brodersen, J., Nilsson, P. A., Skov, C., Hansson, L. A., et al. (2012). Partial migration in fishes: Causes and consequences. *Journal of Fish Biology, 81,* 456–478. doi:10.1111/j.1095-8649.2012.03342.x

Colléter, M., & Brown, C. (2011). Personality traits predict hierarchy rank in male rainbowfish social groups. *Animal Behaviour, 81,* 1231–1237. doi:10.1016/j.anbehav.2011.03.011

Collins, A. B., Heupel, M. R., & Motta, P. J. (2007). Residence and movement patterns of cownose rays *Rhinoptera bonasus* within a south-west Florida estuary. *Journal of Fish Biology, 71,* 1159–1178. doi:10.1111/j.1095-8649.2007.01590.x

Conrad, J. L., Weinersmith, K. L., Brodin, T., Saltz, J. B., & Sih, A. (2011). Behavioural syndromes in fishes: A review with implications for ecology and fisheries management. *Journal of Fish Biology, 78,* 395–435. doi:10.1111/j.1095-8649.2010.02874.x

Cote, J., Fogarty, S., & Sih, A. (2012). Individual sociability and choosiness between shoal types. *Animal Behaviour, 83,* 1469–1476. doi:10.1016/j.anbehav.2012.03.019

Cote, J., Fogarty, S., Weinersmith, K., Brodin, T., & Sih, A. (2010). Personality traits and dispersal tendency in the invasive mosquitofish (*Gambusia affinis*). *Proceedings of the Royal Society of London B: Biological Sciences*. doi:10.1098/rspb.2009.2128

Cotton, C. F., & Grubbs, R. D. (2015). Biology of deep-water chondrichthyans: Introduction. *Deep Sea Research Part II: Topical Studies in Oceanography, 115,* 1–10

Croft, D. P., Madden, J. R., Franks, D. W., & James, R. (2011). Hypothesis testing in animal social networks. *Trends in Ecology and Evolution, 26,* 502–507. doi:10.1016/j.tree.2011.05.012

Crow, G. L., & Hewitt, J. D. (1988). Longevity records for captive Tiger sharks *Galeocerdo cuvier* with notes on behaviour and management. *International Zoo Yearbook, 27,* 237–240. doi:10.1111/j.1748-1090.1988.tb03216.x

Dibattista, J. D., Feldheim, K. A., Gruber, S. H., & Hendry, A. P. (2007). When bigger is not better: Selection against large size, high condition and fast growth in juvenile lemon sharks. *Journal of Evolutionary Biology, 20,* 201–212. doi:10.1111/j.1420-9101.2006.01210.x

Dingemanse, N. J., Bouwman, K. M., van de Pol, M., van Overveld, T., Patrick, S. C., Matthysen, E., et al. (2012). Variation in personality and behavioural plasticity across four populations of the great tit *Parus major*. *Journal of Animal Ecology, 81,* 116–126. doi:10.1111/j.1365-2656.2011.01877.x

Dingemanse, N. J., Wright, J., Kazem, A. J. N., Thomas, D. K., Hickling, R., & Dawnay, N. (2007). Behavioural syndromes differ predictably between 12 populations of three-spined stickleback. *Journal of Animal Ecology, 76,* 1128–1138. doi:10.1111/j.1365-2656.2007.01284.x

Dulvy, N. K., Fowler, S. L., Musick, J. A., Cavanagh, R. D., Kyne, P. M., Harrison, L. R., et al. (2014). Extinction risk and conservation of the world's sharks and rays. *eLife, 3,* e00590. doi:10.7554/eLife.00590

Espinoza, M., Heupel, M. R., Tobin, A. J., & Simpfendorfer, C. A. (2016). Evidence of partial migration in a large coastal predator: Opportunistic foraging and reproduction as key drivers? *PLoS ONE, 11,* e0147608. doi:10.1371/journal.pone.0147608

Farine, D. R., Montiglio, P.-O., & Spiegel, O. (2015). From individuals to groups and back: The evolutionary implications of group phenotypic composition. *Trends in Ecology and Evolution, 30,* 609–621. doi:10.1016/j.tree.2015.07.005

Feldheim, K. A., Gruber, S. H., DiBattista, J. D., Babcock, E. A., Kessel, S. T., Hendry, A. P., et al. (2014). Two decades of genetic profiling yields first evidence of natal philopatry and long-term fidelity to parturition sites in sharks. *Molecular Ecology, 23,* 110–117. doi:10.1111/mec.12583

Ferretti, F., Worm, B., Britten, G. L., Heithaus, M. R., & Lotze, H. K. (2010). Patterns and ecosystem consequences of shark declines in the ocean. *Ecology Letters, 13,* 1055–1071

Field, I. C., Meekan, M. G., Speed, C. W., White, W., & Bradshaw, C. J. A. (2010). Quantifying movement patterns for shark conservation at remote coral atolls in the Indian Ocean. *Coral Reefs, 30,* 61–71. doi:10.1007/s00338-010-0699-x

Finger, J. S., Dhellemmes, F., Guttridge, T. L., Kurvers, R. H. J. M., Gruber, S. H., & Krause, J. (2016). Rate of movement of juvenile lemon sharks in a novel open field, are we measuring activity or reaction to novelty? *Animal Behaviour, 116,* 75–82. doi:10.1016/j.anbehav.2016.03.032

Fisher, R. A., Call, G. C., & Grubbs, R. D. (2011). Cownose ray (*Rhinoptera bonasus*) predation relative to bivalve ontogeny. *Journal of Shellfish Research, 30,* 187–196

Fuss, T., Bleckmann, H., & Schluessel, V. (2013). Place learning prior to and after telencephalon ablation in bamboo and coral cat sharks (*Chiloscyllium griseum* and *Atelomycterus marmoratus*). *Journal of Comparative Physiology A, 200,* 37–52. doi:10.1007/s00359-013-0859-x

Gallagher, A. J., Serafy, J. E., Cooke, S. J., & Hammerschlag, N. (2014). Physiological stress response, reflex impairment, and survival of five sympatric shark species following experimental capture and release. *Marine Ecology Progress Series, 496,* 207–218. doi:10.3354/meps10490

Gallagher, A. J., Vianna, G. M. S., Papastamatiou, Y. P., Macdonald, C., Guttridge, T. L., & Hammerschlag, N. (2015). Biological effects, conservation potential, and research priorities of shark diving tourism. *Biological Conservation, 184,* 365–379. doi:10.1016/j.biocon.2015.02.007

Garamszegi, L. Z., Markó, G., & Herczeg, G. (2013). A meta-analysis of correlated behaviors with implications for behavioral syndromes: Relationships between particular behavioral traits. *Behavioral Ecology.* doi:10.1093/beheco/art033

Gleiss, A. C., Wright, S., Liebsch, N., Wilson, R. P., & Norman, B. (2013). Contrasting diel patterns in vertical movement and locomotor activity of whale sharks at Ningaloo Reef. *Marine Biology, 160,* 2981–2992. doi:10.1007/s00227-013-2288-3

Griffiths, A. M., Jacoby, D. M. P., Casane, D., McHugh, M., Croft, D. P., Genner, M. J., et al. (2011). First analysis of multiple paternity in an oviparous shark, the small-spotted catshark (*Scyliorhinus canicula* L.). *Journal of Heredity.* doi:10.1093/jhered/esr112

Groothuis, T. G. G., & Carere, C. (2005). Avian personalities: Characterization and epigenesis. *Neuroscience and Biobehavioral Reviews, 29,* 137–150. doi:10.1016/j.neubiorev.2004.06.010

Grubbs, R. D. (2010). Ontogenetic shifts in movements and habitat use. In J. C. Carrier, J. A. Musick, & M. R. Heithaus (Eds.), *Sharks and their relatives II: biodiversity, adaptive physiology, and conservation* (pp. 319–350). Boca Raton, FL: CRC Press

Grubbs, R. D., Kraus, R. T., Breed, M. D., & Moore, J. (2010). Fish migration. *Encyclopedia of Animal Behavior, 1,* 715–724

Gruber, S. H., de Marignac, J. R. C., & Hoenig, J. M. (2001). Survival of juvenile lemon sharks at Bimini, Bahamas, estimated by mark–depletion experiments. *Transactions of the American Fisheries Society, 130,* 376–384. doi:10.1577/1548-8659(2001)130<0376:SOJLSA>2.0.CO;2

Gruber, S. H., & Keyes, R. S. (1981). Keeping sharks for research. In A. D. Hawkins (Ed.), *Aquarium systems,* (pp.373–402). New York, NY: Academic Press.

Guttridge, T. L., & Brown, C. (2013). Learning and memory in the Port Jackson shark, *Heterodontus portusjacksoni. Animal Cognition, 17,* 415–425. doi:10.1007/s10071-013-0673-4

Guttridge, T. L., Gruber, S. H., DiBattista, J. D., Feldheim, K. A., Croft, D. P., Krause, S., et al. (2011). Assortative interactions and leadership in a free-ranging population of juvenile lemon shark *Negaprion brevirostris. Marine Ecology Progress Series, 423,* 235–245

Guttridge, T. L., Gruber, S. H., Franks, B., Kessel, S., Gledhill, K., Uphill, J., et al. (2012). Deep danger: Intra-specific predation risk influences habitat use and aggregation formation of juvenile lemon sharks *Negaprion brevirostris. Marine Ecology Progress Series, 445,* 279–291

Guttridge, T. L., Gruber, S. H., Gledhill, K., Croft, D., Sims, D., & Krause, J. (2009). Social preferences of juvenile lemon sharks, *Negaprion brevirostris. Animal Behaviour, 78,* 543–548

Guttridge, T. L., Gruber, S. H., Krause, J., & Sims, D. W. (2010). Novel acoustic technology for studying free-ranging shark social behaviour by recording individuals' interactions. *PLoS ONE, 5,* e9324. doi:10.1371/journal.pone.0009324

Hammerschlag, N., Martin, R. A., Fallows, C., Collier, R. S., & Lawrence, R. (2012). Investigatory behavior toward surface objects and nonconsumptive strikes on seabirds by white sharks, *Carcharodon carcharias*, at Seal Island, South Africa (1997–2010). In M. L. Domeier (Ed.), *Global perspectives on the biology and life history of the white shark* (pp. 91–103). Boca Raton, FL: CRC Press

Harcourt, J. L., Sweetman, G., Johnstone, R. A., & Manica, A. (2009). Personality counts: The effect of boldness on shoal choice in three-spined sticklebacks. *Animal Behaviour, 77,* 1501–1505. doi:10.1016/j.anbehav.2009.03.004

Haulsee, D. E., Fox, D. A., Breece, M. W., Brown, L. M., Kneebone, J., Skomal, G. B., et al. (2016). Social network analysis reveals potential fission-fusion behavior in a shark. *Scientific Reports.* doi:10.1038/srep34087

Heithaus, M. R., & Dill, L. M. (2002). Food availability and tiger shark predation risk influence bottlenose dolphin habitat use. *Ecology, 83,* 480–491. doi:10.1890/0012-9658(2002)083[0480:FAATSP]2.0.CO;2

Heithaus, M. R., Dill, L., Marshall, G., & Buhleier, B. (2002). Habitat use and foraging behavior of tiger sharks (*Galeocerdo cuvier*) in a seagrass ecosystem. *Marine Biology, 140,* 237–248

Heithaus, M. R., Frid, A., Wirsing, A. J., & Worm, B. (2008). Predicting ecological consequences of marine top predator declines. *Trends in Ecology and Evolution, 23,* 202–210. doi:10.1016/j.tree.2008.01.003

Herborn, K. A., Macleod, R., Miles, W. T., Schofield, A. N., Alexander, L., & Arnold, K. E. (2010). Personality in captivity reflects personality in the wild. *Animal Behaviour, 79,* 835–843

Heupel, M. R., Carlson, J. K., & Simpfendorfer, C. A. (2007). Shark nursery areas: Concepts, definition, characterization and assumptions. *Marine Ecology Progress Series, 337,* 287–297. doi:10.3354/meps337287

Heupel, M. R., Simpfendorfer, C. A., Espinoza, M., Smoothey, A. F., Tobin, A., & Peddemors, V. (2015). Conservation challenges of sharks with continental scale migrations. *Marine Megafauna, 2,* 12. doi:10.3389/fmars.2015.00012

Highfill, L. E., & Kuczaj, S. A. (2007). Do bottlenose dolphins (*Tursiops truncatus*) have distinct and stable personalities? *Aquatic Mammals, 33,* 380

Hussey, N. E., DiBattista, J. D., Moore, J. W., Ward, E. J., Fisk, A. T., Kessel, S., et al. (2017). Risky business for a juvenile marine predator? Testing the influence of foraging strategies on size and growth rate under natural conditions. *Proceedings of the Royal Society of London B: Biological Sciences, 284,* 20170166. https://doi.org/10.1098/rspb.2017.0166

Hussey, N. E., Kessel, S. T., Aarestrup, K., Cooke, S. J., Cowley, P. D., Fisk, A. T., et al. (2015). Aquatic animal telemetry: A panoramic window into the underwater world. *Science, 348,* 1255642. doi:10.1126/science.1255642

Huveneers, C., Rogers, P. J., Beckmann, C., Semmens, J. M., Bruce, B. D., & Seuront, L. (2013). The effects of cage-diving activities on the fine-scale swimming behaviour and space use of white sharks. *Marine Biology, 160,* 2863–2875. doi:10.1007/s00227-013-2277-6

Irving, E., & Brown, C. (2013). Examining the link between personality and laterality in a feral guppy *Poecilia reticulata* population. *Journal of Fish Biology, 83,* 311–325. doi:10.1111/jfb.12165

Jacoby, D. M. P., Croft, D. P., & Sims, D. W. (2012). Social behaviour in sharks and rays: Analysis, patterns and implications for conservation. *Fish and Fisheries, 13,* 399–417. doi:10.1111/j.1467-2979.2011.00436.x

Jacoby, D. M. P., Fear, L. N., Sims, D. W., & Croft, D. P. (2014). Shark personalities? Repeatability of social network traits in a widely distributed predatory fish. *Behavioral Ecology and Sociobiology, 68,* 1995–2003. doi:10.1007/s00265-014-1805-9

Jacoby, D. M. P., & Freeman, R. (2016). Emerging Network-Based Tools in Movement Ecology. *Trends in Ecology and Evolution, 31,* 301–314. doi:10.1016/j.tree.2016.01.011

Kaplan, D. M., Chassot, E., Amandé, J. M., Dueri, S., Demarcq, H., Dagorn, L., et al. (2014). Spatial management of Indian Ocean tropical tuna fisheries: Potential and perspectives. *ICES Journal of Marine Science: Journal Du Conseil, 71,* 1728–1749. doi:10.1093/icesjms/fst233

Ketchum, J. T., Hearn, A., Klimley, A. P., Espinoza, E., Peñaherrera, C., & Largier, J. L. (2014). Seasonal changes in movements and habitat preferences of the scalloped hammerhead shark (*Sphyrna lewini*) while refuging near an oceanic island. *Marine Biology, 161,* 755–767. doi:10.1007/s00227-013-2375-5

Kim, S. L., Tinker, M. T., Estes, J. A., & Koch, P. L. (2012). Ontogenetic and among-individual variation in foraging strategies of northeast pacific white sharks based on stable isotope analysis. *PLoS ONE, 7,* e45068. doi:10.1371/journal.pone.0045068

King, J. E., & Figueredo, A. J. (1997). The five-factor model plus dominance in chimpanzee personality. *Journal of Research in Personality, 31,* 257–271. doi:10.1006/jrpe.1997.2179

Kiszka, J. J., Mourier, J., Gastrich, K., & Heithaus, M. R. (2016). Using unmanned aerial vehicles (UAVs) to investigate shark and ray densities in a shallow coral lagoon. *Marine Ecology Progress Series, 560,* 237–242. doi:10.3354/meps11945

Klimley, A. P. (1985). Schooling in *Sphyrna lewini*, a species with low risk of predation: A non-egalitarian state. *Zeitschrift Für Tierpsychologie, 70,* 297–319. doi:10.1111/j.1439-0310.1985.tb00520.x

Klimley, A. P., & Nelson, D. R. (1984). Diel movement patterns of the scalloped hammerhead shark (*Sphyrna lewini*) in relation to El Bajo Espiritu Santo: A refuging central-position social system. *Behavioral Ecology and Sociobiology, 15,* 45–54. doi:10.1007/BF00310214

Krause, J. (1994). Differential fitness returns in relation to spatial position in groups. *Biological Reviews, 69,* 187–206. doi:10.1111/j.1469-185X.1994.tb01505.x

Krause, J., Krause, S., Arlinghaus, R., Psorakis, I., Roberts, S., & Rutz, C. (2013). Reality mining of animal social systems. *Trends in Ecology and Evolution, 28,* 541–551. doi:10.1016/j.tree.2013.06.002

Krause, S., Mattner, L., James, R., Guttridge, T. L., Corcoran, M. J., Gruber, S. H., et al. (2009). Social network analysis and valid Markov chain Monte Carlo tests of null models. *Behavioral Ecology and Sociobiology, 63,* 1089–1096. doi:10.1007/s00265-009-0746-1

Laroche, R. K., Kock, A. A., Dill, L. M., & Oosthuizen, W. H. (2007). Effects of provisioning ecotourism activity on the behaviour of white sharks *Carcharodon carcharias*. *Marine Ecology Progress Series, 338,* 199–209. doi:10.3354/meps338199

Lowry, D., & Motta, P. J. (2008). Relative importance of growth and behaviour to elasmobranch suction-feeding performance over early ontogeny. *Journal of the Royal Society Interface, 5,* 641–652. doi:10.1098/rsif.2007.1189

Martin, R. A. (2007). A review of shark agonistic displays: Comparison of display features and implications for shark–human interactions. *Marine and Freshwater Behaviour and Physiology, 40*(1), 3–34. doi:10.1080/10236240601154872

Martin, R. A., Hammerschlag, N., Collier, R. S., & Fallows, C. (2005). Predatory behaviour of white sharks (*Carcharodon carcharias*) at Seal Island, South Africa. *Journal of the Marine Biological Association of the United Kingdom, 85,* 1121–1135

Martin, J. G. A., & Réale, D. (2008). Temperament, risk assessment and habituation to novelty in eastern chipmunks, *Tamias striatus*. *Animal Behaviour, 75,* 309–318

Matich, P., Heithaus, M. R., & Layman, C. A. (2011). Contrasting patterns of individual specialization and trophic coupling in two marine apex predators. *Journal of Animal Ecology, 80,* 294–305. doi:10.1111/j.1365-2656.2010.01753.x

McFarlane, G. A., & King, J. R. (2003). Migration patterns of spiny dogfish (*Squalus acanthias*) in the North Pacific Ocean. *Fishery Bulletin, 101,* 358–367

McGhee, K. E., & Travis, J. (2010). Repeatable behavioural type and stable dominance rank in the bluefin killifish. *Animal Behaviour, 79,* 497–507. doi:10.1016/j.anbehav.2009.11.037

McKibben, J. N., & Nelson, D. R. (1986). Patterns of movement and grouping of gray reef sharks,

Carcharhinus amblyrhynchos, at Enewetak, Marshall Islands. *Bulletin of Marine Science, 38,* 89–110

Meyer, C. G., Papastamatiou, Y. P., & Holland, K. N. (2010). A multiple instrument approach to quantifying the movement patterns and habitat use of tiger (*Galeocerdo cuvier*). *Marine Biology, 157,* 1857–1868. doi:10.1007/s00227-010-1457-x

Mittelbach, G. G., Ballew, N. G., Kjelvik, M. K., & Fraser, D. (2014). Fish behavioral types and their ecological consequences. *Canadian Journal of Fisheries and Aquatic Sciences, 71,* 927–944

Morgan, A., & Burgess, G. (2007). At-vessel fishing mortality for six species of sharks caught in the northwest Atlantic and gulf of Mexico. *Gulf and Caribbean Research, 19*(2), 123–129. doi:10.18785/gcr.1902.15

Morrissey, J. F., & Gruber, S. H. (1993a). Habitat selection by juvenile lemon sharks, *Negaprion brevirostris*. *Environmental Biology of Fishes, 38,* 311–319. doi:10.1007/BF00007524

Morrissey, J. F., & Gruber, S. H. (1993b). Home range of juvenile lemon sharks, Negaprion brevirostris. *Copeia*. doi:10.2307/1447141

Mourier, J., Vercelloni, J., & Planes, S. (2012). Evidence of social communities in a spatially structured network of a free-ranging shark species. *Animal Behaviour, 83,* 389–401. doi:10.1016/j.anbehav.2011.11.008

Mucientes, G. R., Queiroz, N., Sousa, L. L., Tarroso, P., & Sims, D. W. (2009). Sexual segregation of pelagic sharks and the potential threat from fisheries. *Biology Letters*. doi:10.1098/rsbl.2008.0761

Myrberg, A. A., & Gruber, S. H. (1974). The behavior of the bonnethead shark, *Sphyrna tiburo*. *Copeia, 1974,* 358–374. doi:10.2307/1442530

Nelson, D. R. (1977). On the field study of shark behavior. *Integrative and Comparative Biology, 17,* 501–507. doi:10.1093/icb/17.2.501

Nosal, A. P., Cartamil, D. C., Long, J. W., Lührmann, M., Wegner, N. C., & Graham, J. B. (2012). Demography and movement patterns of leopard sharks (*Triakis semifasciata*) aggregating near the head of a submarine canyon along the open coast of southern California, USA. *Environmental Biology of Fishes, 96,* 865–878. doi:10.1007/s10641-012-0083-5

Ortega, L. A., Heupel, M. R., Beynen, P. V., & Motta, P. J. (2009). Movement patterns and water quality preferences of juvenile bull sharks (*Carcharhinus leucas*) in a Florida estuary. *Environmental Biology of Fishes, 84,* 361–373. doi:10.1007/s10641-009-9442-2

Øverli, Ø., Korzan, W. J., Höglund, E., Winberg, S., Bollig, H., Watt, M., et al. (2004). Stress coping style predicts aggression and social dominance in rainbow trout. *Hormones and Behavior, 45,* 235–241. doi:10.1016/j.yhbeh.2003.12.002

Papastamatiou, Y. P., Cartamil, D. P., Lowe, C. G., Meyer, C. G., Wetherbee, B. M., & Holland, K. N. (2011). Scales of orientation, directed walks and movement path structure in sharks. *Journal of Animal Ecology, 80,* 864–874. doi:10.1111/j.1365-2656.2011.01815.x

Papastamatiou, Y. P., DeSalles, P. A., & McCauley, D. J. (2012). Area-restricted searching by manta rays and their response to spatial scale in lagoon habitats. *Marine Ecology Progress Series, 456,* 233–244. doi:10.3354/meps09721

Papastamatiou, Y. P., Friedlander, A. M., Caselle, J. E., & Lowe, C. G. (2010). Long-term movement patterns and trophic ecology of blacktip reef sharks (*Carcharhinus melanopterus*) at Palmyra Atoll. *Journal of Experimental Marine Biology and Ecology, 386,* 94–102. doi:10.1016/j.jembe.2010.02.009

Papastamatiou, Y. P., & Lowe, C. G. (2012). An analytical and hypothesis-driven approach to elasmobranch movement studies. *Journal of Fish Biology, 80,* 1342–1360. doi:10.1111/j.1095-8649.2012.03232.x

Papastamatiou, Y. P., Meyer, C. G., Carvalho, F., Dale, J. J., Hutchinson, M. R., & Holland, K. N. (2013). Telemetry and random-walk models reveal complex patterns of partial migration in a large marine predator. *Ecology, 94,* 2595–2606. doi:10.1890/12-2014.1

Petelle, M. B., McCoy, D. E., Alejandro, V., Martin, J. G. A., & Blumstein, D. T. (2013). Development of boldness and docility in yellow-bellied marmots. *Animal Behaviour, 86,* 1147–1154. doi:10.1016/j.anbehav.2013.09.016

Pough, F. H., Janis, C. M., & Heiser, J. B. (1999). *Vertebrate life* (Vol. 733). Upper Saddle River, NJ: Prentice Hall

Réale, D., Garant, D., Humphries, M. M., Bergeron, P., Careau, V., & Montiglio, P.-O. (2010). Personality and the emergence of the pace-of-life syndrome concept at the population level. *Philosophical Transactions of the Royal Society of London B: Biological Sciences, 365,* 4051–4063. doi:10.1098/rstb.2010.0208

Réale, D., Reader, S. M., Sol, D., McDougall, P. T., & Dingemanse, N. J. (2007). Integrating animal temperament within ecology and evolution. *Biological Reviews, 82,* 291–318

Rodríguez-Prieto, I., Martín, J., & Fernández-Juricic, E. (2011). Individual variation in behavioural plasticity: Direct and indirect effects of boldness, exploration and sociability on habituation to predators in lizards. *Proceedings of the Royal Society of London B: Biological Sciences, 278,* 266–273. doi:10.1098/rspb.2010.1194

Rohner, C. A., Pierce, S. J., Marshall, A. D., Weeks, S. J., Bennett, M. B., & Richardson, A. J. (2013). Trends in sightings and environmental influences on a coastal aggregation of manta rays and whale sharks. *Marine Ecology Progress Series, 482,* 153–168. doi:10.3354/meps10290

Rooney, N., McCann, K., Gellner, G., & Moore, J. C. (2006). Structural asymmetry and the stability of diverse food webs. *Nature, 442,* 265–269. doi:10.1038/nature04887

Schluessel, V., & Bleckmann, H. (2005). Spatial memory and orientation strategies in the elasmobranch *Potamotrygon motoro*. *Journal of Comparative Physiology A, 191,* 695–706. doi:10.1007/s00359-005-0625-9

Shepard, E. L. C., Wilson, R. P., Quintana, F., Laich, A. G., Liebsch, N., Albareda, D. A., et al. (2008). Identification of animal movement patterns using tri-axial accelerometry. *Endangered Species Research, 10,* 47–60. doi:10.3354/esr00084

Sih, A., Cote, J., Evans, M., Fogarty, S., & Pruitt, J. (2012). Ecological implications of behavioural syndromes. *Ecology Letters, 15,* 278–289. doi:10.1111/j.1461-0248.2011.01731.x

Sims, D. W. (2003). Tractable models for testing theories about natural strategies: Foraging behaviour and habitat selection of free-ranging sharks. *Journal of Fish Biology, 63,* 53–73. doi:10.1111/j.1095-8649.2003.00207.x

Sims, D. W., Nash, J., & Morritt, D. (2001). Movements and activity of male and female dogfish in a tidal sea lough: Alternative behavioural strategies and apparent sexual segregation. *Marine Biology, 139,* 1165–1175. doi:10.1007/s002270100666

Sims, D. W., Southall, E. J., Richardson, A. J., Reid, P. C., & Metcalfe, J. D. (2003). Seasonal movements and behaviour of basking sharks from archival tagging: No evidence of winter hibernation. *Marine Ecology Progress Series, 248,* 187–196

Smith, M., Warmolts, D., Thoney, D., & Hueter, R. (2004). *The elasmobranch husbandry manual: Captive care of sharks, rays and their relatives*. Columbus, OH: Special Publication of the Ohio Biological Survey

Stamps, J. A. (2007). Growth-mortality tradeoffs and "personality traits" in animals. *Ecology Letters, 10,* 355–363

Stamps, J. A., & Groothuis, T. G. G. (2010). The development of animal personality: Relevance, concepts and perspectives. *Biological Reviews, 85,* 301–325. doi:10.1111/j.1469-185X.2009.00103.x

Stevens, J. E. (1994). The delicate constitution of sharks. *BioScience, 44,* 661–664. doi:10.2307/1312508

Stevens, J., Bonfil, R., Dulvy, N., & Walker, P. (2000). The effects of fishing on sharks, rays, and chimaeras (chondrichthyans), and the implications for marine ecosystems. *ICES Journal of Marine Science: Journal Du Conseil, 57,* 476–494

Sundström, L. F., Gruber, S. H., Clermont, S. M., Correia, J. P. S., de Marignac, J. R. C., Morrissey, J. F., et al. (2001). Review of elasmobranch behavioral studies using ultrasonic telemetry with special reference to the lemon shark, *Negaprion Brevirostris*, around Bimini islands, Bahamas. *Environmental Biology of Fishes, 60,* 225–250. doi:10.1023/A:1007657505099

Tinker, M. T., Costa, D. P., Estes, J. A., & Wieringa, N. (2007). Individual dietary specialization and dive behaviour in the California sea otter: Using archival time–depth data to detect alternative foraging strategies. *Deep Sea Research Part II: Topical Studies in Oceanography, 54,* 330–342. doi:10.1016/j.dsr2.2006.11.012

Toscano, B. J., Gownaris, N. J., Heerhartz, S. M., & Monaco, C. J. (2016). Personality, foraging behavior and specialization: Integrating behavioral and food web ecology at the individual level. *Oecologia, 182,* 55–69. doi:10.1007/s00442-016-3648-8

Towner, A. V., Leos-Barajas, V., Langrock, R., Schick, R. S., Smale, M. J., Kaschke, T., et al. (2016). Sex-specific and individual preferences for hunting strategies in white sharks. *Functional Ecology, 30,* 1397–1407. doi:10.1111/1365-2435.12613

Urszán, T. J., Garamszegi, L. Z., Nagy, G., Hettyey, A., Török, J., & Herczeg, G. (2015). No personality without experience? A test on *Rana dalmatina* tadpoles. *Ecology and Evolution, 5,* 5847–5856. doi:10.1002/ece3.1804

Uusi-Heikkilä, S., Wolter, C., Klefoth, T., & Arlinghaus, R. (2008). A behavioral perspective on fishing-induced evolution. *Trends in Ecology and Evolution, 23,* 419–421. doi:10.1016/j.tree.2008.04.006

Vandeperre, F., Aires-da-Silva, A., Fontes, J., Santos, M., Santos, R. S., & Afonso, P. (2014). Movements of blue sharks (*Prionace glauca*) across their life history. *PLoS ONE, 9,* e103538. doi:10.1371/journal.pone.0103538

Vaudo, J. J., Wetherbee, B. M., Harvey, G., Nemeth, R. S., Aming, C., Burnie, N., et al. (2014). Intraspecific variation in vertical habitat use by tiger sharks (*Galeocerdo cuvier*) in the western North Atlantic. *Ecology and Evolution, 4,* 1768–1786. doi:10.1002/ece3.1053

Ward, A. J., Thomas, P., Hart, P. J., & Krause, J. (2004). Correlates of boldness in three-spined sticklebacks (*Gasterosteus aculeatus*). *Behavioral Ecology and Sociobiology, 55*(6), 561–568

Ward-Paige, C. A., Britten, G. L., Bethea, D. M., & Carlson, J. K. (2015). Characterizing and predicting essential habitat features for juvenile coastal sharks. *Marine Ecology, 36,* 419–431. doi:10.1111/maec.12151

Whitney, N. M., Papastamatiou, Y. P., Holland, K. N., & Lowe, C. G. (2007). Use of an acceleration data logger to measure diel activity patterns in captive whitetip reef sharks, *Triaenodon obesus*. *Aquatic Living Resources, 20,* 299–305. doi:10.1051/alr:2008006

Whitney, N. M., Pyle, R. L., Holland, K. N., & Barcz, J. T. (2012). Movements, reproductive seasonality, and fisheries interactions in the whitetip reef shark (*Triaenodon obesus*) from community-contributed photographs. *Environmental Biology of Fishes, 93,* 121–136. doi:10.1007/s10641-011-9897-9

Wilson, A. D. M., Brownscombe, J. W., Krause, J., Krause, S., Gutowsky, L. F. G., Brooks, E. J., et al. (2015). Integrating network analysis, sensor tags, and observation to understand shark ecology and behavior. *Behavioral Ecology*. doi:10.1093/beheco/arv115

Wilson, R. P., White, C. R., Quintana, F., Halsey, L. G., Liebsch, N., Martin, G. R., et al. (2006). Moving towards acceleration for estimates of activity-specific metabolic rate in free-living animals: The case of the cormorant. *Journal of Animal Ecology, 75,* 1081–1090. doi:10.1111/j.1365-2656.2006.01127.x

Wolf, M., & Krause, J. (2014). Why personality differences matter for social functioning and social structure. *Trends in Ecology and Evolution, 29,* 306–308. doi:10.1016/j.tree.2014.03.008

Wolf, M., van Doorn, G. S., Leimar, O., & Weissing, F. J. (2007). Life-history trade-offs favour the evolution of animal personalities. *Nature, 447,* 581–584. doi:10.1038/nature05835

Wolf, M., & Weissing, F. J. (2012). Animal personalities: Consequences for ecology and evolution. *Trends in Ecology and Evolution, 27,* 452–461

Worm, B., Davis, B., Kettemer, L., Ward-Paige, C. A., Chapman, D., Heithaus, M. R., et al. (2013). Global catches, exploitation rates, and rebuilding options for sharks. *Marine Policy, 40,* 194–204

Yopak, K. E. (2012). Neuroecology of cartilaginous fishes: The functional implications of brain scaling. *Journal of Fish Biology, 80,* 1968–2023. doi:10.1111/j.1095-8649.2012.03254.x

Young, F. A., Kajiura, S. M., Visser, G. J., Correia, J. P. S., & Smith, M. F. L. (2002). Notes on the long-term transport of the scalloped hammerhead shark (*Sphyrna lewini*). *Zoo Biology, 21,* 243–251. doi:10.1002/zoo.10019

Yuen, C. H., Pillay, N., Heinrichs, M., Schoepf, I., & Schradin, C. (2016). Personality traits are consistent when measured in the field and in the laboratory in African striped mice (*Rhabdomys pumilio*). *Behavioral Ecology and Sociobiology, 70,* 1235–1246. doi:10.1007/s00265-016-2131-1

Personality and Individuality in Reptile Behavior

8

R. Mark Waters, Bonnie B. Bowers
and Gordon M. Burghardt

Abstract

Despite the taxonomic, behavioral, and lifestyle diversity among reptile species, behavioral consistency in reptiles has not been examined to the extent that it has been in fish, birds, and mammals. Careful use of terms such as individuality, temperament, personality, and behavioral syndromes is needed as they carry overlapping connotations and varying dangers of being applied in an anthropomorphic fashion. The majority of research on such phenomena in reptiles has utilized snakes and lizards. Studies on antipredator behavior in natricine snakes, primarily in the genus *Thamnophis*, has produced strong evidence of individual consistency of behavior over time, some evidence for consistency across situations, clear evidence for a heritable basis for individual differences in antipredator behavior, and limited evidence linking individual variation with fitness-related outcomes. Research in lizards has mirrored findings reported for other vertebrates, focusing on one or more of five traits: shyness–boldness, exploration-avoidance, activity, sociability, conspecific aggression, and possible relationships among them. We review the methodology and statistical analyses used to study these traits in lizards and the relationships of these traits to morphology, reproduction, hormones, fitness, life history, and other factors. The common lizard (*Zootoca vivipara*) has become a model species, for example, in studies assessing the pace-of-life hypothesis, with results differing from those found in other vertebrates. Studies on turtles and crocodilians are also presented that further illustrate the comparative and methodological value of reptilian studies.

R.M. Waters
Ohio University Eastern Campus,
St. Clairsville, OH 43950, USA
e-mail: watersr1@ohio.edu

B.B. Bowers
Department of Psychology, Hollins University,
Roanoke, VA 24020, USA
e-mail: bbowers@hollins.edu

G.M. Burghardt (✉)
Departments of Psychology and Ecology &
Evolutionary Biology, University of Tennessee,
Knoxville, TN 37996, USA
e-mail: gburghar@utk.edu

J. Vonk et al. (eds.), *Personality in Nonhuman Animals*,
DOI 10.1007/978-3-319-59300-5_8

Personality and Individuality in Reptile Behavior

Nonavian reptiles are perhaps the most diverse vertebrates in terms of their morphology, lifestyles, ecology, and behavioral adaptations. The early history of reptiles paints an even more diverse history with true flight in pterosaurs and relatives, seafaring dolphin-like ichthyosaurs and mosasaurs, theraspids—so-called mammal-like reptiles (Hotton et al. 1986), and dinosaurs—the ancestors of birds. So, one would expect much behavioral variation between orders, families, genera, and species. However, until recently, the diversity in reptile behavior, although acknowledged, was, with few exceptions, not a focus of research to the extent found with fishes, birds, and mammals. This relative degree of neglect was true whether the research focused on ethology, comparative psychology, natural history, or laboratory experimentation. This was also true of descriptions of intraspecific variation, which were primarily anecdotal. For example, in the Reptile Ethology Laboratory at the University of Tennessee, anyone interacting with iguanas in the colony room knew which were docile and which were aggressive; one iguana was so tame that she would raise her head to be scratched and rubbed (Burghardt 2000a), whereas another would consistently perch by the door and slap unsuspecting caretakers in the face with her tail. In addition, numerous anecdotal accounts of individuality of turtles, alligators, lizards, and snakes interacting with humans have been described (Bowers and Burghardt 1992). However, in a review of studies of animal personality, Gosling (2001) included research on only two reptile species (two species of gartersnakes) out of the 65 animal species reviewed. Similarly, Bell et al. (2009) reviewed studies of repeatability of behavior and included data from only six reptile species (one snake, one lizard, and four turtle species) out of 98 total species. Even more recently, the review by Weiss and Altschul (2017) referenced no studies of reptiles in their authoritative overview of the field. We hope to illustrate the value of studying nonavian reptiles in this chapter.

Without systematic observation of individuals, it is impossible to separate variation among conspecifics from factors such as observational conditions, testing methods, motivation, sex, age, previous experiences including rearing and parental care, nutritional state, illness, and so forth. Nonetheless, starting in the 1960s, studies of differences among species in display behavior, reproduction, diet, foraging, and antipredator behavior entered the field, as shown in a definitive early review by Carpenter and Ferguson (1977). Carpenter was one of the first researchers to focus on the head-bob and push-up communicative displays in lizards, and developed a method of graphical recording of head, dewlap, and forelimb movements in amplitude, duration, and temporal sequencing, using Display Action Pattern (DAP) graphs. He studied and compared dozens of species, focusing on species and population typical behavior rather than individual differences. However, Dugan (1982a, b) documented consistent individual differences in the signature head-bob displays of male green iguanas (*Iguana iguana*) that could provide information to both males and females in courtship and mating territories.

Species and individual differences in temperament as expressed in antipredator ("defensive") behavioral repertoires and strategies were also studied. Such recognition was especially important for reptile curators and keepers in zoos, as awareness of differences in proneness to attack, strike, constrict, or flee was essential when working with large, dangerous, or venomous reptiles, especially crocodilians and snakes.

Reptilian taxa also differ in sociality, foraging, diet, reproduction, parental care, habitat, and in other ways. Although early work, to be described below, addressed species differences in these traits, and, later, in stable intraspecific differences distinct from sex, age, population, health, etc., the term personality was not used to describe these differences. Thus, although there is a literature on reptile individuality going back decades, this literature has largely gone unrecognized in the field of animal personality, as terms such as temperament and personality were not used (but see Burghardt 2000b for an exception). In this chapter, we review this early work, especially on snakes, review recent work that builds on current approaches to studying animal personality, and discuss some methodological and terminological issues. We focus on snakes and lizards, which represent about 95% of extant reptiles and which are the focus of the majority of relevant research. We begin with terminology and advocate greater use of the term "temperament," as "personality" has quite anthropomorphic connotations due to its origins in research on humans and subsequently in nonhuman primates (Weiss et al. 2011; Weiss, this volume). Although it is important to examine evolutionary continuity across human and nonhuman behavior, and personality is a time-honored term used by psychologists as well as farmers, animal breeders, and pet owners, recent work, much described in this volume, highlights some of the problems in deploying the term and its referents clearly and consistently when applied across wider taxonomic chasms (c. f., chapters in Weiss et al. 2011; Zeigler-Hill and Highfill, this volume).

Terminology and Food Preferences

Individual differences, temperament, and personality are terms that, while seemingly clear, are sometimes problematic in how they are used. On one level, personality can be viewed as a way of dealing with individual differences by grouping similar individuals or aspects of their behavior into "types" or categories. If done in this way, then it is not so different from grouping individuals into types based on dietary preferences, foraging styles, and courtship strategies. Of course, empirically measured variation is continuous on most dimensions and how such behavioral differences are measured, grouped, and deployed is often decided pragmatically and statistically, as well as theoretically (Weiss et al. 2011; Zeigler-Hill and Highfill, this volume), as shown in the voluminous work on dog personality (reviewed in Fratkin, this volume).

Temperament was a term used in ancient medicine to characterize different humans well before personality, let alone psychology, arose as a field. The four humors (sanguine, melancholic, choleric, phlegmatic), popularized by Galen and other early pioneers of medicine were used to classify people, and even animals, as belonging to a type, often using physical appearance as well as psychological propensities (Lavater 1798). Animals and people could be viewed as irritable or placid, for example. Temperament may be a less anthropomorphic term for some of the measures being labeled personality. Also, some researchers emphasize social interactions, as reflected in the interpersonal circumplex approach being proposed as an alternative or supplement to personality measures derived from the "Big Five" model (Zeigler-Hill and Highfill, this volume). Although individuality implies stable variation, both personality and temperament have been applied to entire species, as when some lizard or snake species are characterized as being high-strung or irritable, aggressive and prone to bite or attack, or shy and retiring.

But there are other problems with using the term personality rather than more neutral terms, such as behavioral style or syndrome. One is its origins in human personality research and the loose application of the term. Are animals persons? This is still an area of discussion (Rowlands 2016 and commentaries/response). If they are not, then personality may be a misleading term. Of course, the recent popularity of the term has helped to increase interest in the neglected area of individual differences, as shown by hundreds of recent studies, but care is needed. Some research on animals, especially nonhuman primates, uses scales, terminology, and methods derived from human personality research (Weiss and Altschul 2017; Weiss, this volume). Weiss and Altschul (2017) provide a useful glossary of terms used within the field that may be consulted for a somewhat different viewpoint. For example, temperament is viewed as those aspects of personality that are innate.

There have been several attempts to apply variants of personality traits uncovered in research on humans to other mammals including chimpanzees, horses, and elephants (Latzman et al. 2014; Lee and Moss 2012; Morris et al. 2002; Weiss et al. 2011). However, applying such an anthropocentric lens to other species using questionnaires may not accurately incorporate or reflect the ethology of the target species, although some careful and promising recent comparative studies suggest a path forward (Robinson et al. 2016). With reptiles, these concerns may loom larger than with mammals such as dogs and nonhuman primates, as reading their emotions, likes, and social interactions accurately is less obvious. In any event, let animals be themselves and let us follow their lead. One study that highlights the difficulty in applying personality-trait terms to snakes used 22 owner-rated descriptors of the behavior of six royal pythons, *Python regius* (Dutton and Anderson 2002). Using multidimensional scaling analysis (MDS), two dimensions of behavior were identified—one with opposite poles of sociability and anxiety, the other, incorporating terms like "prompt," "inefficient," and "complex," being harder to interpret, perhaps due to idiosyncratic labeling by the owner (Dutton and Anderson 2002). The small sample size (six individuals), lack of interrater reliability measures (all traits were assessed by the owner), and use of somewhat anthropomorphic traits such as "trusting" and "bashful" limit the conclusions about snake personality that can be drawn from this study.

If the term personality is not limited to traditional human personality derived measures, then what does get included in the study of consistent individual differences in behavior? Consider one of the first reptilian studies, in gartersnakes, to document within-family stable individual differences that could be grouped into classes akin to personality types (Burghardt 1975). It has long been known that snakes have prey preferences that are mediated by chemical cues received via vomerolfaction, the tongue-vomeronasal system (Cooper and Burghardt 1990; Halpern and Frumin 1979; Kubie and Halpern 1979). Snakes will approach, tongue-flick at, and even attack cotton swabs dipped in clear aqueous solutions containing chemical cues from the surface of prey. Furthermore, naïve newly born or hatched snakes show such chemosensory prey preferences and such preferences are species typical, heritable, and vary by population (review in Burghardt 1993). However, stable individual chemosensory prey preferences, as between fish and earthworms, can be shown even without the snakes ingesting food. For example, when neonatal common gartersnakes (*Thamnophis sirtalis*), a prey generalist, were tested on chemical fish and worm cues daily for seven days, most individuals had statistically significant worm or fish preferences. In fact, snakes fell into three groups: those having a strong worm preference, a mild worm preference, or a fish preference. So, not only do different individuals in the same litter exhibit individual differences in prey preference, they can be grouped into types prior to any feeding experiences. From an evolutionary point of view, given yearly changes in prey availability, a mother can hedge her bets by having offspring with different prey preferences. Regardless of the initial and often highly specific preferences, however, both feeding experience and maturational change can alter them (Burghardt 1993; Waters and Burghardt 2005, 2013), including learned discriminations among food classes (small fish, earthworms) that initially were responded to about equally, one trial illness-induced food aversions in gartersnakes, and a maturational shift from dragonfly larvae to crayfish in crayfish snakes in the genus *Regina*. Other research on snakes has demonstrated individual differences in social aggregation and social preferences as a function of diet and social interactions (e.g., Lyman-Henley and Burghardt 1994; Yeager and Burghardt 1991). Individuality in food preferences, then, was not only an early demonstration of consistent individual variability in reptiles, but can also be linked to social interactions.

Individuality and Antipredator Behavior in Snakes

We begin by summarizing the strength of evidence in snakes for (a) consistency in individual differences across time and situations, (b) heritability of individual variation, and (c) individual differences being related to fitness measures, thus putting the snake research into the conceptual framework reviewed by Bell et al. (2009) on diverse taxa. The major focus of this section is on antipredator behavior in snakes because there are numerous relevant studies.

Snakes exhibit a variety of antipredator behaviors, such as death-feigning in eastern hognose snakes (*Heterodon platirhinos,* Burghardt 1991), vigorous striking in Mexican black-bellied gartersnakes (*Thamnophis melanogaster*), and balling in royal pythons (*Python regius*) that are observable in the field and in captivity (Greene 1988). Most species display a limited repertoire of antipredator behaviors, so the frequency of specific behaviors can be used as a measure. Hognose snakes, however, have a complex strategy involving bluff strikes followed by writhing, defecating, blood oozing from the mouth, and then cessation of all movement, including breathing. Newly hatched snakes can perform this immediately and consistent differences are seen when repeatedly tested within hours and days of age (Burghardt 1991). Some snakes bluffed only and others both bluffed and feigned. Some quickly dropped the bluff and feigned death (always turning over on their back) when approached.

Several factors affect individual variation in the type and amount of antipredator behavior exhibited, including species (Bowers et al. 1993; Scudder and Burghardt 1983), population (Burghardt and Schwartz 1999; Herzog and Schwartz 1990), litter (Herzog et al. 1989a), skin pattern (Brodie 1992, 1993a), age (Herzog et al. 1992), sex (Herzog and Burghardt 1986; Scudder and Burghardt 1983), size (Mayer et al. 2016), internal state (e.g., reproductive status, Maillet et al. 2015; hormones, King 2002), environmental factors (e.g., habitat, Greene 1979; temperature, Mori and Burghardt 2001), and eliciting stimuli (Herzog et al. 1989b). Individual consistency of behavior has been demonstrated in several species of snakes, and there is evidence for a genetic basis for at least some of the variability detected.

Much of the quantitative research with standardized testing has used North-American natricine snakes as subjects. These snakes include watersnakes (genus *Nerodia*), gartersnakes and ribbonsnakes (genus *Thamnophis*), brown and red-bellied snakes (genus *Storeria*), and crayfish snakes (genus *Regina*). All natricine snakes in North America give live birth, do not provide parental care, and produce neonates that are highly precocial. Neonates can be tested as early as the day after birth and can be housed individually under identical conditions, controlling for effects of experience on behavior. The first year of life is crucial to neonatal natricine snakes, as there is low survivorship over this time span (Gangloff et al., in press); many of the studies conducted on this group fall within this timeframe.

An initial study was carried out by Scudder and Burghardt (1983) with three species of lab-reared 14-month-old watersnakes that had been individually housed under similar conditions as part of a growth study. Most watersnake species are noted for their propensities to strike, bite, and express obnoxious smelling cloacal secretions. An escalating series of tests with humans approaching, touching, and eventually grabbing the snakes at identically timed intervals led to more antipredator behavior (coiling, body flattening, striking) as the "threat" increased, and these behaviors differed in frequency among species and also by sex—with males more prone to strike compared to females. The 43 snakes tested from these three species were from single litters born to wild-caught females and, thus, the differences found could not be definitively attributed to species. However, this study showed that different defensive temperaments among snakes could be measured in a controlled setting. In a subsequent study using escalating threat with a somewhat different methodology, in which snakes were tapped on the head rather than grasped, four species of lab-born and reared gartersnakes and ribbonsnakes showed even more dramatic species differences with all species but one represented by multiple litters (Bowers et al. 1993). Species differed in what could be termed defensive temperament in their propensity to coil, hide their head, tail wave, escape with reversals, or strike; individual differences within each species were also noted.

Other studies in the Burghardt lab were spearheaded by Herzog and colleagues, who conducted an extensive comparative research program on antipredator behavior in natricine snakes, primarily on species of gartersnakes and ribbonsnakes; some topics explored were species differences (Herzog and Burgardt 1986), ontogeny of antipredator behavior (Herzog et al. 1992), effects of recent feeding (Herzog and Bailey 1987), effects of experience (Herzog 1990), stimulus control (Herzog et al. 1989b), habituation (Herzog et al. 1989a), and stability of individual and litter differences over time (Herzog and Burghardt 1988). Many of these experiments used a common methodology in which snakes, tested in a 51 cm × 59 cm × 25 cm glass-walled arena, were given a 30 s acclimation period followed by the experimenter bringing a nonmoving finger within 2 cm of the snake's snout. After this 60 s "nonmoving trial," there was another 30 s undisturbed period and then a 60 s "moving finger trial" in which the finger was oscillated 3 to 4 times per second. The distinction between the nonmoving and moving stimulus conditions is important as patterns of striking can differ between these two conditions, especially when snakes are tested repeatedly over time (e.g., Herzog et al. 1989a). Unlike the previous studies described, snakes were never physically touched during the testing. Using a human hand as a stimulus has been shown to be as effective as using predator models in eliciting antipredator behavior and has excellent inter-tester reliability, $r(24) = 0.86$, $p < .001$ (Herzog et al. 1989b).

The primary dependent variable used in this series of studies was number of strikes at the stimulus; an unambiguous behavior that can be modified by experience (e.g., Herzog 1990) as well as maturation (Bowers 1992; Herzog et al. 1992). The number of strikes observed has an interobserver reliability of over 0.99 (Herzog and Burgardt 1986). Individual variation in striking has been found in all studies, with consistent litter differences (e.g., Herzog and Burghardt 1988) but with few reported sex differences (e.g., Herzog and Burghardt 1988). In many of these studies, individual snakes have shown consistency of striking behavior over time, and litter means have tended to be consistent in their relative rank. Consistent litter differences found in snakes with identical rearing conditions support a genetic origin of the stable individual differences. There has been limited research relating individual differences in behavior to ecological and fitness-related variables (but see Brodie 1992, summarized below). If the type of predator and/or intensity of predation varies from year to year, then individual differences in antipredator behavior may be adaptive, especially in species with no parental care (Herzog et al. 1989b).

Consistency of Antipredator Behavior Over Time

Studies that test many animals just once and find significant differences are not an adequate means to assess individuality, temperament, or personality because of the limited sample of behavior and potential sensitivity of behavior to transient physiological and environmental effects (Burghardt and Schwartz 1999). Studies that retest animals repeatedly within a short interval (hours, days) allow for significant differences to be demonstrated, as in the studies of chemical prey preference polymorphism noted above (Burghardt 1975); however, demonstrating stable differences across longer periods (weeks, months, or years) or across settings are even more convincing demonstrations.

In a pioneering study, Arnold and Bennett (1984) used several methods (e.g., tail-tapping) in several contexts, to elicit antipredator behavior from neonatal plains gartersnakes, *T. radix*. Antipredator behavior was assessed by deriving a score on a continuum from defensive to offensive behavior using characteristics displayed in the snakes' head, body, and tail. They found that the snakes varied in their displays but also showed individual consistency over time and test condition with phenotypic correlations ranging between 0.55 and 0.73. Table 8.1 summarizes multiple studies that clearly show both short-term and long-term consistency of antipredator

Table 8.1 Individual consistency over time in antipredator behavior of natricine snakes; studies are arranged from shortest to longest inter-test interval(s)

Species (*N; # litters*)	Age at initial test	Research focus (Testing location)	Stimulus/behavior	Inter-test interval(s)	Evidence of consistency[a]	Reference
Eastern gartersnake *Thamnophis sirtalis* (19)	Wild-caught adults	Effects of sex and body size; behavioral consistency (Lab)	Model osprey foot/Defensive behavior score	1 day (5 tests)	Percent of measurement error, ME = 65.9%	Maillet et al. (2015)
T. sirtalis (248; 46 litters)	Neonate	Inheritance of speed, stamina, and antipredator behavior (Lab)	Prodded on tail while on treadmill/antipredator score	1 day	0.68***	Garland (1988)
Northwestern gartersnake *T. ordinoides* Tenmile [TM] population (721, 126 litters) CCQ population (393, 77 litters)	Neonate	Homogeneity of genetic variance-covariance matrix (Lab)	Tapping on tail/locomotor behavior —crawling speed, distance, reversals	1 day	Repeatability Speed TM 0.85*** CCQ 0.80*** Distance TM 0.82*** CCQ 0.80*** Reversals TM 0.73*** CCQ 0.65***	Brodie (1993b)
Mexican black-bellied gartersnake *Thamnophis melanogaster* (12)	Wild-caught adult females	Species differences (Lab)	Nonmoving and moving human hand/# strikes	3 days	0.84** (NM) 0.72** (M)	Herzog and Burgardt (1986)
T. sirtalis (25; 2 litters)	68-70 days	Effects of recent feeding (Lab)	Nonmoving and moving human hand/# strikes	5 days	0.56 ** (M)	Herzog and Bailey (1987)
Keelback snake (Australia) *Tropidonophis mairii*	Neonate	Boldness and body size (Lab)	Emergence from shelter	2–7 days	Head out: $r_s = 0.43$** Repeatability: 0.39 Body out: $r_s = 0.40$** Repeatability: 0.45	Mayer et al. (2016)
T. melanogaster (38; 4 litters)	Neonate	Habituation (Lab)	Nonmoving and moving human hand/# strikes	~10 days	0.70* (Mean correlation)	Herzog et al. (1989a)
Butler's gartersnake *T. butleri*(29; 3 litters)	Neonate	Habituation (Lab)	Nonmoving and moving human hand/# strikes	~10 days	0.44 (Mean correlation)	Herzog et al. (1989a)
T. sirtalis (56; 7 litters)	4 months	Effects of experience (Lab)	Model of bird head/# strikes	35 days	Ranged from 0.62* to 0.83**	Herzog (1990)
Plains gartersnake *T. radix* (18; 10 litters)	3 years	Species comparisons; habituation (Lab)	"Finger" model (FM) "Ball" model (BL)	35 days	FM-BL: 0.70** FM-HK: 0.60** BL-HK 0.70**	Bowers (1992)

(continued)

Table 8.1 (continued)

Species (*N; # litters*)	Age at initial test	Research focus (Testing location)	Stimulus/behavior	Inter-test interval(s)	Evidence of consistency[†]	Reference
			"Hawk" model (HK)/# strikes			
T. melanogaster (36; 12 litters)	3 years	Species comparisons; habituation (Lab)	"Finger" model (FM) "Ball" model (BL) "Hawk" model (HK)/# strikes	35 days	BL-HK 0.63***	Bowers (1992)
Western terrestrial gartersnake *T. elegans* (21)	Wild-caught adult females	Variation in antipredator behavior and physiology effects on offspring fitness (Lab)	Foam-tipped rod/# tongue flicks and escape latency	Median: 63 days, range: 51-77 days	Tongue flicks $W = 0.48$** Escape latency $W = 0.50$***	Gangloff et al. (in press)
T. melanogaster (25; 3 litters)	Neonate	Ontogeny of habituation (Lab)	Model of hawk head/# strikes and individual regression parameters	Tested at ages: 1 day, 61 days, 300 days	Day1/Day 61 0.88*** (TOT) 0.87*** (Intercepts) 0.76** (Slopes) Day 1/Day 300 0.86*** (TOT) 0.77 ** (Intercepts) 0.61* (Slopes) Day 61/Day 300 0.66 ** (TOT) 0.83*** (Intercepts) 0.70*** (Slopes)	Bowers (1992)
T. melanogaster (38; 4 litters)	Neonate	Consistency over time of antipredator behavior (Lab)	Nonmoving and moving human hand/# strikes	Tested at ages: 1 day, 7 weeks, 16 weeks, 31 weeks, 54 weeks	0.54*** (NM, intercorrelation) 0.63*** (M, intercorrelation)	Herzog and Burghardt (1988)
T. ordinoides (74; 50 litters – both lab and field study)	Neonate	Consistency over time of antipredator behavior and color pattern (Lab and Field)	Tapping on tail/locomotor behavior —crawling speed, distance, reversals	Tested at ages: 3 days, 4 months, 8 months, 20 months (lab); 1 year and 2 year groups in field	Lab group: Speed: W = 0.52*** Distance: W = 0.60*** Reversals: W = 0.54*** 1-year-field group: Reversals: .65*** 2-year-field group: Speed: $r_s = 0.47$* Distance: $r_s = 0.51$* Reversals: $r_s = 0.55$*	Brodie (1993a, b)

Note NM = nonmoving human hand, M = moving hand, TOT = total strikes

[†]Results from Pearson *r* correlation coefficients unless otherwise noted. W = Kendall coefficient of concordance; r_s = Spearman rank-order correlation coefficients

* $p < .05$; ** $p < .01$; *** $p < .001$

Table 8.2 Pearson *r* correlation coefficients (sample size) between retests at five different ages of *T. melanogaster* on number of strikes directed at a nonmoving and moving human hand during 60 s trials

Age	Nonmoving stimulus				Moving stimulus			
	7 weeks	16 weeks	31 weeks	54 weeks	7 weeks	16 weeks	31 weeks	54 weeks
1 day	0.56** (36)	0.39* (30)	0.54** (22)	0.48** (25)	0.38* (36)	0.47** (30)	0.53** (22)	0.40* (25)
7 weeks		0.61** (29)	0.63** (21)	0.54** (24)		0.72** (29)	0.69** (21)	0.67** (24)
16 weeks			0.54** (22)	0.53** (25)			0.84** (22)	0.68** (25)
31 weeks				0.61** (22)				0.79** (22)

Note: Recreated from Herzog and Burghardt (1988), p. 254,* $p < .05$; ** $p < .01$ (one-tailed test)

behaviors in natricine snakes; in some of the studies, e.g., Herzog and Burghardt (1988) and Brodie (1993a, b), snakes show consistency of behavior across months or even years.

Long-Term Developmental Trait Consistency

Two studies of gartersnakes focused specifically on long-term consistency of behavior. In an in-depth assessment of long-term stability of snake behavior, Herzog and Burghardt (1988) assessed consistency of antipredator behavior over the first year of life in the Mexican black-bellied gartersnake (*T. melanogaster*). This species is found in central Mexico near water sources and feeds primarily on aquatic prey (Rossman et al. 1996). Both neonates and older individuals are aggressive, i.e., when approached they strike readily and frequently compared to other species of *Thamnophis* (Bowers 1992; Bowers et al. 1993; Herzog et al. 1989a, 1992; Herzog and Burgardt 1986). In Herzog and Burghardt's (1988) experiment, snakes were tested the day after they were born and subsequently at 7, 16, 31, and 54 weeks. The lifespan of *T. melanogaster* in the wild is not known, but in the laboratory, we have kept individuals alive for over 14 years. Highly consistent individual and litter differences were found at each age (see Table 8.2). In addition, litter ranks were remarkably consistent over the first year of life, Kendall's coefficient of concordance, $W = 0.81$, $p < .01$ (nonmoving hand) and $W = 0.94$, $p < .01$ (moving hand). These stable litter differences over the testing span are illustrated in Fig. 8.1.

Brodie (1993a) similarly tested northwestern gartersnakes, *T. ordinoides*, to assess ontogenetic change in antipredator behaviors (maximum escape speed, total distance crawled, number of reversals) and color pattern, both in the laboratory and in the field. The laboratory group was tested at 3 days of age and then at 4, 8, and 20 months. The field groups were tested as neonates, marked, and released, with one subset being captured and retested after one year and the other subset being captured and retested after two years. The coefficients of concordance for the laboratory group and Spearman correlations for the field groups, all adjusted for size, are shown in Table 8.1. Most of the tests of relationship were significant, indicating that these snakes showed individual consistency in their antipredator behavior. It is notable both that laboratory and field results were similar (as rearing and testing conditions may affect an individual's behavior), and that behavior was consistent over a lengthy time period (two years).

Experiential Effects

Experience can modify striking behavior in gartersnakes (Herzog 1990; Herzog et al. 1989a;

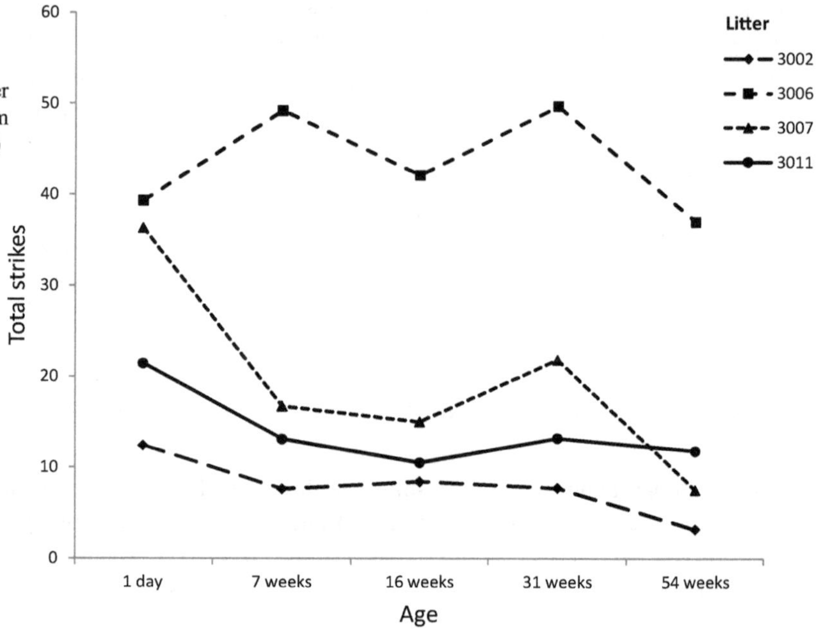

Fig. 8.1 Consistency of antipredator behavior (mean total strikes given) by four litters of *T. melanogaster* over the first year of life (data from Herzog and Burghardt 1988)

Herzog and Burghardt 1988). The number of strikes will increase under conditions when the snakes are harassed repeatedly (Herzog 1990) or will decrease when repeated stimulus presentation has no aversive consequences (Bowers 1992; Herzog et al. 1989a; Herzog and Burghardt 1988). Bowers (1992) investigated the ontogeny of habituation in *T. melanogaster* using 25 neonates from 3 litters tested in a split-litter design. Twelve snakes were given a test the day after birth (day 1) in which a model of a hawk head was presented for 10 30 s trials with a 30 s inter-stimulus interval. The number of strikes given and other antipredator behaviors were measured in each trial. On day 61, all 25 of the snakes were given an identical test; 20 snakes (10 from each group) were also tested on day 300. Regression analyses were performed on the data for each snake (methodology described by Petrinovich and Widaman 1984), so that an intercept and slope were generated for each snake. The intercept provides a measure of responsiveness in terms of striking but is highly affected by slope, which indicates the rate of habituation. Total number of strikes over all trials was calculated as an additional measure of responsiveness. Individual consistency over the 60-day period was found, indicated by strong positive correlations for all measures between all test days (see Table 8.1). Number of strikes over trials on day 1 and on day 61 for three snakes are shown in Fig. 8.2; these individuals show consistency not only in their total responsiveness but also in their habituation profiles.

Stamps and Biro (2016) underscore the importance of measuring personality differences and their influence on individual differences in behavioral plasticity. For example, individual differences in propensity to habituate are associated with fitness-related outcomes in Iberian wall lizards, *Podarcis hispanica*; the body conditions of individuals that habituate readily to low-risk predators tended to improve, presumably because they can spend more time foraging and/or experience less of a physiological stress response than those that do not habituate readily (Rodríguez-Prieto et al. 2010). Thus, individual differences in plasticity may have implications for fitness. And, whereas experience can interact with behavior in complex ways, it can also alter, indeed increase, the heritability of a trait when measured at different ages and after certain experiences (Burghardt et al. 2000). In this study, the responsiveness to chemical cues from prey

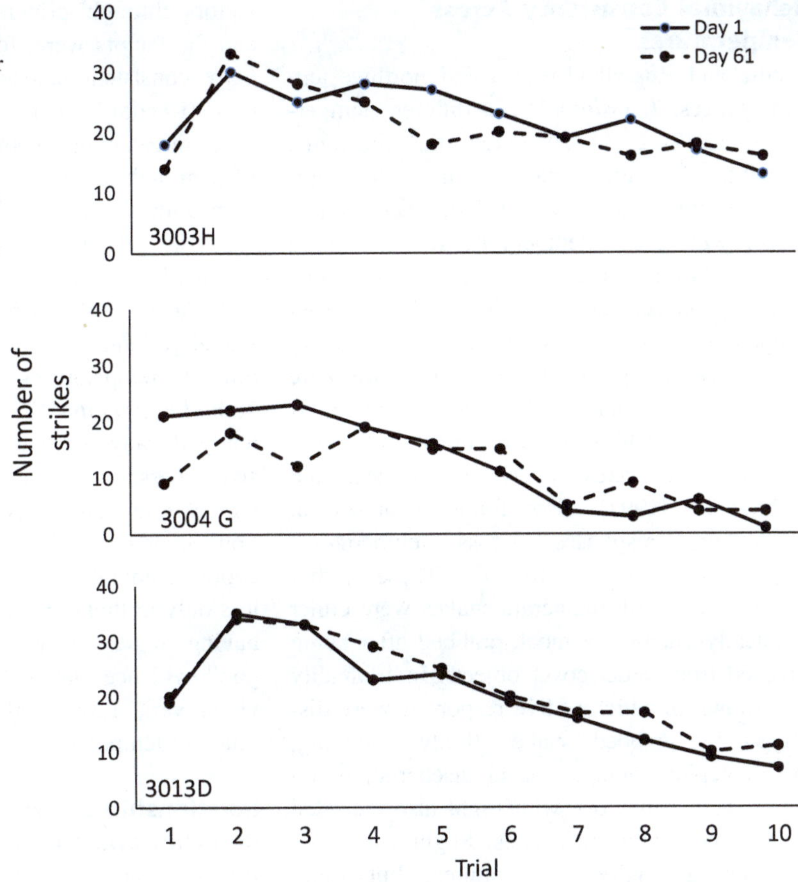

Fig. 8.2 Consistency of habituation profiles of strikes given in each of ten trials for three individual Mexican black-bellied gartersnakes, *Thamnophis melanogaster* from three different litters. These snakes were tested the day after birth (Day 1) and again 60 days later (Day 61)

was assessed in neonatal *T. sirtalis* before feeding experience and after 12 meals on fish. Heritability (h^2) of response to fish stimuli increased from 0 in the initial pre-feeding test to 0.323 ($p < .05$) after feeding experience; similarly, the heritability of response to worm stimuli increased from 0 to 0.497 ($p < .005$). Interestingly, the change in fish response (indicating behavioral plasticity) was itself significantly heritable ($h^2 = 0.226, p < .05$), whereas the heritability for change in worm response was not significant (Burghardt et al. 2000). In another twist, those snakes that had a greater preference for fish after the 12 fish meals gained more weight and length, in spite of all animals receiving the exact same size meals. Thus, individual differences in feeding "personality" and plasticity are mirrored in other physiological traits in snakes.

The studies reviewed in this section have demonstrated clear evidence for individual consistency of antipredator behavior over time in natricine snakes. Individual consistency was found in different species, under different testing conditions, and for different antipredator behaviors, all facts that strengthen these conclusions.

Consistency of Behavior Across Situations

Another marker of temperament/personality is consistency of behavior across situations. There has been less research on behavioral consistency across situations in snakes, and the evidence for this consistency is mixed, as will be reviewed below.

Behavioral Consistency Across Temperatures

Brodie and Russell (1999) tested northwestern gartersnakes, *T. ordinoides*, at different temperatures (15, 22.2, 30 °C) representing the range encountered by these snakes in the wild. Temperature affected distance traveled, reversals, and speed, with cooler temperature resulting in a decrease in all measures; however, snakes were relatively consistent in their rankings across temperatures. Specifically, snakes with relatively faster crawling speeds at one temperature were likely to have faster speeds at the other temperatures. In a field study of almost 200 adult Eastern gartersnakes, *T. sirtalis*, Passak and Gillingham (1997) also found changes in antipredator responses across temperatures ranging from 6 to 40° C. These rather cold-tolerant north temperate snakes were either physically grabbed or mock grabbed after being exposed from under cover objects. Individuality was shown in which of five responses were displayed by grabbed snakes (body flattening, mouth gaping, biting, cloacal discharge, or no antipredator behavior), with flight also recorded for the mock-grabbed snakes. Flight was more common at higher temperatures, but other behaviors such as body flattening and cloacal discharge, which could occur singly or in combination, did not vary by temperature. Although these snakes were not retested, there were individual differences in behavior not explained solely by temperature. Mori and Burghardt (2001) tested 24 adult Japanese tiger keelback snakes, *Rhabdophis tigrinus*, under three temperature conditions (14, 22, 30 °C). In the tests, snakes were briefly pinned with a snake hook, and the occurrence of eight different antipredator behaviors was measured; some of the behaviors (e.g., "neck arch") exposed nuchal glands that excrete toxic secretions. Temperature affected antipredator behavior overall, with snakes exhibiting more passive behaviors and behaviors that exposed the nuchal glands at the lowest temperature; when the snakes were tested under higher temperatures, they, as with *T. sirtalis* above, showed more active fleeing from the stimulus. The behaviors were sorted into three factors through principal component analysis, and the factors were, for the most part, individually consistent across the temperature conditions (Kendall's coefficient of concordance: PC1 static threatening response. $W = 0.75$, $p < .001$; PC2 immobility $W = 0.48$, $p = .09$; PC3 active threatening response, $W = 0.55$, $p < .05$). An example of behavioral consistency of a non-natricine snake was described by Citadini and Navas (2013) who found consistent individual variation of antipredator behaviors across different temperatures in the neotropical snake, *Tomodon dorsatus* (Dipsadidae). Snakes consistently displayed behaviors classified as aggressive, passive, or evasive in the different temperature conditions; these initial classifications interacted with behavioral tendency, i.e., cooler temperatures increased aggressive behaviors only in those snakes previously classified as having aggressive dispositions. Thus, there is good evidence that snakes show consistent individual variability in antipredator behavior across different temperatures.

Relationship Between Antipredator Behavior and Open-Field exploration

It may be expected that antipredator behavior and open-field exploration would be negatively correlated, because less placid snakes may be more inhibited in an open-field test. Tongue flick rates and squares traversed in an open-field test have been shown to be consistent over time in eight plains gartersnakes, *T. radix* (Chiszar and Carter 1975). Herzog and Burgardt (1986) compared open-field behavior and antipredator responses in three gartersnake species (*T. melanogaster*, *T. sirtalis*, and *T. butleri*). The only significant phenotypic correlations were found in eastern gartersnakes, *T. sirtalis*, between number of strikes in the moving hand condition and a) latency to leave the center square of the open field, $r(88) = 0.25$, $p < .05$ and b) ambulation time in the open field, $r(88) = -0.32$, $p < .01$. The correlations between antipredator behavior and open-field measures were near zero for the other two species. Maillet et al. (2015) reported no correlations between exploratory behavior and defensive behavior in *T. sirtalis*, although their methodology differed from that of Herzog and Burgardt (1986). The open-field tests were roughly similar in the two studies, but Herzog and Burghardt measured number of strikes given as the major dependent variable and used a human hand as the threatening stimulus. Maillet et al. (2015), on the other hand, classified snake behavior along a continuum of defensive to offensive behaviors and elicited antipredator behavior using a model claw. Furthermore, Maillet et al. (2015) found no evidence of a behavioral syndrome between exploratory, boldness, and defensive behaviors, although they did report an interaction between sex and body size with longer males exploring more than shorter males, but shorter females exploring more than longer females. In addition, males and nongravid females were more likely to show fleeing behavior than were gravid females. Thus, there is only weak and limited evidence of a relationship between antipredator behavior and exploration.

Relationship Between Antipredator Behavior and Responses to Chemical Prey Extracts

The relationship between antipredator behavior and responses in a prey preference test may be inverse because more defensive, easily irritable snakes could be stressed during prey extract testing and less motivated by food cues. Burghardt and Schwartz (1999) found genetic and phenotypic correlations between behaviors from a battery of tests run on snakes from two populations of *T. sirtalis* from Wisconsin and Michigan. Snakes that struck most were less responsive in the chemical prey test than were those that struck fewer times. Similarly, Bowers (unpublished data) found a significant negative phenotypic correlation between strike scores of *T. sirtalis* at 15 days of age and overall mean responsiveness to chemical prey extracts 10 months later, $r(36) = -0.37$, $p = .02$.

Measuring antipredator behaviors and responsiveness to chemical prey cues over multiple trials in a habituation test revealed a different pattern of results (Bowers 1992). Individual intercepts and slopes were calculated for responses over ten trials (tongue flick—attack scores) to worm and fish extract as well as for antipredator behavior to three different threatening stimuli. The only significant correlations between chemical prey extract and antipredator responses were with the control stimulus in the antipredator test, a finger-shaped foam model, which has been shown to be relatively ineffective at eliciting antipredator behavior (Herzog et al. 1989b); typically, only the most aggressive snakes would strike at this stimulus. Significant phenotypic Pearson correlations were found between strikes to the control stimulus and the following: *T. radix*, earthworm extract test/intercepts, (0.49), *T. melanogaster*, fish extract test/intercepts (0.42) and slopes (0.48), and worm extract test/intercepts (0.39). The persistence of striking at a mildly threatening stimulus or responding to a prey extract may tap into a general reactivity dimension in snakes (Burghardt and Schwartz 1999).

Heritability of Antipredator Behavior

Heritable individual differences can be raw material for, as well as the result of, natural selection (Careau and Garland 2012), and there is good evidence of heritability for antipredator behaviors in gartersnakes. Arnold and Bennett (1984) found that the heritability of a composite antipredator behavior score in plains gartersnakes (*T. radix*) was ~0.37 when based on single trials and ~0.45 for averages of two trials. In common gartersnakes (*T. sirtalis*), Garland (1988; 1994) studied the genotypic and phenotypic correlations of antipredator behavior with sprint (escape) speed and physiological endurance and found significant phenotypic, but not genetic, linkages among these measures. His heritability estimate for antipredator behavior was 0.41, very similar to that of Arnold and Bennett (1984). Brodie (1989) examined genetic linkages between color/pattern and antipredator behaviors in northwestern gartersnakes, *T. ordinoides*, and found statistically significant heritabilities for all behaviors in each of the four populations he tested. Further tests in two populations of

T. ordinoides found significant heritabilities between 0.49 and 0.66 for all three antipredator behaviors (Brodie 1993b). Finally, Schwartz and Herzog (unpublished data) found significant full sibling heritability estimates (h^2) for number of strikes given by three species of day-old gartersnakes: *T. melanogaster* (0.75), *T. butleri* (0.80, males; 0.47, females), *T. sirtalis* (0.62, Michigan population; 0.52, Wisconsin population). In the latter study, the snakes from the Michigan population had a mean number of strikes per test ($M = 25.9$) almost five times higher than that of the Wisconsin snakes ($M = 5.8$) (Burghardt and Schwartz 1999). The heritability estimates are roughly in the same range as in previous studies. The moderate heritability estimates for antipredator behaviors in gartersnakes indicate that there is not strong directional selection on the traits measured for the given populations (Garland 1994). This may be due to the existence of alternative individually consistent strategies within the same population (Careau and Garland 2012) as well as to genetic linkages between color/pattern and behavior (Brodie 1993a, b). The converging results from multiple laboratories using different species and different methodologies underscore the robustness of these findings. The recognition of extensive multiple paternity in many reptiles (e.g., Burghardt and Schwartz 1999) would increase h^2 estimates somewhat.

Fitness Consequences of Inter-Individual Variation

The relationship between inter-individual variation and fitness variables has been relatively unexplored in snakes. Gangloff et al. (in press) investigated how "energetic phenotypes" relate to females' resource allocation to offspring in a population of Western terrestrial gartersnake, *T. elegans*. They sampled 21 pregnant snakes from a "slow-paced life history" population and performed both physiological and behavioral assays to assess maternal energetic phenotype effects on offspring fitness variables. For the behavioral tests, females were tested four times in a simulated predatory attack; total tongue flicks given (assessing exploratory activity) and latency to escape the test arena (assessing boldness towards a predator) were measured and found to be consistent over two months of testing (see Table 8.1). Three blood plasma assays of corticosterone and glucose concentrations were also taken; glucose level was only marginally significantly consistent over time, whereas corticosterone showed no consistency. Principle component analysis sorted the behavioral and physiological traits of the snakes into "high reactive" phenotypes, with a high number of tongue flicks and short escape latency and "low reactive" phenotypes with the opposite pattern of behaviors; these phenotypes were said to be extremes on a single behavioral axis of general activity. Females with consistent behavioral and physiological phenotypes (high activity–high hormone levels; low activity–low hormone levels) had offspring with better body conditions, resulting in more rapid growth and better survivorship, than did those with a mismatch, establishing that individual variation can affect fitness in snakes.

Brodie (1992) investigated correlational selection between color pattern and antipredator behavior in *T. ordinoides*, a species that is noted for the variability of skin color and pattern found in natural populations (Brodie 1993a). In general, striped snakes tend to rely on flight to escape predators due to an optical illusion of longitudinal stripes making detection of motion and judgement of speed difficult, whereas banded or blotched-patterned snakes tend to rely on crypsis or aggression in predatory encounters (Brodie 1992). Individual differences in antipredator displays and color pattern have been shown to be relatively consistent over the first two years of life in this species (Brodie 1993a; Table 8.1). The unique feature of Brodie's 1992 study was that, after neonatal snakes were tested for antipredator behavior (number of reversals/evasive maneuvers, speed of crawling, and distance crawled before resorting to antipredator displays) and assessed for color/pattern in the lab, they were released into the wild. The effects of having different combinations of skin pattern and antipredator behavior on survivorship, i.e., the fitness consequences of individual variability, were assessed. The only combination of traits that affected survivorship was the negative relationship between number of reversals and striping. Striped snakes should favor direct flight from a predator over evasive movements; thus, the combination of striping and exhibiting a high number of reversals was selected against. Snakes high in one of these traits and low in the other tended to have the best survivorship (Brodie 1992). It was concluded that the maintenance of the heritable variation found in the population was due to selection for specific combinations of traits possibly in conjunction with frequency-dependent section (Brodie 1992).

In sum, the studies reviewed above provide evidence of consistency of individual behavior over time in several snake species, with fewer studies of consistency of behavior over situations. Antipredator behavior has been shown to be heritable in gartersnakes but does not seem to be under strong directional selection; the variability found in populations has in some cases (e.g., Brodie 1992), been linked with fitness.

Sociality

Although in-depth studies of social behavior in natricine snakes are not yet available, neonates have a strong tendency to aggregate, preferring conspecifics (Burghardt 1983), and nearest neighbor distances can be heritable (Burghardt and Schwartz 1999).

Rattlesnakes (genus *Crotalus*) exhibit complex social behaviors, including parental care, in communal dens. Technological developments such as miniaturized telemetry equipment with GPS capacity, improved molecular genetic methods for identifying kin relationships, and time-lapse photography, now make it possible to study the behavior of individual snakes over time (Schuett et al. 2016). The rich behavioral repertoire of rattlesnakes has been underappreciated, and this is an exciting branch of research for studying reptile personality. For example, Amarello (2012) studied Arizona black rattlesnakes (*Crotalus cerberus*) at communal basking sites (areas not associated with aggregations due to overwintering or with reproduction) in two populations. She used time-lapse photography to follow association patterns of snakes individually identified by natural markings in April and May over two years. There was individual variability in gregariousness, with some snakes consistently preferring to associate in either small or large groups. In addition, adult snakes were selective as to the individuals with which they associated. Social behavior in reptiles is an area ripe for further investigation of personality/temperament, as has been demonstrated in the studies of lizards reviewed below.

Individual Differences and Behavioral Syndromes in Lizards

Most recent research on reptile behavioral variation has focused on the other major group of squamate reptiles, lizards. Several lizard species have become popular model animals for studying animal personality, temperament, and consistent individual differences and behavioral syndromes. This may be partly due to the ease with which lizards are caught and maintained in large numbers in laboratories or observed in the field (McEvoy et al. 2015). Furthermore, the density of lizards in the field and the frequency with which they can be observed, even nocturnal ones such as geckos, is often remarkable. In fact, it seems easier to collect extensive data on lizards than it is for most other terrestrial vertebrate species.

Researchers studying lizards typically embed their work in personality terminology and we will follow such use in this discussion. Several species have been studied and personality traits and/or behavioral syndromes have been demonstrated in most (see Table 8.3), with the common lizard *Zootoca* (*Lacerta*) *vivipara* receiving the most attention.

Table 8.3 Studies that have examined personality traits or behavioral syndromes in lizards

Species	Age (location tested)	Personality Trait(s)	Behavioral Syndrome	Correlative traits	Reference
Zootoca (Lacerta) vivipera Common Lizard	Adult (enclosure and lab)	*Activity*[a], *Sociability*, Exploration,		Mating behavior and female mate choice under different levels of predation risk	Teyssier et al. (2014)
	Adult (enclosure and lab)	*Activity, Aggression, Boldness* Sociability	No correlation between traits	Standard metabolic rate (SMR); corticosterone levels	Mell et al. (2016)
	Captive born Neonate to 1-year and adult (enclosure)	*Sociability*		Dispersal; Population density	Cote and Clobert (2007)
	Captive born Neonates to 1-year (enclosure).	*Sociability, Boldness*	Positive correlation between Boldness and Sociability	Survival, body growth and fecundity at different population densities	Cote et al. (2008)
	Captive born neonates to juvenile, adult (Lab and enclosure)	*Exploration*[a]		Resting metabolic rate (RMR); maximal sprint speed (MSS); endurance capacity. Survival and growth selection	Le Galliard et al. (2013)
	Yearlings and adults (enclosures)	*Activity, Boldness, Sociability* (short term) *Activity, Boldness*, Sociability (long term)	Weak evidence of negative correlation between activity and boldness at birth	Juvenile survival; body growth rate and reproduction at different population densities	Le Galliard et al. (2015)
Iberolacerta (Lacerta) monticola Iberian/Rock Lizard	Wild-caught adults (lab)	*Boldness*		Body size, head size and body condition and T-cell immunocompetence levels	López et al. (2005)
Lacerta viridis European Green Lizard	Wild-caught adults (lab)	*Boldness; Exploration*[a]	Correlation between boldness and exploration in adults	Body size and parasite load	Bajer et al. (2015)
Podarcis hispamica Iberian Wall Lizard	Wild-caught adults (Outdoor cages and enclosures)	*Boldness*		Habituation to none threatening predation risk; Body condition	Rodríguez-Prieto et al. (2010)
	Wild-caught adults (Outdoor cages)	*Boldness, Exploration; Sociability*		Direct and indirect effects of exploration, sociability, boldness and sex on habitation.	Rodriguez-Prieto et al. (2011)
Egernia whittii White's Skink	Wild-caught adult females (lab)	*Aggression* (female)		First year growth and survival of offspring	Sinn et al. (2008)
	Wild-caught adult females (lab and field)	*Aggression* (female)		Body size, territory size, social density and paternity acquisition (within- vs extra pair paternity)	While et al. (2009)

(continued)

Table 8.3 (continued)

Species	Age (location tested)	Personality Trait(s)	Behavioral Syndrome	Correlative traits	Reference
	Wild-caught adult male and females (lab and field)	*Aggression* (male and female)		Testosterone levels	While et al. (2010)
	Wild-caught adult male and females (lab)	*Aggression, Boldness, Exploration,* Activity, Sociability	No evidence of a behavioral syndrome		McEvoy et al. (2015)
Lampropholis delicata Delicate Skink	Wild caught (lab and field)	*Activity, Aggression, Boldness, Exploration; Sociability*	Correlations between activity, exploration and sociability	Trapping techniques	Michelangeli et al. (2016)
	Wild-caught adult males (lab)	*Activity, exploration, boldness, foraging activity*	Correlation between activity and exploration	Degree of urbanization.	Moule et al. (2016)
Eulamprus heatwolei Southern Water skink	Adult males and females (field).		Behavior syndrome relating to territorial, exploratory and predator avoidance behavior (floaters vs. territorial)	Mating system	Stapley and Keogh (2005)
Agama planiceps Namibian Rock Agama	Adult males (field)	*Boldness*	Correlations between Boldness, Basking and Movement	Tail loss; home range size; feeding	A. J. Carter et al. (2010)
	Adult males (field)	*Boldness*		Dry vs. rainy seasons; basking, moving, thigmothermy, sitting in shade, hidden, signaling behavior	Carter et al. (2012a)
	Adult males (field)	*Boldness*		Trapping technique	Carter et al. (2012b)
	Adult males (field)	*Boldness*		Intraindividual variation (IIV) of boldness; Seasons (wet vs. dry)	Highcock and Carter (2014)
Tiliqua rugose Sleepy Lizard	Adult males and females (field)		Male aggression and level of association with females	Size; body condition and head width; Position in social network	Godfrey et al. (2012)
Anolis segrei Brown anole	Captive reared adult males (Lab)	*Boldness*		Tail autotomy Food availability; body size	Kuo et al. (2015)

Traits in italics were found to be repeatable within individuals and showed inter-individual variation
[a]Indicates heritability estimated

Measuring Behavioral Traits

An important aspect of any behavioral study is to define and measure the behavior(s) of interest. Although the use of different measures for a behavioral type may limit our ability to compare species or groups, the choice of measure is often a function of the environmental conditions under which the study was conducted or idiosyncrasies of the species studied. Here we summarize the methods used to measure the behavioral types commonly studied in lizards.

Boldness or risk-taking was the most common personality trait studied (see Table 8.3). This trait was assessed using a variety of methods. A popular method employed in field or outdoor enclosure studies was to have an observer approach the focal lizard and measure variables such as flight initiation distance (Carter et al. 2010; Taylor and Lattanzio 2016), position relative to a refuge (Cooper 2012; López et al. 2005; Rodríguez-Prieto et al. 2010), and/or times associated with a refuge (Mell et al. 2016). The use of a human to simulate a predator approach is well established and was originally used to study lizard antipredator behavior (Burger and Gochfeld 1990; Cooper 1997a, b). It is only recently that the focus of such studies has shifted to understanding individual differences in those behaviors (e.g., Cooper 2009, and references cited above).

Le Galliard et al. (2015) used a different approach to assessing risk-taking. In their study, common lizards were placed in a plastic box with a cardboard shelter at one end and a basking light at the other. To simulate a predatory attack, they touched the lizard repeatedly with a paintbrush until it retreated into the refuge. The level of boldness was determined by measuring how long the lizard spent hidden before emerging from the refuge and how long it took before the lizard resumed normal basking behavior (Le Galliard et al. 2015). This methodology provides a suitable alternative technique for measuring boldness when field or outdoor enclosure studies are not possible.

Another frequently assessed behavioral trait is sociability, which, in lizards, can be measured by exposing individuals to conspecific odors in a choice test (Cote and Clobert 2007; McEvoy et al. 2015; Rodriguez-Prieto et al. 2011). This test typically involved presenting paper scented with a conspecific odor or an unscented control in a shelter (Cote and Clobert 2007) or under a basking rock (McEvoy et al. 2015). Individual lizards were then released into the arena and the time they spent located by or on the odor source was recorded. This duration was subsequently used as a metric of social tolerance (Cote and Clobert 2007; McEvoy et al. 2015). In one study, live conspecifics were used instead of odor scented paper and the time a lizard chose to bask with conspecifics or alone was measured (Michelangeli et al. 2016).

Another trait, exploration, was typically measured by introducing a lizard to a novel area (Le Galliard et al. 2013; Rodriguez-Prieto et al. 2011) or to novel objects (McEvoy et al. 2015). In the former method, lizards were introduced into one compartment (familiar or home area) and then, after a period of acclimatization, offered the opportunity to enter an unfamiliar compartment (Le Galliard et al. 2013; Teyssier et al. 2014). Various measurements were then taken, such as latency to enter the novel enclosure, time spent in the unfamiliar compartment, time spent walking, and number of visits to the novel compartment (Le Galliard et al. 2013; Teyssier et al. 2014). When lizards were introduced to novel objects, latency to explore (touch first object) and number of times a lizard approached a novel object were scored (McEvoy et al. 2015).

The trait of activity is similar to exploration and the two can be difficult to separate (Le Galliard et al. 2015). Activity was distinguished from exploration by testing lizards in a *familiar* environment. The use of an environment in which a lizard has lived reduces the likelihood that movements observed are due to the lizard exploring because the animal has previously explored the area and is familiar with its structures, odors, lighting, and temperature. For example, McEvoy et al. (2015) measured activity by observing movement of lizards within their home cage, while Michelangeli et al. (2016) and

Moule et al. (2016) monitored activity using an open-field apparatus with which the lizards were familiar. In their study on the role of personality types and predation risk on mate choice, Teyssier et al. (2014) used a different technique to estimate the level of activity of lizards. In this study, they determined levels of exploration and sociability using the novel compartment and conspecific odor methods described previously. The level of general activity was then assessed by estimating the time lizards spent walking during each of these trials.

Conspecific aggression has also been studied in both males (Godfrey et al. 2012; McEvoy et al. 2015; While et al. 2010), and females (While et al. 2010, 2009). The aggressiveness of individual lizards was assessed using several different measurements. For example, Godfrey et al. (2012) used two methods to assess male aggression. First, they monitored lizards every two weeks for scale damage. Lizards with fresh scale damage were recorded as being more aggressive, although the authors do point out that nonaggressive lizards may also receive scale damage from attacks by aggressive males. Second, they presented captive lizards with a life-sized open-mouthed lizard model and recorded the subjects' responses. Model lizards were also used by McEvoy et al. (2015) and While et al. (2009, 2010) to assess aggression in males and females. However, an important difference was that their models were enhanced by the scent of either a male or female.

Potential Biases in Research Methods

The need to capture animals to collect biometric information, provide marking for individual identification, and collection for captivity studies, introduces potential sampling biases. Capture bias may be introduced as a result of factors that affect the detectability of individuals and the methods used to capture lizards (Foster 2012; Rodda 2012). But observational and capture methods themselves can alter subsequent behavior, such as movement patterns (Rodda et al. 1988). The choice of capture and observational techniques is therefore a critical component of any study, and especially those using wild-caught animals. The potential significance of capture technique when studying lizards is highlighted by two studies on the association between lizard boldness on trapability. On the one hand, Carter et al. (2012b) found that boldness was associated with a greater probability of capture by passive, baited clap traps. On the other hand, Michelangeli et al. (2016) found that hand trapping, mealworm fishing, and pitfall traps did not produce a bias in the type of lizard captured, because these methods did not rely on an individual finding and inspecting a trap, thus minimizing the likelihood of being rejected or avoided by risk averse individuals. There are other often-ignored, but especially pertinent problems in behavioral research, such as the possible effects of biases in labeling and characterizing animals. This can be mitigated by double-blinding, by computing interobserver reliabilities, and by evaluating video recordings of behavior (Burghardt et al. 2012).

Statistical Analysis of Behavioral Traits and Variables of Interest

To determine whether a lizard species demonstrated consistent individual differences or personality types, researchers, in the majority of studies, measured each behavioral trait two or more times per individual. Recall, however, that in the prey preference study (Burghardt 1975) snakes were tested enough times to be able to show statistically significant effects both within as well as between individuals at a given time, something not all that common in the lizard literature. Regardless, in most lizard studies, each behavioral trait was analyzed as a single measure or, if there were correlations among multiple traits, principal component analysis (PCA) was used to reduce these measures to one or more components (Bajer et al. 2015; Le Galliard et al. 2013; Teyssier et al. 2014; While et al. 2010, 2009).

Behavioral consistency or repeatability was most frequently assessed by computing an intraclass correlation coefficient for each behavior or component. Several statistical techniques can be used to derive intraclass correlation coefficients (see Nakagawa and Schielzeth 2010). McEvoy et al. (2015) raise an important issue about determining whether a behavior, especially one created by aggregating several different behavioral scores, can be considered a personality trait. Such behaviors should not only demonstrate temporal stability, but also structural consistency (the consistency of relationships between behavioral components over time). Although studies almost always report temporal stability, few document structural consistency (Carter et al. 2010; McEvoy et al. 2015).

Once a trait's stability across time has been determined, its relationship to other variables of interest is assessed to identify correlations and possible behavioral syndromes. Various statistical methods are used to determine these relationships, including repeated-measures ANOVA (While et al. 2010), general linear models (Mell et al. 2016; Rodriguez-Prieto et al. 2011; Sinn et al. 2008), general linear mixed models (Bajer et al. 2015; Cote and Clobert 2007; Le Galliard et al. 2013), PCA (Mell et al. 2016; Michelangeli et al. 2016; Moule et al. 2016) and Spearman rank correlations (Bajer et al. 2015; López et al. 2005). For example, Carter et al. (2010) identified a behavioral syndrome in male Namibian rock agamas (*Agama planiceps*) that related boldness to basking duration and movement. Bolder individuals spent more time basking and moving around their home ranges than shy individuals. As a consequence, they also had larger home ranges and fed more, but they were more susceptible to predation.

Rodriguez-Prieto et al. (2011) took a novel approach to analyzing relationships between behavioral types and other potential factors in lizards. They used path analysis to understand the direct and indirect effects of behavioral type on habituation. In addition to boldness, sociability, and exploration, they included body size, sex, exposure to low-risk predation, and habituation index in their model. The analysis showed that, whereas exploration and sex had direct effects on habituation, boldness, sociability, sex indirectly affected habituation through exposure (Fig. 8.3, from Rodriguez-Prieto et al. 2011). This study is one of only a few animal personality studies to take advantage of path analysis (see Konečná et al. 2012, for an example with primates). The use of path analysis provides an opportunity to develop a richer understanding of both direct and indirect associations between traits that are not elucidated using other statistical approaches.

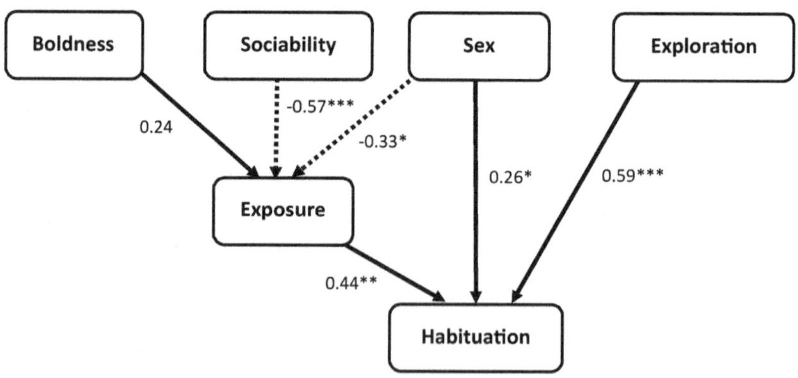

Fig. 8.3 Path analysis diagram and coefficients for direct and indirect predictors of habituation by Iberian wall lizards (redrawn from Rodriguez-Prieto, et al. 2011). Sex was coded as males = 1 and females = 2. Direct relationships are represented by solid lines; inverse relationships by dashed lines. $^*p < 0.05$; $^{**}p < 0.01$; $^{***}p < 0.001$

Heritability and Development of Behavioral Types in Lizards

The heritability of personality traits in lizards was estimated in three studies. Bajer et al. (2015) estimated narrow-sense heritabilities using father–offspring regressions and found no significant heritability for risk-taking and exploratory behaviors in European green lizards (*Lacerta viridis*). However, they did find that their risk-taking and exploration measure were repeatable. Le Galliard et al. (2013), using linear mixed models with maternal identity as a factor, estimated a broad-sense heritability of 0.32 for exploratory behavior in common lizards. Finally, Teyssier et al. (2014) used microsatellite markers to identify the paternity of common lizards born in captivity and then estimated the narrow-sense heritability of activity as 0.41 using a restricted maximum likelihood animal model. They also found that mid-offspring activity was significantly related to the fathers' activity, but not the mothers'. Overall, these studies suggest that exploration and activity are heritable in at least one species of lizard and may be most strongly influenced by the father.

Studies have explored how personality traits/behavioral syndromes change during the course of an individual's life cycle (Bajer et al. 2015; Le Galliard et al. 2015), and are linked to fitness (Bajer et al. 2015), life history (Cote and Clobert 2007), and physiological traits (Le Galliard et al. 2013). For example, Bajer et al. (2015) studied personality in European green lizards in relation to body size and external parasite load at different ontogenetic stages. They found that risk-taking and exploratory behaviors were repeatable within both adult and juvenile individuals, and therefore, may be considered personality traits. However, these behaviors were significantly more repeatable in juveniles (Spearman's rho between 0.60 and 0.84) than in adults (Spearman's rho between 0.23 and 0.63), suggesting an ontogenetic effect, such as maturation or learning, acting on the behaviors. Interestingly, they found significant correlations only between risk-taking behaviors and exploration in adults (Spearman's rho between 0.34 and 0.49) suggesting the existence of a behavioral syndrome in adults, but not juveniles. Bajer et al. (2015) conclude that European green lizards have personalities that are innate, and that behavioral syndromes may develop through experiences affecting different behaviors in concert. These personalities were associated with fitness measures, including body size and parasite load: adult lizards that were large or had low parasite loads were more exploratory than lizards that were small or had a large parasite load. Bajer et al. (2015) interpreted this relationship as either the linking of personality and fitness during development or as the result of the survival of specific phenotypes in the face of environmental challenges, such as parasite load.

Evidence for Hormone-Dependent Personality Trait Expression

Although the relationships between personality traits and levels of hormones have been reported in other taxa, there have been few studies examining this association in lizards. An exception was a study by While et al. (2010) who examined the effect of testosterone on aggression. They found that aggression in male and female White's skink (*Egernia whitii*) were repeatable and stable over 18-months, as were baseline testosterone levels. In addition, a negative correlation existed between aggression and testosterone concentrations in males, but they were not correlated in females. Of particular interest was the finding that the association between testosterone and aggression is context specific and decoupled across the sexes. In male lizards, testosterone and aggression were positively correlated during the mating season, but became negatively correlated towards the end of the mating season. No such change in this association was observed in females. Thaker et al. (2009) provided indirect support of the association between hormone levels and boldness. In a study of the ornate tree lizard (*Urosaurus ornatus*) these authors found that corticosterone levels were positively correlated with boldness measures, such as flight initiation distance and time spent in a refuge when exposed to a live collared lizard (*Crotaphytus nebrius*), a natural predator of small lizards. Unfortunately, these authors did not take repeated measurements of these behaviors and, therefore, they did not document personalities or behavioral syndromes. Similarly, Vitousek and Romero (2013) found that the number of males assessed by female Galapagos marine iguanas (*Amblyrhynchus cristatus*) was negatively correlated with corticosterone levels. Together, these studies suggest that, in lizards, hormone levels are associated with personality, although the underlying mechanisms still need to be elucidated.

Research Programs Studying Personality in Lizards

Namibian Rock Agama

In a series of studies, Carter and colleagues (Carter et al. 2010, 2012a, b; Highcock and Carter 2014) studied personality and behavior syndromes in the Namibian rock agama. The individuals they studied showed consistency in flight initiation distance (their metric for boldness), and they identified a behavioral syndrome in which bolder individuals tended to spend more time basking and moving around their environment (i.e., they were more conspicuous) than their shyer counterparts (Carter et al. 2010). This behavioral syndrome was associated with fitness trade-offs. Although bolder individuals benefited from having larger home ranges and feeding more frequently, they incurred a cost of tail loss due to greater predation risk. Carter et al. (2012a) expanded on their previous study (Carter et al. 2010) and investigated whether personality may exhibit plasticity in response to environmental variability (dry versus wet season). Although they found no evidence for plasticity in flight initiation distance, they did find that the time lizards spent conspicuous was dependent on season and that there was an interaction between individual and season. Given these observations, they concluded that, in predictable environments, selection for plasticity over consistency may be favored. Highcock and Carter (2014) studied intraindividual variability of boldness (flight initiation distance was again used as the metric of boldness) in the same system. They found that individual differences in intraindividual variability were repeatable across seasons and suggest that intraindividual variation could therefore be considered a distinct trait. Additionally, there was a strong correlation between intraindividual variation and boldness. Shy individuals had lower intraindividual variability in boldness than bolder individuals. Highcock and Carter (2014) suggest that, because bolder individuals are at greater risk of predation, individuals with variable flight initiation distances are more likely to escape predators because their escape behavior is less predictable. Furthermore, Highcock and Carter (2014) argued that, given the intraindividual variability in behavioral responses, such as flight initiation distance, the use of a single measure to characterize such behaviors is inadequate. They proposed that behaviors with intraindividual variability would be better represented by a matrix of values that indicate personality, degree of behavioral plasticity and variability of the behavior.

Common Lizard

Extensive work on behavioral types has been done with the common lizard, which occurs throughout most of Europe and the north of Asia. Here we highlight the different topics that have been studied in relation to personality.

Personality, Population Density and Fitness The likelihood of dispersal from natal areas in response to increasing density may increase or decrease. Some individuals may be more likely to disperse with increased density (dispersers) whereas others may be more likely to remain (non-dispersers). Dispersers and non-dispersers also differ in their morphology, physiology, and in their response to olfactory cues (Cote and Clobert 2007). Given these consistent differences, Cote and Clobert (2007) hypothesized that dispersing and non-dispersing individuals might differ in sociability, with dispersers from high density populations being asocial and dispersers from low-density populations being social. To test this hypothesis, they measured the time that lizards associated with conspecific male odor (a measure of social tolerance) at birth and one year later. They found that individual responses were consistent over time and independent of population and density. Significantly, Cote and Clobert (2007) found that the social personality of juveniles determined their dispersal response at different population densities. Social individuals were more likely to become resident in dense populations. They concluded that social personality influences dispersal decisions in common lizards. In a follow-up study, Cote et al. (2008) investigated the fitness consequences of maintaining variability in sociability across population densities. At birth, sociability was independent of body length and body mass, but was positively correlated with boldness. Following hibernation, the probability of survival was negatively correlated with sociability in low-density populations, and growth rates were positively correlated with sociability at both density levels. Furthermore, female reproductive success was positively correlated with sociability, but not density. Cote et al. (2008) concluded from these results that the increased survival of asocial lizards at low density was due to their being less likely to engage in aggressive interactions with adult males, thus avoiding bites and stress, while facing less competition for resources. On the other hand, social females had higher fecundity than asocial females regardless of population density, because they interacted more with males. The maintenance of social personality types in this species may therefore be due to the spatiotemporal variation in environmental conditions faced (Cote et al. 2008).

Le Galliard et al. (2015) expanded on the work of Cote and colleagues by examining density-dependent selection on activity, boldness, and sociability for juvenile survival, body growth rate, and reproduction, these being life-history traits. They found consistent individual differences for all behaviors in the short term, but within a year, behavioral consistency had weakened and was significant only for measures of activity and boldness. Behavioral consistency in sociability was lost by the end of the first year, which contrasts with the findings of Cote and Clobert (2007). Le Galliard et al.'s. (2015) results suggest significant ontogenetic changes in ranking of behavioral scores during the first year of life, which, they note, runs counter to the assumption that personality trait levels remain consistent over time. They also found density-dependent selection on sociability for growth rate, and on level of activity at birth for juvenile survival, but not for boldness. For example, sociability was positively correlated with growth rates at low densities, but not at high densities, and activity level was negatively correlated with juvenile survival at low densities. Le Galliard et al. (2015) concluded that density-dependent selection is important in the maintenance of individual differences in personality traits, such as exploration activity and sociability.

Female Mate Preference Based on Male Personality and Female Predation Risk

Teyssier et al. (2014) studied the role of male and female personality traits on mating behavior and mate choice in female common lizards under different levels of predation risk to the female. In their study, the odor from green whip snakes (*Hierophis viridiflavus*) was used as a predatory stimulus. Activity and sociability showed moderate to high behavioral consistency, while exploration did not. They found that the probability of mating depended on female predation risk and male activity levels. Females that were not exposed to high predation risk preferred more active males, whereas females that had been exposed to predation before mating preferred less active males. Teyssier et al. (2014) suggest that, although high activity in males increases competitive advantage and survival, it may be detrimental when predator risk is high. Therefore, a female's preference for males with a specific personality may be modified by the environmental context in which they find themselves (i.e., high versus low predation levels).

Personality and the Pace-of-Life Syndrome

Réale et al. (2010) expanded the pace-of-life hypothesis to include an animal's personality in addition to its physiology and life history; the pace-of-life syndrome. Le Galliard et al. (2013) studied newborn common lizards to investigate the relationship between exploratory behavior, resting metabolic rate, and locomotor performance. According to the pace-of-life syndrome hypothesis, exploratory behavior should be positively correlated with resting metabolic rate and locomotor performance (Le Galliard et al. 2013). Although their study provided evidence for consistent individual differences in exploratory behavior, support for the pace-of-life hypothesis was mixed. Resting metabolic rate and locomotor performance were not correlated with exploratory behavior in juveniles or yearlings. Although Le Galliard et al. (2013) found evidence for significant correlations between growth rate and offspring survival involving exploration and resting metabolic rate, the direction of those correlations was opposite to that predicted by the pace-of-life hypothesis. Lizards were most likely to survive their first year of life if they had high resting metabolic rate and low exploration, or low resting metabolic rate and high exploration. Furthermore, lizards with low exploration at birth exhibited a positive correlation between resting metabolic rate and growth, while lizards with high exploration exhibited a negative correlation between resting metabolic rate and growth. Le Galliard et al. (2013) suggested that these results may reflect a trade-off between investment in body maintenance (indicated by resting metabolic rate) and energy invested in exploration, or that high exploration represents thoroughness of exploration and such lizards have a slower pace-of-life, as indicated by low relative metabolic rate.

Mell et al. (2016) built on Le Galliard et al.'s (2013) work on personality and pace-of-life syndrome by looking for correlations among the various activity, aggressiveness, risk-taking, and sociability, and physiological variables, including standard metabolic rate and glucocorticoid response. Although they found significant short-term stability in activity, aggressiveness, and risk-taking, but not sociability, in adult common lizards, they found only weak or nonsignificant correlations between these behaviors. They point out that independence between personality traits and a lack of a behavioral syndrome has been documented in other studies on lizards (Le Galliard et al. 2015; McEvoy et al. 2015). In addition, although they found that metabolic rate was negatively correlated with sociability and activity, it was not correlated with risk-taking. They concluded that the link between inter-individual variation in physiology and behavior was trait-dependent in common lizards, suggesting that these behaviors may be free to evolve independently of physiology.

Correlated Variation in Morphology and Behavior

As mentioned earlier, strong flight escape in snakes is often associated with body shape and patterns, such as striping. In lizards, where sexual pattern and coloration dimorphisms are more pronounced, more socially relevant behavioral variation exists. In some lizard species (for example, the tree lizard, *Urosaurus ornatus*, and side-blotched lizards, *Uta stansburiana*), polymorphisms and correlated behavioral variation occur between individuals within one sex (typically males) and are associated with alternative reproductive strategies (Hews et al. 1994; Sinervo and Lively 1996). Sinervo and Lively (1996) developed an evolutionary stable strategy model to explain the maintenance of such alternative male reproductive strategies. Some populations of side-blotched lizard have three distinct male reproductive strategy phenotypes. Males with orange throats are more aggressive and defend larger territories, while dark blue-throated males are less aggressive, defend smaller territories, and exhibit mate guarding. The last phenotypic type, yellow throated individuals do not defend territories and are sneakers, mimicking female behavior and morphology. Sinervo and Lively (1996) proposed that these male reproductive phenotypes were maintained in a rock–paper–scissors game in which each phenotype has an advantage over one of the others, causing the frequency of each to oscillate from year to year. Furthermore, these phenotypic differences appear to be due to the effects of steroid hormones, primarily progesterone (Moore et al. 1998) and testosterone (Hews et al. 1994; Hews and Moore 1996; Sinervo et al. 2000). The discovery of individuals that can transition between two reproductive phenotypes added an interesting dimension to this system. In the studies of side-blotched lizard, yellow throated individuals sometimes transitioned into blue-throated individuals within the breeding season (Sinervo et al. 2000). This transition typically occurred when a yellow throated lizard occupied a territory vacated by an orange throated male (normally through death) and coincided with an elevation in plasma testosterone levels. Yellow throated males who transitioned to blue-throated males no longer engaged in female mimicry or sneaker behavior (Sinervo et al. 2000).

A similar system is found in the tree lizard, *Urosaurus ornatus*, in which several alternative male reproductive phenotypes have also been identified. Individuals with a blue dewlap (Thompson and Moore 1991) or an orange dewlap with a blue spot (Knapp et al. 2003) tend to be territorial and highly aggressive, whereas individuals with an orange dewlap are typically less aggressive, and nomadic (Knapp et al. 2003; Thompson and Moore 1991). A third phenotype, with a yellow dewlap may also occur. Individuals with a yellow dewlap are like the orange morph in their level of aggressiveness, but employ a satellite tactic in association with blue males (Taylor and Lattanzio,2016). Plasticity in reproductive strategy has also been demonstrated in tree lizards. Knapp et al. (2003) found that in drier years, orange males tended to be nomadic, whereas in wetter years they were more territorial. This shift in response to seasonal weather patterns was also correlated with differences in testosterone levels. During dry years, nomadic males had lower testosterone levels than territorial males. In both species of lizard (Knapp et al. 2003; Sinervo et al. 2000), behavioral change occurred in response to changes in the level of interaction with the more aggressive phenotype and the effects of hormones on transitional (plastic) reproductive strategies (Knapp et al. 2003; Moore et al. 1998; Sinervo et al. 2000).

Recently, Taylor and Lattanzio (2016) reported that, in tree lizards, different reproductive phenotypes also differed in their behavioral response to a simulated predator attack. They found that yellow individuals were bolder (less likely to flee or have shorter flight initiation distances) than blue individuals, even though the blue individuals were characteristically more dominant during paired interactions. Taylor and Lattanzio (2016) suggest that the interplay between dominance, aggression and boldness is likely environmentally dependent and may explain both ecological and phenotypic variation in the species.

Studies of Turtles

Although not well represented in the literature, there has been work on individuality in turtles. The righting response of neonates has been the most frequently studied behavior, and has been used as a measure of fitness. Although several measures have been used to represent the righting response, individual differences in righting have been reported in some studies (Carter et al. 2016; Delmas et al. 2007; Ibáñez et al. 2013, 2014), but not others (e.g., Davy et al. 2014). Additionally, righting has been linked to temperament, if not personality.

Ibáñez et al. (2013) studied the effect of boldness on conspecific avoidance in male Spanish terrapin (*Mauremys leprosa*). Boldness was determined by measuring the time an individual spent motionless and protected by their shells before righting themselves. Ibáñez et al. (2013) predicted that bolder males would spend more time in pools with familiar conspecific odors than in pools with unfamiliar males, thus reducing the risk of an antagonist encounter. On the other hand, shy turtles would reduce the risk of antagonist encounters by avoiding chemicals of both familiar, but dominant males, and unfamiliar conspecifics. Ibáñez et al.'s (2013) experiments supported these predictions and they concluded that behavioral type (along the shy-bold axis) may influence male–male interactions both during and following agonistic interactions and the outcome of these interactions then determined how individuals responded to familiar and unfamiliar conspecific chemicals.

Carter et al. (2016) investigated the effect of estrone sulfate, corticosterone, and thermal fluctuations during incubation on righting response and exploratory behavior in the red-eared slider (*Trachemys scripta elegans*). Although these treatments did not affect righting responses or latencies to begin exploration in neonates, these behaviors were repeatable and positively correlated with each other. Individuals that quickly righted themselves were also quicker to engage in exploratory behavior. However, these behavioral types did not predict neonate dispersal time and survival in naturalistic settings. This finding is contrary to the interpretation that righting speed reflects dispersal time and subsequent survival (e.g., Delmas et al. 2007), and Carter et al. (2016) urge caution about using the righting response as a proxy for dispersal ability.

On a different note, two studies found evidence for behavioral individuality in food competition (Froese and Burghardt 1974) and food acquisition (Davis and Burghardt 2007). Froese and Burghardt (1974) studied the formation of food-competitive hierarchies in snapping turtles, *Chelydra serpentina*. In a series of trials, they paired turtles together in an arena and allowed them to compete for a single food item (a piece of carp). The turtle that successfully consumed the food was the winner. By calculating the ratio of total food wins to total food losses for each turtle, Froese and Burghardt (1974) constructed a food dominance hierarchy. Importantly, they found that this hierarchy was stable over time, as is true of many animals living in social groups. This study suggests stable social hierarchies may be indicators of individuality as well as individual recognition, especially when not confounded with size, age, or sex. In a more recent study from the same laboratory, Davis and Burghardt (2007) reported that identically reared red-bellied turtles (*Pseudemys nelsoni*) had different strategies in a learned food acquisition task. Turtles were trained to emerge from the water and displace a bottle covering a barely visible food item versus a bottle not containing food. Some animals were "impulsive" and made decisions rapidly, but with frequent errors, while others were "deliberate" and made the choice decision more slowly but were more successful in selecting the correct bottle. Taken together, these studies suggest that, like snakes and lizards, turtles exhibit behavioral, even psychological, individuality.

Studies of Crocodilians

As with turtles, there have been few studies of personality in crocodilians. Still, similarity in the expression of aggression in neonates and adults suggests that agonistic behaviors are innate. Although common in multiple species, variation in the type and level of antagonistic behavior across species may reflect divergence in morphology, ecology, and life history (Brien et al. 2013a). In studies of aggression in freshwater (Brien et al. 2013b) and saltwater (Brien et al. 2013c) crocodiles, individuals that participate in an encounter were classified as aggressive or nonaggressive. During an antagonistic encounter either one or both individuals may be aggressive. Unfortunately, the studies did not specify whether individuals were consistent in their level of aggression during these encounters or whether aggression levels differed between encounters. However, in hatchling saltwater crocodiles, clutch differences in the frequency of antagonistic interactions were reported. Zoo and fieldworkers engaged with adult crocodilians often report great consistent individuality in their behavior (Burghardt, pers. obs.).

Conclusions

Although studies on individuality, temperament, and personality in reptiles have not figured prominently in the personality literature, they provide intriguing, perhaps crucially important, material for our attempts to understand the development, evolution, mechanisms, and adaptive function of these phenomena. This is because the diversity within even closely related reptilian taxa dwarfs that of birds and mammals. Viviparous and oviparous species are common in snakes and lizards, sometimes within the same genus and species. The often large clutches and litters make experimental studies tractable and the lack of parental care in many, but not all, species, can facilitate developmental studies by eliminating confounding effects on neonatal behavior due to postnatal care, especially when offspring are reared together in clutches or litters. For example, parent birds and siblings can differentially respond to nestlings in ways that may affect later behavioral syndromes. On the other hand, the variation in their sex determination mechanisms allows exploration of topics not addressed in mammal and avian studies. Squamate reptiles can both lay eggs and give live birth. Sex determination can also be temperature dependent as well as genetic. Recently it has been found that bearded dragon lizards (*Pogona vitticeps*), which normally have genetic sex determination, can show sex reversal when eggs with the male (ZZ) genotype are incubated at high temperatures and become permanent functional females, a phenomenon found in the field and not just the laboratory. These females can actually produce more offspring than genetic females, but also show more male-like morphology and behavior (Li et al. 2016). Furthermore, they found that sex reversed females differ from sex concordant males and females in showing greatly increased exploration, boldness, and neophobia in some measures, which may have both adaptive and deleterious consequences for fitness depending on ecological conditions such as level of predation. The authors relate their findings to evolutionary trajectories that, of course, include climate change. Individual difference and personality research in reptiles can thus have far reaching effects and can identify potential phenomena of great importance less evident in other taxa and whose effects are not yet appreciated.

We also want to emphasize the need to use blind coding and/or rater reliability measures (Burghardt et al. 2012) in reptile studies and also the need to do studies in both field and captive conditions whenever possible, as both have strengths and weaknesses (Snowdon and Burghardt 2017). This chapter's goal was to introduce the rich variety of intriguing findings and opportunities that studies of reptile individuality offer to researchers interested in comparative studies of animal personality and individuality.

Acknowledgments We thank the many students and colleagues who have contributed to the studies reported here, and Hal Herzog and the editors for review and comments on early drafts. Much of our research was support by decades of grant support from the National Science Foundation, National Institutes Health (NIMH, NICHD), the University of Tennessee, and others.

References

Amarello, M. (2012). *Social snakes? Non-random association patterns detected in a population of Arizona black rattlesnakes (Crotalus cerebrus)*. Unpublished thesis. Arizona State University, Tempe, Arizona.

Arnold, S. J., & Bennett, A. F. (1984). Behavioural variation in natural populations. III: Antipredator displays in the gartersnake *Thamnophis radix*. *Animal Behaviour, 32,* 1108–1118. doi:10.1016/S0003-3472(84)80227-4

Bajer, K., Horváth, G., Molnár, O., Török, J., Garamszegi, L. Z., & Herczeg, G. (2015). European green lizard (*Lacerta viridis*) personalities: Linking behavioural types to ecologically relevant traits at different ontogenetic stages. *Behavioural Processes, 111,* 67–74. doi:10.1016/j.beproc.2014.11.020

Bell, A. M., Hankison, S. J., & Laskowski, K. L. (2009). The repeatability of behaviour: A meta-analysis. *Animal Behaviour, 77,* 771–783. doi:10.1016/j.anbehav.20

Bowers, B. B. (1992). *Habituation of antipredator behaviors and responses to chemical prey extracts in four species of gartersnakes, Thamnophis (Serpentes: Colubridae)*. Unpublished dissertation. University of Tennessee. Knoxville, Tennessee.

Bowers, B. B., Bledsoe, A. E., & Burghardt, G. M. (1993). Responses to escalating predatory threat in garter and ribbon snakes (*Thamnophis*). *Journal of Comparative Psychology, 107,* 25–33. doi:10.1037//0735-7036.107.1.25

Bowers, B. B., & Burghardt, G. M. (1992). The scientist and the snake: Relationships with reptiles. In H. Davis & D. Balfour (Eds.), *The inevitable bond: Examining scientist–animal interactions* (pp. 250–263). Cambridge, England: Cambridge University Press.

Brien, M. L., Lang, J. W., Webb, G. J., Stevenson, C., & Christian, K. A. (2013a). The good, the bad, and the ugly: Agonistic behaviour in juvenile crocodilians. *PLoS ONE, 8,* e80872. doi:10.1371/journal.pone.0080872

Brien, M. L., Webb, G. J., Lang, J. W., & Christian, K. A. (2013b). Intra- and interspecific agonistic behaviour in hatchling Australian freshwater crocodiles (*Crocodylus johnstoni*) and saltwater crocodiles (*Crocodylus porosus*). *Australian Journal of Zoology, 61,* 196. doi:10.1071/zo13035

Brien, M. L., Webb, G. J., Lang, J. W., McGuinness, K. A., & Christian, K. A. (2013c). Born to be bad: Agonistic behaviour in hatchling saltwater crocodiles (*Crocodylus porosus*). *Behaviour, 150,* 737–762. doi:10.1163/1568539x-00003078

Brodie, E. D., III. (1989). Genetic correlations between morphology and antipredator behaviour in natural populations of the gartersnake *Thamnophis ordinoides*. *Nature, 342,* 542–543. doi:10.1038/342542a0

Brodie, E. D., III. (1992). Correlational selection for color pattern and antipredator behavior in the gartersnake *Thamnophis ordinoides*. *Evolution, 46,* 1284–1298. doi:10.2307/2409937

Brodie, E. D., III. (1993a). Consistency of individual differences in anti-predator behaviour and colour pattern in gartersnake, *Thamnophis ordinoides*. *Animal Behaviour, 45,* 851–861. doi:10.1006/anbe.1993.1106

Brodie, E. D., III. (1993b). Homogeneity of the genetic variance-covariance matrix for antipredator traits in two natural populations of the gartersnake *Thamnophis ordinoides*. *Evolution, 47,* 844–854. doi:10.2307/2410188

Brodie, E. D., III, & Russell, N. H. (1999). The consistency of individual differences in behaviour: Temperature effects on antipredator behaviour in gartersnakes. *Animal Behaviour, 57,* 445–451. doi:10.1006/anbe.1998.0990

Burger, J., & Gochfeld, M. (1990). Risk discrimination of direct versus tangential approach by basking black iguanas (*Ctenosaura similis*): Variation as a function of human exposure. *Journal of Comparative Psychology, 104,* 388–394. doi:10.1037/0735-7036.104.4.388

Burghardt, G. M. (1975). Chemical prey preference polymorphism in newborn gartersnakes *Thamnophis sirtalis*. *Behaviour, 52,* 202–225. doi:10.2307/4533647

Burghardt, G. M. (1983). Aggregation and species discrimination in newborn snakes. *Zeitschrift für Tierpsychologie, 61,* 89–101. doi:10.1111/j.1439-0310.1983.tb01330.x

Burghardt, G. M. (1991). Cognitive ethology and critical anthropomorphism: A snake with two heads and hognose snakes that play dead. In C. A. Ristau (Ed.), *Cognitive ethology: The minds of other animals: Essays in honor of Donald R. Griffin*. Hillsdale, NJ: Lawrence Erlbaum Associates Inc.

Burghardt, G. M. (1993). The comparative imperative: Genetics and ontogeny of chemoreceptive prey responses in natricine snakes. *Brain, Behavior and Evolution, 41,* 138–146. doi:10.1159/000113831

Burghardt, G. M. (2000a). Staying close. In M. Bekoff (Ed.), *The smile of the dolphin: Remarkable accounts of animal emotions* (pp. 162–165). New York, NY: Discovery Books.

Burghardt, G. M. (2000b). Temperature and temperament. In M. Bekoff (Ed.), *The smile of the dolphin: Remarkable accounts of animal emotions* (pp. 202–203). New York, NY: Discovery Books.

Burghardt, G. M., Bartmess-LeVasseur, J. N., Browning, S. A., Morrison, K. E., Stec, C. L., Zachau, C. E., et al. (2012). Perspectives—Minimizing observer bias in behavioral studies: A review and recommendations. *Ethology, 118,* 511–517. doi:10.1111/j.1439-0310.2012.02040.x

Burghardt, G. M., Layne, D. G., & Konigsberg, L. (2000). The genetics of dietary experience in a restricted natural population. *Psychological Science, 11,* 69–72. doi:10.1111/1467-9280.00217

Burghardt, G. M., & Schwartz, J. M. (1999). Geographic variations on methodological themes in comparative ethology: A natricine snake perspective. In S. A. Foster & J. A. Endler (Eds.), *Geographic variation in behavior: Perspectives on evolutionary mechanisms* (pp. 69–94). Oxford, England: Oxford University Press.

Careau, V., & Garland, T., Jr. (2012). Performance, personality, and energetics: Correlation, causation, and mechanism. *Physiological and Biochemical Zoology, 85,* 543–571. doi:10.1086/666970

Carpenter, C. C., & Ferguson, G. W. (1977). Variation and evolution of stereotyped behavior in reptiles. In C. Gans & D. W. Tinkle (Eds.), *Biology of the reptilia* (Vol. 7, pp. 335–554). New York: Academic Press.

Carter, A. J., Goldizen, A. W., & Heinsohn, R. (2012a). Personality and plasticity: Temporal behavioural reaction norms in a lizard, the Namibian rock agama. *Animal Behaviour, 84,* 471–477. doi:10.1016/j.anbehav.2012.06.001

Carter, A. J., Goldizen, A. W., & Tromp, S. A. (2010). Agamas exhibit behavioral syndromes: Bolder males bask and feed more but may suffer higher predation. *Behavioral Ecology, 21,* 655–661. doi:10.1093/beheco/arq036

Carter, A. J., Heinsohn, R., Goldizen, A. W., & Biro, P. A. (2012b). Boldness, trappability and sampling bias in wild lizards. *Animal Behaviour, 83,* 1051–1058. doi:10.1016/j.anbehav.2012.01.033

Carter, A. W., Paitz, R. T., McGhee, K. E., & Bowden, R. M. (2016). Turtle hatchlings show behavioral types that are robust to developmental manipulations. *Physiology & Behavior, 155,* 46–55. doi:10.1016/j.physbeh.2015.11.034

Chiszar, D., & Carter, T. (1975). Reliability of individual differences between gartersnakes (*Thamnophis radix*) during repeated exposures to an open field. *Bulletin of the Psychonomic Society, 5,* 507–509. doi:10.3758/BF03333315

Citadini, J. M., & Navas, C. A. (2013). Inter-individual variation and temperature-dependent antipredator behavior in the snake *Tomodon dorsatus* (Dipsadidae). *Behavioral Processes, 97,* 11–17. doi:10.1016/j.beproc.2013.03.008

Cooper, W. E., Jr. (1997a). Factors affecting risk and cost of escape by the broad-headed skink (*Eumeces laticeps*): Predator speed, directness of approach, and female presence. *Herpetologica, 53,* 464–474. http://www.jstor.org/stable/3893260

Cooper, W. E., Jr. (1997b). Threat factors affecting antipredatory behavior in the broad-headed skink (*Eumeces laticeps*): Repeated approach, change in predator path, and predator's field of view. *Copeia,* 613–619. doi:10.2307/1447569

Cooper, W. E., Jr. (2009). Variation in escape behavior among individuals of the striped plateau lizard *Sceloporus virgatus* may reflect differences in boldness. *Journal of Herpetology, 43,* 495–502. doi:10.1670/08-197R1.1

Cooper, W. E., Jr. (2012). Risk, escape from ambush, and hiding in the lizard, *Sceloporus virgatus*. *Herpetologica, 68,* 505–513. doi:10.1655/HERPETOLOGICA-D-11-00069

Cooper, W. E., Jr., & Burghardt, G. M. (1990). Vomerolfaction and vomodor. *Journal of Chemical Ecology, 16,* 103–105. doi:10.1007/BF01021271

Cote, J., & Clobert, J. (2007). Social personalities influence natal dispersal in a lizard. *Proceedings of the Royal Society B, 274,* 383–390. doi:10.1098/rspb.2006.3734

Cote, J., Dreiss, A., & Clobert, J. (2008). Social personality trait and fitness. *Proceedings of the Royal Society B, 275,* 2851–2858. doi:10.1098/rspb.2008.0783

Davis, K. M., & Burghardt, G. M. (2007). Training and long-term memory of a novel food acquisition task in a turtle (*Pseudemys nelsoni*). *Behavioral Processes, 75,* 225–230. doi:10.1016/j.beproc.2007.02.021

Davy, C. M., Paterson, J. E., & Leifso, A. E. (2014). When righting is wrong: Performance measures require rank repeatability for estimates of individual fitness. *Animal Behaviour, 93,* 15–23. doi:10.1016/j.anbehav.2014.04.013

Delmas, V., Baudry, E., Girondot, M., & Prevot-Julliard, A.-C. (2007). The righting response as a fitness index in freshwater turtles. *Biological Journal of the Linnean Society, 91,* 99–109. doi:10.1111/j.1095-8312.2007.00780.x

Dugan, B. A. (1982a). A field study of the headbob displays of male green iguanas (*Iguana iguana*): Variation in form and context. *Animal Behaviour, 30,* 327–338. doi:10.1016/S0003-3472(82)80043-2

Dugan, B. A. (1982b). The mating behavior of the green iguana, *Iguana iguana*. In G. M. Burghardt & A. S. Rand (Eds.), *Iguanas of the world: Their behavior, ecology, and conservation* (pp. 320–341). Park Ridge, NJ: Noyes Publications.

Dutton, D., & Anderson, M. (2002). Personality in royal pythons and the human–animal relationship. *Anthrozoos, 15,* 243–250.

Foster, M. S. (2012). Standard techniques for inventory and monitoring. In R. W. McDiarmid, M. S. Foster, C. Guyer, W. Gibbond, & N. Chernoff (Eds.), *Reptile biodiversity: Standard methods for inventory and monitoring* (pp. 205–271). Berkley, CA: University California Press.

Froese, A. D., & Burghardt, G. M. (1974). Food competition in captive juvenile snapping turtles, *Chelydra serpentina*. *Animal Behaviour, 22,* 735–740. doi:10.1016/S0003-3472(74)80025-4

Gangloff, E. J., Sparkman, A. M., & Bronikowski, A. M. (in press). Individual heterogeneity in behaviour and physiology affects fitness in a model ectothermic amniote, the gartersnake, *Thamnophis elegans*. *Oikos*.

Garland, T., Jr. (1988) Genetic basis of activity metabolism. I. Inheritance of speed, stamina, and antipredator displays in the gartersnake *Thamnophis sirtalis*. *Evolution, 42,* 335–350. doi:10.2307/2409237

Garland, T., Jr. (1994). Quantitative genetics of locomotor behavior and physiology in a gartersnake. In C. R. Boake, (Ed.), *Quantitative genetic studies of behavioral evolution* (pp. 251–277). Chicago: University of Chicago Press.

Godfrey, S. S., Bradley, J. K., Sih, A., & Bull, C. M. (2012). Lovers and fighters in sleepy lizard land: Where do aggressive males fit in a social network? *Animal Behaviour, 83,* 209–215. doi:10.1016/j.anbehav.2011.10.028

Gosling, S. D. (2001). From mice to men: What can we learn about personality from animal research? *Psychological Bulletin, 127,* 45–86. doi:10.1037/0033-2909.127.1.45

Greene, H. W. (1979). Behavioral convergence in the defensive displays of snakes. *Experientia, 35,* 747–748. doi:10.1007/BF01968221

Greene, H. W. (1988). Antipredator mechanisms in reptiles. In C. Gans & R. B. Huey (Eds.), *Biology of the reptilia* (Vol. 16, pp. 1–152). New York: Alan R. Liss Inc.

Halpern, M., & Frumin, N. (1979). Roles of the vomeronasal and olfactory systems in prey attack and feeding in adult gartersnakes *Thamnophis sirtalis parietalis*. *Physiology & Behavior, 22,* 1183–1190. doi:10.1016/0031-9384(79)90274-9

Herzog, H. A., Jr. (1990). Experiential modification of defensive behaviors in gartersnakes (*Thamnophis sirtalis*). *Journal of Comparative Psychology, 104,* 387–389. doi:10.1037/0735-7036.104.4.334

Herzog, H. A., Jr., & Bailey, B. D. (1987). Development of antipredator responses in snakes: II. Effects of recent feeding on defensive behaviors of juvenile gartersnakes (*Thamnophis sirtalis*). *Journal of Comparative Psychology, 101,* 387–389. doi:10.1037/0735-7036.101.4.387

Herzog, H. A., Jr., Bowers, B. B., & Burghardt, G. M. (1989a). Development of antipredator responses in snakes: IV. Interspecific and intraspecific differences in habituation of defensive behavior. *Developmental Psychobiology, 22,* 489–508. doi:10.1002/dev.420220507

Herzog, H. A., Jr., Bowers, B. B., & Burghardt, G. M. (1989b). Stimulus control of antipredator behavior in newborn and juvenile gartersnakes (*Thamnophis*).

Journal of Comparative Psychology, 103, 233–242. doi:10.1037//0735-7036.103.3.233
Herzog, H. A., Jr., Bowers, B. B., & Burghardt, G. M. (1992). Development of antipredator responses in snakes: V. Species differences in ontogenetic trajectories. *Developmental Psychobiology, 25,* 199–211. doi:10.1002/dev.420250305
Herzog, H. A., Jr., & Burgardt, G. M. (1986). Development of antipredator responses in snakes: I. Defensive and open-field behaviors in newborns and adults of three species of gartersnakes (*Thamnophis melanogaster, T. sirtalis, T. butleri*). *Journal of Comparative Psychology, 100,* 372–379. doi:10.1037/0735-7036.100.4.372
Herzog, H. A., Jr., & Burghardt, G. M. (1988). Development of antipredator responses in snakes: III. Stability of individual and litter differences over the first year of life. *Ethology, 77,* 250–258. doi:10.1111/j.1439-0310.1988.tb00208.x
Herzog, H. A., Jr., & Schwartz, J. M. (1990). Geographic variation in the antipredator behaviour of neonate gartersnakes (*Thamnophis s sirtalis*). *Animal Behaviour, 40,* 597-598. doi:10.1016/S0003-3472(05)80541-X
Hews, D. K., Knapp, R., & Moore, M. C. (1994). Early exposure to androgens affects adult expression of alternative male types in tree lizards. *Hormones and Behavior, 28,* 96–115. doi:10.1006/hbeh.1994.1008
Hews, D. K., & Moore, M. C. (1996). A critical period for the organization of alternative male phenotypes of tree lizards by exogenous testosterone? *Physiology & Behavior, 60,* 425–429. doi:10.1016/S0031-9384(96)80014-X
Highcock, L., & Carter, A. J. (2014). Intraindividual variability of boldness is repeatable across contexts in a wild lizard. *Plos One, 9.* doi:10.1371/journal.pone.0095179.g001
Hotton, N., III, Maclean, P. D., Roth, J. J., & Roth, E. C. (Eds.). (1986). *The ecology and biology of mammal-like reptiles*. Washington, DC: Smithsonian Institution Press.
Ibáñez, A., López, P., Martín, J., & Wright, J. (2014). Inter-individual variation in antipredator hiding behavior of spanish terrapins depends on sex, size, and coloration. *Ethology, 120,* 742–752. doi:10.1111/eth.12245
Ibáñez, A., Marzal, A., López, P., & Martín, J. (2013). Boldness and body size of male Spanish terrapins affect their responses to chemical cues of familiar and unfamiliar males. *Behavioral Ecology and Sociobiology, 67,* 541–548. doi:10.1007/s00265-012-1473-6
King, R. B. (2002). Family, sex and testosterone effects on gartersnake behaviour. *Animal Behaviour, 64,* 345–359. doi:10.1006/anbe.2002.3046
Knapp, R., Hews, D. K., Thompson, C. W., Ray, L. E., & Moore, M. C. (2003). Environmental and endocrine correlates of tactic switching by nonterritorial male tree lizards (*Urosaurus ornatus*). *Hormones and Behavior, 43,* 83–92. doi:10.1016/s0018-506x(02)00018-1
Konečná, M., Weiss, A., Lhota, S., & Wallner, B. (2012). Personality in Barbary macaques (*Macaca sylvanus*): Temporal stability and social rank. *Journal of Research in Personality, 46,* 581–590. doi:10.1016/j.jrp.2012.06.004
Kubie, J. L., & Halpern, M. (1979). Chemical senses involved in gartersnake prey trailing. *Journal of Comparative and Physiological Psychology, 93,* 648–667.
Kuo, C.-Y., Irschick, D. J., Lailvaux, S. P., & Van Damme, R. (2015). Trait compensation between boldness and the propensity for tail autotomy under different food availabilities in similarly aged brown anole lizards. *Functional Ecology, 29,* 385–392. doi:10.1111/1365-2435.12324
Latzman, R. D., Hopkins, W. D., Keebaugh, A. C., & Young, L. J. (2014). Personality in chimpanzees (*Pan troglodytes*): Exploring the hierarchical structure and associations with the vasopressin V1A receptor gene. *PLoS ONE, 9*(4), e95741. doi:10.1371/journal.pone.0095741
Lavater, J. C. (1798). *Essays on physiognomy*. London, UK: John Murray.
Lee, P. C., & Moss, C. J. (2012). Wild female African elephants (*Loxodonta africana*) exhibit personality traits of leadership and social integration. *Journal of Comparative Psychology, 126,* 224–232. doi:10.1037/a0026566
Le Galliard, J.-F., Paquet, M., Cisel, M., Montes-Poloni, L., & Franklin, C. (2013). Personality and the pace-of-life syndrome: Variation and selection on exploration, metabolism and locomotor performances. *Functional Ecology, 27,* 136–144. doi:10.1111/1365-2435.12017
Le Galliard, J.-F., Paquet, M., & Mugabo, M. (2015). An experimental test of density-dependent selection on temperament traits of activity, boldness and sociability. *Journal of Evolutionary Biology, 28,* 1144–1155. doi:10.1111/jeb.12641
Li, H., Holleley, C. E., Elphick, M., Georges, A., & Shine, R. (2016). The behavioural consequences of sex reversal in dragons. *Proceedings of the Royal Society B, 283,* 20160217. doi:10.1098/rspb.2016.0217
López, P., Hawlena, D., Polo, V., Amo, L., & Martín, J. (2005). Sources of individual shy–bold variations in antipredator behaviour of male Iberian rock lizards. *Animal Behaviour, 69,* 1–9. doi:10.1016/j.anbehav.2004.05.010
Lyman-Henley, L. P., & Burghardt, G. M. (1994). Opposites attract—Effects of social and dietary experience on snake aggregation behavior. *Animal Behaviour, 47,* 980–982. doi:10.1006/anbe.1994.1131
Maillet, Z., Halliday, W. D., & Blouin-Demers, G. (2015). Exploratory and defensive behaviours change with sex and body size in eastern gartersnakes (*Thamnophis sirtalis*). *Journal of Ethology, 33,* 47–54. doi:10.1007/s10164-014-0416-2
Mayer, M., Shine, R., & Brown, G. P. (2016). Bigger babies are bolder: Effects of body size on personality

in hatchling snakes. *Behaviour*. doi:10.1163/1568539X-00003343

McEvoy, J., While, G. M., Sinn, D. L., Carver, S., & Wapstra, E. (2015). Behavioural syndromes and structural and temporal consistency of behavioural traits in a social lizard. *Journal of Zoology, 296,* 58–66. doi:10.1111/jzo.12217

Mell, H., Josserand, R., Decencière, B., Artacho, P., Meylan, S., & Le Galliard, J.-F. (2016). Do personalities co-vary with metabolic expenditure and glucocorticoid stress response in adult lizards? *Behavioral Ecology and Sociobiology, 70,* 951–961. doi:10.1007/s00265-016-2117-z

Michelangeli, M., Wong, B. B. M., & Chapple, D. G. (2016). It's a trap: Sampling bias due to animal personality is not always inevitable. *Behavioral Ecology, 27,* 62–67. doi:10.1093/beheco/arv123

Moore, M. C., Hews, D. K., & Knapp, R. (1998). Hormonal control and evolution of alternative male phenotypes: Generalizations of models for sexual differentiation. *American Zoologist, 38,* 133–151. http://www.jstor.org/stable/4620127

Mori, A., & Burghardt, G. M. (2001). Temperature effects on anti-predator behaviour in *Rhabdophis tigrinus*, a snake with toxic nuchal glands. *Ethology, 107,* 795–811. doi:10.1046/j.1439-0310.2001.00706.x

Morris, P. H., Gale, A., & Duffy, K. (2002). Can judges agree on the personality of horses? *Personality and Individual Differences, 33,* 67–81. doi:10.1016/S0191-8869(01)00136-2

Moule, H., Michelangeli, M., Thompson, M. B., & Chapple, D. G. (2016). The influence of urbanization on the behaviour of an Australian lizard and the presence of an activity-exploratory behavioural syndrome. *Journal of Zoology, 298,* 103–111. doi:10.1111/jzo.12288

Nakagawa, S., & Schielzeth, H. (2010). Repeatability for Gaussian and non-Gaussian data: A practical guide for biologists. *Biological Reviews of the Cambridge Philosophical Society, 85,* 935–956. doi:10.1111/j.1469-185X.2010.00141.x

Passak, K. M., & Gillingham, J. C. (1997). Thermal influence on defensive behaviours of the Eastern gartersnake, *Thamnophis sirtalis*. *Animal Behaviour, 54,* 629–633. doi:10.1006/anbe.1996.0458

Petrinovich, L., & Widaman, K. F. (1984). An evaluation of statistical strategies to analyze repeated-measures data. In H. V. S. Peeke & L. Petrinovich (Eds.), *Habituation, sensitization, and behavior* (pp. 156–201). Orlando, FL: Academic Press.

Réale, D., Garant, D., Humphries, M. M., Bergeron, P., Careau, V., & Montiglio, P. O. (2010). Personality and the emergence of the pace-of-life syndrome concept at the population level. *Philosophical Transactions of the Royal Society of London. Series B, Biological Sciences, 365,* 4051–4063. doi:10.1098/rstb.2010.0208

Robinson, L. M., Morton, F. B., Gartner, M. C., Widness, J., Paukner, A., Essler, J. L., et al. (2016). Divergent personality structures of brown (*Sapajus apella*) and white-faced capuchins (*Cebus capucinus*). *Journal of Comparative Psychology, 130,* 305–312. doi:10.1037/com0000037

Rodda, G. H. (2012). Statistical properties and techniques and validation. In R. W. McDiarmid, M. S. Foster, C. Guyer, W. Gibbond, & N. Chernoff (Eds.), *Reptile biodiversity: Standard methods for inventory and monitoring* (pp. 197–203). Berkley, CA: University of California Press.

Rodda, G. H., Bock, B. C., Burghardt, G. M., & Rand, A. S. (1988). Techniques for identifying lizard individuals at a distance reveal influences of handling. *Copeia, 1988,* 905–913. doi:10.2307/1445713

Rodriguez-Prieto, I., Martin, J., & Fernandez-Juricic, E. (2011). Individual variation in behavioural plasticity: Direct and indirect effects of boldness, exploration and sociability on habituation to predators in lizards. *Proceedings of the Royal Society of Biological Sciences, 278,* 266–273. doi:10.1098/rspb.2010.1194

Rodríguez-Prieto, I., Martín, J., & Fernández-Juricic, E. (2010). Habituation to low-risk predators improves body condition in lizards. *Behavioral Ecology and Sociobiology, 64,* 1937–1945. doi:10.1007/s00265-010-1004-2

Rossman, D. A., Ford, N. B., & Seigal, R. A. (1996). *The gartersnakes: Evolution and ecology*. Norman, OK: University of Oklahoma Press.

Rowlands, M. (2016). Are animals persons? *Animal Sentience, 2016,* 101.

Schwartz, J. M., & Herzog, H. A., Jr. (unpublished). Heritability of antipredator behavior in three gartersnakes species (*Thamnophis butleri, T. melanogaster and T sirtalis*) in nature.

Schuett, G. W., Clark, R. W., Repp, R. A., Amarello, M., Smith, C. F., & Greene, H. W. (2016). Social behavior in rattlesnakes: A shifting paradigm. In G. W. Schuett, M. J. Feldner, C. F. Smith, & R. S. Reiserer (Eds.), *Rattlesnakes of Arizona. Conservation, behavior, venom, and evolution* (Vol. 2, pp. 161–244). Rodeo, New Mexico: ECO Publishing.

Scudder, R. M., & Burghardt, G. M. (1983). A comparative study of defensive behavior in three sympatric species of water snakes (*Nerodia*). *Zeitschrift für Tierpsychologie, 63,* 17–26. doi:10.1111/j.1439-0310.1983.tb00737.x

Sinervo, B., & Lively, C. M. (1996). The rock-paper-scissors game and the evolution of alternative male strategies. *Nature, 380,* 240–243. doi:10.1038/380240a0

Sinervo, B., Miles, D. B., Frankino, W. A., Klukowski, M., & DeNardo, D. F. (2000). Testosterone, endurance, and Darwinian fitness: Natural and sexual selection on the physiological bases of alternative male behaviors in side-blotched lizards. *Hormones and Behavior, 38,* 222–233. doi:10.1006/hbeh.2000.1622

Sinn, D. L., While, G. M., & Wapstra, E. (2008). Maternal care in a social lizard: Links between female aggression and offspring fitness. *Animal Behaviour, 76,* 1249–1257. doi:10.1016/j.anbehav.2008.06.009

Snowdon, C. T. & Burghardt, G. M. (2017). Studying animal behavior: Integration of field and laboratory approaches. In J. Call (Ed.), G. M. Burghardt, I. M. Pepperberg, C. T. Snowdon, & T. Zentall (Assoc. Eds.), *APA handbook of comparative psychology: Vol 1. Basic concepts, methods, neural substrate, and behavior* (pp. 39–63). Washington, DC: American Psychological Association.

Stamps, J. A., & Biro, P. A. (2016). Personality and individual differences in plasticity. *Current Opinion in Behavioral Sciences, 12,* 18–23. doi:10.1016/j.cobeha.2016.08.008

Stapley, J., & Keogh, J. S. (2005). Behavioral syndromes influence mating systems: Floater pairs of a lizard have heavier offspring. *Behavioral Ecology, 16,* 514–520. doi:10.1093/beheco/ari019

Taylor, J. N., & Lattanzio, M. S. (2016). Boldness, dominance, and territoriality in the color polymorphic tree lizard, *Urosaurus ornatus. Ethology, 122,* 1–10. doi:10.1111/eth.12534

Teyssier, A., Bestion, E., Richard, M., & Cote, J. (2014). Partners' personality types and mate preferences: Predation risk matters. *Behavioral Ecology, 25,* 723–733. doi:10.1093/beheco/aru049

Thaker, M., Lima, S. L., & Hews, D. K. (2009). Alternative antipredator tactics in tree lizard morphs: Hormonal and behavioural responses to a predator encounter. *Animal Behaviour, 77,* 395–401. doi:10.1016/j.anbehav.2008.10.014

Thompson, C. W., & Moore, M. C. (1991). Throat colour reliably signals status in male tree lizards, *Urosaurus ornatus. Animal Behaviour, 42,* 745–753. doi:10.1016/S0003-3472(05)80120-4

Vitousek, M. N., & Romero, L. M. (2013). Stress responsiveness predicts individual variation in mate selectivity. *General and Comparative Endocrinology, 187,* 32–38. doi:10.1016/j.ygcen.2013.03.009

Waters, R. M., & Burghardt, G. M. (2005). The interaction of food motivation and experience in the ontogeny of chemoreception in crayfish snakes. *Animal Behaviour, 69,* 363–374. doi:10.1016/j.anbehav.2004.03.014

Waters, R. M., & Burghardt, G. M. (2013). Prey availability influences the ontogeny and timing of chemoreception-based prey shifting in the striped crayfish snake, *Regina alleni. Journal of Comparative Psychology, 127,* 49–55. doi:org.hollins.idm.oclc.org/10.1037/a0029267

Weiss, A., & Altschul, D. M. (2017). Methods and applications of animal personality research. In J. Call (Ed.), G. M. Burghardt, I. M.Pepperberg, C. T. Snowdon, & T. Zentall (Assoc. Eds.), *APA handbook of comparative psychology: Vol 1. Basic concepts, methods, Neural substrate, and behavior* (pp. 179–200). Washington, DC: American Psychological Association.

Weiss, A., King, J. E., & Murray, L. (Eds.). (2011). *Personality and temperament in nonhuman primates.* New York, NY: Springer.

While, G. M., Isaksson, C., McEvoy, J., Sinn, D. L., Komdeur, J., Wapstra, E., et al. (2010). Repeatable intra-individual variation in plasma testosterone concentration and its sex-specific link to aggression in a social lizard. *Hormones and Behavior, 58,* 208–213. doi:10.1016/j.yhbeh.2010.03.016

While, G. M., Sinn, D. L., & Wapstra, E. (2009). Female aggression predicts mode of paternity acquisition in a social lizard. *Proceedings of the Royal Society B: Biological Sciences, 276,* 2021–2029. doi:10.1098/rspb.2008.1926

Yeager, C. P., & Burghardt, G. M. (1991). Effect of food competition on aggregation: Evidence for social recognition in the plains gartersnake (*Thamnophis radix*). *Journal of Comparative Psychology, 105,* 380–386. doi:10.1037/0735-7036.105.4.380

Personality in Swine

9

Kristina M. Horback

Abstract

There are approximately 977 million domestic swine (*Sus scrofa domesticus*) that live on farms across the world. Knowledge of individual variation in behavior, morphology and productivity has been the basis for artificial selection of this domestic species for thousands of years. Determining personality traits among group living swine may help to identify certain individuals that possess an increased ability to cope with, or a heightened sensitivity to, environmental and/or social stress. There have been approximately 48 studies that investigated behavioral variation in commercial swine. The majority of these studies have been limited to animals aged seven months or younger. Experimental trials are the predominant method of measuring behavioral differences among swine and include such methods as physical restraint, open field tests, human approach tests, novel object tests, and feeding competitions. Previous research on swine personality has reported the existence of multiple traits such as aggressive, exploratory or fearful, as well as dichotomous coping styles (e.g., proactive vs. reactive). Despite excellent work on themes such as environmental influences and trait development, scholars examining personality in swine have not yet fully explored proximate and ultimate explanations of swine personality traits. Considerable effort has gone into identifying personality traits and coping styles in swine in order to evaluate the relationship between behavior, health, and production (e.g., mothering skills, weight gain, lean meat percentage). Knowledge of individual personality traits among domestic swine may help the pork industry to adjust husbandry practices and housing conditions, which are currently under transition due to market-driven welfare concerns.

K.M. Horback (✉)
Department of Animal Science, University of California, Davis, Davis, CA, USA
e-mail: km.horback@gmail.com

J. Vonk et al. (eds.), *Personality in Nonhuman Animals*,
DOI 10.1007/978-3-319-59300-5_9

Personality in Domestic Farm Animals

There are many potential sources of environmental and social stress for domestic farm animals, such as housing conditions and flooring (Meunier-Salaun et al. 1987), transport (Von Borell 2001), integration with unfamiliar conspecifics (Jensen and Yngvesson 1998), and handling procedures (Hemsworth and Barnett 1991). A better understanding of personality differences among livestock may help farmers to identify individuals that possess an increased ability to cope with environmental and/or social stress. Research on animal personality suggests that animals that are highly resilient to stress may experience a difference in HPA reactivity (Koolhaas 2008; Ruis et al. 2000), immunity strength (Bolhuis et al. 2003), productivity (e.g., milk yield, weight gain) and health (e.g., disease resistance; Capitanio et al. 2008; Capitanio et al. 1999; Carere et al. 2010; Koolhaas and Van Reenen 2016). Therefore, increasing our knowledge of the ontogeny and maintenance of personality traits in farm animals can potentially impact both animal welfare and productivity.

Knowledge of individual variation in behavior and morphology is the basis for artificial selection and breeding of farm animals. Humans have selectively bred animals for production traits (e.g., milk yield or weight gain) and behavioral traits, such as a reduced fear of humans, to increase ease of handling (Price 2002). Due to the fact that one gene can influence two or more seemingly unrelated phenotypic traits, known as pleiotropy, selection for certain behavioral traits may result in unintentional selection for particular biological traits (Wright 2015). For example, research indicates that dairy cows that are calmer in response to human handling produce higher milk yields (Breuer et al. 2000; Rushen et al. 1999), and ewes that are calmer in response to restraint and isolation produce milk with more proteins (Sart et al. 2004).

Research on farm animal personality has shown that individual differences in stress response can be phenotypically associated with biological welfare measures. For example, response to physical restraint in piglets is related to differences in growth, lean meat percentage (Hessing et al. 1994; Geverink et al. 1998; Van Erp-Van Der Kooij et al. 2000), reproductive success (Janczak et al. 2003a), and the occurrence of stereotypic behaviors (Geverink et al. 2002). In addition, researchers have found phenotypic correlations between the duration of struggling during restraint among beef cattle and differences in growth and meat quality (Petherick et al. 2009; Voisinet et al. 1997), immune function (Fell et al. 1999), and pregnancy rate (Cooke et al. 2009). And finally, individual differences in both neuroendocrine and behavioral stress response patterns in poultry have been reported to be related to differences in growth (Marin et al. 2003), egg production (Uitdehaag et al. 2008), and the frequency of conspecific-directed feather pecking (Rodenburg et al. 2004). From an economic standpoint, knowledge of personality traits among farm animals may improve both animal health and production, which may potentially increase profits for the farmer.

Assessing Personality in Domestic Farm Animals

A prerequisite for conducting animal personality research is to establish the test–retest reliability or repeatability of trait assessments. Like other forms of reliability, test–retest reliability supports the idea that reactivity to environmental and social stimuli is mediated by underlying temperamental characteristics. Assessing animal personality can be accomplished through two methods: trait ratings and behavioral codings (Gosling 2001; Vazire et al. 2007). Rating personality traits requires animal caretakers to use their intimate knowledge of each individual to place the animal on a continuum of a trait (e.g., 1 [timid]—5 [bold]). The number of response categories can vary according to the species of interest. This method may be applied in a farm setting by inviting the animal care staff to share their knowledge of individual behavior patterns based on their daily interactions with the animals. There is an opportunity to examine which factors (e.g., housing system, duration of animal interaction) may influence the accuracy of personality and welfare assessment by stockpersons (Hemsworth et al. 1989; Highfill et al. 2010).

The second method of assessing personality traits involves behavioral coding. Given that personality is most clearly expressed in novel and unpredictable situations (Gosling 2001), behavioral tests are used to elicit high-arousal responses (Table 9.1). These experiments measure an animal's behavioral response to a range of relevant situations, using operational definitions for the constructs of interest. For example, behaviors obtained in the open field test (e.g., sniff, lick, walk, or freeze), which involves placing an animal in a novel arena, are measures of activity and/or fear (Walsh and Cummins 1976). Interpretation of the behaviors displayed by the individual in the open field test is contingent upon the species of interest. Exposure without cover from predators can be fear-inducing for a domestic chicken; however, for species that evolved for open ground foraging, like swine, the open field test can elicit behaviors related to activity and exploration. In addition, behaviors expressed in experiments that are assumed to elicit similar reactions (e.g., boldness in novel object tests and predation-threat tests) may not correlate with one another (Carter et al. 2012). Therefore, researchers must be careful to not incorrectly assign behaviors to certain personality traits based on the assumption that the behaviors are similar (i.e., jingle-jangle fallacy; Block 1995).

Personality traits can also be assessed through coding of naturally occurring behaviors (i.e., ethological coding) in a given environment or context such as feeding or socializing. Researchers create ethograms; catalogs of behaviors that can be displayed by the animal, to be species-specific. Ethograms for pigs, for example, may include social grooming, grazing, wallowing, and interacting with environmental stimuli (Vazire and Gosling 2004). Behavioral data are then collected using common sampling techniques for either singular events and/or sustained states of the focal animal (Altmann 1974). Behavior events are relatively short in duration, such as discrete body movements or vocalizations while behavior states are relatively long in duration, such as resting or fighting, or the body posture of an animal, such as stand or sit. Researchers may use instantaneous sampling to record the behavior of an animal at a single point

Table 9.1 Example of behavioral measurements employed in animal personality research

Behavioral experiments	Description	Behaviors recorded
Restraint test	Limit the movement of an animal for a brief period of time, for example by tethering, or by close confinement in a weighing crate	Vocalization, freezing, and escape attempts
Open field test	Separate animal from its pen or herd mates and place it in a novel arena for brief time	Vocalization, exploration, freezing, and activity
Novel object test	Animal is presented a species-relevant object in home pen/pasture or experimental arena	Latency to approach, duration of contact, and average distance from object
Feeding competition	Measure aggression and social rank by placing a single highly valued food source in a pen/pasture with a certain number of animals	Aggression, duration/rate of food access, latency to approach food resource, and displacement at food resource
Resident–Intruder Test	Single animal is confronted with unknown conspecific in their home pen/pasture	Aggression, tactile contact, and inter-animal distance
Human approach test	Animal is placed in solitary arena with an unfamiliar human. The location of experiment can be familiar (home pen) or unfamiliar (open field test)	Latency to approach a human (Voluntary Approach), or the response to an approach by a human (Forced Approach), and frequency/duration of tactile contact

in time (e.g., every 5 min) or they may use continuous sampling to record all occurrences of a behavior within a given time period (e.g., all bites initiated by an animal in 1 h).

Personality Data Analysis

Researchers may use variable reduction techniques, like principal components analysis (PCA) or exploratory factor analysis (EFA), to cluster many behaviors into composite traits. Principal components analysis is a procedure for reducing a larger set of behavioral variables into a smaller set of artificial variables called principal components, which are based on the strength of the correlation coefficients among the behavioral variables (Gorsuch 1983). An observed behavioral variable "loads" on a component if it is highly correlated with the component. The goal of exploratory factor analysis, on the other hand, is to explain the structure of data by determining the shared variance among the variables, called communality. Factor analysis summarizes data into a few dimensions by condensing a large number of variables into a smaller set of variables, called factors. Behaviors that are clustered using PCA or EFA share a significant proportion of variance in the data (Budaev and Zworykin 2002; Sih et al. 2004). For example, behaviors such as "bite", "chase", and "hit" may have high loadings on a single component, and this component may be labeled "aggression".

Prior to running the data reduction analysis, researchers must consider their sample size. There are many subject-to-variable ratios recommended, such as 100:1, 10:1, or 5:1 (Bryant and Yarnold 1995; Gorsuch 1983). For example, a subject-to-variables ratio of 5:1 means that for every five animals in the study, one behavior can be placed in the analysis. This recommendation of a minimum sample size is not universal, and there are reports of successful PCA and EFA with small samples sizes (de Winters et al. 2009; Guadagnoli and Velicer 1988; MacCallum et al. 1999).

After reducing many behavioral variables into smaller sets of components or factors, researchers must determine the number of components or factors to retain. A researcher may use the K1 method to decide which components or factors to retain based on the strength of variance explained, or, the eigenvalue (Kaiser 1960). According to this rule, only the components or factors that have eigenvalues greater than one are retained for interpretation. Researchers must be aware of the threat of retaining too few or too many factors or components, as it is difficult to justify retaining a component with an eigenvalue of 1.01, but rejecting a component with an eigenvalue of 0.99 (Zwick and Velicer 1986). Another popular approach is based on the Cattell's scree test (Cattell 1966), which involves the visual exploration of the eigenvalues in graph form. In this method, a researcher retains the components or factors in the steep curve before the "cliff" of flat line trend (Cattell and Jaspers 1967). Parallel analysis (PA) is another technique for determining the number of components to retain when using PCA on a correlation matrix (Franklin et al. 1995; Horn 1965). This method compares the observed eigenvalues derived from the correlation matrix to the eigenvalues obtained from uncorrelated normal variables. Parallel analysis adjusts for the effect of sampling error and is a sample-based as opposed to the population-based method of K1 approach (Zwick and Velicer 1986).

Once a researcher has identified the number of factors or components to retain, he or she may wish to identify an individual's placement or ranking on the trait (Gorsuch 1983). Knowledge of individual rankings on each trait allows a researcher to determine how factor or component scores differ between groups or to investigate correlations between traits and other physiological measures (e.g., rank on bold–shy trait and body weight). There are many methods of calculating trait scores, such as the least squares regression approach, Bartlett's approach, or Anderson–Rubin approach, with each method having different strengths and weaknesses (DiStefano et al. 2009). One simple way to estimate trait scores for each individual involves summing raw scores corresponding to all behavioral variables loading on a trait. If a behavior has a negative loading, the raw score is subtracted rather than added because the behavior is negatively related to the trait (DiStefano et al. 2009).

Coping Styles

Assessing animal personality assumes that there may be different traits that explain the variance in behavior among a group of animals. Alternatively, researchers may categorize animals into one of two categories based on the occurrence of specific stress-related behaviors and physiological reactions (Koolhaas et al. 1999). These categories are called coping styles. Individual animals may be characterized as having a proactive or reactive coping style (Benus et al. 1991; Koolhaas et al. 1999). Proactive animals are bold (e.g., shorter approach/attack latencies), aggressive, and actively cope with challenges in their environment. Reactive individuals are cautious (e.g., longer approach/attack latencies) and less aggressive. Proactive animals use consistent behavior patterns when investigating a change in the environment, whereas reactive animals are more adaptive to change and develop diverse behavioral patterns (Benus et al. 1991; Koolhaas et al. 1999; Sih et al. 2004). In addition, the reactive coping style is associated with high hypothalamus-pituitary-adrenal axis responsiveness (e.g., cortisol production), whereas the proactive coping style is associated with high sympathetic reactivity (e.g., adrenaline production; Koolhaas 2008). Given this disparity in physiological stress response, the coping style of domesticated farm animals may impact the immunocompetence of individuals.

Swine Natural Behavior

Approximately 977 million domestic swine (*Sus scrofa domesticus*) live on farms across the world (FAOSTAT 2013). Although they have been domesticated for almost 9000 years, swine maintain instinctual behaviors of rooting, exploring, and social fighting. In the wild, groups of two to four adult females, called sows, will travel together with their piglets in groups called "sounders" (D'Eath and Turner 2009). Each sounder is led by a dominant sow, while the adult males, called boars, roam throughout non-overlapping home ranges in bachelor herds (Graves 1984). Through frequent nose-to-body and nose-to-nose contact, swine use olfactory cues (i.e., pheromones) to recognize other individuals and gauge social dominance (Kristensen et al. 2001; Mendl et al. 2002; Stookey and Gonyou 1998). The establishment of social rank among a sounder is mediated through both wins and losses of fighting, with younger animals fighting more frequently and thus sustaining more lesions (Arey 1999; Hodgkiss et al. 1998). The main fighting stance among swine is called the inverse parallel press (Jensen 1980). During this agonistic interaction, the animals face head on and push their shoulders against each other, throwing the head against the neck and flanks of the other and biting the other's neck and shoulders. The fight ends when one animal turns away to flee and is not pursued by the displacing animal (Jensen 1980).

Swine are opportunistic omnivores and explore their environments by digging and rooting up objects to test for edibility (Studnitz et al. 2007; Wood-Gush and Vestergaard 1989). Swine will also manipulate objects using the snout or front limbs when performing behaviors such as biting, sniffing, pushing, kicking, licking, and chewing (Blackshaw et al. 1997). Piglets are able to discriminate odors from their nest (e.g., sow's urine, udders) as early as one day old and can identify their mother's vocalizations soon after (Horrell and Hodgson 1992; Walser 1986). Piglets will engage in antagonistic tussling within minutes of being born while teat order is disputed (Algers 1993; Fraser and Jones 1975). This early social interaction often determines future dominance and aggression levels among the litter, with piglets positioned at the anterior teats being more dominant and those in the middle teats fighting more frequently (Ruis et al. 2000; Sundman 2011). This relationship between teat rank and social dominance highlights how rearing experience may influence the development of certain personality traits.

Domestic Swine Welfare

The swine farming industry can be divided into categories based on the age of the animals. Farmers that operate gestation barns care for pregnant sows and their piglets. Over 75% of gestation barns in the United States house their sows in gestation stalls for approximately 110 days after artificial insemination (Schulz and Tonsor 2015). Standard gestation stalls are 0.64 m wide and 1.73 m long with metal side bars and a trough for food and water. While the original goal of gestation stalls was to prevent fighting among the sows and to ensure individual nutrition, this housing system precludes the sow from turning around and/or performing many natural behaviors. In many developed countries, there is increasing public interest in the quality of life for gestating sows. Due to both legislative and market-driven initiatives, swine farmers in nine US states, the entire EU, Australia, New Zealand, and South Africa have been asked to transition their gestating sows from individual stalls to group pens (EU Council Directive 2001/88/EC; Schulz and Tonsor 2015; Weaver and Morris 2004).

The maintenance of gestating animals in pens can be difficult due to the multifaceted considerations of pen size, group composition, feeding system, and individual health care and nutrition requirements (Bench et al. 2013). Group housing of gestating sows may result in increased fighting while the hierarchy is established (Broom et al. 1995), and can thus lead to minor or serious injuries (Anil et al. 2005; Hodgkiss et al. 1998), loss of productivity (McGlone et al. 2004), and may decrease sow longevity (Anil et al. 2005). These issues create economic losses for the producer and welfare concerns for the sows. Knowledge of personality traits may help identify individuals who are more susceptible to stress, as well as allow animal care staff to gauge social compatibility among a pen of sows.

Once the sow is ready to farrow, or give birth, she is moved into a single farrowing crate or pen. In modern swine husbandry, the ability of a sow to raise large litters and to meet the piglets' needs for warmth and milk provision is essential. To reduce piglet losses and to facilitate human intervention, farrowing crates are used to limit the sow's movements during parturition and lactation (Edwards and Fraser 1997). Farrowing crates are similar in size to gestation stalls (0.64 × 1.73 m) and include guardrails to prevent the sow from crushing her piglets when she is in the process of lying down. While the goal of limiting the sow's movement is to increase the piglets' welfare and survivability, this restriction prevents the sow from performing species-specific behaviors at a crucial time.

To date, just Sweden, Switzerland, and Norway have banned farrowing crate use completely, with current bans in New Zealand (NAWAC 2010) and Australia implementing bans in 2017 (PISC 2008). There is also voluntary industry uptake of loose-farrowing alternatives (e.g., UK, Denmark, and Australia), with a number of different systems being developed and tested (Arey 1997; Edwards and Fraser 1997; Johnson and Marchant-Forde 2009; Baxter et al. 2012). It may be beneficial to understand the impact of personality traits on the survivability of piglets in both standard and alternative farrowing environments. Consistent differences in maternal behavior among sows could impact piglet welfare as well as productivity for the farmer.

In most commercial farms, piglets are weaned any time between 20 and 35 days of age, whereas free-ranging sows wean their piglets at around 60–137 days of age (Jensen and Recen 1989; Newberry and Wood-Gush 1985). Weaning can be highly stressful for piglets due to the transport away from the farrowing environment, adjustment to an unfamiliar pen, and an increase in aggression as the dominance hierarchy of unfamiliar piglets is established (Held and Mendl 2001). After weaning, piglets are often transported to another farm that specializes in caring for the animals until they reach the market weight of 250 lb. By understanding the differences in personality or coping ability among piglets, producers may be able to adjust husbandry practices, housing conditions, transportation procedures, and social groupings that may impact the health and welfare of the animals.

Swine Personality Research

There have been approximately 48 published reports of behavioral variation in commercial swine. The majority of these studies have been limited to animals aged seven months or younger and only seven focused on behavioral differences among gestating sows (Table 9.5). This discrepancy could be due to the logistic difficulties of implementing standard behavioral experiments with animals ranging from 300 to 700 lb. Experimental trials were the predominant method used to measure behavioral differences and included such methods as physical restraint, open field tests, human approach tests, novel object tests, and feeding competitions. While there is ample information on behavioral differences in pen housed piglets, there is a lack of information on the individual differences of group-housed gestating sows. Given the current focus of gestating sow welfare in the swine industry, the lack of research addressing behavioral variation among these mature, pregnant animals is surprising.

Although PCA is frequently used in animal personality research as an unbiased approach to cluster behaviors into composite traits, only twelve of the 48 published studies on swine personality or coping styles utilized this analytical method (Table 9.5). Although many studies on swine personality have large sample sizes, many simply present multiple correlations among the behaviors recorded and do not attempt to ascertain specific traits. Many studies on swine personality have identified multiple traits, instead of a dichotomous coping style, which may explain the variance observed in behavioral data (Donald et al. 2011; Forkman et al. 1995; Horback and Parsons 2016).

Forkman et al. (1995) analyzed behavioral responses of 45 piglets to novel objects, open fields, resident-intruder agonistic interactions, and restraint tests to assess personality, and identified three factors (aggression, sociability, and exploration) that explained the behavioral covariance. In human personality research, there is a popular model that identifies five key traits that explain the variance in behavior among humans. This is known as the Five Factor Model and corresponds to openness to experience, conscientiousness, extroversion, agreeableness, and neuroticism (Goldberg 1990). Gosling and John (1999) suggested that the three traits identified in the Forkman et al. (1995) study (aggression, sociability, and exploration) may be analogous to the human dimensions of agreeableness (inversely), extraversion, and openness.

This description of swine personality as being comprised of multiple traits is contrary to some reports of a dichotomous proactive/reactive coping style in pigs (Bolhuis et al. 2003; Hessing et al. 1993; Ruis et al. 2000; van Erp-van der Kooij et al. 2002). The likely reason for this difference is that many studies attempt to categorize piglets based on early stress or restraint tests (e.g., tonic immobility) in an effort to determine behavioral predictors of coping styles. For example, Hessing et al. (1993, 1994) performed non-social (i.e., back-test) and social challenges (i.e., pen mixing) with young piglets and found that those that were more active and responsive in the non-social test were more aggressive in the social test. In contrast, personality studies using PCA evaluate individual differences based on a continuum of many traits, rather than assigning a low versus high category. In a critical review of research addressing coping styles in pigs, Jensen et al. (1995b) argue that researchers have determined these dichotomous behavioral strategies "by choosing arbitrary cut-off points when splitting their individuals into categories rather than performing adequate tests of correlation in the unbiased raw data", and therefore, "great care is needed in the handling of correlational results" (p. 250).

As noted previously, much of the research on behavioral differences among piglets is focused on the strength of response to handling, social isolation, and social aggression. On the other hand, research on sows is focused on differences in mothering skills or fear/aggression toward humans. An exception is a recent study of personality traits among group-housed gestating sows (Horback and Parsons 2016). In this study, 46 sows from a commercial genetic line were observed for behaviors occurring during human approach tests, ease of handling tests, open field tests, novel object tests and for social behaviors during pen mixing. The behavioral tests were repeated five months later to assess the consistency in individual response. Three components, which accounted for over 60% of the variance, were uncovered using principal component analysis. Based on the behaviors that encompassed them, these components were labeled aggressive/dominant, avoidant of humans, and active/exploratory (Table 9.2). This was the first description of personality traits in commercial gestating sows housed in large groups. In addition, individual component scores were significantly correlated between replicates demonstrating their temporal stability (Horback and Parsons 2016).

Behavioral Experiments in Swine Personality Research

Open Field Test

The open field test, originally developed for laboratory rodents (Walsh and Cummins 1976), has mostly been used in swine behavioral research to measure activity (Forkman et al. 1995), whereas a few studies use behaviors collected in an open field as a measure of fear of novelty (e.g., Spoolder et al. 1996). In an open field test, an animal is placed alone in an unfamiliar arena covered in grid lines and behaviors are recorded, which, in the case of Spoolder et al. (1996), included locomotor activity (number of gridlines crossed and duration of time to explore walls or ground), eliminative behavior (defecation, urination) and vocalizations (squeals, grunts). Due to constraints in arena size, this approach is most often used to measure fear or exploration in pigs younger than five months.

In addition to differences in personality traits, the variance in behaviors displayed during open field tests in swine has been attributed to breed (Mormède et al. 1984), genetics (Fabrega et al. 2004; Shea-Moore 1998), rearing (Beattie et al. 1995), and age (Wood-Gush et al. 1990).

Table 9.2 Loading scores of the behavioral measurements collected on group-housed sows for each principle component (Horback and Parsons 2016)

Behavioral measurements	PC1[1]	PC2[1]	PC3[1]
Pen mixing			
Nose/smell other sow (#/h)	−0.35	0.09	**0.76**
Displace other sow (#/h)	**0.86**	0.03	0.10
Retreat from other sow (#/h)	**−0.54**	−0.14	−0.06
Initiate aggressive behavior [bite, hite] (#/h)	**0.83**	−0.06	−0.16
Chase other sow (#/hr)	**0.80**	−0.15	−0.03
Open field test			
Number of lines crossed (10 min)	0.03	−0.04	**0.76**
Ease of Handling Test (scale 1–5)	−0.07	**0.87**	−0.10
Human Approach Test (scale 1–5)	0.06	**0.84**	0.15
Eigenvalue	2.30	1.50	1.20
Variance explained	28.40%	19.00%	15.20%

[1]PC1: aggressive/dominant, PC2: avoidant of humans, PC3: active/exploratory
Component loadings ≥ 0.5 (bold) were considered to have significant weight and were clustered to create composite personality traits

For example, piglets with a genetic mutation on the Halothane RYR(1) gene, which is associated with a susceptibility to physiological stress, were significantly less active in an open field test than RYR(1) negative piglets (Fabrega et al. 2004). General activity and vocalizations displayed in open field tests with piglets of various ages tends to be repeatable; with test–retest intervals ranging from days to months. In addition, open field studies of swine indicate that, unlike rodents, defecation does not seem to be a fear response, but more of a consequence of increasing activity (Andersen et al. 2000; Forkman et al. 2007). Personality traits derived from behaviors measured in open field tests include activity (Andersen et al. 2000; Horback and Parsons 2016) and response to novelty (Giroux et al. 2000; Ruis et al. 2000).

Novel Object Test

Due to their omnivorous diet, swine instinctively dig up and gnaw on objects in their environment in search of edible materials (Studnitz et al. 2007). Swine explore their environments to evaluate both the extrinsic value of their surroundings (i.e., shelter, food, water source) and the intrinsic value (i.e., social group, pen size; Studnitz et al. 2007; Wood-Gush and Vestergaard 1989). The novel object test allows researchers to assess differences in the interest in and exploration of unfamiliar, inedible objects. In this test, an individual is placed either in their home pen or in an experimental arena (e.g., open field) and observed within a group or alone and is presented a species-relevant object, such as a colorful ball for a pig. Some behaviors commonly recorded during novel object tests include latency, frequency or duration of contact, exploration, and orientation to object without physical contact. The consistency of behavior toward a novel object depends on the time interval between tests and whether the animal was tested within a group. Personality traits observed in novel object tests include exploration (Thodberg et al. 1999) and curiosity (Spake et al. 2012).

Social Challenge

In order to assess an individual's response to a social challenge, or their degree of social aggression, researchers may use one of two approaches: resident–intruder tests with young piglets or social behaviors at pen mixing with older pigs. In resident–intruder tests, the focal resident piglets are observed for behaviors toward an intruder, often a piglet that is younger than or the same weight as the resident. This test may take place in the resident's home pen or unacquainted piglets may be placed in a separate experimental arena. Behaviors measured in the resident–intruder test include the rate of aggressive behaviors initiated and received, the duration of vocalizations and the latency to approach, attack, and/or make physical contact with the intruder piglet. Researchers will often terminate the test after the first highly aggressive attack to prevent severe lesions or injury.

Group housing of gestating sows often results in increased physical aggression as the social hierarchy is established. Researchers investigating swine personality can take advantage of this high-arousal context to collect behavior data related to both affiliative and agonistic social behavior (Table 9.3). Data collected at resident–intruder tests and at pen mixing allow researchers to create records of the wins and losses during agonistic encounters, and an individual's role as initiator and recipient. Personality traits assessed using social challenge tests include aggression and dominance (Horback and Parsons 2016; Løvendahl et al. 2005; Spoolder et al. 1996).

Back-Test

In many species, animals become immobile as a fear–response when threatened (i.e., tonic immobility; Gallup 1977). Behavioral responses

Table 9.3 Example ethogram for data collection of social behaviors during mixing of sows

Behavior	Operational definition
Head-to-head knock	Sow uses snout and side of face to hit another sow's head with force
Head-to-body knock	Sow uses snout and side of face to hit another sow's body with force
Bite	Sow uses teeth to clamp down or scrape at another sow's head and body
Lever	Sow places snout under the limbs or torso or another sow and lifts up
Displace other	Focal sow forces another sow to move away from current location. May or may not be at the end of a fight
Retreat from other	Focal sow moves away from another sow, at a walking or running pace. May or may not be at the end of a fight
Chase other	Focal sow pursues another sow at a fast pace, trying to reduce the distance between the dyad
Nose other	Focal sow places snout near the snout, head or body of another sow. Relatively short physical contact (<5 s) is possible

measured in tonic immobility tests can be categorical (becoming immobile or not) or continuous (duration of immobility). Research on piglet personality evaluates individual differences in tonic immobility response using "back-tests". The back-test can be carried out in two ways: the experimenter may place the piglet in a supine position in a wood or plastic v-shaped structure and immobilize the piglet by placing a light weight on its chest (Erhard et al. 1999), or the researcher places the piglet on its back on a flat surface and the experimenter uses their hands to press firmly on the piglet's chest and hind legs (Hessing et al. 1993). The variables measured in this test are latency to the first escape attempt, number of escape attempts, duration or strength of struggling, and duration of vocalizing. Often, each wriggle or bout of struggling made by a piglet is counted as one escape attempt. Researchers use the back-test as a non-social measure of stress response in order to classify piglets as "high-resistant/reactive" or "low-resistant/reactive" based on predetermined cut points in the behaviors that are measured. Researchers tend to associate high-responding piglets as demonstrating proactive coping styles, and low-responding piglets as demonstrating reactive coping styles. The validity of the back-test in piglet personality research is currently debated as researchers are not certain which motivational system the back-test challenges (Jensen et al. 1995b; Spake et al. 2012).

Response to Humans

A piglet's or sow's response to handling or approach by a human is an important factor when investigating swine personality. Handling tests involve common husbandry tasks such as an experimenter physically holding a piglet in a prone position, the restraint of a piglet in a weighing chute or snout snare, or the movement of a piglet during transportation out of the home pen and/or onto a loading truck (Bolhuis et al. 2003; Forkman et al. 1995). The most common behaviors recorded during handling tests include the ease of transit (e.g., physical resistance to moving) and responsiveness to restraint (e.g., physical struggling and vocalizations).

Another method to assess individual differences in response to humans among swine is the human approach test. In this test, the behavior of a piglet or sow is recorded in either a familiar (home pen) or unfamiliar arena with a familiar or unfamiliar human. The human approach test can be voluntary (i.e., measure latency of animal to approach human) or forced (i.e., measure animal's response to an approach by a human). Additional measurements in the voluntary human approach

Table 9.4 Operationally definition scale for evaluating sow's response to a forced human approach

Score	Operational definition
1	Initiates physical contact to experimenter standing 1 m from head, does not show avoidance of human hand proximity or touch
2	Tolerant of human standing 1 m from head, initiates physical contact to human when hand is presented, does not show avoidance of human hand contact
3	Tolerant of human standing 1 m from head and tolerant of hand proximity to face, but intolerant of hand to snout contact
4	Tolerant of human standing 1 m from head, but avoidant of hand proximity to face and hand to snout contact
5	Complete avoidance of human standing in proximity (1 m from head)

test include the duration of physical contact, average distance from a human, and reaction to a human changing posture. In the forced human approach test, the experimenter first records behaviors in response to a human's presence and then gradually moves toward the animal and eventually initiates physical contact (Table 9.4). Personality traits obtained from tests involving human experimenters include fear of humans (Brown et al. 2009; Lensink et al. 2009a, b)

and response/avoidance to human proximity or approach (Giroux et al. 2000; Horback and Parsons 2016) (Table 9.5).

Discussion

When dealing with the behavior of domestic swine, researchers must keep in mind that swine evolved in an environment different from the current agricultural setting and their perceptual systems are the product of evolutionary development and artificial selection. Despite excellent work on themes such as environmental influences and trait development, scholars examining personality in swine have not yet fully explored proximate and ultimate explanations of swine

Table 9.5 Summary table of swine personality research

Study	Sample size	Age	Behavioral measurements	Traits
*Andersen et al. (2000)	84	2.5 weeks	TI	Fear of novelty, Activity
		6 weeks	Elevated puzzle maze	
		7 weeks	Light/dark exploration test	
		8 weeks	OFT	
Bolhuis et al. (2003)	38	10, 17 days	TI (back-test)	Low/high resisting
		9 weeks	Lymphocyte stimulation test	
Bolhuis et al. (2004)	76	10–17 days	TI (back-test)	Low/high resisting
	64	8 weeks	T-maze	
Bolhuis et al. (2005)	60	10, 17 days	TI (back-test)	Low/high resisting
		5, 9, 11, 15, 19 weeks	Activity/posture Social behaviors Ingestive/comfort behaviors Exploratory behaviors	
Bolhuis et al. (2006)	120	10, 17 days	TI (back-test)	Low/high resisting
		15, 19 weeks	Behavior in enriched/barren pens	

(continued)

Table 9.5 (continued)

Study	Sample size	Age	Behavioral measurements	Traits
Brown et al. (2009)	118	22–23 weeks	HAT NOT ODT Feeding competition	Fear
D'Eath (2002)	125	44, 47 days	RI	Aggressiveness
	112	49–51 days	Aggression in group mixing	
D'Eath and Burn (2002)	176	3, 9 days	TI	Low/High responding
	128	42, 44 days	RI	
D'Eath and Pickup (2002)	125	47 days	RI	Aggressiveness
D'Eath and Lawrence (2004)	125	37, 39 days	RI	Aggressiveness
D'Eath et al. (2009)	1663	10 weeks	Aggression in group mixing	Aggressiveness, response to handling
	1620	22 weeks	Response to handling	
	1212	11–13 weeks	Inactivity in pen	
de Sevilla et al. (2009)	119	4 weeks	TI	Active/passive coping style
	99	11–13 weeks	OFT NOT	
		15–16 weeks	HAT	
Donald et al. (2011)	24	5–6 weeks	OFT (pharmacological manipulation)	Emotionality (cautious to exploratory, arousal level)
	20	5–6 weeks	OFT (social manipulation)	
	12	5–6 weeks	OFT (novelty manipulation)	
Erhard and Mendl (1997)	218	7, 11 weeks	RI	Aggressiveness
Erhard and Mendl (1999)	29	3 weeks	TI Emergence test	Active/passive behavioral strategies
Erhard et al. (1999)	219	2.5 weeks	TI	Active/passive behavioral strategies
	109	4 weeks	Handling test	
	70	10 weeks	Speed of movement	
Fraser (1974)	35	8–10 weeks	OFT	Activity, vocalization
*Forkman et al. (1995)	65	2 weeks	TI	Aggression, sociability, exploration
	45	1–5 weeks		
		8 weeks	Extinction learning	
		9 weeks	NOT, RI	
		10 weeks	Social dependence	
Geverink et al. (2002)	72	10, 17 days	Back-test	Low/high responding
		10 months	Nose-sling test	
*Giroux et al. (2000)	252	20–25 days	OFT HAT Feeding competition	Reaction to humans, active/passive stress response, rank order

(continued)

Table 9.5 (continued)

Study	Sample size	Age	Behavioral measurements	Traits
Hellbrügge et al. (2008)	1327	lactating sow	Separation from litter at birth	"aggressive in group", "maternal abilities"
	1279		Separation from litter at day 21	
	1453		Reaction to screaming piglet and music	
	1220	gestating sow	Aggression in group mixing	
Hemsworth et al. (1981)	1225	gestating sow	HAT (forced approach)	Response to humans
	480		HAT (voluntary approach)	
Hessing et al. (1993)	218	1 week	Social confrontation	Aggressive, resistant
		1–3 weeks	TI (back-test)	
		10, 15 weeks	Aggression in group mixing	
*Horback and Parsons (2016)	46	gestating sow	HAT (forced approach) Ease of handling OFT NOB Social behavior in group mixing	Aggressive/dominant, active/exploratory, avoidant of humans
Ison et al. (2015)	24	20, 23, 25 weeks	HAT (voluntary approach) Startle object test Behavior at farrowing	"maternal abilities"
Janczak et al. (2003a)	92	3 weeks	TI	Fear, anxiety, aggression
		8, 24 weeks	RI HAT (voluntary) NOT	
*Janczak et al. (2003b)	89	8 weeks	NOT HA (voluntary)	Fear of humans, aovelty induced anxiety
	30	first farrow	Maternal behavior	
*Jensen et al. (1995a)	45	birth, 12–24 h, 3 and 5 weeks	Behavior at farrowing OFT Social challenge test	Activity, suckling behavior
Lawrence et al. (1991)	62	30 weeks	Response to handling Ease of transit HAT (forced) Resistance to restraint	Low/high responding
		34 weeks	NOT Group feeding competition	
Lensink et al. (2009a)	75	26 weeks	HAT (forced)	"fear of humans", "nervousness"
			NOT	
			Response to weighing crate	
		first farrow	Behavior in farrowing crate	
Lensink et al. (2009b)	100	sow	HAT (forced approach) Reaction to transfer Reaction to piglet handling	"fear of humans"
Løvendahl et al. (2005)	835	sows	Aggression in group mixing	Aggression
	1076		Maternal behavior	

(continued)

Table 9.5 (continued)

Study	Sample size	Age	Behavioral measurements	Traits
Marchant-Forde (2002)	62	gilt	HAT (voluntary)	Shy–bold, aggression
			Food competition test	
			Aggression to human in farrowing	
Melotti et al. (2011)	128	15 days	TI (back-test)	Low/high responding; aggression
		30 days	Aggressive, explorative and manipulative behavior at weaning	
Mendl et al. (1992)	37	primiparous sows	Behavior 1st and 3rd week in pen	Social dominance (Low, high, no success)
Reimert et al. (2014a)	72	10 days	TI (back-test)	Low/high resisting
		11 weeks	Social support test	
*Reimert et al. (2014b)	480	2 weeks	TI (back-test)	Low/high resisting, locomotion, calm
		6 weeks	NOT	
	478	7 weeks	HAT (voluntary)	
	467	13 weeks	OFT (novel environment) NOT	
Ruis et al. (2000)	128	2–4 days	TI (back-test)	Low/high resisting, aggression
		3–4 weeks	Teat order	
		10, 24 weeks	Group feeding competition test	
		10, 24 weeks	OFT (novel environment)	
Ruis et al. (2002)	96	2–4 days	TI (back-test)	Low/high resisting, social dominance
		7 weeks	Behavior at pen mixing	
		8 weeks	OFT (novel environment) NOT	
Scheffler et al. (2016)	1382	12–19 days	TI (back-test)	Low/high resisting, aggression
	1318	6–9 weeks	HAT	
	272	22 weeks	HAT (voluntary)	
	1111	weaning	Aggression at group mixing	
	446	growing		
	279	gilt		
Spake et al. (2012)	575	6, 13 days	TI (back-test)	Low/high resisting, aggression, fear/curiosity
	120	5, 6 weeks	RI NOT	
*Spoolder et al. (1996)	208	13–17 weeks	NOT Food motivation Social dominance General activity	Fearful, restlessness, food motivation, social dominance, active
*Sundman (2011)	63	9–31 days	Teat order Suckling behavior NOT Straw test	Exploration, playfulness, food interest, straw interest

(continued)

Table 9.5 (continued)

Study	Sample size	Age	Behavioral measurements	Traits
*Svennson (2011)	63	9–31 days	Teat order Suckling behavior NOT Straw test	Curiosity, playful, food interest, straw interest
*Thodberg et al. (1999)	56	17–19 weeks	OFT/NOT HAT (voluntary approach) Social test Undisturbed behavior Food competition test	Immobility, exploratory, inactivity, object exploration
*van Erp-van der Kooij et al. (2002)	315	3, 10, 17 days	TI (back-test)	Coping style
		5–7, 10–12 weeks	HAT (forced) NOT ODT	
von Borell and Ladewig (1992)	24	18, 22, 27 weeks	Behavior in home pen	Activity, stress response
		10, 28 weeks	OFT	
		16, 23, 33 weeks	ACTH challenge	

TI tonic immobility, *OFT* open field test, *NOT* novel object test, *RI* resident-intruder test, *HAT* human approach test, *ODT* open door test

personality traits. Without such an understanding, we are left with an incomplete analysis that can lead to ill-informed policy decisions. By combining traditional measurements of biological welfare (e.g., health, weight gain, fecundity) with metrics of psychological welfare (i.e., cognitive bias testing; Douglas et al. 2012), future research could incorporate an ethological approach to personality assessment. This approach can increase our understanding of the genetic and phenotypic relationship between production, welfare, and personality traits which may improve breeding and husbandry practices.

Ongoing animal welfare science research is directed toward enhancing engineering and production standards in housing and care on farms, such as appropriate bedding, shelter, allowance for natural self-maintenance (e.g., grooming, defecating), and opportunity to exercise. An important factor to keep in mind while establishing attainable, science-based, auditable standards is the influence of temperament. A wealth of research has demonstrated that individual differences can affect a wide range of biological mechanisms and processes, such as the formation of social bonds (Massen and Koski 2014), immunity strength (Capitanio et al. 1999, 2008), individual fitness (ability to survive and reproduce) (Smith and Blumstein 2008), the ability to cope with physiological stress (Carere et al. 2010), the performance of abnormal or stereotypic behaviors (Cussen and Mench 2015; Ijichi et al. 2013), and the expression of pain (Ijichi et al. 2014). Therefore, the personality of an animal may influence individual ability to cope with environmental stressors (e.g., air/light quality, housing structure, and stocking density) and thus influence their health and overall welfare.

Implications

Considerable effort has gone into identifying personality traits and coping styles in piglets in order to evaluate the relationship between personality, health, and production variables (e.g., mothering skills, weight gain, lean meat percentage).

For example, studies have reported that piglets that were more aggressive toward litter mates had lower susceptibility to tonic immobility in the back-test, while other studies correlate high response to the back-test to a higher percentage of carcass protein at slaughter (Hartsock et al. 1977; Hessing et al. 1993; Ruis et al. 2000; Van Erp-van der Kooij et al. 2000). Future research may identify traits associated with susceptibility or resilience to stress, which can impact the biological and psychological welfare of swine (Koolhaas and Van Reenen 2016). Whereas domestic swine have been bred for traits related to tameness and production, it may soon be common practice to breed for personality traits or coping styles as well. Given consumers' increasing interest in the welfare of farm animals, and corresponding legal requirements in animal care and housing, the swine industry could benefit from identifying personality traits associated with an increase in animal welfare and productivity in various social and environmental housing conditions.

References

Algers, B. (1993). Nursing in pigs: Communicating needs and distributing resources. *Journal of Animal Science, 71*(10), 2826–2831.

Altmann, J. (1974). Observational study of behavior: Sampling methods. *Behaviour, 49*(3), 227–266.

Andersen, I. L., Bøe, K. E., Fœrevik, G., Janczak, A. M., & Bakken, M. (2000). Behavioural evaluation of methods for assessing fear responses in weaned pigs. *Applied Animal Behaviour Science, 69,* 227–240.

Anil, L., Anil, S., Deen, J., Baidoo, S., & Wheaton, J. (2005). Evaluation of well-being, productivity, and longevity of pregnant sows housed in groups in pens with an electronic sow feeder or separately in gestation stalls. *American Journal of Veterinary Research, 66,* 1630–1638.

Arey, D. S. (1997). Behavioural observations of peri-parturient sows and the development of alternative farrowing accommodation: a review. *Animal Welfare, 6,* 217–229.

Arey, D. S. (1999). Time course for the formation and disruption of social organization in group-housed sows. *Applied Animal Behavior Science, 62,* 199–207.

Baxter, E. M., Lawrence, A. B., & Edwards, S. A. (2012). Alternative farrowing accommodation: welfare and economic aspects of existing farrowing and lactation systems for pigs. *Animal, 6,* 96–117.

Beattie, V. E., Walker, N., & Sneddon, I. A. (1995). Effect of rearing environment and change of environment on the behavior of gilts. *Applied Animal Behaviour Science, 46,* 57–65.

Bench, C. J., Rioja-Lang, F. C., Hayne, S. M., & Gonyou, H. W. (2013). Group gestation sow housing with individual feeding—II: How space allowance, group size and composition, and flooring affect sow welfare. *Livestock Science, 152,* 218–227.

Benus, R. F., Bohus, B., Koolhaas, J. M., & Van Oortmerssen, G. A. (1991). Heritable variation for aggression as a reflection of individual coping strategies. *Cellular and Molecular Life Sciences, 47*(10), 1008–1019.

Blackshaw, J. K., Swain, A. J., Blackshaw, A. W., Thomas, F. J. M., & Gillies, K. J. (1997). The development of playful behaviour in piglets from birth to weaning in three farrowing environments. *Applied Animal Behaviour Science, 55*(1), 37–49.

Block, J. (1995). A contrarian view of the five-factor approach to personality description. *Psychological Bulletin, 117,* 187.

Bolhuis, J. E., Parmentier, H. K., Schouten, W. G., Schrama, J. W., & Wiegant, V. M. (2003). Effects of housing and individual coping characteristics on immune responses of pigs. *Physiology & Behavior, 79,* 289–296.

Bolhuis, J. E., Schouten, W. G. P., de Leeuw, J. A., Schrama, J. W., & Wiegant, V. M. (2004). Individual coping characteristics, rearing conditions and behavioural flexibility in pigs. *Behavioral Brain Research, 152,* 351–360.

Bolhuis, J. E., Schouten, W. G., Schrama, J. W., & Wiegant, V. M. (2005). Behavioural development of pigs with different coping characteristics in barren and substrate-enriched housing conditions. *Applied Animal Behaviour Science, 93,* 213–228.

Bolhuis, J. E., Schouten, W. G., Schrama, J. W., & Wiegant, V. M. (2006). Effects of rearing and housing environment on behaviour and performance of pigs with different coping characteristics. *Applied Animal Behaviour Science, 101,* 68–85.

Breuer, K., Hemsworth, P. H., Barnett, J. L., Matthews, L. R., & Coleman, G. J. (2000). Behavioural response to humans and the productivity of commercial dairy cows. *Applied Animal Behaviour Science, 66,* 273–288.

Broom, D. M., Mendl, M., & Zanella, A. (1995). A comparison of the welfare of sows in different housing conditions. *Animal Science, 61,* 369–385.

Brown, J. A., Dewey, C., Delange, C. F., Mandell, I. B., Purslow, P. P., Robinson, J. A., et al. (2009). Reliability of temperament tests on finishing pigs in group-housing and comparison to social tests. *Applied Animal Behaviour Science, 118,* 28–35.

Bryant, F. B., & Yarnold, P. R. (1995). Principal components analysis and exploratory and confirmatory factor analysis. In L. G. Grimm & R. R. Yarnold (Eds.), *Reading and understanding multivariale statistics* (pp. 99–136). Washington, DC: American Psychological Association.

Budaev, S. V., & Zworykin, D. D. (2002). Individuality in fish behavior: Ecology and comparative psychology. *Journal of Ichthyology, 42,* S189–S195.

Capitanio, J. P., Mendoza, S. P., & Baroncelli, S. (1999). The relationship of personality dimensions in adult male rhesus macaques to progression of simian immunodeficiency virus disease. *Brain, Behavior, and Immunity, 13,* 138–154.

Capitanio, J. P., Abel, K., Mendoza, S. P., Blozis, S. A., McChesney, M. B., Cole, S. W., et al. (2008). Personality and serotonin transporter genotype interact with social context to affect immunity and viral set-point in simian immunodeficiency virus disease. *Brain, Behavior, and Immunity, 22,* 676–689.

Carere, C., Caramaschi, D., & Fawcett, T. (2010). Covariation between personalities and individual differences in coping with stress: Converging evidence and hypotheses. *Current Zoology, 56,* 728–740.

Carter, A. J., Marshall, H. H., Heinsohn, R., & Cowlishaw, G. (2012). How not to measure boldness: novel object and antipredator responses are not the same in wild baboons. *Animal Behaviour, 84,* 603–609.

Cattell, R. B. (1966). The scree test for the number of factors. *Multivariate Behavioral Research, 1,* 245–276.

Cattell, R. B., & Jaspers, J. (1967). A general plasmode for factor analytic exercises and research. *Multivariate Behavioral Research Monographs, 3,* 1–212.

Cooke, R. F., Arthington, J. D., Araujo, D. B., & Lamb, G. C. (2009). Effects of acclimation to human interaction on performance, temperament, physiological responses, and pregnancy rates of Brahman-crossbred cows. *Journal of Animal Science, 87*(12), 4125–4132.

Cussen, V., & Mench, J. (2015). The relationship between personality dimensions and resiliency to environmental stress in orange-winged amazon parrots, as indicated by the development of abnormal behaviors. *PLoS ONE, 10,* e0126170.

D'Eath, R. B. (2002). Individual aggressiveness measured in a resident-intruder test predicts the persistence of aggressive behaviour and weight gain of young pigs after mixing. *Applied Animal Behaviour Science, 77,* 267–283.

D'Eath, R. B., & Burn, C. C. (2002). Individual differences in behaviour: A test of'coping style'does not predict resident-intruder aggressiveness in pigs. *Behaviour, 139,* 1175–1194.

D'Eath, R. B., & Pickup, H. E. (2002). Behaviour of young growing pigs in a resident-intruder test designed to measure aggressiveness. *Aggressive Behavior, 28,* 401–415.

D'Eath, R. B., & Lawrence, A. B. (2004). Early life predictors of the development of aggressive behaviour in the domestic pig. *Animal Behaviour, 67,* 501–509.

D'Eath, R. B., Roehe, R., Turner, S. P., Ison, S. H., Farish, M., Jack, M. C., et al. (2009). Genetics of animal temperament: Aggressive behaviour at mixing is genetically associated with the response to handling in pigs. *Animal, 3,* 1544–1554.

de Sevilla, X. F., Casellas, J., Tibau, J., & Fàbrega, E. (2009). Consistency and influence on performance of behavioural differences in Large White and Landrace purebred pigs. *Applied Animal Behaviour Science, 117,* 13–19.

de Winter, J. C. F., Dodou, D., & Wieringa, P. A. (2009). Exploratory factor analysis with small sample sizes. *Multivariate Behavioral Research, 44,* 147–181.

DiStefano, C., Zhu, M., & Mîndrilă, D. (2009). Understanding and using factor scores: Considerations for the applied researcher. *Practical Assessment, Research & Evaluation, 14*(20), 1–11.

Donald, R. D., Healy, S. D., Lawrence, A. B., & Rutherford, K. (2011). Emotionality in growing pigs: Is the open field a valid test? *Physiology & Behavior, 104,* 906–913.

Douglas, C., Bateson, M., Walsh, C., Bédué, A., & Edwards, S. A. (2012). Environmental enrichment induces optimistic cognitive biases in pigs. *Applied Animal Behaviour Science, 139,* 65–73.

Edwards, S. A., & Fraser, D. (1997). Housing systems for farrowing and lactation. *Pig Journal, 39,* 77–89.

Erhard, H. W., & Mendl, M. (1997). Measuring aggressiveness in growing pigs in a resident-intruder situation. *Applied Animal Behaviour Science, 54,* 123–136.

Erhard, H. W., Mendl, M., & Christiansen, S. B. (1999). Individual differences in tonic immobility may reflect behavioural strategies. *Applied Animal Behaviour Science, 64,* 31–46.

Fabrega, E., Diestre, A., Font, J., Carrion, D., Velarde, A., Ruiz-de-la-Torre, J. L., et al. (2004). Differences in open field behavior between heterozygous and homozygous negative gilts for the RYR (1) gene. *Journal of Applied Animal Welfare Science, 7,* 83–93.

Fell, L. R., Colditz, I. G., Walker, K. H., & Watson, D. L. (1999). Associations between temperament, performance and immune function in cattle entering a commercial feedlot. *Animal Production Science, 39,* 795–802.

Food and Agriculture Organization of the United Nations, FAOSTAT database (FAOSTAT, 2013). http://faostat3.fao.org/home

Forkman, B., Furuhang, I. L., & Jensen, P. (1995). Personality, coping patterns and aggression inpiglets. *Applied Animal Behaviour Science, 45,* 31–42.

Forkman, B., Boissy, A., Meunier-Salaün, M. C., Canali, E., & Jones, R. B. (2007). A critical review of fear tests used on cattle, pigs, sheep, poultry and horses. *Physiology & Behavior, 92,* 340–374.

Franklin, S. B., Gibson, D. J., Robertson, P. A., Pohlmann, J. T., & Fralish, J. S. (1995). Parallel analysis: A method for determining significant principal components. *Journal of Vegetation Science, 6,* 99–106.

Fraser, D. (1974). The vocalizations and other behaviour of growing pigs in an 'open field' test. *Applied Animal Ethology, 1,* 3–16.

Fraser, D., & Jones, R. M. (1975). The 'teat order'of suckling pigs: I. Relation to birth weight and subsequent growth. *Journal of Agricultural Science, 84*(03), 387–391.

Gallup, G. G. (1977). Tonic immobility: The role of fear and predation. *The Psychological Record, 27,* 41.

Geverink, N. A., Kappers, A., Van de Burgwal, E., Labooij, E., Blokhuis, J., & Wiegant, V. (1998). Effects of regular moving and handling on the behavioral and physiological responses of pigs to pre-slaughter treatment and consequences for meat quality. *Journal of Animal Science, 76,* 2080–2085.

Geverink, N., et al. (2002). Individual differences in behavioral and physiological responses to restraint stress in pigs. *Physiology & Behavior, 77,* 451–457.

Giroux, S., Martineau, G. P., & Robert, S. (2000). Relationships between individual behavioural traits and post-weaning growth in segregated early-weaned piglets. *Applied Animal Behaviour Science, 70,* 41–48.

Goldberg, L. R. (1990). An alternative description of personality: The big-five factor structure. *Journal of Personality and Social Psychology, 59,* 1216–1229.

Gorsuch, R. L. (1983). *Factor analysis* (2nd ed.). Hillsdale, NJ: Erlbaum.

Gosling, S. D. (2001). From mice to men: What can we learn about personality from animal research? *Psychological Bulletin, 127,* 45–86.

Gosling, S. D., & John, O. P. (1999). Personality dimensions in nonhuman animals a cross-species review. *Current Directions in Psychological Science, 8,* 69–75.

Graves, H. B. (1984). Behavior and ecology of wild and feral swine (*Sus scrofa*). *Journal of Animal Science, 58,* 482–492.

Guadagnoli, E., & Velicer, W. F. (1988). Relation to sample size to the stability of component patterns. *Psychological Bulletin, 103,* 265.

Hartsock, T. G., Graves, H. B., & Baumgardt, B. R. (1977). Agonistic behavior and the nursing order in suckling piglets: Relationships with survival, growth and body composition. *Journal of Animal Science, 44,* 320–331.

Held, S., & Mendl, M. (2001). *Behaviour of the young weaner pig* (pp. 273–298). The Weaner Pig, Oxford: CAB International.

Hellbrügge, B., Tölle, K. H., Bennewitz, J., Henze, C., Presuhn, U., & Krieter, J. (2008). Genetic aspects regarding piglet losses and the maternal behaviour of sows. Part 2. Genetic relationship between maternal behaviour in sows and piglet mortality. *Animal: An International Journal of Animal Bioscience, 2,* 1281–1288.

Hessing, M. J. C., Hagelso, A. M., Van Beek, J. A. M., Wiepkema, P. R., Schouten, W. G. P., & Krukow, R. (1993). Individual behavioral and physiological strategies in pigs. *Applied Animal Behavior Science, 55,* 39–46.

Hessing, M. J. C., Schouten, W. G. P., Wiepkema, P. R., & Tielen, M. J. M. (1994). Implications of individual behavioural characteristics on performance in pigs. *Livestock Production Science, 40,* 187–196.

Hemsworth, P. H., Barnett, J. L., Coleman, G. J., & Hansen, C. (1989). A study of the relationships between the attitudinal and behavioural profiles of stockpersons and the level of fear of humans and reproductive performance of commercial pigs. *Applied Animal Behaviour Science, 23,* 301–314.

Hemsworth, P. H., Brand, A., & Willems, P. J. (1981). The behavioural response of sows to the presence of human beings and their productivity. *Livestock Production Science, 8,* 67–74.

Hemsworth, P. H., & Barnett, J. L. (1991). The effects of aversively handling pigs, either individually or in groups, on their behaviour, growth and corticosteroids. *Applied Animal Behaviour Science, 30,* 61–72.

Highfill, L., Hanbury, D., Kristiansen, R., Kuczaj, S., & Watson, S. (2010). Rating vs. coding in animal personality research. *Zoo Biology, 29,* 509–516.

Hodgkiss, N. J., Eddison, J. C., Brooks, P. H., & Bug, P. (1998). Assessment of the injuries sustained by pregnant sows housed in groups using electronic feeders. *Veterinary Record, 143,* 604–607.

Horback, K. M., & Parsons, T. D. (2016). Temporal stability of personality traits in group-housed gestating sows. *Animal, 10*(08), 1351–1359.

Horn, J. L. (1965). A rationale and test for the number of factors in factor analysis. *Psychometrika, 32,* 179–185.

Horrell, I., & Hodgson, J. (1992). The bases of sow-piglet identification. 2. Cues used by piglets to identify their dam and home pen. Applied Animal *Behaviour Science, 33*(4), 329–343.

Ijichi, C., Collins, L., & Elwood, R. (2013). Evidence for the role of personality in stereotypy predisposition. *Animal Behavior, 85,* 1145–1151.

Ijichi, C., Collins, L., & Elwood, R. (2014). Pain expression is linked to personality in horses. *Applied Animal Behavior Science, 152,* 38–43.

Ison, S. H., Wood, C. M., & Baxter, E. M. (2015). Behaviour of pre-pubertal gilts and itsrelationship to farrowing behaviour in conventional farrowing crates and loose-housed pens. *Applied Animal Behaviour Science, 170,* 26–33.

Janczak, A. M., Pedersen, L. J., Rydhmer, L., & Bakken, M. (2003a). Relation between early fear- and anxiety-related behaviour and maternal ability in sows. *Applied Animal Behaviour Science, 82,* 121–135.

Janczak, A. M., Pedersen, L. J., & Bakken, M. (2003b). Aggression, fearfulness and coping styles in female pigs. *Applied Animal Behaviour Science, 81*(1), 13–28.

Jensen, P. (1980). An ethogram of social interaction patterns in group-housed dry sows. *Applied Animal Ethology, 6,* 341–350.

Jensen, P., & Recén, B. (1989). When to wean—Observations from free-ranging domestic pigs. *Applied Animal Behaviour Science, 23,* 49–60.

Jensen, P., Forkman, B., Thodberg, K., & Köster, E. (1995a). Individual variation and consistency in piglet behaviour. *Applied Animal Behaviour Science, 45,* 43–52.

Jensen, P., Rushen, J., & Forkman, B. (1995b). Behavioural strategies or just individual variation in behaviour? A lack of evidence for active and passive piglets. *Applied Animal Behaviour Science, 43,* 135–139.

Jensen, P., & Yngvesson, J. (1998). Aggression between unacquainted pigs—Sequential assessment and effects of familiarity and weight. *Applied Animal Behaviour Science, 58,* 49–61.

Johnson, A. K., & Marchant-Forde, J. N. (2009). Welfare of pigs in the farrowing environment. In J. N. Marchant-Forde (Ed.), *The welfare of pigs* (pp. 141–188). New York, NY: Springer.

Kaiser H. F. (1960). The application of electronic computers to factor analysis. *Educational and Psychological Measurement, 20*(1), 141–151.

Koolhaas, J. M. (2008). Coping style and immunity in animals: Making sense of individual variation. *Brain, Behavior, and Immunity, 22,* 662–667.

Koolhaas, J. M., Korte, S., De Boer, S., Van Der Vegt, B., Van Reenen, C., Hopster, H., et al. (1999). Coping styles in animals: Current status in behavior and stress-physiology. *Neuroscience Biobehavioral Review, 23,* 925–935.

Koolhaas, J. M., & Van Reenen, C. G. (2016). Animal behavior and well-being symposium: Interaction between coping style/personality, stress, and welfare: Relevance for domestic farm animals. *Journal of Animal Science, 94,* 2284–2296.

Kristensen, H. H., Jones, R. B., Schofield, C. P., White, R. P., & Wathes, C. M. (2001). The use of olfactory and other cues for social recognition by juvenile pigs. *Applied Animal Behaviour Science, 72,* 321–333.

Lawrence, A. B., Terlouw, E., & Illius, A. W. (1991). Individual differences in behavioural responses of pigs exposed to non-social and social challenges. *Applied Animal Behaviour Science, 30,* 73–86.

Lensink, B. J., Leruste, H., Le Roux, T., & Bizeray-Filoche, D. (2009a). Relationship between the behaviour of sows at 6 months old and the behaviour and performance at farrowing. *Animal, 3,* 128–134.

Lensink, B. J., Leruste, H., De Bretagne, T., & Bizeray-Filoche, D. (2009b). Sow behaviour towards humans during standard management procedures and their relationship to piglet survival. *Applied Animal Behavior Science, 119,* 151–157.

Løvendahl, P., Damgaard, L. H., Nielsen, B. L., Thodberg, K., Su, G., & Rydhmer, L. (2005). Aggressive behaviour of sows at mixing and maternal behaviour are heritable and genetically correlated traits. *Livestock Production Science, 93,* 73–85.

MacCallum, R. C., Widaman, K. F., Zhang, S., & Hong, S. (1999). Sample size in factor analysis. *Psychological Methods, 4,* 84.

Marchant-Forde, J. N. (2002). Piglet-and stockperson-directed sow aggression after farrowing and the relationship with a pre-farrowing, human approach test. *Applied Animal Behaviour Science, 75,* 115–132.

Marin, R. H., Satterlee, D. G., Castille, S. A., & Jones, R. B. (2003). Early T-maze behavior and broiler growth. *Poultry Science, 82,* 742–748.

Massen, J., & Koski, S. (2014). Chimps of a feather sit together: Chimpanzee friendships are based on homophily in personality. *Evolution and Human Behavior, 35,* 1–8.

McGlone, J. J., Vines, B., Rudine, A. C., & DuBois, P. (2004). The physical size of gestating sows. *Journal of Animal Science, 82,* 2421–2427.

Melotti, L., Oostindjer, M., Bolhuis, J. E., Held, S., & Mendl, M. (2011). Coping personality type and environmental enrichment affect aggression at weaning in pigs. Applied Animal Behaviour Science, 133 (3), 144–153.

Mendl, M., Zanella, A. J., & Broom, D. M. (1992). Physiological and reproductive correlates of behavioral strategies in female domestic pigs. *Animal Behaviour, 44,* 1107–1121.

Mendl, M., Randle, K., & Pope, S. (2002). Young female pigs can discriminate individual differences in odours from conspecific urine. *Animal Behavior, 64,* 97–101.

Meunier-Salaun, M. C., Vantrimponte, M. N., Raab, A., & Dantzer, R. (1987). Effect of floor area restriction upon performance, behavior and physiology of growing finishing pigs. *Journal of Animal Science, 64,* 1371–1377.

Morméde, P., Dantzer, R., Bluthe, R. M., & Caritez, J. C. (1984). Differences in adaptive abilities of three breeds of Chinese pigs. *Genetique, Selection et Evolution, 16,* 85–102.

National Animal Welfare Advisory Committee. (2010). *Animal welfare (pigs) code of welfare 2010. Animal welfare directorate.* Wellington, New Zealand: MAF Biosecurity New Zealand.

Newberry, R. C., & Wood-Gush, D. G. (1985). The suckling behaviour of domestic pigs in a semi-natural environment. *Behaviour, 95,* 11–25.

Petherick, J., Doogan, V. J., Venus, B. K., Holroyd, R. G., & Olsson, P. (2009). Quality of handling and holding yard environment, and beef cattle temperament: 2. Consequences for stress and productivity. *Applied Animal Behaviour Science, 120,* 28–38.

Primary Industries Standing Committee. (2008). *Model code of practice for the welfare of animals: Pigs* (3rd ed). PISC Report 92. Victoria, Australia: CSIRO Publishing.

Price, E. O. (2002). *Animal domestication and behaviour.* Wallingford: CABI Publishing.

Reimert, I., Bolhuis, J. E., Kemp, B., & Rodenburg, T. B. (2014a). Social support in pigs with different coping styles. *Physiology & Behavior, 129,* 221–229.

Reimert, I., Rodenburg, T. B., Ursinus, W. W., Kemp, B., & Bolhuis, J. E. (2014b). Responses to novel situations of female and castrated male pigs with divergent social breeding values and different backtest classifications in barren and straw-enriched housing. *Applied Animal Behaviour Science, 151,* 24–35.

Rodenburg, T. B., Buitenhuis, A. J., Ask, B., Uitdehaag, K. A., Koene, P., Van Der Poel, J. J., et al. (2004). Genetic and phenotypic correlations between feather pecking and open-field response in laying hens at two different ages. *Behavior Genetics, 34,* 407–415.

Ruis, M. A., te Brake, J. H., van de Burgwal, J. A., de Jong, I. C., Blokhuis, H. J., & Koolhaas, J. M. (2000). Personalities in female domesticated pigs: Behavioural and physiological indications. *Applied Animal Behaviour Science, 66,* 31–47.

Ruis, M. A., te Brake, J. H., Engel, B., Buist, W. G., Blokhuis, H. J., & Koolhaas, J. M. (2002). Implications of coping characteristics and social status for welfare and production of paired growing gilts. *Applied Animal Behaviour Science, 75*(3), 207–231.

Rushen, J., De Passille, A. M. B., & Munksgaard, L. (1999). Fear of people by cows and effects on milk yield, behavior, and heart rate at milking. *Journal of Dairy Science, 82,* 720–727.

Sart, S., Bencini, R., Blache, D., & Martin, G. B. (2004). Calm ewes produce milk with more protein than nervous ewes. *Animal Production in Australia, 25,* 307.

Scheffler, K., Stamer, E., Traulsen, I., & Krieter, J. (2016). Relationship between behavioural tests and agonistic interactions at different age levels in pigs. *Applied Animal Behaviour Science, 177,* 19–24.

Schulz, L. L., & Tonsor, G. T. (2015). The U.S. gestational stall debate. *Choices, 30,* 1–4.

Shea-Moore, M. M. (1998). The effect of genotype on behavior in segregated early-weaned pigs tested in an open field. *Journal of Animal Science, 76,* 100.

Sih, A., Bell, A., & Johnson, J. C. (2004). Behavioral syndromes: An ecological and evolutionary overview. *Trends in Ecology & Evolution, 19,* 372–378.

Smith, B. R., & Blumstein, D. T. (2008). Fitness consequences of personality: A meta-analysis. *Behavioral Ecology, 19,* 448–455.

Spake, J. R., Gray, K. A., & Cassady, J. P. (2012). Relationship between backtest and coping styles in pigs. *Applied Animal Behaviour Science, 140,* 146–153.

Spoolder, H., Burbidge, J., Edwards, S., Lawrence, A., & Simmins, P. (1996). Social recognition in gilts mixed into a dynamic group of 30 sows. *Animal Science, 62,* 630.

Stookey, J. M., & Gonyou, H. W. (1998). Recognition in swine: Recognition through familiarity or genetic relatedness? *Applied Animal Behaviour Science, 55,* 291–305.

Sundman, A. (2011). *Personality in piglets: Is there a difference in personality traits between pigs from different teat order positions?* (Unpublished bachelors thesis). Linköping University, Linköping, Sweden.

Studnitz, M., Jensen, M. B., & Pedersen, L. J. (2007). Why do pigs root and in what will they root? A review on the exploratory behaviour of pigs in relation to environmental enrichment. *Applied Animal Behaviour Science, 107,* 183–197.

Svennson, M. (2011). *Personality types and teat order in piglets.* (Unpublished bachelors thesis). Linköping University, Linköping, Sweden.

Thodberg, K., Jensen, K. H., & Herskin, M. S. (1999). A general reaction pattern across situations in prepubertal gilts. *Applied Animal Behaviour Science, 63,* 103–119.

Uitdehaag, K., Rodenburg, T. B., Komen, H., Kemp, B., & Van Arendonk, J. A. M. (2008). The association of response to a novel object with subsequent performance and feather damage in adult, cage-housed, pure-bred Rhode Island Red laying hens. *Poultry Science, 87,* 2486–2492.

Van Erp-van der Kooij, E., Kuijpers, A. H., Schrama, J. W., Ekkel, E. D., & Tielen, M. J. M. (2000). Individual behavioural characteristics in pigs and their impact on production. *Applied Animal Behaviour Science, 66,* 171–185.

Van Erp-van der Kooij, E. E. V., Kuijpers, A. H., Schrama, J. W., Van Eerdenburg, F. J. C. M., Schouten, W. G. P., & Tielen, M. J. M. (2002). Can we predict behaviour in pigs?: Searching for consistency in behaviour over time and across situations. *Applied Animal Behaviour Science, 75*(4), 293–305.

Vazire, S., & Gosling, S. D. (2004). Personality and temperament: A comparative perspective. In M. Bekoff (Ed.), *Encyclopedia of animal behavior* (pp. 818–822). Westport, CT: Greenwood Publishing Group.

Vazire, S., Gosling, S. D., Dickey, A. S., & Schaprio, S. J. (2007). Measuring personality in nonhuman animals. In R. W. Robins, R. C. Fraley, & R. Krueger (Eds.), *Handbook of research methods in personality psychology* (pp. 190–206). New York: Guilford.

Voisinet, B. D., Grandin, T., Tatum, J. D., O'Connor, S. F., & Struthers, J. J. (1997). Feedlot cattle with calm temperaments have higher average daily gains than cattle with excitable temperaments. *Journal of Animal Science, 75,* 892–896.

Von Borell, E. (2001). The biology of stress and its application to livestock housing and transportation assessment. *Journal of Animal Science, 79,* E260–E267.

von Borell, E., & Ladewig, J. (1992). Relationship between behaviour and adrenocortical response pattern in domestic pigs. *Applied Animal Behaviour Science, 34,* 195–206.

Walsh, R. N., & Cummins, R. A. (1976). The open-field test: A critical review. *Psychological Bulletin, 83,* 482.

Walser, E. E. S. (1986). Recognition of the sow's voice by neonatal piglets. *Behaviour, 99*(3), 177–188.

Weaver, S. A., & Morris, M. C. (2004). Science, pigs, and politics: A New Zealand perspective on the phase-out of sow stalls. *Journal of Agricultural and Environmental Ethics, 17,* 51–66.

Wood-Gush, D. G. M., & Vestergaard, K. S. (1989). Exploratory behavior and the welfare of intensively kept animals. *Journal of Agricultural Ethics, 2,* 161–169.

Wood-Gush, D. G. M., Vestergaard, K., & Petersen, H. V. (1990). The significance of motivation and environment in the development of exploration in pigs. *Biology of Behaviour, 15,* 39–52.

Wright, D. (2015). The genetic architecture of domestication in animals. *Bioinformatics and Biology Insights, 9,* 11.

Zwick, W. R., & Velicer, W. F. (1986). Comparison of five rules for determining the number of components to retain. *Psychological Bulletin, 99,* 432.

Personality in Dogs

10

Jamie L. Fratkin

Abstract

Dogs have an ever-increasing presence in our lives from being a child's best friend to detecting roadside bombs to guiding people with visual impairments. Examining the personalities of these animals can help us to better understand them. This chapter reviews many topics related to dog personality, beginning by examining the structure of dog personality. Studies have found that dog personality consists of two to more than five dimensions with little consensus as to the optimal number of dimensions needed to best describe dog personality. This chapter then examines the reliability and validity of dog personality assessments. Finally, this chapter reviews the evidence for various genetic, biological, and environmental factors related to dog personality and concludes that what we currently know is largely mixed. Future studies on dog personality should further examine several personality dimensions and factors likely associated with dog personality.

Personality in Dogs

In the last 20 years, dog personality has become a subject of interest for dog owners and dog practitioners. Potential dog owners want to understand dog personality to be able to select a dog that is suitable for their own lifestyle (e.g., an active dog for an active lifestyle). Dog owners want to understand their dog's personality to be able to understand the best way to train their dog or to provide the type of environment their dog needs. Dogs are now utilized for a variety of purposes, such as guiding people who are visually impaired (e.g., Goddard and Beilharz 1986), detecting drugs and explosives (e.g., Maejima et al. 2007), helping children with autism or veterans with post-traumatic stress disorder (e.g., Hall et al. 2016), and providing therapy to children and adults who are ill (e.g., Haubenhofer and Kirchengast 2007). With these ever-increasing working roles created for dogs,

J.L. Fratkin (✉)
Department of Psychology, The University of Texas at Austin, 108 E. Dean Keeton Stop A8000, Austin, TX 78712, USA
e-mail: fratkijl@gmail.com

J. Vonk et al. (eds.), *Personality in Nonhuman Animals*,
DOI 10.1007/978-3-319-59300-5_10

dog practitioners want to better understand dog personality so that they can select dogs that are most suitable for these working roles. This chapter examines the work to date in the field of dog personality to better understand these practical applications as well as to better understand dog personality more generally. Specifically, this chapter covers the structure of dog personality, different methods to assess dog personality, and the reliability and validity of those assessments. In addition, this chapter reviews studies that have begun to explore different factors that that may be associated with dog personality.

The Structure of Dog Personality

In humans, there is a general consensus among researchers that personality can be described by five broad dimensions, known as the Big Five: extraversion, agreeableness, conscientiousness, neuroticism, and openness to experience (McCrae and Costa 2008). In dogs, there is less consensus about the number of dimensions required to describe personality: some studies find five dimensions (De Palma et al. 2005; Ledger and Baxter 1997; Ley et al. 2009; Svartberg and Forkman 2002), but others find only three (Arata et al. 2010) or four dimensions (Dowling-Guyer et al. 2011; Draper 1995; Mirkó et al. 2012; Sinn et al. 2010; Wahlgren and Lester 2003). Some find more than five dimensions, with some indicating up to 22 dimensions (Hsu and Serpell 2003; Ruefenacht et al. 2002; Goodloe and Borchelt 1998; Serpell and Hsu 2001).

Whereas many dog personality studies use factor analysis to examine the structure of personality, other methods have also been used to be able to understand how already existing traits should be categorized. One other method that has been used in dog personality research is an expert-sorting method, which helps to evaluate the points of similarity and difference across past structural studies (Jones and Gosling 2005). From this method, Jones and Gosling found seven dimensions that consistently emerged across studies (reactivity, fearfulness, activity, sociability, responsiveness to training, submissiveness/dominance, and aggression). In addition, researchers found a none/other dimension that was not considered to be related to personality (e.g., barking, pain sensitivity) or related to an additional aspect of personality that was not part of the seven broad dimensions (see Table 10.1).

Although the exact number of dimensions relating to dog personality varies from study to study, many studies find similar dimensions of dog personality (see Table 10.2). Common dog personality dimensions are activity/excitability, aggression, confidence/fear, extraversion/friendliness/sociability, and obedience/trainability. Less common dimensions that also appear in dog personality studies are chase, distraction/focus, drive, motivation, and playfulness. Even though many dimensions of dog personality appear to be the same and share similar names, the procedures for behaviorally defining those dimensions may differ. For example, Svartberg (2006) defined curiosity/fearfulness as ranging from high fearfulness (e.g., avoidance, flight) and a low tendency to explore in potentially threatening nonsocial situations, to low fearfulness and a high tendency to explore in potentially threatening nonsocial situations. On the other hand, Dowling-Guyer et al. (2011) defined fearfulness as indicated by avoidance and flight behaviors, such as running and looking away, as well as cowering, trembling, tail tucking, and lip licking. In another example, Svartberg (2006) defined aggressiveness as ranging between a low tendency to threaten and act aggressively in unfamiliar and potentially threatening situations, to a high tendency to threaten and act aggressively in unfamiliar and potentially threatening situations. On the other hand, Dowling-Guyer et al. (2011) defined aggressiveness as growling, biting, snapping, and showing teeth as well as hard eye, stiff body, stiff tail, and barking. These differences in definitions led Diederich and Giffroy (2006) to propose a need for consensus and standardization in dog personality assessments and the dimensions generated from those assessments. Although some dog personality assessments today are widely used to assess general dog personality (e.g., the Dog Mentality

Table 10.1 Description of the Jones and Gosling (2005) personality framework

	Dimension description
Reactivity	Indexed by such behaviors and repeated approach/avoidance of novel objects, raised hackles, and increased activity in novel situations. Assessed through procedures such as presenting a novel object or series of novel objects to a puppy and recording is subsequent behavior. Includes traits labeled as 'excitability', 'sound reaction', and 'heart reactivity'
Fearfulness	Exhibited by signs of excitement, pacing, or running around, avoidance of social stimuli, and barking. Also includes shaking and a tendency to avoid novel stimuli without approaching them. Includes trait labels 'courage', 'confidence', 'self-confidence', 'apprehension', 'dog-directed fear or aggression', and 'timidity'
Activity	Often assessed by placing a puppy or dog in an empty arena with gridlines on the floor and seeing how many times the puppy or dog crosses the lines. Includes traits labeled as 'activity', 'locomotor activity,' and 'general activity'
Sociability	Indexed by such behaviors as initiating friendly interactions with people and other dogs. Primarily assessed in meetings between dogs and an unfamiliar person. Includes traits labeled 'extraversion', 'affection demand', and 'affability'
Responsiveness to training	Indexed by such behaviors as working with people, learning quickly in new situations, playfulness, and overall reaction to the environment. Related to a dog's tendency to stay focused on engaged in a given activity. Normally assessed through giving dogs puzzles to solve, willingness to work with a person, and retrieval tests. Includes traits labeled 'distractibility' 'focus', 'problem solving', 'willingness to work', and 'cooperative'
Submissiveness	The opposite of dominance. Dominance can be judged by observing which dogs bully others, and which guard food areas and feed first. Submission can also be reflected by such behaviors as urination upon greeting people
Aggression	Indexed by behaviors such as biting, growling, and snapping at people or other dogs. Often assessed through having strangers approach the dog in a threatening manner. Includes traits labeled as 'stranger-directed fear or aggression', 'owner-directed aggression', 'dog-directed fear or aggression', 'sharpness', and the 'willingness to bite a human being'

Assessment; Svartberg and Forkman 2002), there still remains a wide variety of personality assessments used to assess dog personality with varying personality dimensions examined from these assessments. More work needs to be done to create dog personality assessments that assess similar dimensions with similar definitions in terminology, so as in human personality studies, comparisons can be made across studies and populations.

Some dog personality assessments are beginning to emerge as frequently used instruments. Perhaps the most frequently used is the Canine Behavioral Assessment and Research Questionnaire (C-BARQ; Hsu and Serpell 2003), which consists of 11 dimensions that were derived through factor analysis: stranger-directed aggression, owner-directed aggression, stranger-directed fear, nonsocial fear, dog-directed fear or aggression, separation-related behavior, attachment or attention-seeking behavior, trainability, chasing, excitability, and pain sensitivity. Another common instrument for measuring dog personality is the Dog Mentality Assessment (Svartberg and Forkman 2002), which was developed with the Swedish Working Dog Association and consists of five dimensions (playfulness, curiosity/fearfulness, chase-proneness, sociability, and aggressiveness). A more recently developed instrument is the Monash Canine Personality Questionnaire (Ley et al. 2009), which consists of five dimensions (extraversion, motivation, training focus, amicability, and neuroticism). Future dog personality studies may want to focus on using personality assessments that are already used frequently, so comparisons can be made across studies.

Table 10.2 Summary of names of dimensions found across studies examining the structure of dog personality

Dimension	Name of dimension from study	Studies
Activity/excitability	Active, distraction, energy level, excitability, intraspecific dominance-activity, reactivity	Arata et al. (2010), De Palma et al. (2005), Draper (1995), Hsu and Serpell (2003), Ledger and Baxter (1997), Mirkó, et al. (2012), Serpell and Hsu (2001), Wahlgren and Lester (2003)
Aggression	Aggression, aggression to family member, aggression to strangers, aggression to unfamiliar dogs, dog-directed fear or aggression, owner-directed aggression, sharpness, stranger-directed aggression, subordination/aggressiveness	De Palma et al. (2005), Dowling-Guyer et al. (2011), Draper (1995), Hsu and Serpell (2003), Ledger and Baxter (1997), Mirkó et al. (2012), Ruefenacht et al. (2002), Serpell and Hsu (2001), Sinn et al. (2010), Svartberg and Forkman (2002)
Chase	Chase-proneness, chasing	Hsu and Serpell (2003), Serpell and Hsu (2001), Svartberg and Forkman (2002)
Confidence/fear	Curiosity/fearlessness, dog-directed fear or aggression, fearfulness, fear or avoidance of strangers, hardness, nerve stability, nonsocial fear, neuroticism, self-confidence, sensitivity, stranger-directed fear, timidity	Arata et al. (2010), Dowling-Guyer et al. (2011), Hsu and Serpell (2003), Ledger and Baxter (1997), Ley et al. (2009), Ruefenacht et al. (2002), Serpell and Hsu (2001), Svartberg and Forkman (2002), Wahlgren and Lester (2003)
Distraction/focus	Human focus, object focus, search focus	Sinn et al. (2010)
Drive	Defense drive, fighting drive	Ruefenacht et al. (2002)
Extraversion/ friendliness/ sociability	Affiliation, amicability, anxiety-sociability towards dogs, attachment, extraversion, friendliness, sociability, sociability toward humans, stranger-directed sociability	De Palma et al. (2005), Dowling-Guyer et al. (2011), Hsu and Serpell (2003), Ley et al. (2009), Mirkó et al. (2012), Serpell and Hsu (2001), Svartberg and Forkman (2002), Wahlgren and Lester (2003)
Motivation	Motivation	Ley et al. (2009)
Obedience/trainability	Docility, obedience, training focus	Arata et al. (2010), Draper (1995), Hsu and Serpell (2003), Ledger and Baxter (1997), Ley et al. (2009), Mirkó et al. (2012), Serpell and Hsu (2001), Wahlgren and Lester (2003)
Playfulness	Interest, investigation, playfulness	De Palma et al. (2005), Dowling-Guyer et al. (2011), Draper (1995), Svartberg and Forkman (2002)

Methods for Assessing Dog Personality

Various methods have been used to assess dog personality, including test batteries, observational tests, ratings of individual dogs, and expert ratings of breed prototypes (Jones and Gosling 2005). The most common method for assessing dog personality is using a test battery (Jones and Gosling 2005). Test batteries often involve presenting various stimuli to a dog and observing the dog's reaction to the stimuli. For example, one study used a test battery to examine genetic and non-genetic effects on personality traits for German Shepherds (Ruefenacht et al. 2002).

The test battery consisted of eight parts and eight traits (self-confidence, nerve stability, reaction to gunfire, temperament, hardness, sharpness, defense drive, fighting drive) that were scored from the test, along with an overall score that indicated the dog's overall performance.

Observational tests are another method and involve assessing and describing traits seen in the dog's naturally occurring behavior. Observational tests take place in dogs' natural environments rather than in the controlled environments used in test batteries. For example, one study examined how dogs in training to become explosive detection dogs reacted to different environments, including a general store, a woodshop, an airport cargo area, and an airport, with 10 traits (confidence, concentration, responsiveness, initiative, excitability, hearing sensitivity, body sensitivity, chase retriever, independent possession, physical possession) being assessed from the test (Fratkin et al. 2015).

Another method, ratings of individual dogs, involves gathering information about a dog's behaviors from an informant, often an owner or someone else familiar with the dog. For example, the C-BARQ involves having dog owners rate their dog on a series of traits based on how the dog typically responds to various stimuli. The ratings are based on an overall impression of the dog, rather than a dog's reaction during a short testing situation (Hsu and Serpell 2003).

In addition to these methods of assessing the personalities of individual dogs, expert ratings of breed prototypes involve experts (e.g., veterinarians, dog trainers) describing breeds of dogs. For example, one study examined how dog show judges, professional dog handlers, and veterinarians rated 56 different breeds on 13 traits (Hart and Hart 1985). This method is used to examine overall impressions of breeds of dogs, rather than examining a specific individual dog.

Each of these methods has advantages and disadvantages. Test batteries and observational tests allow for one person to score a large number of dogs because the rater does not have to have any knowledge about a particular dog going into the test. However, in test batteries and observational tests, raters view dogs during only a short period of time and thus lack the aggregation (i.e., summing or averaging multiple observations) that the other methods use, which often helps improve reliability (Epstein 1983). Ratings of individual dogs can be used in situations where informants are familiar with certain dogs and thus dogs can be rated across a number of situations, rather than viewing the dog through a short testing situation. However, there is a limit to the number of dogs each informant can rate because each informant may know only one or two dogs well enough to provide ratings. Expert ratings of breed prototypes can also be used to rate multiple breeds of dogs because experts often know many dogs well. However, expert ratings do not factor in the idea that individuals within breeds may differ from the typical breed prototype. Ultimately, to best determine a dog's personality it may be helpful to use more than one method (e.g., Fratkin et al. 2015; Mirkó et al. 2013).

How Good Are Dog Assessments?

While there are various ways to assess dog personality, for any of these assessments to be useful they need to be both reliable and valid. Below, I discuss results from studies examining reliability and validity of dog personality assessments.

Reliability

Reliability examines the consistency of a measurement and is a prerequisite for validity, so it is important to first examine the reliability of a measure (John and Soto 2007). Many studies before 2005 did not report on the reliability of dog personality measures (Jones and Gosling 2005). Further, many of the studies that reported reliability failed to quantify the reliability assessment or did not report complete reliability results. However, in more recent years the number of studies reporting reliability estimates has increased with 57% of recent studies reporting reliability, compared to 19% of studies prior to 2005 (Gartner 2015). Below, I discuss findings from studies examining reliability.

Test–Retest Reliability

Test–retest reliability examines the consistency of personality over time (John and Soto 2007). A meta-analysis of dog personality studies found evidence for good test–retest reliability, with an estimated weighted mean effect size of 0.43 (Fratkin et al. 2013). However, several factors affected test–retest reliability. First, test–retest reliability was not as good when dogs were first assessed as puppies (<12 months) compared to adults (>12 months), which is similar to humans (Roberts and DelVecchio 2000), but different in other animals (Bell et al. 2009), where there was no overall difference between juveniles and adults. Second, the longer the interval between assessments, the worse test–retest reliability became. Third, in puppies, assessments of aggression and submissiveness were found to be more stable than those of responsiveness to training and reactivity/fearfulness. These results suggest that puppy personality may not be as consistent as adult dog personality, but certain dimensions may be examined with relatively good test–retest reliability in puppies.

Inter-Observer Reliability

Inter-observer reliability examines the consistency of personality across raters or coders. Before 2005, inter-observer reliability was found to differ across studies and dimensions, with estimates ranging from 0.00 to 0.86, with a sample weighted mean correlation of 0.60 (Jones and Gosling 2005). After 2005, estimates ranged from 0.01 to 1.0 with a sample weighted mean correlation of 0.68 (Gartner 2015).

Inter-observer reliability seems to differ based on the dimension that is being assessed. Some studies found inter-observer reliability to be the highest for confidence or fear (Fratkin et al. 2015; Mornement et al. 2014), but one study indicated inter-observer reliability was poorest for self-confidence (Paroz et al. 2008). Several studies found inter-observer reliability was poorer for activity than for other dimensions (Fratkin et al. 2015; Jakuba et al. 2013; Paroz et al. 2008). Other studies found inter-observer reliability to be poor for sociability with other dogs (Diesel et al. 2008), sociability with people (Valsecchi et al. 2011), and friendliness (Mornement et al. 2014), but good for stranger-directed aggression (Jakuba et al. 2013). With so much inconsistency in the findings, whether some dimensions have better inter-observer reliability than others is yet to be clearly established. Many of these studies use different assessments and designs to score each dimension, so it is difficult to directly compare reliability results. However, overall, there is evidence that dog personality assessments can be measured with good inter-observer reliability.

Validity

Validity provides an index of the accuracy of dog assessments or how well an instrument is measuring what it is meant to be measuring. Only 27% of dog personality studies reported convergent validity estimates before 2005, but of those studies, convergent validity was found to be relatively good in comparison to human measures of validity (e.g., Paulhus and Bruce 1992), averaging 0.40 (Jones and Gosling 2005). More recently, 70% of studies of dog personality reported convergent validity findings, averaging between 0.19 and 0.74, depending on the dimension assessed, which will be discussed in further detail below (Gartner 2015).

Convergent Validity

Convergent validity examines how well a measure correlates with other measures of the same construct (e.g., one measure of fearfulness should match with another measure of fearfulness). Many studies found good evidence for convergent validity, but like reliability, convergent validity was better for some dimensions than for others. In one synthesis of past research, activity had the lowest convergent validity, whereas fearfulness had the highest convergent validity (Jones and Gosling 2005). However, in that synthesis of past research, only two studies reported convergent validity for activity, so the estimate for that dimension may not be robust. More recent work suggested the lowest convergent validity was for boldness, whereas the best convergent validity was for trainability (Gartner 2015), but again, few estimates of convergent validity were available from which to calculate validity.

Discriminant Validity

Discriminant validity examines how much one measure correlates with another measure that it should not be related to (e.g., a measure of fearfulness should not correlate with activity). Very few studies have examined discriminant validity in dog personality studies, but from the studies that do report discriminant validity, there is evidence for good discriminant validity (Hsu and Serpell 2003; Serpell and Hsu 2001). More research is needed on discriminant validity in the future. Examining both convergent and discriminant validity altogether allows researchers to evaluate construct validity or how well a measure assesses a construct as they intend (Cronbach and Meehl 1955).

Predictive Validity

An important question about dog personality assessments is in regards to what these measures of dog personality predict. For example, do personality measures predict whether a given dog can become a guide dog? Predictive validity is particularly important to examine dogs because doing so addresses questions posed by dog organizations (e.g., how do working dogs behave in their job or how do shelter dogs behave in their future home?). Approximately half of the dogs that are bred and trained for working roles are found to be unsuitable as working dogs through the training process (Maejima et al. 2007; Sinn et al. 2010; Slabbert and Odendaal 1999; Wilsson and Sundgren 1997), so determining predictors of successful dogs is important. With working dogs potentially costing up to $50,000 or more to breed and train (Wirth and Rein 2011), determining predictors of success may help to increase the 50% success rate and reduce the amount of time and resources spent on dogs that are not suitable as working dogs.

Several studies have examined predictors of success in different types of working dogs, including guide dogs, drug and explosives detector dogs, police dogs, military dogs, service dogs, and shelter dogs. One of the earliest studies found that more fearful dogs were more likely to be rejected from programs that trained them as guide dogs (Goddard and Beilharz 1982a). Later studies found that fearfulness (Tomkins et al. 2011) and stress reactions (Mizukoshi et al. 2008) were negatively related to guide dog success. Some studies found that dogs that were less easily distracted were more likely to succeed as guide dogs (Arata et al. 2010; Kobayashi et al. 2013). Other personality traits predictive of guide dog success have included traits related to trainability, such as the latency to sit (Tomkins et al. 2011) and pulling on leash (Duffy and Serpell 2012). Greater sociability (Asher et al. 2013) and lower aggression (Duffy and Serpell 2012; Takeuchi et al. 2009a, b) also predicted guide dog success.

Other studies focused on detector, police, and military working dog success. Successful police dogs tend to be aggressive and tend to do well on a test assessing their ability to retrieve a toy, which is often related to responsiveness to training. In this assessment, dogs were scored based on their focus on retrieving the toy and the steps they used to find and return the toy to the handler (Slabbert and Odendaal 1999). Successful drug detection dogs had a high desire for work, which was based on high scores on items such as concentration, general activity, and anxiety (Maejima et al. 2007). Successful military working dogs had high search focus and sharpness (Sinn et al. 2010) and high confidence (Foyer et al. 2014; Wilsson and Sinn 2012).

Other research examined factors that predict service dog success in dogs from shelters. Early research was not successful in finding predictors of success (Weiss and Greenberg 1997). However, later research found that a dog's sensitivity as measured by their reaction to being lightly pinched, and a dog's activity level positively predicted service dog success (Weiss 2002). Other predictors of success from the same study were based on the dog's reaction to another dog, a stranger, and a stranger staring at the dog for 30 s. In another study, some researchers developed an assessment tool that was successful at identifying shelter dogs that would pass a Delta Society test, which assesses therapy dogs (Lucidi et al. 2005). This assessment tool consisted of assessing dog aggressiveness, initiative, sociability, fearfulness, and trainability and dogs that displayed low aggression and fearfulness and high initiative and trainability passed the assessment the researchers developed.

A key concern for animal shelters is finding the right home for their animals. Some studies have focused on shelter dogs, aiming to predict the dog's behavior in his or her new home. Some research suggested that wariness predicted problem behaviors (Hennessy et al. 2001), but other studies found that behavioral assessments were not associated with later behavior (Christensen et al. 2007; Mornement et al. 2014, 2015). In one study, researchers found evidence suggesting that 40.9% of dogs that passed a behavioral assessment in a shelter displayed aggressive behaviors (e.g., lunging, growling, snapping, and/or biting) after they were placed in their new homes (Christensen et al. 2007). Further research suggested that most traits that were tested did not predict a dog's behavior in its new home, but that fear predicted later fearful/inappropriate toileting behavior (Mornement et al. 2015) and friendliness predicted later friendly behavior (Mornement et al. 2014). Although the reliability of many shelter behavioral assessments is good, their predictive validity often is not (Mornement et al. 2014). Predicting later behavior in shelter dogs may be particularly difficult because the shelter environment may be extremely stressful for dogs. Shelters would like to determine the suitability of dogs as soon as possible, but some research suggests little consistency in scores between assessments at day 0 of intake and day 3 (Bennett et al. 2015). Dogs may need time to acclimate to the shelter before their personality can be observed accurately.

Overall, studies have examined the predictive validity of dog personality assessments and have determined that several traits are associated with later outcomes. These assessments can help us determine the successful personality profiles of many types of working dogs as well as companion and shelter dogs. Knowing personality traits that make successful working dogs allows working dog organizations to focus on breeding and training more dogs with those particular traits. Knowing the personality of shelter dogs can help us to better match shelter dogs to owners. Eventually, the success rates of working dogs should increase and there should be more successful matches between shelter dogs and new owners, which means more dogs trained to do tasks that help keep us safe (e.g., detecting explosives) and less dogs in shelters.

What Factors Are Associated with Dog Personality?

Many studies examined how factors relating to biology and the dog's environment may be related to the dog's personality. Many studies have been conducted for a variety of purposes, from better understanding the genetics of dogs being trained as guide dogs (Goddard and Beilharz 1982a, b), to examining personality differences across different breeds (Turcsán et al. 2011), to examining how characteristics of dog owners are associated with dog personality (Kubinyi et al. 2009). To get a sense of the factors associated with dog personality, I reviewed the literature, organizing the findings using the framework set out in Jones and Gosling's (2005) review of the dog personality literature (see Table 10.2). Traits related to fearfulness and traits related to reactivity often overlap, so to make the findings clearer and more concise, these dimensions will be combined.

Biological Factors

Heredity

Many studies have examined the heritability of personality, seeking to explain the proportion of variance related to genetic effects. Heritability studies often involve examining one or two large samples of purebred dogs. German Shepherds are one of the most frequently studied breeds in genetic studies (Hradecká et al. 2015); other breeds studied include Belgian Shepherd Dogs, Border Collies, Boxers, Brittany Spaniels, Dachshunds, English Cocker Spaniels, English Setters, Finnish Spitz, Flat Coated Retrievers, German Shorthaired Pointers, German Wirehaired Pointers, Golden Retrievers, Hovawarts, Labrador Retrievers, Rottweilers, Rough Collies, and Shiba Inus.

A meta-analysis of studies examining heritability of dog personality found that they were low with the overall weighted mean between 0.09 (for play) and 0.15 (for environment and hunting) (Hradecká et al. 2015). In addition, there were no differences in heritability based on sex or breed. Other research dividing personality along different personality dimensions suggests that some personality dimensions are more heritable than others. Reactivity/fearfulness (Goddard and Beilharz 1982a, b; Meyer et al. 2012; Strandberg et al. 2005) and activity (Lindberg et al. 2004; Wilsson and Sundgren 1998) appear to be highly heritable with estimates ranging from 0.13 for courage in Labrador Retrievers to 0.56 for gun shyness in Labrador Retrievers (van der Waaij et al. 2008) and from 0.17 for hunting eagerness in German Wirehaired Pointers (Brenøe et al. 2002) to 0.53 for activity in German Shepherds (Wilsson and Sundgren 1998). Sociability does not appear to be as strongly heritable with estimates ranging from 0.03 for affability in Labrador Retrievers to 0.38 for affability in German Shepherds (van der Waaij et al. 2008). Some estimates of heritability for traits related to responsiveness to training and aggression are high too, such as for cooperation in Brittany Spaniels (0.90) (Brenøe et al. 2002) and dog-directed aggression in Golden Retrievers (0.88) (Liianmo et al. 2007), but others are lower, such as for willingness to cooperate in Labrador Retrievers (0.35) (Wilsson and Sundgren 1997) and sharpness in German Shepherds (0.09) (Meyer et al. 2012). Heredity studies allow some basis to understand dog personality, but other biological factors also account for personality differences.

Breed

The American Kennel Club currently lists 189 breeds of dogs (American Kennel Club 2016) and other estimates suggest that there are more than 400 breeds of dogs (Clutton-Brock 1995). Dog breeds have existed since around 3000–4000 years ago (Clutton-Brock 1995) or even earlier (Miklósi 2007) likely being created through human selection based on certain tasks that humans needed dogs for (e.g., hunting or herding). Humans sought dogs that had characteristics or traits that made them successful in these tasks. Dogs with those traits were then treated favorably and received food and other resources that helped them to survive and reproduce (Serpell and Duffy 2014). These dogs likely formed the basis for breeds, with different breeds displaying different personalities based on the specific needs humans used them for.

With the emergence of dog shows and kennel clubs in the mid-nineteenth century (Sampson and Binns 2006), modern dog breeding began to focus on physical appearance, with less of a focus on behavioral traits in breed standards (McGreevy and Nicholas 1999). Even though the focus of dog breeding shifted to focusing on physical appearance, dog breeding still likely led to some personality differences between dog breeds. In one of the most comprehensive studies of breed differences, Serpell and Duffy (2014) examined 30 of the most popular American Kennel Club registered breeds and found differences between breeds on C-BARQ scores (see Table 10.3). Specifically, toy breeds (e.g., Chihuahuas) scored high on dimensions similar to reactivity/fearfulness, activity, and aggression, which is further supported by other research indicating that toy breeds were the most fearful dogs (Temesi et al. 2014). In contrast, Golden Retrievers (Serpell and Duffy 2014), Labrador Retrievers (Goddard and Beilharz 1985; Serpell and Duffy 2014; Svartberg 2006; Wilsson and Sundgren 1997), and Siberian Huskies (Serpell and Duffy 2014) tended to score low on fearfulness. Serpell and Duffy (2014) also found that herding breeds (e.g., Australian Shepherds) and sporting breeds (e.g., Golden Retrievers) scored

Table 10.3 Summary of breed differences of 30 popular dog breeds examined by Serpell and Duffy (2014) using the C-BARQ dimensions

	Stranger-directed aggression	Owner-directed aggression	Dog-directed aggression	Dog rivalry	Attachment/attention-seeking	Predatory chasing	Trainability
Dog breed scoring high	Chihuahua, Dachshund, GSD, Maltese, Miniature Schnauzer, Yorkshire Terrier	Chihuahua, Dachshund, GSD, Miniature Schnauzer, Yorkshire Terrier	Beagle, Chihuahua, French Bulldog, Maltese, Pomeranian, Shih Tzu, Yorkshire Terrier	Beagle, Chihuahua, French Bulldog, Yorkshire Terrier	Toy Poodle, Yorkshire Terrier	German Shorthaired Pointer, Miniature Schnauzer, Siberian Husky	Australian Shepherd, Doberman Pinscher, Golden Retriever, Labrador Retriever, Standard Poodle, Rottweiler, Shetland Sheepdog
Dog breeds scoring low	Bulldog, Cavalier King Charles Spaniel, Golden Retriever, Labrador Retriever, Pug, Siberian Husky	Cavalier King Charles Spaniel, Golden Retriever, Labrador Retriever, Pug	Australian Shepherd, Boxer, Doberman Pinscher, Golden Retriever, Great Dane, Labrador Retriever	Golden Retriever, Great Dane, Havanese, Labrador Retriever, Pug	Bulldog, Siberian Husky	Bulldog, Chihuahua, English Mastiff, Pug, Shih Tzu	Beagle, Dachshund, Pug, Yorkshire Terrier

	Stranger-directed fear	Dog-directed fear	Nonsocial fear	Separation-related problems	Touch sensitivity	Excitability	Energy
Dog breed scoring high	Chihuahua, Dachshund, Yorkshire Terrier	Beagle, Chihuahua, Dachshund, Maltese, Shih Tzu, Toy Poodle, Yorkshire Terrier	Beagle, Chihuahua, Dachshund, Maltese, Shih Tzu, Toy Poodle, Yorkshire Terrier	Beagle, Dachshund, Maltese, Toy Poodle, Yorkshire Terrier	Beagle, Chihuahua, Maltese, Shih Tzu, Toy Poodle, Yorkshire Terrier	Boston Terrier, Maltese, Miniature Schnauzer, Yorkshire Terrier	Australian Shepherd, Boxer, Doberman Pinscher, German Shorthaired Pointer
Dog breeds scoring low	French Bulldog, Golden Retriever, Labrador Retriever, Rottweiler, Siberian Husky	Doberman Pinscher, Rottweiler, Siberian Husky	Doberman Pinscher, GSD, Labrador Retriever, Rottweiler, Siberian Husky	Australian Shepherd, Bulldog, Golden Retriever, Labrador Retriever, Rottweiler	Boxer, Cavalier King Charles Spaniel, English Mastiff, Golden Retriever, Rottweiler	Bulldog, English Mastiff, Siberian Husky	Bulldog, English Mastiff, Great Dane

Note The dog breeds listed in each category represent the breeds that scored high or low for the dimension and do not include every breed assessed. High and low scorings are based on either being below or above the overall average for that dimension high on traits related to responsiveness to training. Other studies found similar results (Ley et al. 2009; Turcsán et al. 2011), particularly in Labrador Retrievers (Serpell and Hsu 2005). Finally, guard breeds (e.g., Doberman Pinschers, German Shepherds) scored high on stranger-directed aggression (Serpell and Duffy 2014).

Today, most dogs are companions and may not perform the role that they were originally bred for (King et al. 2011). In addition, dogs also perform a variety of working roles for people. As a result, along with differences between breeds, it is important to note that there are also significant personality differences within breeds (Mehrkam and Wynne 2014; Serpell and Hsu 2014). These within-breed differences may occur based on the newer roles certain dog breeds are bred for. For example, German Shepherds are often bred for both police dog and guide dog work. While successful police dogs are often aggressive (Slabbert and Odendaal 1999), successful guide dogs are not often aggressive (Duffy and Serpell 2012; Takeuchi et al. 2009a, b), but are confident (Goddard and Beilharz 1982a). In addition, even within specific populations of dogs (e.g., military working dogs) there appear to be substantial personality differences. For example, Foyer et al. (2014) examined German Shepherds trained as military working dogs and found personality differences within that population of German Shepherds. Breed differences and heredity alone cannot account for all of the personality differences between dogs, so it is important to examine other biological and environmental factors associated with dog personality.

Morphology

Some research has shown that size, height, and coat color are associated with dog personality (see Table 10.4). Smaller dogs were found to be more neurotic (Arhant et al. 2010; Ley et al. 2009), fearful (Arhant et al. 2010; Stone et al. 2016), less sociable (Wahlgren and Lester 2003) and less obedient (Wahlgren and Lester 2003) than bigger dogs. Shorter dogs were found to be more aggressive than taller dogs (Stone et al. 2016). However, these results are likely confounded due to the breed of the dog. Many toy breeds also score high on reactivity/fearfulness (Serpell and Duffy 2014) and toy breeds tend to be small and short. However, not all small breeds score high on reactivity/fearfulness (e.g., Pugs), so other factors may be associated with personality differences based on size.

Other personality differences have been found based on the coat color of the dog. In Labrador Retrievers, black and yellow dogs displayed more noise sensitivity than chocolate dogs, but chocolate dogs were more excitable and less trainable than black dogs (Lofgren et al. 2014). In English Cocker Spaniels, solid color dogs were more aggressive than particolored dogs and golden colored dogs were more aggressive than black dogs (Podberscek and Serpell 1997) or other colored dogs (Amat et al. 2009). Differences based on coat color are likely related to genetics. In the 1960s, researchers selectively bred foxes based on tameness and found a connection between behavior and morphology from that population of foxes (Trut 1999). Specifically, similar to aggression differences in particolored versus solid colored English Cocker Spaniels, researchers found that tamer foxes had a piebald pattern whereas less tame foxes had a solid color. This work suggests an association between coat color and personality.

Sex

Several studies have reported sex differences in dog personality (Amat et al. 2009; Döring et al. 2016; Goddard and Beilharz 1982a; Goodloe and Borchelt 1998; Hsu and Sun 2010; Kubinyi et al. 2004, 2009; Lindberg et al. 2004; Plutchik 1971; Podberscek and Serpell 1997; Reuterwall and Ryman 1972; Rooney and Bradshaw 2003; Roth and Jensen 2015; Starling et al. 2013; Strandberg et al. 2005; Takeuchi et al. 2009a, b; Temesi et al. 2014; Wells and Hepper 2000; Wilsson and Sundgren 1997). For some dimensions, there appear to be consistent sex differences, but for other dimensions, the differences are less clear (see Table 10.4). Two dimensions with relatively consistent sex differences are reactivity/fearfulness and aggression. All studies found that females scored higher than males on reactivity/fearfulness (Döring et al. 2016;

Table 10.4 Summary of findings of characteristics associated with dog personality

	Reactivity/ fearfulness	Activity	Sociability	Responsiveness to training	Aggression
Sex					
Male		Evidence		Evidence	Strong evidence
Female	Strong evidence	Strong evidence	Evidence	Evidence	Evidence
No difference		Strong evidence	Evidence	Strong evidence	Strong evidence
Sexual status					
Intact		Evidence		Evidence	Evidence
Neutered	Evidence	Evidence		Evidence	Evidence
No difference				Evidence	Evidence
Age					
Younger	Strong evidence	Strong evidence	Evidence	Evidence	Evidence
Older	Evidence			Evidence	Evidence
No difference	Evidence			Evidence	Evidence
Human gender					
Male					
Female		Evidence	Evidence	Evidence	Evidence
Human age					
Younger		Evidence	Evidence		
Older					
Human personality					
Extraversion			Evidence		
Agreeableness					
Conscientiousness					
Emotional stability			Evidence		
Openness					
Experience					
No experience	Evidence	Evidence			Evidence
Experience				Evidence	
Country					
USA	Evidence				Evidence
Hungary					
Exercise/ training/play					
Little/none	Strong evidence				Strong evidence
More/some			Evidence	Strong evidence	Evidence
No difference		Evidence			
Weight					
Heavier		Evidence	Evidence	Strong evidence	
Lighter	Evidence	Evidence			

(continued)

Table 10.4 (continued)

	Reactivity/ fearfulness	Activity	Sociability	Responsiveness to training	Aggression
Height					
Taller					
Shorter					Evidence
Coat color					
Black (LR)	Evidence			Evidence	
Yellow (LR)	Evidence				
Chocolate (LR)		Evidence			
Solid (ECS)					Evidence
Golden (ECS)					Evidence

Note Strong evidence = 3 or more studies have found dogs with that characteristic scored higher than the comparison characteristic, evidence = 1 or 2 studies have found dogs with that characteristic scored higher than the comparison characteristic, Blank cells = no studies reporting dogs with characteristic scored higher than the comparison characteristic, *LR* Labrador Retriever, *ECS* English Cocker Spaniel

Goddard and Beilharz 1982a; Kubinyi et al. 2009; Plutchik 1971; Starling et al. 2013; Strandberg et al. 2005; Temesi et al. 2014; Wilsson and Sundgren 1997), which includes fear relating to avoiding objects (Plutchik 1971) and being low in boldness (Kubinyi et al. 2009; Starling et al. 2013; Strandberg et al. 2005). For aggression, male dogs were more aggressive than female dogs (Amat et al. 2009; Borchelt 1983; Hsu and Sun 2010; Podberscek and Serpell 1997; Reuterwall and Ryman 1972; Strandberg et al. 2005), though some studies failed to find evidence for sex differences for aggression (Marder et al. 2013; Meyer et al. 2012; Schneider et al. 2013). Hormonal differences are one possible explanation for the difference between male and female aggression. Male dogs have higher levels of testosterone than females and, because testosterone is linked to aggression in dogs (Borchelt 1983), males may be more likely to display aggression than females.

There was mixed evidence for sex differences relating to activity, sociability, and responsiveness to training with several studies even finding no sex differences for these dimensions (Gazzano et al. 2008; Schneider et al. 2013; Seksel et al. 1999; Serpell and Hsu 2005; Vas et al. 2007; Wahlgren and Lester 2003; Wright et al. 2011). For example, female Labrador Retrievers training to become guide dogs scored higher on activity than male Labrador Retrievers (Takeuchi et al. 2009a, b), but male Flat Coated Retrievers were more excitable than female Flat Coated Retrievers (Lindberg et al. 2004). Although there are clear sex difference findings for reactivity/fearfulness and aggression, further work needs to be done to better understand sex differences for activity, sociability, and responsiveness to training.

Reproductive Status

Although many assumptions have been made about reproductive status (i.e., whether dogs are spayed/neutered) and dog personality, few studies have reported differences between intact and neutered dogs (see Table 10.4). Some research examining reproductive status found that intact male and female dogs scored lower on reactivity/fearfulness (Lofgren et al. 2014; Tonoike et al. 2015), but higher on activity (Lofgren et al. 2014) than spayed female dogs, though other research has found that the opposite is true for activity (Kubinyi et al. 2009; Tonoike et al. 2015). Personality differences based on reproductive status may vary based on breed. For example, intact Shetland Sheepdogs and Rottweilers scored lower on responsiveness to training than altered dogs of these breeds, but there were no differences based on reproductive status for trainability for the nine other breeds that were examined (Serpell and Hsu 2005).

Research on reproductive status has most commonly focused on whether neutering results in reduced aggression in male dogs. The idea is that aggression may be a result of hormonal differences and, when male dogs are neutered, gonadal steroid hormones decrease (Landsberg et al. 2012). Decreasing gonadal steroid hormones should decrease aggression, but findings have been mixed in this regard. Some studies found that neutering reduces aggression (Borchelt 1983; Hart and Eckstein 1997; Nielson et al. 1997), but not in all dogs (Hart and Eckstein 1997). Some studies even found that neutered dogs were more aggressive than non-neutered dogs (Fadel et al. 2016; Podberscek and Serpell 1997), so the effects of reproductive status of neutered dogs still need to be examined further.

Age

Several studies have found personality differences based on the age of the dog (see Table 10.4). For example, researchers found that younger dogs scored higher than older dogs on traits relating to reactivity/fearfulness (Starling et al. 2013; Tiira and Lohi 2015; Wells and Hepper 2000) (except for one study: Döring et al. 2016), activity (Kubinyi et al. 2009; Mirkó e al. 2012; Vas et al. 2007), and sociability (Kubinyi et al. 2009; Wahlgren and Lester 2003). Age-related differences in personality could be a result of developmental changes in dogs as dogs reach sexual maturity at around 6 months of age and thus hormonal changes may influence behavior at around this age (Lindsay, 2001).

Age-related differences are less clear for responsiveness to training and aggression. Some research suggests that younger dogs are more responsive to training (Kubinyi et al. 2009) and aggressive (Kubinyi et al. 2004) than older dogs, whereas other research suggests that older dogs are more responsive to training (Roth and Jensen 2015) and aggressive (Bennett and Rolhf 2007; Hsu and Sun 2010; Wells and Hepper 2000) than younger dogs. Some research suggests no age-related difference for responsiveness to training (Schneider et al. 2013) or aggression (Meyer et al. 2012; Schneider et al. 2013). Age differences for aggression may be related to the type of aggression. Older dogs displayed more owner-directed aggression (Hsu and Sun 2010) and dog-directed aggression (Wells and Hepper 2000) than younger dogs, but younger dogs barked and growled more than older dogs (Kubinyi et al. 2004). These results suggest that older dogs may be more physically aggressive, whereas younger dogs may be more vocally aggressive.

Environmental Factors

Several characteristics related to the dog's environment appear to be related to dog personality. Many of these characteristics are related to the dog's caregiver, including the caregiver's gender, age, personality, experience, and environmental factors within the caregiver's control (e.g., the number of other people the dog lives with) (see Table 10.4). First, the gender of the caregiver appears to be important. Male dog owners tended to have dogs that were calmer (Kubinyi et al. 2009), less obedient (Bennett and Rohlf 2007), less social (Kotrschal et al. 2009) and less aggressive (Hsu and Sun 2010) than did female dog owners. Second, the age of the owner appears to be important. Dog owners who were under 18 years of age tended to have calmer dogs than did older dog owners (Kubinyi et al. 2009), and older dog owners tended to have dogs that were less extraverted/social than did younger dog owners (Ley et al. 2009). Third, the personality of the caregiver seems to be associated with the dog's personality. Compared to the dogs of less extraverted dog owners, dogs of more extraverted owners looked more often at an experimenter (Kis et al. 2012). In addition, owners who were higher in neuroticism had dogs that took longer to sit on command than dogs whose owners were lower in neuroticism (Kis et al. 2012). Fourth, the experience of the caregiver is associated with dog personality. The dogs of first-time owners scored higher on noise sensitivity (Tiira and Lohi 2015), overexcitement (Jagoe and Serpell 1996), and disobedience (Bennett and Rohlf 2007; Jagoe and Serpell 1996) than dogs belonging to more experienced owners. Fifth, training, socialization and play appear to be associated with personality. These studies show that dogs that did not receive these experiences were more fearful (Fuchs et al. 2005; Schneider et al. 2013; Tiira and Lohi 2015; Tóth et al. 2008), less trainable (Kubinyi et al. 2009; Kutsumi et al. 2013; Mirkó et al. 2012), and less aggressive (Jagoe and Serpell 1996; Schneider et al. 2013) than dogs that did receive these experiences. Sixth, the home environment of the dog appears to be associated with personality. Dogs living in environments where more dogs were present tended to be calmer than dogs living in environments in which fewer dogs were present (Kubinyi et al. 2009). Finally, there may be cultural effects because the country the dog (and owner) lives in appears to be related to personality. American owners of German Shepherds tended to report their dogs were more confident and aggressive than dogs of Hungarian German Shepherd owners (Wan et al. 2009).

While many environmental characteristics appear to be related to dog personality, it is unclear as to why these variables are associated with dog personality. Research suggests dog owners do not simply project their own personality onto their dogs (Kwan et al. 2008), so dog personality is not just a reflection of the owner's personality. One possible reason for why these variables are associated with dog personality is that people may choose dogs with certain characteristics that they find desirable (e.g., activity/energy). Personality differences may then be a result of the owner's preference when selecting the dog (e.g., selecting an active/energetic dog), rather than the owner influencing their dog to display those desirable characteristics. Supporting this idea, one study found a similarity between dogs and owners for most personality dimensions (Turcsán et al. 2012). This study found that the length of ownership did not affect the similarity between the dog and owner, which may suggest that dog owners selected dogs similar to themselves to begin with. Dogs and owners did not grow more similar over time, so the similarity was there to begin with. Alternatively, dogs could possibly influence the behavior of the owner. For example, an active/energetic dog may encourage a less active owner to eventually become more active over time. Further research is needed to better understand how environmental factors are related to dog personality.

Summary

Dog personality research has expanded over the past 20 years, which has allowed for our understanding of dogs to also increase tremendously. There is now considerable evidence that the methods we use to examine dog personality are sound, displaying good levels of reliability and validity. There is also evidence suggesting that biological and environmental factors are associated with dog personality. However, there are still areas in which more work is needed to better understand dog personality.

One area in which dog personality research could use some work is in understanding the structure of dog personality. Whereas much work has been done to examine the structure of dog personality, there is little consensus as to the optimal number of dimensions that describe dog personality. In humans, there is much more consensus, which is beneficial because most human personality studies can be categorized together and compared with one another. As Costa and McCrae (1993) state the five-factor model, "is the Christmas tree on which findings of stability, heritability, consensual validation, cross-cultural invariance, and predictive utility are hung like ornaments" (p. 302). Having a better consensus may allow us to be able to build upon what we know about dog personality and better apply this knowledge.

Although some research shows that biological and environmental factors relate to dog personality, much more research needs to be done on these topics, especially in regards to environmental factors. A dog's environment from the house the dog lives into the type of food the dog eats to the amount of exercise the dog gets is selected by the dog's human caregiver. The dog's caregiver may select these environmental aspects based on his/her own personality. Understanding more about how the environment relates to dog personality can help us to understand how to care for and train a dog, whether the dog is a companion or a working dog. Certain dog personality dimensions also need to be studied in more depth (e.g., sociability and responsiveness to training), where there are conflicting reports as to how biological factors relate to these dimensions. These studies may need to focus on how environmental factors as well as biological factors relate to these dimensions.

Learning about dog personality helps us to better examine an animal that is such a huge part of our lives and that often shares our food and our homes. By further exploring dog personality, we can learn more about topics such as what dogs make the best companions for people with certain personalities. These studies may even help us to deepen our bond with dogs. In addition, because dogs often live alongside humans, studying dogs may allow us to examine topics that are difficult to examine in other species. Understanding dog personality is useful for many applied reasons. Continuing to research this topic can help humans develop better relationships with an animal so entwined with our society.

References

Amat, M., Manteca, X., Mariotti, V. M., de la Torre, J. L., & Fatjó, J. (2009). Aggressive behavior in the English cocker spaniel. *Journal of Veterinary Behavior, 4,* 111–117.

American Kennel Club. (2016). *Breeds recognized by year*. Retrieved from http://www.akc.org/press-center/facts-stats/page-3/

Arata, S., Momozawa, Y., Takeuchi, Y., & Mori, Y. (2010). Important behavioral traits for predicting guide dog qualification. *Journal of Veterinary Medical Science, 72,* 539–545.

Arhant, C., Bubna-Littitz, H., Bartels, A., Futschik, A., & Troxler, J. (2010). Behaviour of smaller and larger dogs: Effects of training methods, inconsistency of owner behaviour and level of engagement in activities with the dog. *Applied Animal Behaviour Science, 123,* 131–142.

Asher, L., Blythe, S., Roberts, R., Toothill, L., Craigon, P. J., Evans, K. M., et al. (2013). A standardized behavior test for potential guide dog puppies: Methods and association with subsequent success in guide dog training. *Journal of Veterinary Behavior, 8,* 431–438.

Bell, A. M., Hankison, S. J., & Laskowski, K. L. (2009). The repeatability of behaviour: A meta-analysis. *Animal Behaviour, 77,* 771–783.

Bennett, P. C., & Rohlf, V. I. (2007). Owner-companion dog interactions: Relationships between demographic variables, potentially problematic behaviours, training engagement and shared activities. *Applied Animal Behaviour Science, 102,* 65–84.

Bennett, S. L., Weng, H., Walker, S. L., Placer, M., & Litster, A. (2015). Comparison of SAFER behavior assessment results in shelter dogs at intake and after a 3-day acclimation period. *Journal of Applied Animal Welfare Science, 18,* 153–168.

Borchelt, P. L. (1983). Aggressive behavior of dogs kept as companion animals: Classification and influence of sex, reproductive status and breed. *Applied Animal Ethology, 10,* 45–61.

Brenøe, U. T., Larsgard, A. G., Johannessen, K., & Uldal, S. H. (2002). Estimates of genetic parameters for hunting performance traits in three breeds of gun hunting dogs in Norway. *Applied Animal Behaviour Science, 77,* 209–215.

Christensen, E., Scarlett, J., Campagna, M., & Houpt, K. A. (2007). Aggressive behavior in adopted dogs that passed a temperament test. *Applied Animal Behaviour Science, 106,* 85–95.

Clutton-Brock, J. (1995). Origins of the dog: Domestication and early history. In J. Serpell (Ed.), *The domestic dog, its evolution, behaviour and interactions with people* (pp. 8–20). Cambridge: Cambridge University Press.

Costa, P. T., & McCrae, R. R. (1993). Bullish on personality psychology. *The Psychology, 6,* 302–303.

Cronbach, L. J., & Meehl, P. E. (1955). Construct validity in psychological tests. *Psychological Bulletin, 52,* 281–302.

De Palma, C., Viggiano, E., Barillari, E., Palme, R., Dufour, A. B., Fantini, C., et al. (2005). Evaluating the temperament in shelter dogs. *Behaviour, 142,* 1307–1328.

Diederich, C., & Giffoy, J. (2006). Behavioural testing in dogs: A review of methodology in search for standardization. *Applied Animal Behaviour Science, 97,* 51–72.

Diesel, G., Brodbelt, D., & Pfeiffer, D. U. (2008). Reliability of assessment of dogs' behavioural responses by staff working at a welfare charity in the UK. *Applied Animal Behaviour Science, 115,* 171–181.

Döring, D., Haberland, B. E., Ossig, A., Küchenhoff, H., Dobenecker, B., Hack, R., et al. (2016). Behavior of laboratory beagles: Assessment in a standardized behavior test using novel stimuli and situations. *Journal of Veterinary Behavior, 11,* 18–25.

Fadel, F. R., Driscoll, P., Pilot, M., Wright, H., Zulch, H., & Mills, D. (2016). Differences in trait impulsivity indicate diversification of dog breeds into working and show lines. *Scientific Reports, 6,* 22162.

Dowling-Guyer, S., Marder, A., & D'Arpino, S. (2011). Behavioral traits detected in shelter dogs by a behavior evaluation. *Applied Animal Behaviour Science, 130,* 107–114.

Draper, T. W. (1995). Canine analogs of human personality factors. *The Journal of General Psychology, 122* (3), 241–252.

Duffy, D. L., & Serpell, J. A. (2012). Predictive validity of a method for evaluating temperament in young guide and service dogs. *Applied Animal Behaviour Science, 138,* 99–109.

Epstein, S. (1983). Aggregation and beyond: Some basic issues on the prediction of behavior. *Journal of Personality, 51,* 360–392.

Foyer, P., Bjällerhag, N., Wilsson, E., & Jensen, P. (2014). Behaviour and experiences of dogs during the first year of life predict the outcome in a later temperament test. *Applied Animal Behaviour Science, 155,* 93–100.

Fratkin, J. L., Sinn, D. L., Patall, E. A., & Gosling, S. D. (2013). Personality consistency in dogs: A meta-analysis. *PLoS ONE, 8,* e54907.

Fratkin, J. L., Sinn, D. L., Thomas, S., Hilliard, S., Olson, Z., & Gosling, S. D. (2015). Do you see what I see?: Can non-experts with minimal training reproduce expert ratings in behavioral assessments of working dogs. *Behavioural Processes, 110,* 105–115.

Fuchs, T., Gaillard, C., Gebhardt-Henrich, S., Ruefenacht, S., & Steiger, A. (2005). External factors and reproducibility of the behaviour test in German shepherd dogs in Switzerland. *Applied Animal Behaviour Science, 94,* 287–301.

Gartner, M. C. (2015). Pet personality: A review. *Personality and Individual Differences, 75,* 102–113.

Gazzano, A., Mariti, C., Sighieri, C., Ducci, M., Ciceroni, C., & McBride, E. A. (2008). Survey of undesirable behaviors displayed by potential guide dogs with puppy walkers. *Journal of Veterinary Behavior, 3,* 104–113.

Goddard, M. E., & Beilharz, R. G. (1982a). Genetic and environmental factors affecting the suitability of dogs as guide dogs for the blind. *Theoretical and Applied Genetics, 62,* 97–102.

Goddard, M. E., & Beilharz, R. G. (1982b). Genetics of traits which determine the suitability of dogs as guide-dogs for the blind. *Applied Animal Ethology, 9,* 299–315.

Goddard, M. E., & Beilharz, R. G. (1985). A multivariate analysis of the genetics of fearfulness in potential guide dogs. *Behavior Genetics, 33,* 1338–1342.

Goddard, M. E., & Beilharz, R. G. (1986). Early prediction of adult behaviour in potential guide dogs. *Applied Animal Behaviour Science, 15,* 247–260.

Goodloe, L. P., & Borchelt, P. L. (1998). Companion dog temperament traits. *Journal of Applied Animal Welfare Science, 1*(4), 303–338.

Hall, S. S., Wright, H. F., & Mills, D. S. (2016). What factors are associated with positive effects of dog ownership in families with children with autism spectrum disorder? The development of the Lincoln Autism Pet Dog Impact Scale. *PLoS ONE, 11*(2), e0149736.

Hart, B. L., & Eckstein, R. A. (1997). The role of gonadal hormones in the occurrence of objectionable behaviours in dogs and cats. *Applied Animal Behaviour Science, 52,* 331–344.

Hart, B. L., & Hart, L. A. (1985). Selecting pet dogs on the basis of cluster analysis of breed behavior profiles and gender. *Journal of the American Veterinary Medical Association, 186,* 1181–1185.

Haubenhofer, D. K., & Kirchengast, S. (2007). Dog handlers' and dogs' emotional and cortisol secretion responses associated with animal-assisted therapy sessions. *Society and Animals, 15,* 127–150.

Hennessy, M. B., Voith, V. L., Mazzei, S. J., Buttram, J., Miller, D. D., & Linden, F. (2001). Behavior and cortisol levels of dogs in a public animal shelter, and an exploration of the ability of these measures to predict problem behavior after adoption. *Applied Animal Behaviour Science, 73,* 217–233.

Hradecká, L., Bartoš, L., Svobodová, I., & Sales, J. (2015). Heritability of behavioural traits in domestic dogs: A meta-analysis. *Applied Animal Behaviour Science, 170,* 1–13.

Hsu, Y., & Serpell, J. A. (2003). Development and validation of a questionnaire for measuring behavior and temperament in pet dogs. *Journal of the American Veterinary Medical Association, 229,* 1293–1300.

Hsu, Y., & Sun, L. (2010). Factors associated with aggressive responses in pet dogs. *Applied Animal Behaviour Science, 123,* 108–123.

Jagoe, A., & Serpell, J. (1996). Owner characteristics and interactions and the prevalence of canine behaviour problems. *Applied Animal Behaviour Science, 47,* 31–42.

Jakuba, T., Polcová, Z., Fedákova, D., Kottferová, J., Mareková, J., Fejsakova, M., et al. (2013). Differences in evaluation of a dog's temperament by individual members of the same household. *Society & Animals, 21,* 582–589.

John, O. P., & Soto, C. J. (2007). The importance of being valid: Reliability and the process of construct validation. In R. W. Robins, R. C. Fraley, & R. F. Krueger (Eds.), *Handbook of research methods in personality psychology* (pp. 461–494). New York: Guilford Press.

Jones, A. C., & Gosling, S. D. (2005). Temperament and personality in dogs (Canis familiaris): A review and evaluation of past research. *Applied Animal Behaviour Science, 95,* 1–53.

King, T., Marston, L. C., & Bennett, P. C. (2011). Breeding dogs for beauty and behaviour: Why scientists need to do more to develop valid and reliable

behaviour assessments for dogs kept as companions. *Applied Animal Behaviour Science, 137,* 1–12.
Kis, A., Turcsán, B., Miklósi, A., & Gásci, M. (2012). The effect of the owner's personality on the behaviour of owner-dog dyads. *Interaction Studies, 13,* 373–385.
Kobayashi, N., Arata, S., Hattori, A., Kohara, Y., Kiyokawa, Y., Takeuchi, Y., et al. (2013). Association of puppies' behavioral reaction at five months of age assessed by questionnaire with their later 'distraction' at 15 months of age, an important behavioral trait for guide dog qualification. *Journal of Veterinary Medical Science, 75*(1), 63–67.
Kortschal, K., Schöberl, I., Bauer, B., Thibeaut, A., & Wedl, M. (2009). Dyadic relationships and operational performance of male and female owners and their male dogs. *Behavioral Processes, 81,* 383–391.
Kubinyi, E., Miklósi, A., Kaplan, F., Gásci, M., Topál, J., & Csányi, V. (2004). Social behaviour of dogs encountering AIBO, an animal-like robot in a neutral and in a feeding situation. *Behavioural Processes, 65,* 231–239.
Kubinyi, E., Turcsán, B., & Miklósi, A. (2009). Dog and owner demographic characteristics and dog personality trait associations. *Behavioural Processes, 81,* 392–401.
Kutsumi, A., Nagasawa, M., Ohta, M., & Ohtani, N. (2013). Importance of puppy training for future behavior of the dog. *Journal of Veterinary Mecial Science, 75*(2), 141–149.
Kwan, V. S. Y., Gosling, S. D., & John, O. P. (2008). Anthropomorphism as a special case of social perception: A cross-species social relations model analysis of humans and dogs. *Social Cognition, 26,* 129–142.
Ledger, R. A., & Baxter, M. R. (1997). The development of a validated test to assess the temperament of dogs in a rescue shelter. In D. S. Mills, S. E. Heath, L. J. Harrington (Eds.), *Proceedings of the first international conference on veterinary behaviour medicine*.
Ley, J., Bennett, P., & Coleman, G. (2009). A refinement and validation of the Monash Canine Personality Questionnaire (MCPQ). *Applied Animal Behaviour Science, 116,* 220–227.
Landsberg, G., Hunthausen, W., & Ackerman, L. (2012). *Behavior problems of the dog and cat*. Edinburgh: Saunders.
Liinamo, A., van den Berg, L., Leegwater, P. A. J., Schilder, M. B. H., van Arendonk, J. A. M., & van Oost, B. A. (2007). Genetic variation in aggression-related traits in Golden Retriever dogs. *Applied Animal Behaviour Science, 104,* 95–106.
Lindberg, S., Strandberg, E., & Swenson, L. (2004). Genetic analysis of hunting behaviour in Swedish Flatcoated Retrievers. *Applied Animal Behaviour Science, 88,* 289–298.
Lindsay, S. R. (2001). *Handbook of applied dog behaviour and training*. Ames, Iowa: Blackwell Publishing.
Lofgren, S. E., Wiener, P., Blott, S. C., Sanchez-Molano, E., Woolliams, J. A., Clements, D. N., et al. (2014). Management and personality in Labrador Retriever dogs. *Applied Animal Behaviour Science, 156,* 44–53.
Lucidi, P., Bernabò, N., Panunzi, M., Dalla Villa, P., & Mattioli, M. (2005). Ethotest: A new model to identify (shelter) dogs' skills as service animals or adoptable pets. *Applied Animal Behaviour Science, 95,* 103–122.
Maejima, M., Inoue-Murayama, M., Tonosaki, K., Matsuura, N., Kato, S., Saito, Y., et al. (2007). Traits and genotypes may predict the successful training of drug detection dogs. *Applied Animal Behaviour Science, 107,* 287–298.
Marder, A. R., Shabelansky, A., Patronek, G. J., Dowling-Guyer, S., & D'Arpino, S. S. (2013). Food-related aggression in shelter dogs: A comparison of behavior identified by a behavior evaluation in the shelter and owner reports after adoption. *Applied Animal Behaviour Science, 148,* 150–156.
McCrae, R. R., & Costa, P. T. (2008). The five-factor theory of personality. In O. P. John, R. W. Robins, & L. A. Pervin (Eds.), *Handbook of personality: Theory and research* (3rd ed., pp. 159–181). New York: Guildford.
McGreevy, P. D., & Nicholas, F. W. (1999). Some practical solutions to welfare problems in dog breeding. *Animal Welfare, 8,* 329–341.
Mehrkam, L. R., & Wynne, C. D. L. (2014). Behavioral differences among breeds of dogs (Canis lupus familiaris): Current status of the science. *Applied Animal Behaviour Science, 155,* 12–27.
Meyer, F., Schawalder, P., Gaillard, C., & Dolf, G. (2012). Estimation of genetic parameters for behavior based on results of German Shepherd Dogs in Switzerland. *Applied Animal Behaviour Science, 140,* 53–61.
Miklosi, A. (2007). *Dog behaviour, evolution and cognition*. Oxford: Oxford University Press.
Mirkó, E., Dóka, A., & Miklósi, A. (2013). Association between subjective rating and behaviour coding and the role of experience in making video assessments on the personality of the domestic dog (Canis familiaris). *Applied Animal Behaviour Science, 149,* 45–54.
Mirkó, E., Kubinyi, E., Gásci, M., & Miklósi, A. (2012). Preliminary analysis of an adjective-based dog personality questionnaire developed to measure some aspects of personality in the domestic dog (Canis familiaris). *Applied Animal Behaviour Science, 138,* 88–98.
Mizukoshi, M., Kondo, M., & Nakamura, T. (2008). Evaluation of the potential suitability of guide dog candidates by continuous observation during training. *Journal of Veterinary Behavior, 3,* 193–198.
Mornement, K. M., Coleman, G. J., Toukhsati, S., & Bennett, P. C. (2014). Development of the behavioural assessment for re-homing K9's (B.A.R.K.) protocol. *Applied Animal Behaviour Science, 151,* 75–83.
Mornement, K. M., Coleman, G. J., Toukhsati, S. R., & Bennett, P. C. (2015). Evaluation of the predictive validity of the behavioural assessment for re-homing K9's (B.A.R.K.) protocol and owner satisfaction with

adopted dogs. *Applied Animal Behaviour Science, 167,* 35–42.

Neilson, J. C., Eckstein, R. A., & Hart, B. L. (1997). Effects of castration on problem behaviors in male dogs with reference to age and duration of behavior. *Journal of the American Veterinary Medical Association, 211,* 180–182.

Paroz, C., Gebhardt-Henrich, S. G., & Steiger, A. (2008). Reliability and validity of behaviour tests in Hovawart dogs. *Applied Animal Behaviour Science, 115,* 67–81.

Paulhus, D. L., & Bruce, M. N. (1992). The effect of acquaintanceship on the validity of personality impressions: A longitudinal study. *Journal of Personality and Social Psychology, 63,* 816–824.

Plutchik, R. (1971). Individual and breed differences in approach and withdrawal in dogs. *Behaviour, 40,* 302–311.

Podberscek, A. L., & Serpell, J. A. (1997). Environmental influences on the expression of aggressive behaviour in English Cocker Spaniels. *Applied Animal Behaviour Science, 52,* 215–227.

Reuterwall, C., & Ryman, N. (1972). An estimate of the magnitude of additive genetic variation of some mental characters in Alsatian dogs. *Hereditas, 73,* 277–284.

Roberts, B. W., & DelVecchio, W. F. (2000). The rank-order consistency of personality traits from childhood to old age: A quantitative review of longitudinal studies. *Psychological Bulletin, 126,* 3–25.

Rooney, N. J., & Bradshaw, J. W. S. (2003). Links between play and dominance and attachment dimensions of dog–human relationships. *Journal of Applied Animal Welfare Science, 6,* 67–94.

Roth, L. S. V., & Jensen, P. (2015). Assessing companion dog behavior in a social setting. *Journal of Veterinary Behavior, 10,* 315–323.

Ruefenacht, S., Gebhardt-Henrich, S., Miyake, T., & Gaillard, C. (2002). A behaviour test on German Shepherd dogs: Heritability of seven different traits. *Applied Animal Behaviour Science, 79,* 113–132.

Sampson, J., & Binns, M. M. (2006). The kennel club and the early history of dogs shows and breed clubs. In E. A. Ostranger, U. Giger, & K. Lindblad-Toh (Eds.), *The dog and its genome* (pp. 19–30). Cold Spring Harbor, NY: Cold Spring Harbor Laboratory Press.

Schneider, L. A., Delfabbro, P. H., & Burns, N. R. (2013). Temperament and lateralization in the domestic dog (Canis familiaris). *Journal of Veterinary Behavior, 8,* 124–134.

Seksel, K., Mazurski, E. J., & Taylor, A. (1999). Puppy socialisation programs: Short and long term behavioral effects. *Applied Animal Behaviour Science, 62,* 335–349.

Serpell, J. A., & Duffy, D. L. (2014). Dog breeds and their behaviour. In A. Horowitz (Ed.), *Domestic dog cognition and behavior* (pp. 31–57). Berlin, Heidelberg: Springer.

Serpell, J. A., & Hsu, Y. (2001). Development and validation of a novel method for evaluating behavior and temperament in guide dogs. *Applied Animal Behaviour Science, 72,* 347–364.

Serpell, J. A., & Hsu, Y. (2005). Effects of breed, sex, and neuter status on trainability in dogs. *Anthrozoös, 18* (3), 196–207.

Sinn, D. L., Gosling, S. D., & Hilliard, S. (2010). Personality and performance in military working dogs: Reliability and predictive validity of behavioral tests. *Applied Animal Behaviour Science, 127,* 51–65.

Slabbert, J. M., & Odendaal, J. S. J. (1999). Early prediction of adult police dog efficiency: A longitudinal study. *Applied Animal Behaviour Science, 64,* 269–288.

Starling, M. J., Branson, N., Thomson, P. C., & McGreevy, P. D. (2013). Age, sex and reproductive status affect boldness in dogs. *The Veterinary Journal, 197,* 868–872.

Stone, H. R., McGreevy, P. D., Starling, M. J., & Forkman, B. (2016). Associations between domestic-dog morphology and behaviour scores in the dog mentality assessment. *PLoS ONE, 11*(2), e014903.

Strandberg, E., Jacobsson, J., & Saetre, P. (2005). Direct genetic, maternal and litter effects on behaviour in German shepherd dogs in Sweden. *Livestock Production Science, 93,* 33–42.

Svartberg, K. (2006). Breed-typical behaviour in dogs-historical remnants or recent constructs? *Applied Animal Behaviour Science, 96,* 293–313.

Svartberg, K., & Forkman, B. (2002). Personality traits in the domestic dog (Canis familiaris). *Applied Animal Behaviour Science, 79,* 133–156.

Takeuchi, Y., Hashizume, C., Arata, S., Inoue-Murayama, M., Maki, T., Hart, B. L., et al. (2009a). An approach to canine behavioural genetics employing guide dogs for the blind. *Animal Genetics, 40,* 217–224.

Takeuchi, Y., Kaneko, F., Hashizume, K., Masuda, K., Ogata, N., Maki, T., et al. (2009b). Association analysis between canine behavioural traits and genetic polymorphisms in the Shiba Inu breed. *Animal Genetics, 40,* 616–622.

Temesi, A., Turcsán, B., & Miklósi, A. (2014). Measuring fear in dogs by questionnaires: An exploratory study toward a standardized inventory. *Applied Animal Behaviour Science, 161,* 121–130.

Tiira, K., & Lohi, H. (2015). Early life experiences and exercise associate with canine anxieties. *PLoS ONE, 10*(11), e0141907.

Tomkins, L. M., Thomson, P. C., & McGreevy, P. D. (2011). Behavioral and physiological predictors of guide dog success. *Journal of Veterinary Behavior, 6,* 178–187.

Tonoike, A., Nagasawa, M., Mogi, K., Serpell, J. A., Ohtsuki, H., & Kikusui, T. (2015). Comparison of owner-reported behavioral characteristics among genetically clustered breeds of dog (*Canis familiaris*). *Scientific Reports, 5,* 17710.

Tóth, L., Gásci, M., Topál, J., & Miklósi, A. (2008). Playing styles and possible causative factors in dogs' behaviour when playing with humans. *Applied Animal Behaviour Science, 114,* 473–484.

Trut, L. N. (1999). Early canid domestication: The farm-fox experiment. *American Science, 87,* 160–169.

Turcsàn, B., Kubinyi, E., & Miklósi, A. (2011). Trainability and boldness differ between dog breed clusters based on conventional breed categories and genetic relatedness. *Applied Animal Behaviour Science, 132,* 61–70.

Turcsán, B., Range, F., Virányi, Z., Miklósi, A., & Kubinyi, E. (2012). Birds of a feather flock together? Perceived personality matching in owner-dog dyads. *Applied Animal Behaviour Science, 140,* 154–160.

Valsecchi, P., Barnard, S., Stefanini, C., & Normando, S. (2011). Temperament test for re-homed dogs validated through direct behavioral observation in shelter and home environment. *Journal of Veterinary Behavior, 6,* 161–177.

van der Waaij, E. H., Wilsson, E., & Strandberg, E. (2008). Genetic analysis of results of a Swedish behavior test on German Shepherd Dogs and Labrador Retrievers. *Journal of Animal Science, 86,* 2853–2861.

Vas, J., Topál, J., Péch, E., & Miklósi, A. (2007). Measuring attention deficit and activity in dogs: A new application and validation of a human ADHD questionnaire. *Applied Animal Behaviour Science, 103,* 105–117.

Wahlgren, K., & Lester, D. (2003). The big four: Personality in dogs'. *Psychological Reports, 92,* 828.

Wan, M., Kubinyi, E., Miklósi, A., & Champagne, F. (2009). A cross-cultural comparison of reports by German Shepherd owners in Hungary and the United States of America. *Applied Animal Behaviour Science, 121,* 206–213.

Weiss, E. (2002). Selecting shelter dogs for service dog training. *Journal of Applied Animal Welfare Science, 5,* 43–62.

Weiss, E., & Greenberg, G. (1997). Service dog selection tests: Effectiveness for dogs from animal shelters. *Applied Animal Behaviour Science, 53,* 297–308.

Wells, D. L., & Hepper, P. G. (2000). Prevalence of behaviour problems reported by owners of dogs purchased from an animal rescue shelter. *Applied Animal Behaviour Science, 69,* 55–65.

Wilsson, E., & Sinn, D. L. (2012). Are there differences between behavioral measurement methods? A comparison of the predictive validity of two ratings methods in a working dog program. *Applied Animal Behaviour Science, 141,* 158–172.

Wilsson, E., & Sundgren, P. (1997). The use of a behaviour test for selection of dogs for service and breeding. II: Heritability for tested parameters and effect of selection based on service dog characteristics. *Applied Animal Behaviour Science, 54,* 234–241.

Wilsson, E., & Sundgren, P.-E. (1998). Behaviour test for eight-week old puppies–heritabilities of tested behaviour traits and its correspondence to later behaviour. *Applied Animal Behaviour Science, 58,* 151–162.

Wirth, K. E., & Rein, D. B. (2011). The economic costs and benefits of dog guides for the blind. *Ophthalmic Epidemiology, 15,* 92–98.

Wright, H. F., Mills, D. S., & Pollux, P. M. J. (2011). Development and validation of a psychometric tool for assessing impulsivity in the domestic dog (*Canis familiaris*). *International Journal of Comparative Psychology, 24,* 210–225.

Felid Personality and Its Implications 11

Marieke Cassia Gartner

Abstract

Animal personality is becoming more of a focus in a variety of fields, and whereas there is still a bias toward primates, it is important to assess personality in other species. There is little work done with felids, yet felids represent a population of animals that could benefit from such research. Most of the 37 species of felids are endangered, with as few as 60 of some species left in the wild (e.g., the Amur leopard). In addition, due to their large home range sizes, they face distinct challenges in zoos. Although steps are being taken to address both conservation and welfare of these animals, understanding the causes and consequences of felid personality has the potential to aid in these efforts, with the goal of improving the lives of individual animals, and ensuring the future of each species. This chapter considers the measurement of felid personality, its relationship to health and life outcomes, and its potential role in conservation and welfare. Much more research focused on felids is needed, and personality research can add greatly to the field.

Felid Personality and Its Implications

The last few decades have seen an exponential increase in nonhuman animal personality studies. However, there is a bias toward studying species more closely related to humans, or more useful to humans in terms of understanding health outcomes. Among mammals, this has led to a large literature on personality in nonhuman primates (210 articles as indicated in Freeman and Gosling 2010). There is also a growing literature on dog (*Canis lupus familiaris*) personality (95 articles as reviewed in Gartner 2015 and Jones and Gosling 2005; there are very few published studies on other canid species). By comparison, there are only 24 articles on personality in cat species (for a review, see Gartner 2015 and Gartner and Weiss 2013a). By far, the most studied cat species, in terms of personality, is the domestic cat (*Felis silvestris catus*), with 87.5%

M.C. Gartner (✉)
Philadelphia Zoo, Philadelphia, PA, USA
e-mail: mariekecassia@gmail.com

J. Vonk et al. (eds.), *Personality in Nonhuman Animals*,
DOI 10.1007/978-3-319-59300-5_11

of studies on felids focused on that species. This emphasis is mostly due to owner participation in studies, and also, to a lesser extent, availability of domestic cats in laboratories, veterinary clinics, and shelters. Similarly, the focus of canid personality research is on domestic dogs (often working dogs—for example, police and guide dogs) and shelter animals. Big cats in zoos are more difficult to study in terms of obtaining a large sample size, and coordinating across zoos. Methodologies for studying felid personality in sanctuaries/the wild are difficult to establish, as cats are notoriously elusive. In addition, even though scientists have made substantial progress in establishing nonhuman animal personality as a valid field of study, there is still resistance to the word "personality" when it is applied to animals, especially animals more distantly related to humans. As a result, this chapter will focus on the potential for studying personality in these taxa, with possible implications for domestication, health, welfare, and conservation. Because many definitions for personality persist, for the purpose of this chapter, like other researchers (e.g., Freeman and Gosling 2010), I will define personality as a construct described by a set of behaviors that are consistent across context and time.

The formal study of cat personality began in the 1980s (Feaver et al. 1986; Meier and Turner 1985; Turner et al. 1986), following numerous studies of primates, but humans have been fascinated by the multi-faceted behavior of cats since ancient times. Worshipped as gods in Egypt (Serpell 2000), buried with humans in Cyprus (Vigne et al. 2004), possibly domesticated in China (Hu et al. 2014), symbols of good fortune in Japan (Miller 2010), and revered by Muslims (Glassé 2003), cats are currently one of the most popular pets across the world (2015–2016 APPA National Pet Owners Survey). As a whole, they are an alluring species, and it is very possibly their expression of individual differences, from independence to affection, that draws so many different people to them.

The study of personality, more than a century old in humans, originally extended to nonhuman primates, a natural step considering our close genetic relationship to great apes, in particular. But once it was clear that personality could be measured reliably in primates (for example, King and Figueredo 1997; King et al. 2005; Stevenson-Hinde et al. 1980; Weiss et al. 2006, 2011) and indeed was genetically based (for example, in primates: Adams et al. 2012; Hopkins et al. 2012; and in cats: Arahori et al. 2015), it seemed probable—considering phylogenetic continuity—that other species would exhibit individual differences defined by clusters of traits as well. In 1986, Feaver et al. used a similar methodology—behavioral observations and a survey used in primates (both human and nonhuman)—to assess domestic cat personality. The authors found three factors of personality (broad dimensions comprised of a number of traits); Alert (comprised of the traits active and curious), Sociable (comprised of the traits sociable with people, not fearful of people, not hostile to people, and tense), and Equable (calm with cats). The only personality studies on big cats include single studies of each of the following species: cheetahs (*Acionyx jubatus*, Wielebnowski 1999), clouded leopards (*Neofelis nebulosi*, Wielebnowski et al. 2002), tigers (*Panthera tigris*, Phillips and Peck 2007), African lions (*Panthera leo*, Torgerson-White and Bennett 2014), and snow leopards (*Panthera uncia*, Gartner and Powell 2012), and a mix of African lion, clouded leopards, snow leopards, Scottish wildcats (*Felis silvestris grampia*), and domestic cats (Gartner et al. 2014). Generally speaking, most studies confirm a three-factor structure of personality, with the most frequently emerging factors identified as Sociable, Curious, and Dominant (Gartner and Weiss 2013a). These particular factors are not consistently found together across studies, however. For example, Feaver et al. (1986) found two factors related to Sociable and one to Curious. In contrast, human personality is comprised of five factors (Openness to Experience, Conscientiousness, Extraversion, Agreeableness, and Neuroticism; Digman 1990). Some studies with great apes have revealed five or six factors. Chimpanzees (*Pan troglodytes*), for example, have a similar structure to that found in humans, with the addition of a Dominance factor (King and Figueredo 1997). In domestic dogs, as in cats, the factor structure is much less clear (for a review, see Fratkin, this volume; Freeman and Gosling 2010; Gartner 2015). Most studies rely on behavioral observations, but some also make use of caretaker surveys (for a review, see Gartner and Weiss 2013a). Because of the differences in methodology, species, and the lack of replication, in addition to the omission of breed, neuter/spay status, age, and longitudinal studies in most research, much more work needs to be done before any definitive pronouncements on the nature of felid personality can be made.

The study of personality is important in view of its associations with other aspects of life. Personality is related to health outcomes such as morbidity (e.g., Capitanio 2011), mortality (e.g., Weiss et al. 2012), and subjective well-being (e.g., King and Landau 2003), as well as social behavior (e.g., Capitanio 2002) in nonhuman primates, and outcomes such as learning ability (e.g., Pavlov, in Locurto 2007) and adoption rates in dogs (DePalma et al. 2005). In cats, similar studies have not been widely conducted (but see Fromont et al. 1997; Natoli et al. 2005 for morbidity; Gartner et al. 2016 for subjective well-being; Wielebnowski 1999; Wielebnowski et al. 2002 for stress). For an animal that is commonly kept as a pet, and for big cats that face extinction or special challenges in zoos, this type of information could be vital to their survival and well-being.

Implications of Personality for Felids

Wild Versus Domestic Cats

There are eight phylogenetic lineages of felids (pantherine lineage, 10.8 million years ago (MYA); baycat group, 9.4 MYA; caracal group, 8.5 MYA; ocelot lineage, 8.0 MYA; lynx genus, 7.2 MYA; puma group, 6.7 MYA; leopard cat group, 6.2 MYA; and domestic cat lineage, 3.4 MYA; O'Brien et al. 2008). The domestic cat and the African wildcat probably diverged from a common ancestor approximately 131,000 years ago (Driscoll et al. 2007). A genetic study of domestic cats and five species of wildcat showed that cats were most likely domesticated in the Near East (Israel, Saudi Arabia, Bahrain, and the United Arab Emirates) from the African wildcat, with five founders across the region, about 10,000 years ago, and then brought to Cyprus and Egypt (Driscoll et al. 2007). In comparison, there is evidence of domestic dogs emerging anywhere from 27 to 40,000 years ago. This evidence has led to the assumption that unlike dogs, domestic cats are not very different from their wild counterparts, with evidence that certain species interbreed (Scottish wildcat and domestic cat, for example), although the capability to breed only within the species is no longer a distinct part of the definition of speciation (for example, wolves and coyotes breed successfully [Way et al. 2010]).

Another assumption is that domestic cats can survive in the wild, as opposed to domestic dogs. However, it seems that neither species does particularly well without help from humans (e.g., Boitani and Ciucci 1995; Schmidt et al. 2007). In addition, in opposition to some societies worshipping cats, a dislike and fear of cats is present in some cultures, whereas dislike of dogs is not comparable (Lockwood 2005). There is also a perception that cats are impossible to study (Grimm retrieved June 30, 2016), as compared to dogs, which has been proven untrue. For example, Miklósi et al. (2005) found that both dogs and cats respond in the same way to a human pointing. Both species were pre-trained so that they knew they were being offered food in one of two bowls. During the experiment, the animals could not see in which bowl the experimenter put the food; the owner of each animal held them while the experimenter got the animal's attention, and then pointed at the bowl with the food in it. There were no significant species differences with regard to accurate searching. However, differences were found in attention-getting behavior, which is to be expected, given the differences in the species' social structures. Dogs are a social species, and it is generally believed that many interactions between humans and dogs are due to their longer domestication period than cats. This has been shown in differences in the use of gaze between domestic dogs and wolves, where domestic dogs look to humans for help to solve problems, whereas socialized wolves do not (Miklósi et al. 2003). Because domestic cats are semi-solitary, it makes sense that tests done with dogs may not always work in the same way with cats, which suggests that different tests must be established to test cat abilities.

The relatively recent domestication of cats also seems to have led to a striking similarity in personality across various species of wild cats and domestic cats. In our work on five felid species (Gartner et al. 2014, 2016), three factors of personality emerged in each species (domestic cat, Scottish wildcat, clouded and snow leopards, and African lions). Domestic cat personality factors were labeled Neuroticism (with highest loadings on traits such as anxious and insecure on one end of the scale and stable and affectionate on the other), Impulsiveness (excitable and active; constrained and independent), and Dominance (aggressive and bullying; submissive). Factors related to Impulsiveness, Openness, and Agreeableness were related across species, as were Neuroticism factors. To understand where there might be differences between these species, we then compared the traits that made up the factors across species as well. Most correlated significantly, with the exception of African lions, whose traits did not correlate significantly with other species, and Scottish wildcats and snow leopards whose traits did not correlate with each other. These results indicate that there is something different at the trait level for African lions, especially—perhaps to do with the fact that they are the only purely social species among the felids, with adults living in both female-only and male-only groups. The highest reliabilities in lions, but not in any other species, were for the traits dominant and submissive (which loaded on the factors Dominance and Neuroticism, respectively). However, other felids do sometimes live in social groups. For example, small, male-only groups exist in cheetahs, and female-only groups occur within domestic cat societies (Macdonald et al. 2000). It would be interesting to add cheetahs to this analysis to assess any similarities to lions or domestic cats on the trait level. The similarity on the factor level indicates that overall felid personality structures may have evolved early on in cat species, and that some part of lion personality structure may be derived from behavioral traits (such as social behavior) that are found only in that species. This possible derivation from behavioral traits is an indication of the adaptive nature of personality, which is also seen in its relationship to health and life outcomes across species, from humans to primates to felids and others, as noted earlier. This adaptive nature lends itself well to the study of both proximate and ultimate explanations of behavior (Mayr 1961; Orians 1962), as well as to an understanding of related behaviors that may have co-evolved with personality (Carere and Locurto 2011).

So does domestication, or even captivity, affect personality in felids? The results from our study suggest that the felid personality factors, Neuroticism and Impulsiveness, have not evolved much since modern cats split off from *Carnivora*, although, until genetic analyses are conducted, this is uncertain. It is not clear how either of these traits would be adaptive to the felid predator way of life. It is possible that the results are related to the status of all the species in the study as captive animals, as contemporary evolution (Hendry and Kinnison 1999) may play a role in the strengthening or weakening of traits (McDougall et al. 2006). A study of personality in wild-living felids, then, might be helpful in understanding the results. Studies have been done on groups of free-living domestic cats (e.g., Natoli et al. 2005); however, they are limited to behavioral analyses, not those based on ratings by a familiar caretaker. One possibility would be to examine animals living in large sanctuaries, such as lions in Africa; however, this would still present problems for the current methodology as the knowledge of the caretakers is probably different than the knowledge of zookeepers, who spend more time in close proximity with the animals in their care. In some cases the same problem with free-living domestic cats would be applicable as well. Alternatively, a comparison between sanctuary-living and zoo-living chimpanzees found no difference in factor structure attributable to environment (King et al. 2005). The differences that were found between the zoo- and sanctuary-living chimpanzees were also found in a comparison between two different sets of zoo-living chimpanzees, indicating that personality could be measured reliably across environment.

Coat Color

Another indication that personality is genetic is the link between coat color and personality. There is very little work done on this in felids, but evidence from work with other animals (silver foxes [*Vulpes vulpes*]: Trut et al. 2009; domestic horses [*Equus caballus*]: Brunberg et al. 2013; and rats [*Rattus norvegicus*] and deer mice [*Peromyscus maniculatus*]: Hayssen 1997) shows that such an association is possible in felids as well. The work that has been done in felids shows some interesting results. One study found that black cats are more tolerant of crowding (Todd 1977). Another found that red, cream, or tortoiseshell kittens struggle for a longer time and make more escape attempts when handled by an unfamiliar person compared to other colored kittens (Ledger and O'Farrell 1996). The orange allele may be linked to aggressiveness in males, while lacking the agouti allele (usually black cats) is linked to greater amicability and aggregative tendencies (Robinson 1977). More recently, support for the idea that non-agouti cats are friendlier was found by Morgan (2010) in her dissertation, along with an increased shyness among "wildtype appearance" cats, described as having agouti coloration and no sex-linked orange pigmentation. However, Stelow et al. (2016) found that owners reported orange females (including tortoiseshells, calicos, and torbies), black-and-white, and gray-and-white cats to be more frequently aggressive toward humans. Wilhelmy et al. (2016) found that associations independent of breed included more aggression in agouti cats and prey interest in red cats, decreased stranger-directed aggression in piebald cats, and increased separation anxiety in Siamese and Tonkinese patterned cats. On the other hand, Umbelino (2014) found no association between tameness and coat color. These varying results do not allow for any assumptions to be made between personality and coat color. However, this work would be useful in shelters, and so more work should be done.

Rearing Environment

A question that follows from exploring the role of personality in pet or captive versus wild animals is one of rearing environment. Do stray cats show different personality traits than feral cats, and do zoo-bred cats show different personality traits than wild animals? If living space does affect personality, is it due to growing up in that environment, or because of other life experiences? Although this topic has not been formally studied in felids, comparisons between captive and wild primates do not reveal such a difference, as noted above. However, developmental studies have found that domestic cats show individual differences in motor activity soon after birth (Raihani et al. 2014). In addition, kittens that are handled during their first eight weeks of life are more likely to be bold at four months and, to some extent, a year, but by two years of age, handled kittens are no more likely to be bold than those who weren't handled (Lowe and Bradshaw 2001), indicating that perhaps environment does not play as big of a role as some might think, which is also the case in humans (Tellegen et al. 1988). Similarly, McCune (1995) found that cats sired by a friendly father (one that immediately initiated proximity and/or contact with people), or who were socialized (handled between 2 and 12 weeks of age), were quicker to approach, touch, and rub a test person, were more vocal, and spent more time near both a familiar and unfamiliar test person. These results suggest that there are both genetic and environmental influences on felid personality, but whether environment also affects personality development is unclear as yet.

Health Outcomes

Approximately 59 percent of cat species are endangered or in decline, and their survival may depend on their success in zoos. However, because cat species have large ranges, they often face challenges in zoo enclosures (Clubb and Mason 2003). In small cages in shelters and laboratories, felid welfare is often compromised, and an estimated 71% of cats in shelters in the United States alone are euthanized (Animal Shelter Euthanasia retrieved June 30, 2016). Although various techniques have been used to increase welfare, some species still face challenges in captivity. Knowing an animal's personality has the potential to address some of these issues. Whereas there is some data on the relationship between personality and health outcomes in other species, the amount of data in cat species is meager. What data there is, though, is telling. For example, Wielebnowski (1999) found that cheetahs rated highly on the dimension Tense–Fearful were more likely to be non-breeders, and that females scored higher on this factor than males. She suggests that these cheetahs might have more difficulty in coping with the captive environment, and might therefore need more secluded enclosures and/or more hiding places. Similarly, stress, which can be measured by HPA activity (Tsigos and Chrousos 2002), and is related to disease susceptibility (Elenkov and Chrousos 1999), is also related to personality. Wielebnowski et al. (2002) found that clouded leopards rated as more Fearful/Tense than those rated lower on that scale, and who self-injured, paced, slept, and hid more often, had increased overall, base, and peak fecal corticoid concentrations, indicating chronic stress.

A link has also been found between personality and disease contraction in domestic cats. Natoli et al. (2005) analyzed temperament, social rank, and prevalence of feline immunodeficiency virus (FIV), a lethal disease that is transmitted by biting (Fromont et al. 1997), in three cat colonies in Rome and Lyon. They found one factor, which ranged from proactive to reactive. Proactive included the most aggressive and affiliative males who marked frequently. Reactive described submissive individuals who were rarely aggressive. Proactive males had the highest social rank, and were the oldest, largest, and heaviest animals. Blood sampling of the male cats (who have higher rates of FIV infection than females) revealed that most infected males were high ranking and proactive. Proactive males had increased reproductive success, but they also had increased chances of being infected with FIV. The opposite result is found in domestic cats with feline leukemia virus (FeLV), which is transmitted mainly during affiliative interactions including licking and grooming (Fromont et al. 1997). The prevalence of FeLV in more aggressive cats, then, is lower than that in more socially active cats (Fromont et al. 1997).

Finkler and Terkel (2015) found that boldness in female cats (but not in males) was correlated with reproductive status. That is, spayed females were less bold than intact females. Taken together, this small amount of work in the relationship between health outcomes and personality may have implications for adoption, as different people have different desires in a prospective pet, and allowing pets outside is the norm in some places. If neutering a male cat has no effect on his boldness, then allowing him outside may expose him to FIV more readily. However, a female cat may not face the same challenges. This may be taken into account when deciding whether to let a pet cat outside or not. In addition, personality has been rated as one of the most important factors in adoption in shelters. For example, one study found that friendliness and happiness are rated as the number one and number two (respectively) criteria by potential adopters of cats (Sinn 2016), and another found that personality makes up 94% of the reason for satisfaction/dissatisfaction with cat adoptions (Neidhart and Boyd 2002; see also Gourkow and Fraser 2006; Zito et al. 2015).

Personality is one of the strongest and most consistent predictors of well-being (or happiness) in humans (Diener et al. 1999) and nonhuman primates (e.g., King and Landau 2003). In humans, the relationship is especially strong between well-being and Neuroticism (a negative relationship), and, to a lesser extent, well-being and Extraversion (a positive relationship; Steel et al. 2008). In rhesus macaques (*Macaca mulatta*), higher ratings on the personality factors Confidence and Friendliness, and lower ratings on Anxiety are positively related to well-being (Weiss et al. 2011). In chimpanzees, well-being is positively associated with Dominance, Extraversion, and Dependability (King and Landau 2003), as well as Agreeableness and Openness, and negatively with Neuroticism (Weiss et al. 2009). Higher Extraversion, Agreeableness, and lower Neuroticism are related to well-being in orangutans (*Pongo* sp.; Weiss et al. 2006).

We studied well-being and its relationship to personality in felids (Gartner et al. 2016) to see if this relationship was also evident in felids. We found that subjective well-being was related to Agreeableness/Openness (positive relationship) and Neuroticism (negative relationship) in clouded leopards, Neuroticism in snow leopards (negative relationship), and Impulsiveness (negative relationship) and Neuroticism (negative relationship) in African lions. In Scottish wildcats, well-being is also positively related to the factor Self-Control, which has high loadings on decisive, self-assured, and cool, and includes elements related to those known to comprise Conscientiousness and low Neuroticism in primates (Gartner and Weiss 2013b). These results are similar to some of those found in both human and nonhuman primates. Because of the similarities across species with regard to personality, we attempted to see if we could define a taxa personality structure, and found Neuroticism, Dominance, and Impulsiveness factors (Gartner et al. 2014). Interestingly, none of these factors correlated with well-being (Gartner et al. 2016). To understand this result, we examined the relationship of the individual species personality traits to well-being and found important differences. There was only one trait that correlated with well-being across species: insecure. A combination of the adjectives affectionate, calm, fearful of people, suspicious, tense, and trusting correlated with well-being in some of each of three species. Thirteen traits were found to correlate only in one species. Clouded leopard traits that correlated with well-being include cool, cooperative, friendly to conspecifics, and smart. This relationship indicates that positive interactions—both with conspecifics and human caretakers—are important to the clouded leopard's well-being. Two snow leopard personality traits correlate with well-being—individualistic and predictable. This may mean that for this species, environment is what is important in determining well-being—known situations that either do not change much or that allow for control over that situation. Finally, for African lions, it is unsurprising that the traits that correlate with well-being are indicative of social structure—constrained, dominant, fearful of conspecifics, persevering, stable, submissive, and timid. This indicates that, for African lions, how they are grouped socially is important for their well-being, and that schisms within the grouping may be more challenging for them than for other species, some of which, as noted earlier, are not as solitary as originally thought. This type of information can clearly be used for animal management in zoos, and should be studied further. For example, it is difficult in zoo situations to separate groups easily, due to such challenges as space concerns and Species Survival Plan recommendations. However, if it is known that a lion group is experiencing a split, it may be recommended to try to expedite plans to separate the group (for instance, already having a plan in place should such a situation arise), whereas with other species more time may be available, without affecting well-being. This proposal would clearly need more testing, including assessing the well-being of the group during a normal period and then at the point of the break.

In addition to its applications for felid captive management, well-being is also important to assess due to its possible relationship with longevity, as has been shown in humans (Diener and Chan 2011) and orangutans (Weiss, Adams & King 2011). Well-being may be a good marker for health outcomes, as personality may influence health via subjective well-being. For example, in cynomolgus monkeys (*Macaca fascicularis*), depression, a facet of subjective well-being, mirrors that in humans in terms of physiology, neurobiology, behavior, and health. Depression is associated with increased risk of cardiovascular disease risk and mortality and is subject to individual differences in terms of response to environmental challenges (Willard and Shively 2012). This relationship has yet to be shown in felids, but would be invaluable to know in caring for any of the felid species.

The relationships between health outcomes and personality can be used to inform care. In humans, Deary et al. (2010) argue that there are four major applications of personality to health care and improved well-being: heightened surveillance for those with traits related to earlier mortality; the development of specific, individual intervention strategies; targeted drug treatments; and improved relationships between patients and healthcare practitioners. There is no reason to think that these might not also apply to animals (Gartner and Weiss 2013b). A relationship between longevity and personality has been found in gorillas (*Gorilla gorilla gorilla*), for example, (Weiss et al. 2012), so it makes sense to also study this relationship in other animals. In parallel to the suggestion to help a patient rated low in Conscientiousness, a physician's directive could be accompanied by incentives or regular monitoring (Deary et al. 2010), individual interventions based on personality for felids might include giving shy ones more places to hide, or giving bold ones more novel objects more often to help each type cope with their environment better (Gartner and Powell 2012). Pharmacological treatments are recommended by animal behaviorists and have been used with some success in pet cats (Kaur et al. 2016). Any chance of increasing that success can only help pet retention and maintain lower numbers of homeless pet cats. Relationships between animals and their human caretakers are increasingly being studied—in fact an entire journal (*Anthrozoös*) is dedicated to the investigation of these relationships. It makes sense, then, that a better relationship between an animal and its keeper should improve welfare (Carlstead et al. 1999; Wielebnowski 1999). Findings from each of these areas have the potential to contribute to the overall welfare of an animal, given that each works to decrease specific causes of stress. Whereas some have suggested that personality factors have implications for increasing welfare directly (for example, Wielebnowski 1999), more work is needed. This should be an important next step in furthering the literature on felid personality and welfare.

Welfare and Conservation

There is little existing work on the direct effect that personality may have on captive animal welfare. One study found that chimpanzees rated as higher in Openness were more interested than other chimpanzees in cognitive enrichment in the form of mirror recognition and touchscreen tasks (Herrelko et al. 2012). Several researchers have made suggestions as to possible links between personality and welfare, including the idea mentioned earlier that shy or fearful animals may need more places to hide (snow leopards: Gartner and Powell 2012; cheetahs: Wielebnowski 1999), or that an animal's physical behavior may need further explanation, which may lead to better understanding of welfare. For example, if an animal is alert and active, is it also calm, or is it fearful (Wemelsfelder 2007)? How that animal is handled or managed would depend on the latter assessment, and without knowledge of both the species and the individual, that animal's welfare could be impacted.

To help conserve species, zoos are often called upon to orchestrate captive breeding efforts (Wielebnowski 1998). These efforts require careful consideration of spatial needs, species and individual behavior, as well as genetic differentiation to avoid inbreeding. Whereas some species breed well in captivity, others have more trouble, due to a variety of problems, including aggression (Wielebnowski et al. 2002), disinterest (Powell et al. 2008), or poor health, and/or stress (Clubb and Mason 2003). Several methods have been used to increase the likelihood of successful breeding, including species- or, ideally, individual-specific environmental enrichment and providing an appropriate social environment (Shepherdson 1994), to varying effect. In the past decade, research has shown that, in some species at least, personality can influence how animals interact in captive breeding situations. This makes sense, given the evidence that personality affects fitness in the wild across a number of species, with a clear connection between boldness and reproductive success (at the expense of survival, however) and between exploration and survival (Smith and Blumstein 2008). Because this is the case, personality should be taken into account when conducting captive breeding programs. For example, as noted earlier, Wielebnowski (1999) found that cheetahs who rated highly on the personality factor Tense-Fearful were more likely to be non-breeders.

Captive felids are often kept in multi-animal enclosures. Despite the fact that cat species, excluding lions, are mostly solitary, it is becoming more established that some cat species, such as domestic cats, tigers, cheetahs, and lynx, are semi-solitary, forming social groups in certain situations, often related to food availability, but also family relationships (dispersing sibling groups, long-term parental care; Kitchener 2000). It is unsurprising then, that there is evidence that social interaction may decrease abnormal behaviors and increase natural behaviors in pair-grouped captive tigers, for example (De Rouck et al. 2005). Knowing each animal's personality could potentially increase grouping success. Gorillas (*Gorilla* sp.) who rate higher on the personality factor Understanding maintain more successful groups in captivity (Stoinski et al. 2004). In felids, tigers housed near other tigers show an increase in stereotypic behavior (De Rouck et al. 2005). If personality is taken into consideration when choosing which tigers to house near one another, such behaviors may decrease.

Future Studies

There is clearly a large amount of work that can be done in the field of personality for felids. First and foremost, establishing a validated, replicable structure is paramount, for each of the 38 extant felid species. Because cats are more easily studied in captivity, comparisons across differing facilities—zoos, laboratories, sanctuaries—may offer validation of personality structures. Once this is established, other associations need to be made, starting with longevity/mortality, but also immunity and morbidity. Other ties to welfare and conservation are vital to the health, happiness, and survival of species.

References

Adams, M. J., King, J. E., & Weiss, A. (2012). The majority of genetic variation in orangutan personality and subjective well-being is nonadditive. *Behavior Genetics, 42,* 675–686.

Animal Shelter Euthanasia. Retrieved June 30, 2016 from http://www.americanhumane.org/animals/stop-animal-abuse/fact-sheets/animal-shelter-euthanasia.html

Arahori, M., Hori, Y., Saito, A., Chiijiiwa, H., Takagi, S., Ito, Y., et al. (2015). The oxytocin receptor gene (OXTR) polymorphism in cats (*Felis catus*) is associated with "Roughness" assessed by owners. *Journal of Veterinary Behavior, 11,* 109–112.

Boitani, L., & Ciucci, P. (1995). Comparative social ecology of feral dogs and wolves. *Ethology Ecology & Evolution, 7,* 49–72.

Brunberg, E., Gille, S., Mikko, S., Lindgren, G., & Keeling, L. J. (2013). Icelandic horses with the silver coat colour show altered behaviour in a fear reaction test. *Applied Animal Behaviour Science, 146,* 72–78.

Capitanio, J. P. (2002). Sociability and responses to video playbacks in adult male rhesus monkeys (*Macaca mulatta*). *Primates, 43,* 169–177.

Capitanio, J. P. (2011). Nonhuman primate personality and immunity: Mechanisms of health and disease. In A. Weiss, J. E. King, & L. Murray (Eds.), *Personality and Temperament in Nonhuman Primates* (pp. 233–255). New York: Springer.

Carere, C., & Locurto, C. (2011). Interaction between animal personality and animal cognition. *Current Zoology, 57,* 491–498.

Carlstead, K., Mellen, J., & Kleiman, D. G. (1999). Black rhinoceros (*Diceros bicornis*) in US zoos: I. Individual behavior profiles and their relationship to breeding success. *Zoo Biology, 18,* 17–34.

Clubb, R., & Mason, G. (2003). Captivity effects on wide-ranging carnivores. *Nature, 425,* 473–474. doi:10.1038/425473a

De Palma, C., Viggiano, E., Barillari, E., Palme, R., Dufour, A. B., Fantini, C., et al. (2005). Evaluating the temperament in shelter dogs. *Behaviour, 142,* 1307.

De Rouck, M., Kitchener, A. C., Law, G., & Nelissen, M. (2005). A comparative study of the influence of social housing conditions on the behaviour of captive tigers (*Panthera tigris*). *Animal Welfare, 14,* 229–238.

Deary, I. J., Weiss, A., & Batty, G. D. (2010). Intelligence and personality as predictors of illness and death:

How researchers in differential psychology and chronic disease epidemiology are collaborating to understand and address health inequalities. *Psychological Science in the Public Interest, 11,* 53–79.

Diener, E., & Chan, M. Y. (2011). Happy people live longer: Subjective well-being contributes to health and longevity. *Applied Psychology: Health and Wellbeing, 3,* 1–43. doi:10.1111/j.1758-0854.2010.01045.x

Diener, E., Suh, E. M., Lucas, R. E., & Smith, H. L. (1999). Subjective well-being: Three decades of progress. *Psychological Bulletin, 125,* 276–302.

Digman, J. M. (1990). Personality structure: Emergence of the five-factor model. *Annual Reviews, 41,* 417–440.

Driscoll, C. A., Menotti-Raymond, M., Roca, A. L., Hupe, K., Johnson, W. E., Geffen, E., et al. (2007). The Near Eastern origin of cat domestication. *Science, 317,* 519–523. doi:10.1126/Science.1139518

Elenkov, I. J., & Chrousos, G. P. (1999). Stress hormones, Th1/Th2 patterns, pro/anti-inflammatory cytokines and susceptibility to disease. *Trends in Endocrinology and Metabolism, 10,* 359–368.

Feaver, J., Mendl, M., & Bateson, P. (1986). A method for rating the individual distinctiveness of domestic cats. *Animal Behavior, 34,* 1016–1025.

Finkler, H., & Terkel, J. (2015). The relationship between individual behavioural styles, dominance rank and cortisol levels of cats living in urban social groups. *Applied Animal Behaviour Science, 173,* 22–28.

Freeman, H., & Gosling, S. (2010). Personality in nonhuman primates: A review and evaluation of past research. *American Journal of Primatology, 71,* 1–19.

Fromont, E., Artois, M., Langlais, M., Courchamp, F., & Pontier, D. (1997). Modelling the feline leukemia virus (FeLV) in natural populations of cats (*Felis catus*). *Theoretical Population Biology, 52,* 60–70.

Gartner, M. C. (2015). Pet personality: A review. *Personality and Individual Differences, 75,* 102–113.

Gartner, M. C., & Powell, D. (2012). Personality assessment in snow leopards (*Uncia uncia*). *Zoo Biology, 31,* 151–165. doi:10.1002/zoo.20385

Gartner, M. C., Powell, D. M., & Weiss, A. (2014). Personality structure in five felids (*Felis silvestris catus, Felis silvestris grampia, Neofelis nebulosa, Panthera uncia,* and *Panthera leo*): A comparative study. *Journal of Comparative Psychology, 128,* 414–426.

Gartner, M. C., Powell, D. M., & Weiss, A. (2016). Comparison of subjective well-being and personality assessments in the clouded leopard (*Neofelis nebulosa*), snow leopard (*Panthera uncia*), and African lion (*Panthera leo*). *Journal of Applied Animal Welfare Science, 19,* 1–9.

Gartner, M. C., & Weiss, A. (2013a). Personality in felids: A review. *Applied Animal Behaviour Science, 144,* 1–13.

Gartner, M. C., & Weiss, A. (2013b). Scottish wildcat (*Felis silvestris grampia*) personality and subjective well-being: Implications for captive management. *Journal of Applied Animal Behaviour Science, 147,* 261–267.

Glassé, C. (2003). *The new encyclopedia of Islam* (p. 102). Rowman Altamira.

Gourkow, N., & Fraser, D. (2006). The effect of housing and handling practices on the welfare, behavior and selection of domestic cats (*Felis sylvestris catus*) by adopters in an animal shelter. *Animal Welfare, 15,* 371–377.

Grimm, D. *What are cats thinking?* Retrieved June 30, 2016 from http://www.slate.com/articles/health_and_science/science/2014/04/cat_intelligence_and_cognition_are_cats_smarter_than_dogs.html

Hayssen, V. (1997). Effects of the nonagouti coat-color allele on behavior of deer mice (*Peromyscus maniculatus*): A comparison with Normway rats (*Rattus norvegicus*). *Journal of Comparative Psychology, 111,* 419–423.

Hendry, A. P., & Kinnison, M. T. (1999). The pace of modern life: Measuring rates of contemporary microevolution. *Evolution, 53,* 1637–1653.

Herrelko, E. S., Vick, S.-J., & Buchanan-Smith, H. M. (2012). Cognitive research in zoo-housed chimpanzees: Influence of personality and impact on welfare. *American Journal of Primatology, 74,* 828–840.

Hopkins, W. D., Donaldson, Z. R., & Young, L. J. (2012). A polymorphic indel containing the RS3 microsatellite in the 5′ flanking region of the vasopressin V1a receptor gene is associated with chimpanzee (*Pan troglodytes*) personality. *Genes, Brain and Behavior, 11,* 552–558.

Hu, Y., Hu, S., Wang, W., Wu, X., Marshall, F. B., Chen, X., et al. (2014). Earliest evidence for commensal processes of cat domestication. *Proceedings of the National Academy of Sciences, 111,* 116–120.

Jones, A. C., & Gosling, S. D. (2005). Temperament and personality in dogs (*Canis familiaris*): A review and evaluation of past research. *Applied Animal Behavioiur Science, 95,* 1–53.

Kaur, G., Voith, V. L., & Schmidt, P. L. (2016). The use of fluoxetine by veterinarians in dogs and cats: A preliminary study. *Veterinary Record Open, 3,* e000146. doi:10.1136/vetreco-2015-000146

King, J. E., & Figueredo, A. J. (1997). The five-factor model plus dominance in chimpanzee personality. *Journal of Research in Personality, 31,* 257–271.

King, J. E., & Landau, V. I. (2003). Can chimpanzee (*Pan troglodytes*) happiness be estimated by human raters? *Journal of Research in Personality, 37,* 1–15. doi:10.1016/S0092-6566(02)00527-5

King, J. E., Weiss, A., & Farmer, K. H. (2005). A chimpanzee (*Pan troglodytes*) analogue of cross-national generalization of personality structure: Zoological parks and an African sanctuary. *Journal of Personality, 73,* 389–410.

Kitchener, A. C. (2000). Are cats really solitary? *Lutra, 43,* 1–10.

Ledger, R., & O'Farrell, V. (1996). Factors influencing the reactions of cats to humans and novel objects. In I. J. H. Duncan, T. M. Widowski, & D. B. Haley (Eds.), *Proceedings of the 30th International Congress of the*

ISAE. Guelph: Col. K. L. Campbell Centre for the Study of Animal Welfare.

Lockwood, R. (2005). Cruelty toward cats: Changing perspectives. In D. J. Salem & A. N. Rowan (Eds.), *The state of the animals III: 2005* (pp. 15–26). Washington, DC: Humane Society Press.

Locurto, C. (2007). Individual differences and animal personality. *Comparative Cognition & Behavior Reviews, 2,* 67–78.

Lowe, S. E., & Bradshaw, J. W. S. (2001). Ontogeny of individuality in the domestic cat in the home environment. *Animal Behaviour, 61,* 231–237. doi:10.1006/Anbe.2000.1545

Macdonald, D. W., Yamaguchi, N., & Kerby, G. (2000). Group-living in the domestic cat: Its sociobiology and epidemiology. In D. C. Turner & P. Bateson (Eds.), *The domestic cat: The biology of its behaviour* (2nd ed., pp. 95–118). Cambridge: Cambridge University Press.

Mayr, E. (1961). Cause and effect in biology. *Science, 134,* 1501–1506.

McCune, S. (1995). The impact of paternity and early socialization on the development of cats' behavior to people and novel objects. *Applied Animal Behaviour Science, 45,* 109–124.

McDougall, P. T., Réale, D., Sol, D., & Reader, S. M. (2006). Wildlife conservation and animal temperament: Causes and consequences of evolutionary change for captive, reintroduced, and wild populations. *Animal Conservation, 9,* 39–48.

Meier, M., & Turner, D. C. (1985). Reactions of house cats during encounters with a strange person: Evidence for two personality types. *Journal of the Delta Society, 2,* 45–53.

Miklósi, Á., Kubinyi, E., Topál, J., Gácsi, M., Virányi, Z., & Csányi, V. (2003). A simple reason for a big difference: Wolves do not look back at humans, but dogs do. *Current Biology, 13,* 763–766.

Miklósi, Á., Pongrácz, P., Lakatos, G., Topál, J., & Csányi, V. (2005). A comparative study of the use of visual communicative signals in interactions between dogs (*Canis familiaris*) and humans and cats (*Felis catus*) and humans. *Journal of Comparative Psychology, 119,* 179–186.

Miller, L. (2010). Japan's zoomorphic urge. *ASIANetwork Exchange: A Journal for Asian Studies in the Liberal Arts, 17,* 69–82.

Morgan, R. S. (2010). Personality in domestic cats (*Felis catus*): Identification and description of personality structure, predictive validity, and associations with coat color characteristics. Dissertation, University of California, Davis.

Natoli, E., Say, L., Cafazzo, S., Bonanni, R., Schmid, M., & Pontier, D. (2005). Bold attitude makes male urban feral domestic cats more vulnerable to feline immunodeficiency virus. *Neuroscience and Biobehavioral Reviews, 29*(151–157), 2004. doi:10.1016/J.Neubiorev.06.011

Neidhart, L., & Boyd, R. (2002). Companion animal adoption study. *Journal of Applied Animal Welfare Science, 5,* 175–192.

O'Brien, S. J., Johnson, W., Driscoll, C., Pontius, J., Pecon-Slattery, J., & Menotti-Raymond, M. (2008). State of cat genomics. *Trends in Genetics, 24,* 268–279.

Orians, G. H. (1962). Natural selection and ecological theory. *The American Naturalist, 96,* 257–263.

Phillips, C., & Peck, D. (2007). The effects of personality of keepers and tigers (*Panthera tigris tigris*) on their behaviour in an interactive zoo exhibit. *Applied Animal Behaviour Science, 106*(244–258), 2007. doi:10.1016/J.Applanim.01.007

Powell, D. M., Hong, L., Carlstead, K. J., Kleiman, D. G., Zhang, H., Zhang, G., et al. (2008). Relationships between temperament, husbandry, management, and socio-sexual behavior in captive male and female giant pandas (*Ailuropoda melanoleuca*). *Acta Zoologica Sinica, 54,* 169–175.

Raihani, G., Rodriguez, A., Saldana, A., Guarneros, M., & Hudson, R. (2014). A proposal for assessing individual differences in behaviour during early development in the domestic cat. *Applied Animal Behaviour Science, 154,* 48–56.

Robinson, R. (1977). *Genetics for cat breeders* (2nd ed.). London: Pergamon Press.

Schmidt, P. M., Lopez, R. R., & Collier, B. A. (2007). Survival, fecundity, and movements of free-roaming cats. *Journal of Wildlife Management, 71,* 915–919.

Serpell, J. A. (2000). Domestication and history of the cat. In D. C. Turner & P. Bateson (Eds.), *The domestic cat: The biology of its behaviour* (2nd ed., pp. 179–192). Cambridge: Cambridge University Press.

Shepherdson, D. (1994). The role of environmental enrichment in the captive breeding and reintroduction of endangered species. In P. J. S. Olney, G. M. Mace, & A. T. C. Feistner (Eds.), *Creative conservation: Interactive management of wild and captive animals* (pp. 167–177). London: Chapman & Hall.

Sinn, L. (2016). Factors affecting the selection of cats by adopters. *Journal of Veterinary Behavior*. doi:10.1016/j.jveb.2016.06.001

Smith, B. R., & Blumstein, D. T. (2008). Fitness consequences of personality: A meta-analysis. *Behavioral Ecology, 19,* 448–455.

Steel, P., Schmidt, J., & Shultz, J. (2008). Refining the relationship between personality and subjective well-being. *Psychological Bulletin, 134,* 138–161.

Stelow, E. A., Bain, M. J., & Kass, P. H. (2016). The relationship between coat color and aggressive behaviors in the domestic cat. *Journal of Applied Animal Welfare Science, 19,* 1–15.

Stevenson-Hinde, J., Stillwell-Barnes, R., & Zunz, M. (1980). Subjective assessment of rhesus monkeys over four successive years. *Primates, 21,* 66–82.

Stoinski, T. S., Lukas, K. E., Kuhar, C. W., & Maple, T. L. (2004). Factors influencing the formation and maintenance of all-male gorilla groups in captivity. *Zoo Biology, 2004,* 189–203.

Tellegen, A., Lykken, D. T., Bouchard, T. J., Jr., Wilcox, K. J., Segal, N. L., & Rich, S. (1988). Personality similarity in twins reared apart and together. *Journal of Personality and Social Psychology, 54,* 1031–1039.

Todd, N. B. (1977). Cats and commerce. *Scientific American, 237,* 100–107.
Torgerson-White, L., & Bennett, C. (2014). Rating methodology, personality axes and behavioral plasticity: A case study in African lions. *Animal Behavior & Cognition, 1,* 230–248.
Trut, L., Oskina, I., & Kharlamova, A. (2009). Animal evolution during domestication: The domesticated fox as a model. *BioEssays, 31,* 349–360.
Tsigos, C., & Chrousos, G. P. (2002). Hypothalamic–pituitary–adrenal axis, neuroendocrine factors and stress. *Journal of Psychosomatic Research, 53,* 865–871.
Turner, D. C., Feaver, J., Mendl, M., & Bateson, P. (1986). Variation in domestic cat behaviour towards humans: A paternal effect. *Animal Behaviour, 34,* 1890–1892.
Umbelino, M. T. L. P. (2014). Evaluation of the relation between tameness and coat color in cats. Dissertation, Universidade de Evora.
Vigne, J.-D., Guilaine, J., Debue, K., Haye, L., & Gérard, P. (2004). Early taming of the cat in Cyprus. *Science, 304,* 259.
Way, J. G., Rutledge, L., Wheeldon, T., & White, B. N. (2010). Genetic characterization of eastern "coyotes" ni eastern Massachusetts. *Northeastern Naturalist, 17,* 189–204.
Weiss, A., Adams, M. J., & King, J. E. (2011). Happy orang-utans live longer lives. *Biology Letters, 7,* 872–874. doi:10.1098/Rsbl.2011.0543
Weiss, A., Adams, M. J., Widdig, A., & Gerald, M. S. (2011). Rhesus macaques (*Macaca mulatta*) as living fossils of hominoid personality and subjective well-being. *Journal of Comparative Psychology, 125,* 72–83. doi:10.1037/A0021187
Weiss, A., Gartner, M. C., Gold, K. C., & Stoinski, T. S. (2012). Extraversion predicts longer survival in gorillas: an 18-year longitudinal study. *Proc. R. Soc. B, 280,* 20122231
Weiss, A., Inoue-Murayama, M., Hong, K. W., Inoue, E., Udono, T., Ochiai, T., et al. (2009). Assessing chimpanzee personality and subjective well-being in Japan. *American Journal of Primatology, 71,* 283–292.
Weiss, A., King, J. E., & Perkins, L. (2006). Personality and subjective well-being in orangutans (*Pongo pygmaeus* and *Pongo abelii*). *Journal of Personality and Social Psychology, 90,* 501–511. doi:10.1037/0022-3514.90.3.501
Wemelsfelder, F. (2007). Qualitative behaviour assessment: Application to welfare assessment in extensively managed sheep. In P. Goddard (Ed.), *Proceedings of welfare goals from the perspective of extensively managed sheep* (pp. 57–62). Aberdeen: The Macaulay Institute.
Wielebnowski, N. C. (1998). Contributions of behavioral studies to captive management and breeding of rare and endangered mammals. In T. Caro (Ed.), *Behavioral ecology and conservation biology* (pp. 130–162). New York: Oxford University Press Inc.
Wielebnowski, N. C. (1999). Behavioral differences as predictors of breeding status in captive cheetahs. *Zoo Biology, 18,* 335–349.
Wielebnowski, N. C., Fletchall, N., Carlstead, K., Busso, J. M., & Brown, J. L. (2002). Noninvasive assessment of adrenal activity associated with husbandry and behavioral factors in the North American clouded leopard population. *Zoo Biology, 21,* 77–98. doi:10.1002/Zoo.10005
Wilhelmy, J., Serpell, J., Brown, D., & Siracusa, C. (2016). Behavioral associations with breed, coat type, and eye color in single-breed cats. *Journal of Veterinary Behavior: Clinical Applications and Research, 13,* 80–87.
Willard, S. L., & Shively, C. A. (2012). Modeling depression in adult female cynomolgus monkeys (*Macaca fascicularis*). *American Journal of Primatology, 74,* 528–542.
Zito, S., Paterson, M., Vankan, D., Morton, J., Bennett, P., & Phillips, C. (2015). Determinants of cat choice and outcomes for adult cats and kittens adopted from an Australian animal shelter. *Animals, 5,* 276–314.

What Do We Want to Know About Personality in Marine Mammals?

12

Erin E. Frick, Amber J. de Vere and Stan A. Kuczaj

Abstract

Investigation of personality in a variety of nonhuman animal species has flourished over the past decade. However, personality assessments in marine mammals remain greatly underrepresented. In this chapter, we seek to outline the key areas of interest that warrant further study to improve knowledge of personality in marine mammals. Several definitional challenges associated with personality in marine mammals are examined. An overview of the current marine mammal personality literature is provided, which is limited to trait ratings of bottlenose dolphins (*Tursiops truncatus*) and behavioral axes in grey seals (*Halichoerus grypus*). Possible cross-species comparisons are limited at this time due to the lack of marine species included. With this in mind, it would be useful to assess the personality dimensions of species from each of the major marine mammal taxa: pinnipeds, sirenians, cetaceans, and carnivores. This could provide evidence for the potential impact of marine mammals' aquatic lifestyle on the evolution of species-specific personality traits. Given their aquatic habitat, aspects of typical personality methodology are difficult to apply to marine mammals. Several of these difficulties are discussed, along with proposed solutions to maximize the species-relevance and representativeness of collected data. Finally, the potential contributions of personality to a number of research areas are discussed: social rank and dominance, learning, physiology, and friendships. Although the field of marine mammal personality research is in an early stage of development, this provides huge opportunity for future research.

Introduction

Marine mammals are social animals that exhibit a large behavioral repertoire across various contexts and possess sophisticated cognitive

E.E. Frick (✉) · A.J. de Vere · S.A. Kuczaj
Department of Psychology, University of Southern Mississippi, 118 College Drive #5025, Hattiesburg, MS 39406, USA
e-mail: erin.frick@usm.edu

J. Vonk et al. (eds.), *Personality in Nonhuman Animals*, DOI 10.1007/978-3-319-59300-5_12

abilities. The most frequently studied marine mammal species, the Atlantic bottlenose dolphin (*Tursiops truncatus*), provides a prime example of this behavioral and cognitive range (Hill et al. 2016). Bottlenose dolphins, which exhibit a fission–fusion social structure, engage in social interactions ranging from those that are affiliative or playful (Connor et al. 2000) to those that are agonistic or aggressive (Samuels and Gifford 1997). Dolphins also exhibit highly complex cognitive abilities, such as cooperation (Kuczaj et al. 2015b) and planning/problem solving (Kuczaj et al. 2009). Altruistic (i.e., care-giving) behaviors have also been observed in bottlenose dolphins (Caldwell and Caldwell 1966; Fertl and Shiro 1994; Kuczaj et al. 2015a), rough toothed dolphins (*Steno bredanensis*; De Moura et al. 2009), long-beaked common dolphins (*Delphinus capensis*; Park et al. 2013), and La Plata dolphins (*Pontoporia blainvillei*; Cremer et al. 2006). Furthermore, as in many species (for review see Gosling 2001), dolphins show individual variation in several behavioral states such as foraging (Duffy-Echevarria et al. 2008; Gazda et al. 2005), maternal care (Gibson and Mann 2008; Hill et al. 2007), and group movements (Lewis et al. 2011). Individual variation in the type and frequency of behaviors from each of the aforementioned behavioral states may reflect personality differences (Highfill and Kuczaj 2007, 2010; Kuczaj et al. 2012a).

In this chapter we will provide an overview of the literature on personality in marine mammals, with particular emphasis on bottlenose dolphins —hereafter referred to as dolphins—due to the abundance of available research that has been conducted already with this species compared to other marine mammal species (Hill et al. 2016). In our discussion, other species of marine mammals besides dolphins will be incorporated when appropriate (i.e., there is sufficient literature on a given species to incorporate it into the relevant topic). In doing so, we focus particularly on how well definitions of personality, methods, and techniques are currently applied in marine mammals (in particular, dolphins and pinnipeds) and propose how cross-species comparisons could be made in the future to other mammalian species in which personality studies have been done, such as elephants (e.g., Highfill et al. 2013; Horback et al. 2013; Yasui et al. 2013) and chimpanzees (e.g., King and Figueredo 1997; Koski 2011; Massen and Koski 2014). We also consider the influence and role that personality may serve in diverse aspects of dolphin behavior, including social rank (Aureli et al. 2002; Samuels and Gifford 1997), foraging (Gazda et al. 2005; Highfill and Kuczaj 2010), learning (Kuczaj and Walker 2006; Kuczaj et al. 2009), physiology (Ortiz and Worthy 2000), social associations (Connor et al. 2000; Connor et al. 2006), and play (Kuczaj and Eskelinen 2014; Kuczaj and Highfill 2005; Kuczaj and Makecha 2008; Kuczaj et al. 2006). We offer several suggestions for ways that this field can be expanded to further contribute to our current understanding of personality in dolphins and other marine mammals, as well as the importance of incorporating more species of marine mammals into the personality literature.

Defining Personality for Marine Mammals

Consistent variation in behavioral traits between individuals is known as individual differences or personality (Gosling 2001; Locurto 2007; Sih et al. 2004; Stamps and Groothius 2010a). Individual differences are commonly assessed through questionnaire ratings (Gosling 2001; Highfill and Kuczaj 2007; Kuczaj et al. 2012a), naturalistic observations (Horback et al. 2013; Vazire et al. 2007), or experimental tasks (Eskelinen et al. 2015). Temperament is considered the biological precursor for the development of personality in humans (De Pauw and Mervielde 2010). However, in animal studies, the word "temperament" appears to be often used to avoid referring to "personality" (Gosling 2001; Vazire and Gosling 2004). Subsequently, both temperament and personality have been used interchangeably within the animal personality literature, as have terms such as "coping styles" (Koolhaas et al. 1999) and "behavioral syndromes" (Sih et al. 2004). The common element for all of these labels is the description of individual differences as patterns of behavior that are consistent over time and context/situation (Carlstead et al. 1999; Gosling 2001; Locurto 2007). However, predictions regarding contexts in which individual differences should be consistent are typically lacking. Moreover, authors sometimes differ in whether they distinguish between context and situation (Stamps and Groothius 2010a). It is therefore important to consider how these terms are used to define personality in dolphins and other marine mammals.

"Consistency over time" is a phrase commonly used when defining personality and individual differences (Gosling and John 1999). However, it is not clear how we define "consistency over time." Are multiple time periods of data collection necessary, or is drawing from one's cumulative experience working with an animal (e.g., 5 years as an animal trainer) sufficient for establishing "consistency over time" in personality? For example, personality ratings of bottlenose dolphins have been carried out on two separate occasions over a 2-year time period (Highfill and Kuczaj 2007), as well as a one-time assessment relying on the cumulative experience of the raters that worked with the animals (Kuczaj et al. 2012a). There currently is no existing framework stipulating the time course over which observations would need to be consistent in order to constitute personality.

As noted above, context and situation are often included in definitions of personality, and are sometimes used interchangeably (Stamps and Groothuis 2010a). If/when a distinction is made, context refers to behavioral categories based on the function of the observed behavior, such as foraging or breeding (Sih et al. 2004). Within contexts, situations occur that consist of different external environmental conditions, such as the presence or absence of predators (Sih et al. 2004). Although distinctions between context and situation can be made, we currently do not understand the functions of many behaviors in a variety of animal species, including dolphins. This lack of clarity can result in incorrectly assuming a behavior's function (Stamps and Groothuis 2010b), and therefore potentially incorrectly categorizing the context. In their investigation of the role of context in the personalities of dolphins, Kuczaj et al. (2012a) found that only 4 of 20 dolphins were rated consistently for a personality trait across all measured contexts. These results emphasize not only the importance of considering context in studies of personality, but also stresses that different animals may have individually specific patterns of consistency across contexts. The flexibility of individual differences in behavior across contexts also may serve a role in personality expression in dolphins and other species of marine mammals (Stamps and Groothuis 2010b).

It is also important to consider the scale of a particular context. Contexts are continuums that can vary in time and space (Stamps and Groothuis 2010a). A context could have a duration of as little as a few seconds, or a larger temporal scale such as day versus night. It could have a small spatial scale of only the animal's immediate surroundings or could include an entire social group. These details should always be considered when designing studies, and accompanied by a clear definition that clarifies what the term "context" is referring to. Furthermore, no matter how stringent the definition, we feel it is important to note that all environments are dynamic, and, even if researchers focus on a small number of characteristics, no two contexts will ever be identical. This emphasizes further the importance of clearly operationalizing the contexts of interest, to maintain as much similarity as possible. Finally, we suggest that researchers should strive to explicitly examine the interaction of various forms of context and durations in their assessments of animal personality.

Overall, as do Stamps and Groothuis (2010a), we recommend that context be used as an all-encompassing term that refers to all stimuli external to an individual. When it is possible to make distinctions between contexts, these distinctions should be based on physical, observable conditions rather than the function of behaviors. For example, in marine species, we would advocate using clearly distinctive contexts such as surface versus underwater behaviors. Contexts that are dependent upon human interpretation should be incorporated only if there is a substantial body of research that justifies their use. Furthermore, when they lack information about the functions of behaviors, researchers should collect data on finer scale behaviors, which can be grouped based on the results of principal components or factor analyses. In doing so, researchers avoid assuming that behaviors have the same function when they do not (Carter et al. 2012, 2013).

Overview of Current Marine Mammal Personality Research

Cetaceans: Bottlenose Dolphins

Highfill and Kuczaj (2007) were the first to investigate whether adult bottlenose dolphins could be characterized as having personalities. For this study, they obtained personality ratings of 16 bottlenose dolphins, with one assessment occurring prior to Hurricane Katrina and the second occurring 15 months after Hurricane Katrina. Due to the hurricane, the dolphins were displaced from the MarineLife Oceanarium in Gulfport MS to the Mississippi Sound, rescued, and then relocated to the Atlantis facility in the Bahamas. MarineLife staff conducted ratings for the first assessment and new staff members at the Bahamas resort location completed the second set of ratings 9 months after the relocation. Of these dolphins, 12 exhibited fairly stable and consistent individual differences in personality. This is remarkable in light of the fact that these dolphins underwent significant changes to their environment between ratings.

Kuczaj et al. (2012a) further investigated the importance of context in dolphin personality by assessing certain traits across specific contexts. Twenty dolphins housed at Dolphins Plus in Key Largo, Florida were rated by experienced animal care staff on traits Observant, Timidity, Curiosity, and Playful across three contexts (interactions with the physical world, interactions with other dolphins, and interactions with humans). In a separate assessment targeting social behaviors across contexts, personality traits Aggressiveness, Gentleness, and Cooperative were analyzed over the two social contexts (interactions with humans and interactions with dolphins). Four subjects were stable in all traits across all contexts, while the remaining dolphins' ratings varied across these contexts. This emphasizes the importance of recognizing traits specific for the target species, as well as accounting for context when assessing individual differences in dolphins and other nonhuman animals.

Pinnipeds: Grey Seals (*Halichoerus grypus*)

Due to the lack of studies on personality in any marine mammal species other than the bottlenose dolphin, it is not possible at this time to make personality comparisons between marine mammalian species. However, there is some data on individual differences in a few specific behaviors in grey seals. Active scanning behaviors (i.e., alert behavior) in males have been measured as a possible indicator of the bold–shy axis over two consecutive breeding seasons; individual time spent alert was highly consistent between seasons (Twiss and Franklin 2010). In a further study, across a short re-test interval of 4–12 days, Twiss et al. (2012) used a remote control vehicle (RCV) to test individual differences in pup-checking by females and individual differences in aggressive behaviors of males in response to disturbance by the RCV. The grey seals' responses to the RCV were significantly repeatable and the length of re-test interval had no effect on repeatability. The same protocol was used to test females for the same pup-checking behavior the following year, to test for long-term response repeatability (Twiss et al. 2012). Seven females were tested in both years and showed some degree of repeatability overall, but with a wide range between individual repeatability values. Twiss et al. (2012) attributed individual differences in reaction to the RCV as being indicative of differing positions on a proactive–reactive behavioral axis. However, we suggest that these differences could also indicate individually specific patterns of consistency across contexts, such as was found in bottlenose dolphins (Kuczaj et al. 2012a).

Although these studies support the existence of individual differences in a few behaviors in grey seals, no studies examine a greater proportion of the behavioral repertoire of this species. As a result, it is not known whether individual differences are exhibited in other behaviors, or if any of these relate to each other. However, the data used in Twiss and Franklin's (2010) study were not gathered for the purpose of studying personality. It may therefore be possible to test for a range of further individual differences in behavior by carrying out another re-analysis of these existing datasets (Twiss and Franklin 2010).

Furthermore, many pinniped species are highly faithful in their haul-out sites: locations where individuals temporarily leave the water for activities such as resting and mating (Twiss et al. 1994). This allows identification of individuals in consecutive years in the same location. Although it must be recognized that hauled-out behavior is only one subset of their total behavioral repertoire, certain pinniped populations that exhibit a high degree of haul-out site fidelity may lend themselves more readily to studies of personality compared to other marine mammals.

Cross-Species Comparisons

There is a need for more analyses that attempt to measure individual differences in the behavior of other marine mammal species, as we cannot make any cross-species comparisons at this time. It would therefore be useful to assess at least one marine mammal species from the major groups: pinnipeds (i.e., seals, sea lions, walruses), sirenians (i.e., dugongs, manatees), cetaceans (i.e., dolphins, whales), and carnivores (i.e., otters, polar bears). Interestingly, the most closely related taxa to each of these groups are terrestrial mammal groups, so comparisons between these relatives would be particularly intriguing. Most phylogenies place hippopotamus as the closest relative of cetaceans (Berta et al. 2006), but there is no personality data on any hippopotamus species. Sirenians are most closely related to proboscidea (elephants; Berta et al. 2006); although personality in African and Asian elephants has been studied (Grand et al. 2012; Highfill et al. 2013; Horback et al. 2013; Lee and Moss 2012; Yasui et al. 2013), no such research has been carried out in sirenian species. Pinnipeds are possibly closely related to arctoid carnivores such as raccoons and weasels (Springer et al. 2004), but are most commonly grouped with ursids, including the polar bear (*Ursus maritimus*). Although there are no existing studies examining any individual differences in polar bears, it would be interesting to compare their personality with that of pinnipeds and terrestrial bears.

Future research should therefore be directed at expanding personality studies to more species of marine mammals. Cross-species comparisons of species typical traits would allow for more clarity on the evolution of a given personality trait (Gosling 2001). In their review of personality dimensions across species, Gosling and John (1999) conclude that Five-Factor Model derived traits of Extraversion, Neuroticism, and Agreeableness were present across a wide variety of taxa, which suggests there may be a biological foundation for such factors in nonhuman animals. This review also observed that the Conscientious factor was present only in chimpanzees and humans. No personality studies from dolphins or any marine mammal including all dimensions were available to include in this review, but Highfill and Kuczaj (2007) did find the Conscientious component to be present in their study on dolphin personality. We would predict that, due to the cognitive abilities demonstrated by dolphins, that cross-species comparisons of personality factors would further demonstrate the similarity of dolphin personality to that of chimpanzees and humans on factors such as Conscientiousness. The impact of marine mammals' aquatic lifestyle, compared to features shared with their closest terrestrial relatives on any elucidated factor structure, could provide insight into the evolution of personality.

Refining Methods and Techniques for Marine Mammal Personality Research

Animal personality, like all psychological phenomena, must be assessed by observing behavior (Freeman et al. 2011; Gosling 2001; Locurto 2007). One of the primary methods of assessing personality is the rating method (Carter et al. 2013; Highfill et al. 2010). The rating method asks humans to judge an animal's behavioral tendencies across time and contexts (Highfill and Kuczaj 2007, 2010; Highfill et al. 2010; Kuczaj et al. 2012a). These ratings are then pulled together and analyzed for possible personality dimensions/domains (see, e.g., Dutton et al. 1997; King and Figueredo 1997). The personality questionnaires used in these studies can be designed using a top-down design, a bottom-up design, or some combination of both (Freeman et al. 2013). Top-down designs adapt and apply an already established scale from one species to another, such as the Five-Factor Model (Goldberg 1990), which has been widely applied in the study of human personality. For example, factors homologous to the Five-Factor Model have been found in bottlenose dolphins (Highfill and Kuczaj 2007). Although a top-down design tends to be used more across multiple studies and can facilitate cross-species comparisons (Freeman et al. 2013), it has the potential to miss factors relevant to the target species or to include irrelevant factors (Gosling and John 1999; Kuczaj et al. 2012a). Conversely, a bottom-up rating structure derives traits specific to the target species from their behavioral repertoire (Freeman and Gosling 2010; Freeman et al. 2013; Koski 2011). Advantages of utilizing this approach include the use of factors that are directly relevant to the target species (Uher and Asendorf 2008), but this does decrease the ability to make comparisons across species. For example, certain traits included for an assessment of dolphins may not be relevant to another species. Bottom-up designs have been incorporated into ratings in only a few studies of primates (Dutton et al. 1997; Freeman et al. 2013; Koski 2011), and several species-specific elements and contexts were included in Kuczaj et al.'s (2012a) study on rating assessments of bottlenose dolphin personality.

The coding method assesses personality by recording all instances of behavior (Highfill et al. 2010), and then clustering behaviors into personality traits. This method has been used in several species including African elephants (Horback et al. 2013) and chimpanzees (Koski 2011). Additionally, coding reactions to novel stimuli in experimental settings has revealed information about the bold–shy dimension in several species such as great tits (*Parus major*; Verbeek et al. 1996; Verbeek et al. 1999), pumpkinseed fish (*Leopmis gibbosus*; Coleman and Wilson 1998), and rainbowfish (*Melanotaenia duboulayi*; Colléter and Brown 2011). The coding method has also been used in all studies to date of individual differences in grey seals (Twiss and Franklin 2010; Twiss et al. 2012a, b). For dolphins, several studies have recorded individual differences in response to novel stimuli (e.g., Eskelinen et al. 2015; Kuczaj and Eskelinen 2014; Kuczaj and Yeater 2006; Lopes et al. 2016). However, researchers have yet to apply the coding methodology to the broader repertoire of behavior in dolphins, or any other species of marine mammals.

A currently overlooked issue in animal personality research methodology is emotions. In humans, several factors in the Five-Factor Model of personality contain emotional content, such as anxiety in Neuroticism (Zillig et al. 2002), and correlate strongly to measures of emotions, such as anger with low Agreeableness (Kuppens 2005). As a result, a distinction is often made between state emotion, the actual experiencing of an emotion, and trait emotion, individual tendencies to experience those emotions (Izard et al. 1993). In animal personality research, the rating method frequently involves sourcing trait words from the human Five-Factor Model (Gosling and John 1999). The questionnaires constructed then often include terms referring to emotional experiences. For example, King and Figueredo (1997) include traits such as fearful, irritable, and excitable in their questionnaire assessing chimpanzee personality. The coding method may also benefit from considering emotions. When factor analyses are conducted, the functions of behaviors loading onto factors are interpreted in order to provide an explanation of each factor. As a result, when behaviors are interpreted as indicating an emotional trait, such as cautiousness or aggressiveness, this may assume that the species under study does experience some form of these emotional states.

It is therefore worth asking whether the use of this crossover terminology, from emotion to personality research, is justified—are the animals under study capable of experiencing all of the emotional states inferred by trait words or factor labels such as fearful or nervous? Furthermore, if the answer to this question is not known, is it justifiable to assume that the animals are capable? Several approaches could be used to tackle these issues. Firstly, any trait words inferring emotional states could be avoided in both rating and coding. However, this would remove a huge proportion of traits that are currently used, and would fail to consider the increasing body of evidence suggesting that animals do experience some form of emotions (e.g., de Vere and Kuczaj 2016). Alternatively, animals could be experimentally tested to determine which emotional states they do and do not experience. Already there is overlap between the experimental tests used to assess behavioral indicators of emotions and personality. The most common test uses approach versus avoidance of a stimulus to determine an animal's position on the bold–shy axis, or the relative pleasantness of the stimulus (Paul et al. 2005). However, exhaustively testing a sufficient sample of animals for a range of emotional states is not time efficient, especially as each emotion would likely require its own specific measure (Paul et al. 2005). Finally, it may be possible to use the rating method to determine species-specific emotional repertoires. Personality ratings of many species have proven reliable (Gosling and John 1999), and often contain traits inferring emotional states. Therefore, people familiar with a particular species may be able to reliably rate that species' capacity to experience different emotions (Kuczaj and Horback 2013). Only those trait words corresponding to emotions rated reliably could then be included in personality questionnaires, or used to interpret behavioral factors (de Vere and Kuczaj 2016). For example, if bottlenose dolphins were not rated reliably on an emotional trait such as "jealous," then the trait word "jealous" should not be included in personality assessments of this species.

Keeping these aspects of the methodologies in mind, to improve how we assess personality in marine mammals it is imperative to understand the functions and relevance of a given behavior so as to be sure that the raters and coders assessing personality are interpreting the observed behaviors accurately and as objectively as possible. For example, among dolphins open mouth displays are commonly thought to be aggressive "threat" displays (Samuels and Gifford 1997). However, there is some evidence that open mouth displays serve a communicative purpose in other contexts, such as sexual or affiliative contexts (Kuczaj and Frick 2015). Interpretation of behavior, and prior expectations regarding a behavior's connection to a trait, then, can be incorrect when knowledge of the species' behavioral repertoire is incomplete (Vazire et al. 2007).

Finally, marine mammals engage in the majority of their behavioral repertoire underwater, and so it can be difficult for researchers to observe their full behavioral repertoire and the contexts in which behaviors take place. Surface behaviors are more readily visible in wild populations and in captive groups. Underwater behavior recordings therefore need to be introduced to personality assessments to maximize the information that researchers can extract from marine mammals.

Future Directions

Social Rank and Dominance

For numerous mammalian species, individuals compete with one another for access to resources, and for highly social mammals this can result in the formation of a dominance hierarchy. Some benefits associated with high dominance include preferential access to sexual partners, shelter, and food (Connor et al. 2000). Chase et al. (2002) suggested that an individual's position in the hierarchy is a product of individual differences in physical and behavioral characteristics, and social interactions. Although most research in this area focuses on the physical attributes associated with dominance, some studies have investigated whether personality traits are associated with dominance. For example, Colléter and Brown (2011) found that, among male rainbowfish, higher social position was associated with higher Aggression, Activity, and Boldness. More dominant fish were also more likely to be reproductively successful. In addition, exploratory behavior has also been associated with rank in some species (e.g., in great tits; Dingemanse and De Goede 2004). In contrast, among wild male grey seals, no association has been identified between dominance rank (as assessed by outcomes of male–male aggressive encounters) and individual differences in alert scanning behaviors (Twiss and Franklin 2010).

Bottlenose dolphins exhibit dominance hierarchies (Aureli et al. 2002; Veit and Bojanowski 1996). Social conflict occurs when individual dolphins interact and exhibit aggressive and hostile behaviors (Aureli et al. 2002; Samuels and Gifford 1997). In particular, middle-ranked individuals live in a constant state of instability, which renders more opportunities for social conflict (Samuels and Gifford 1997), namely as these individuals frequently challenge more dominant individuals and defend their position from less dominant challengers (Benus et al. 1991). This instability results in more aggressive behaviors produced by and directed toward other middle-ranked individuals (Scott et al. 2005). Thus, individual differences in the behavioral response that a dolphin exhibits when challenged by a dominant or submissive individual could provide some information to that individual's personality (Highfill and Kuczaj 2010; Samuels and Gifford 1997).

The manner in which dolphins form hierarchies in their social groups is not well known. In wild dolphin groups, aggressiveness is difficult to observe but is nonetheless thought to influence social relationships and the resulting social structure (Scott et al. 2005). Aggression is commonly expressed through postural movement, vocalizations, and behaviors such as biting, ramming, and hitting (Samuels and Gifford 1997). Aggressive behaviors are most often observed from males when competing for access to females (Connor et al. 1996, 2001; Scott et al. 2005). The role of personality in reproductive fitness likely influences the formation of dolphin social networks, but the nature of this influence deserves additional attention. Dolphin societies are made up of a variety of relationships, including higher order alliances (long-term pair bonds; Connor et al. 2000) and cooperative associations (Kuczaj et al. 2015b). Lusseau and Newman (2004) found, through close observation of a bottlenose dolphin social network in Doubtful Sound, New Zealand, that sex- and age-related bonds (i.e., homophily) drove the formation of their social network. Kinship may also influence the emergence of a dolphin hierarchy, as seen in matrilineal societies (Conner et al. 1998). Further understanding the relationship between personality, social relationships, and network formation could pinpoint which individuals broker relationships in a community as well as what factors cause certain individuals to emerge as leaders in a given social group (Lusseau and Newman 2004). This information can then be related to social behaviors where individuals each fulfill a certain role in the interaction. For example, Gazda et al. (2005) describe a specialized foraging technique where certain individuals become "drivers" that use their fluke to slap the water to create a mud wall, which herds fish toward other dolphins. The authors found that the same individuals were consistently being the drivers. It is suggested that personality may influence the role dolphins take in these social coordinated behaviors (Highfill and Kuczaj 2010). Behavioral profiles may prove to be the most direct route to identifying personality factors that influence these types of group-coordinated behaviors, as well as social relationships for dolphins and other marine mammals (Highfill and Kuczaj 2010).

Learning

Pavlov hypothesized over a century ago that animals' learning abilities had potential links to personality. The four personality types that he found in dogs (i.e., Excitable, Lively, Quiet, and Inhibited) appeared to be associated with characteristics of associative learning (Pavlov 1906, 1941). For example, in associative learning tasks, Quiet dogs learned slowly but consistently and Lively dogs learned quickly (Pavlov 1906, 1941). Recently, there has been an increased interest in how animal learning is associated with personality traits, such as Exploration, as well as an individual's standing on the bold–shy behavioral axis. For example, in female rhesus macaques (*Macaca mulatta*) a significant positive correlation was found between individuals categorized as "exploratory" and an individual's likelihood to be trained successfully on a learning task (Coleman et al. 2005). Significant correlations between the speed with which tasks are learned and exploratory tendencies are also found in ravens (*Corvus corvax*); fast explorers acquired two discrimination tasks faster than slower explorers (Range et al. 2006). Matzel et al. (2003) investigated this relationship in mice (*Mus musculus*), and demonstrated that individual tendencies to explore open spaces were correlated with performance on four of five different learning tasks: odor discrimination, accuracy locating a reward in a maze, movement suppression in response to an aversive stimulus, and latency to locate a hidden platform in a pool. Performance in these tasks did not correlate with individual tendencies in fearful responses, pain reactivity, running speed and overall activity levels (Matzel et al. 2006). More recently, Altschul et al. (2016) found that, in rhesus macaques, personality traits of Friendliness and Openness were related to performance on a serial cognition task. Specifically, Friendliness was related to task performance over time while openness was related to an individual's rate of learning the task.

More exploratory individuals may perform better at learning tasks because they are more likely to approach and handle novel stimuli or experimental apparatuses rather than because they possess different learning abilities than less exploratory individuals. This is one possible reason why high exploratory animals perform better in learning tasks than low exploratory individuals. Previous research in avian species revealed that highly neophobic individuals also tend to be slower learners, indicating that exploration may facilitate opportunities for faster learning (review in Sih et al. 2004). Other studies attempted to rule out the influence of neophobia on learning by allowing the animals to become familiar with objects or protocols in the learning tasks prior to experiments. For example, Range et al. (2006) tested ravens that had all seen and touched the exploration task object prior to the experiment, suggesting that their latencies to touch the object were more likely due to exploratory tendencies rather than neophobia.

On the basis of several studies, Kuczaj et al. (2009) concluded that dolphins possess the ability to plan and restructure their behaviors in response to novel tasks and contexts. For one, planning was evident in the rapid generation of novel behaviors that resulted in the successful solutions of novel problems, a process that eliminates the potential for trial and error learning in this instance (Kuczaj and Walker 2006; Kuczaj et al. 2009). The capacity to plan and structure behavior for the successful completion of a task, in addition to the individual differences in planning behaviors, suggest that personality differences could provide insight into which animals exhibit more creative planning abilities (Kuczaj et al. 2009). The bold–shy personality dimension in particular has been linked to innovation in problem solving. For example, Bouchard et al. (2007) found that when pigeons attempted to solve a complex feeding problem by observing a pigeon that had been trained to complete the task, Bold individuals were more innovative and solved the task faster than Shy individuals.

From an early age, dolphins also exhibit individual differences in the bold–shy dimension (Highfill and Kuczaj 2007; Mann 1997). Individuals that are bolder and more curious are more likely to have their behavior mimicked and modeled by others in their social group (Kuczaj et al. 2006, 2012b). Play behaviors may also influence the development and first occurrence of problem solving behaviors (Kuczaj and Horback 2013; Kuczaj et al. 2006; Kuczaj and Yeater 2006). Social cooperation in dolphins can be observed through play behaviors, and personality traits may influence whom a dolphin chooses to cooperate with (Kuczaj and Highfill 2005; Kuczaj et al. 2015b). Thus, models for social learning in dolphins may be partly influenced by personality traits (Kuczaj and Yeater 2006; Yeater and Kuczaj 2010). Future studies of learning in marine mammals that incorporate personality as a covariate could prove useful in determining the factors that affect learning. This could be achieved by the addition of trait ratings, which have already been demonstrated to be reliable when used to assess dolphins (Highfill and Kuczaj 2007; Kuczaj et al. 2012a). Even ratings of just a few traits that are implicated in learning abilities, such as Exploration (e.g., Exnerová et al. 2010; Range et al. 2006) and Boldness (Svartberg and Forkman 2002; Wilson et al. 1994) could explain some variation in learning abilities. To begin to fill the gap in knowledge currently present in dolphin literature regarding the relationship between personality and learning, researchers should incorporate personality assessments into experimental tasks that assess cognitive and learning abilities in dolphins, as well as in other marine mammals.

Physiology

Alongside the study of personality from a behavioral perspective, researchers have been interested in the physiological mechanisms underlying personality traits. Understanding links between personality and physiology may allow another form of quantitative information to be collected about personality in marine mammals.

Much of the current literature on personality and physiology focuses on the physiological profiles of animals with differing coping styles (e.g., Koolhaas et al. 1999, 2010). Animals with a proactive coping style react more aggressively and form routines more easily, whereas animals with a reactive coping style are less aggressive, spend more time motionless, and exhibit greater behavioral flexibility (Koolhaas et al. 1999). Aggressive behaviors have often been a proxy for identifying proactive versus reactive individuals (Koolhaas et al. 2010). This approach may not strictly identify those constructs, but at a minimum, studies using this approach assess physiological differences associated with individual differences in aggressive behaviors. Several physiological correlates of proactive–reactive coping styles and aggression have emerged (see Koolhas et al. 1999). Proactive individuals have lower hypothalamic pituitary adrenal (HPA) axis reactivity in response to stress, whereas reactive individuals show the opposite response. There is also accumulating evidence that dopaminergic system sensitivity is higher in proactive than reactive animals (Koolhaas et al. 1999, 2010). Furthermore, there is support for the notion that proactive and reactive individuals are differently susceptible to cardiovascular disease, ulcers, and stereotypic behaviors (Koolhaas et al. 1999; Suomi 1997). The serotonin system has been identified as a target for future research, citing evidence that highly aggressive male rats possess more sensitive versions of a serotonin receptor subtype (1A autoreceptors) than their less aggressive conspecifics (Koolhaas et al. 2010). This suggestion is supported by Suomi's (1997) findings that impulsive rhesus macaques were more likely to have chronically low levels of serotonin metabolism than less impulsive individuals. If future research continues to support the presence of these physiological differences between individuals differing in coping style, then evaluating marine mammals for their risk of developing such conditions could be achieved by assessing their position on the proactive–reactive axis.

Corticosteroids, such as cortisol, have also been consistently linked to differences in coping styles (review in Koolhaas et al. 1999). One of the few studies to assess the relationship between physiological features and a wider range of personality traits found that baseline cortisol levels and cortisol reactivity correlated with several traits in juvenile tufted capuchins (*Cebus apella*; Byrne and Suomi 2002). Cortisol reactivity was negatively correlated with the factors Confident, Curious, Effective, and Opportunistic, and positively with Apprehensive, Fearful, Insecure, and Tense, whereas baseline levels were positively correlated with strong, and negatively with Submissive (Byrne and Suomi 2002). Exploration tendencies have also been linked to corticosteroid levels in male great tits using non-invasive fecal testing. In response to social stress, only slow exploring birds displayed differences in corticosteroid secretion both immediately and a day after being exposed to a stressor (Carere et al. 2003). The authors interpret these results as an indication of higher HPA reactivity in the slow exploring birds.

Few studies in dolphins, whales, pinnipeds, and other marine mammals have incorporated physiological profiles due to the difficulty in collecting this data for marine mammals. Non-invasive methods for collecting physiological data, such as fecal testing, are difficult in marine mammals due to their aquatic habitat. Pinniped species that haul-out on shore may defecate on land, thus facilitating easier fecal collection in these species. Adrenal function in stellar sea lions (*Eumetopias jubatus*) has been measured via radioimmunoassay of fecal corticosterone, determining that age and sex affect corticosterone concentrations (Mashburn and Atkinson 2004). Such corticosterone differences may reflect individual differences in stress response, and so fecal testing may be another way to measure personality in pinnipeds that have loyal haul-out sites (Twiss and Franklin 2010; Twiss et al. 2012a, b). Collection of fecal matter for cetaceans, while possible in captive settings, is difficult in the wild. In some cases, researchers can control aspects of the environment though capture–release assessments that allow for physiological data to be collected in dolphins and whales. Ortiz and Worthy (2000) measured adrenal steroids and arginine vasopressin in free ranging bottlenose dolphins while looking at how much the capture experience affected cortisol levels. The results indicated that relatively short restraint periods did not induce a significant stress response, but any procedure that has the potential to cause stress to the subjects raises some concerns over both welfare and possible confounding effects. However, fecal data from wild North Atlantic Right whales (*Eubalaena glacialis*) were collected over several years in Bay of Fundy, Canada, and was analyzed to provide information on the stress levels of the whales in response to entanglement, as well as hormonal changes due to pregnancy and lactation (Hunt et al. 2006). Researchers need to be able to identify the animals in some way to apply these data to the study of personality. In marine mammal groups where such data collection and long-term identification of individuals is possible, physiological methods and personality assessments can be conducted together. Developing and incorporating minimally invasive measures in an aquatic environment is necessary to introduce this physiological component into marine mammal personality research.

Ultimately, we are just beginning to understand the relationship between personality and physiology in animals, both human and nonhuman. Given the invasive nature of many tests of physiological profiles, and the challenges posed by the aquatic environment, it is worth examining what these data would add to our understanding of marine mammal personality, as well as investigating non-invasive alternatives. However, in general, it would be useful to increase efforts aimed at relating physiological profiles to a range of personality traits and factors.

Friendships

Recent research shows that nonhuman animals with high levels of association (i.e., friendships) share similar personality traits (Weinstein and Capitanio 2008, 2012). Rhesus macaques are selective in forming affiliative relationships with specific individuals, but these preferences are often attributed to age, sex, and relatedness. However, Weinstein and Capitanio (2008) demonstrated that, when these factors were taken into account, young macaques that initiated affiliative behaviors with each other were significantly similar in Adaptability and Equitability. Personality effects in capuchin relationships were also maintained when age, sex, relatedness and social rank were taken into account; dyads that were more similar in Neuroticism had more affiliative relationships, and overall higher quality relationships existed between capuchins with more similar Sociability scores (Morton et al. 2015). Massen and Koski (2014) defined chimpanzee friendships as close associations, characterized by high rates of contact sitting, a known affiliative behavior (Massen et al. 2010). In friendships between both relatives and non-relatives, individuals had similar scores of Sociality, which is similar to the human Extraversion factor. This was observed for high and low Sociability individuals, and this pattern remained when similarity in sex and age were controlled for. The authors also raised an important question about causality: do chimpanzees preferentially associate with individuals with similar personalities, or do individuals become increasingly similar in personality as they spend more time together?

Several studies have assessed the stability of relationships between animals with high degrees of associations over time. Weinstein and Capitanio (2012) defined rhesus macaque friendships as present between individuals who initiated interactions with each other at levels greater than chance. The more similar scores friends had at 1-year-old on Equitability, the more likely they were to remain friends (i.e., continue to associate at above chance levels) from 1 to 2 years of age. The authors concluded that personality does not play as influential a role in the maintenance of friendships, as other variables become more important as the animals age. This is similar to findings in humans that temperament is more influential in friendships among younger rather than older children (Hartup 1996).

Association patterns are the subject of numerous studies in marine mammals, ranging from affiliative, or affable (Connor et al. 2006) to alliance formation (Wiszniewski et al. 2012). Although several factors play a role in these associations, some variation remains unexplained. For example, in one population of bottlenose dolphins, the sex of both actor and recipient was important in the occurrence of contact swimming, an affiliative behavior. This affiliative behavior predominantly occurred between females, but the factors affecting which females associated with one other were unclear (Connor et al. 2006). In Indo-Pacific bottlenose dolphins (*Tursiops aduncus*), twelve pairs of females associated with one another at greater than chance levels, but not all of these associations could be explained by relatedness (Möller et al. 2006). Stable alliances have been observed lasting up to 8 years between unrelated male dolphins, suggesting that relatedness does not appear to play a role in the stability of these alliances (Wiszniewski et al. 2012). Although there are other possible factors that contribute to alliance formation or association levels, such as dominance and familiarity, it seems plausible that personality differences may explain some of this variation. For example, it would be interesting to determine whether females who exhibit highly affiliative bonds (Connor et al. 2006) are similar in personality. To date, there have been no published studies showing a link between dolphin personality and social relationships, and we should therefore consider this an interesting topic for future research.

Conclusions

The study of personality in marine mammals is in an earlier stage of development relative to work with other species, such as avians and nonhuman primates. However, this provides us with the opportunity to ask what it is that we really want to know about marine mammal personality. Although the answer to this question will differ among researchers, we wish to understand the personality of these animals for its own sake, and not just how marine mammal personality might relate to that of humans. We also believe that it is important to ask whether the possession of a certain trait has the same relevance for a marine mammal as it does for a human. For example, would it mean the same thing for a dolphin to be conscientious as it would a human?

Further knowledge of marine mammal individual differences, including personality, can improve our knowledge of the ontogeny and evolution of behavior in these species. In particular, species that have large and complex social structures, such as bottlenose dolphins, may serve as a model for investigating the psychological aspects of personality that influence a given social group in different contexts. Further refinement of operational definitions of individual differences, temperament, and personality, as well as assessing the influence of these constructs on the behavior of marine mammals will allow researchers to better understand relationships between personality and behavior. Research in this area will also continue to have practical applications, such as improving management practices of captive groups and wild populations (Carlstead et al. 1999), and informing researchers of how individuals' roles in social behaviors can be predicted (Highfill and Kuczaj 2010). Future research will hopefully allow more experimental studies to move away from exploratory techniques and incorporate explicit testing of aspects of personality that broaden and refine what we already know about personality in dolphins and other marine mammals.

In this chapter, we identified important subject matters that require further investigation through empirical research, including defining personality, modifying current methodologies for dolphins and other marine mammals, and research areas that may benefit from incorporating personality assessments. The study of personality may enable us to further understand its evolutionary significance, as well as its influence in marine mammals' behavior at the individual and group level. Future research should begin to incorporate personality as a variable in new and unique ways to broaden our knowledge in this body of research.

References

Altschul, D. M., Terrace, H. S., & Weiss, A. (2016). Serial cognition and personality in macaques. *Animal Behavior and Cognition, 3,* 46–64.

Aureli, F., Cords, M., & Van Schaik, C. P. (2002). Conflict resolution following aggression in gregarious animals: A predictive framework. *Animal Behaviour, 64,* 325–343.

Benus, R. F., Bohus, B., Koolhaas, J. M., & Oortmerssen, G. A. (1991). Heritable variation for aggression as a reflection of individual coping strategies. *Cellular and Molecular Biology, 47,* 1008–1019.

Berta, A., Sumich, J. L., & Kovacs, K. M. (2006). *Marine mammals: Evolutionary biology* (pp. 51–110). Saint Louis, MO: Academic Press.

Bouchard, J., Goodyer, W., & Lefebvre, L. (2007). Social learning and innovation are positively correlated in pigeons (*Columba livia*). *Animal Cognition, 10,* 259–266.

Byrne, G., & Suomi, S. J. (2002). Cortisol reactivity and its relation to home cage behavior and personality ratings in tufted capuchin (*Cebus apella*) juveniles from birth to six years of age. *Psychoneuroendocrinology, 27,* 139–154.

Caldwell, M. C., & Caldwell, D. K. (1966). Epimeletic (care-giving) behavior in Cetacea. In K. S. Norris (Ed.), *Whales, porpoises and dolphins* (pp. 755–789). Berkeley, CA: University of California Press.

Carere, C., Groothius, T. G., Möstl, E., Daan, S., & Koolhaas, J. M. (2003). Fecal corticosteroids in a territorial bird selected for different personalities: Daily rhythm and response to social stress. *Hormones and Behavior, 43,* 540–548.

Carlstead, K., Mellen, J., & Kleiman, D. G. (1999). Black rhinoceros (*Diceros bicornis*) in US zoos: Individual behavior profiles and their relationship to breeding success. *Zoo Biology, 18,* 17–34.

Carter, A. J., Feeney, W. E., Marshall, H. H., Cowlishaw, G., & Heinsohn, R. (2013). Animal personality: What are behavioural ecologists measuring? *Biological Reviews, 88,* 465–475.

Carter, A. J., Marshall, H. H., Heinsohn, R., & Cowlishaw, G. (2012). Evaluating animal personalities: Do observer assessments and experimental tests measure the same thing? *Behavioral Ecology and Sociobiology, 66,* 153–160.

Chase, J. M., Abrams, P. A., Grover, J. P., Diehl, S., Chesson, P., Holt, R. D., et al. (2002). The interaction between predation and competition: A review and synthesis. *Ecology Letters, 5,* 302–315.

Coleman, K., Tully, L. A., & McMillan, K. L. (2005). Temperament correlates with training success in adult rhesus macaques. *American Journal of Primatology, 65,* 63–71.

Coleman, K., & Wilson, D. S. (1998). Shyness and boldness in pumpkinseed sunfish: Individual differences are context-specific. *Animal Behaviour, 56,* 927–936.

Colléter, M., & Brown, C. (2011). Personality traits predict hierarchy rank in male rainbowfish social groups. *Animal Behaviour, 81,* 1231–1237.

Connor, R. C., Heithaus, M. R., & Barre, L. M. (2001). Complex social structure, alliance stability and mating access in a bottlenose dolphin 'super-alliance'. *Proceedings of the Royal Society of London B: Biological Sciences, 268,* 263–267.

Connor, R. C., Mann, J., Tyack, P. L., & Whitehead, H. (1998). Social evolution in toothed whales. *Trends in Ecology & Evolution, 13,* 228–232.

Connor, R. C., Mann, J., & Watson-Capps, J. J. (2006). A sex-specific affiliative contact behavior in Indian Ocean bottlenose dolphins, *Tursiops* sp. *Ethology, 112,* 631–638.

Connor, R. C., Richards, A. F., Smolker, R. A., & Mann, J. (1996). Patterns of female attractiveness in Indian Ocean bottlenose dolphins. *Behaviour, 133,* 37–69.

Connor, R. C., Wells, R. S., Mann, J., & Read, A. J. (2000). The bottlenose dolphin: Social relationships in a fission–fusion society. In J. Mann, R. C. Connor, P. L. Tyack, & H. Whitehead (Eds.), *Cetacean societies: Field studies of dolphins and whales* (pp. 91–125). Chicago, IL: University of Chicago Press.

Cremer, M. J., Hardt, F. A. S., & Júnior, A. J. T. (2006). Evidence of epimeletic behavior involving a Pontoporia blainvillei calf (*Cetacea pontoporiidae*). *Biotemas, 19,* 83–86.

De Moura, J. F., da Silva, É. R., & Siciliano, S. (2009). Epimeletic behaviour in rough-toothed dolphins (*Steno bredanensis*) on the east coast of Rio de Janeiro State, Brazil. *Marine Biodiversity Records, 2,* e12. doi:10.1017/S1755267208000122

De Pauw, S. S. W., & Mervielde, I. (2010). Temperament, personality and developmental psychopathology: A review based on the conceptual dimensions underlying childhood traits. *Child Psychiatry and Human Development, 41,* 313–329.

de Vere, A. J., & Kuczaj, S. A. (2016). Where are we in the study of animal emotions? *Wiley Interdisciplinary Reviews: Cognitive Science.* doi:10.1002/wcs.1399

Dingemanse, N. J., & De Goede, P. (2004). The relation between dominance and exploratory behavior is context-dependent in wild great tits. *Behavioral Ecology, 15,* 1023–1030.

Duffy-Echevarria, E. E., Connor, R. C., & St Aubin, D. J. (2008). Observations of strand-feeding behavior by bottlenose dolphins (*Tursiops truncatus*) in Bull Creek, South Carolina. *Marine Mammal Science, 24,* 202–206.

Dutton, D. A., Clark, R. A., & Dickins, D. W. (1997). Personality in captive chimpanzees: Use of a novel rating procedure. *International Journal of Primatology, 18,* 539–552.

Eskelinen, H. C., Winship, K. A., & Borger-Turner, J. L. (2015). Sex, age, and individual differences in bottlenose dolphins (*Tursiops truncatus*) in response to environmental enrichment. *Animal Behavior and Cognition, 2,* 241–253.

Exnerová, A., Svádová, K. H., Fučíková, E., Drent, P., & Štys, P. (2010). Personality matters: Individual variation in reactions of naive bird predators to aposematic prey. *Proceedings of the Royal Society of London B: Biological Sciences, 277,* 723–728.

Fertl, D., & Schiro, A. (1994). Carrying of dead calves by free-ranging Texas bottlenose dolphins (*Tursiops truncatus*). *Aquatic Mammals, 20,* 53–56.

Freeman, H. D., Brosnan, S. F., Hopper, L. M., Lambeth, S. P., Schapiro, S. J., & Gosling, S. D. (2013). Developing a comprehensive and comparative questionnaire for measuring personality in chimpanzees using a simultaneous top-down/bottom-up design. *American Journal of Primatology, 75,* 1042–1053.

Freeman, H. D., & Gosling, S. D. (2010). Personality in nonhuman primates: A review and evaluation of past research. *American Journal of Primatology, 72,* 653–671.

Freeman, H. D., Gosling, S. D., & Schapiro, S. J. (2011). Methods for assessing personality in nonhuman primates. In A. Weiss, A. J. King, & L. Murray (Eds.), *Personality and behavioral syndromes in nonhuman primates* (pp. 17–41). New York, NY: Springer.

Gazda, S. K., Connor, R. C., Edgar, R. K., & Cox, F. (2005). A division of labor with role specialization in group–hunting bottlenose dolphins (*Tursiops truncatus*) off Cedar Key, Florida. *Proceedings of the Royal Society B: Biological Sciences, 272,* 135–140.

Gibson, Q. A., & Mann, J. (2008). Early social development in wild bottlenose dolphins: Sex differences, individual variation and maternal influence. *Animal Behaviour, 76,* 375–387.

Goldberg, L. (1990). An alternative "description of personality": The big-five factor structure. *Journal of Personality and Social Psychology, 59,* 1216–1229.

Gosling, S. D. (2001). From mice to men: What can we learn about personality from animal research? *Psychological Bulletin, 127,* 45–86.

Gosling, S. D., & John, O. P. (1999). Personality dimensions in nonhuman animals: A cross-species review. *Current Directions in Psychological Science, 8,* 69–75.

Grand, A. P., Kuhar, C. W., Leighty, K. A., Bettinger, T. L., & Laudenslager, M. L. (2012). Using personality ratings and cortisol to characterize individual differences in African elephants (*Loxodonta africana*). *Applied Animal Behaviour Science, 142,* 69–75.

Hartup, W. W. (1996). The company they keep: Friendships and their developmental significance. *Child Development, 67,* 1–13.

Highfill, L. E., & Kuczaj, S. A. (2007). Do bottlenose dolphins (*Tursiops truncatus*) have distinct and stable personalities? *Aquatic Mammals, 33,* 380–389.

Highfill, L. E., & Kuczaj, S. A. (2010). How studies of wild and captive dolphins contribute to our understanding of individual differences and personality. *International Journal of Comparative Psychology, 23,* 269–277.

Highfill, L. E., Fad, O., Makecha, R., & Kuczaj, S. A. (2013). Brief Report: Asian elephants (*Elephas maximus*) may demonstrate stable personalities. *International Journal of Comparative Psychology, 26,* 233–240.

Highfill, L. E., Hanbury, D., Kristiansen, R., Kuczaj, S. A., & Watson, S. (2010). Rating vs. coding in animal personality research. *Zoo Biology, 29,* 509–516.

Hill, H. M., Greer, T., Solangi, M., & Kuczaj, S. A. (2007). All mothers are not the same: Maternal styles in bottlenose dolphins (*Tursiops truncatus*). *International Journal of Comparative Psychology, 20,* 35–54.

Hill, H. M., Guarino, S., Dietrich, S., & St Leger, J. (2016). An inventory of peer-reviewed articles on killer whales (*Orcinus orca*) with a comparison to bottlenose dolphins (*Tursiops truncatus*). *Animal Behavior and Cognition, 3,* 135–149.

Horback, K. M., Miller, L. J., & Kuczaj, S. A. (2013). Personality assessment in African elephants (*Loxodonta africana*): Comparing the temporal stability of ethological coding versus trait rating. *Applied Animal Behaviour Science, 149,* 55–62.

Hunt, K. E., Rolland, R. M., Kraus, S. D., & Wasser, S. K. (2006). Analysis of fecal glucocorticoids in the North Atlantic right whale (*Eubalaena glacialis*). *General and Comparative Endocrinology, 148,* 260–272.

Izard, C. E., Libero, D. Z., Putnam, P., & Haynes, O. M. (1993). Stability of emotion experiences and their relations to traits of personality. *Journal of Personality and Social Psychology, 64,* 847–860.

King, J. E., & Figueredo, A. J. (1997). The five-factor model plus dominance in chimpanzee personality. *Journal of Research in Personality, 31,* 257–271.

Koolhaas, J. M., Korte, S. M., de Boer, S. F., Van Der Vegt, B. J., Van Reenen, C. G., Hopster, H., et al. (1999). Coping styles in animals: Current status in behavior and stress-physiology. *Neuroscience & Biobehavioral Reviews, 23,* 925–935.

Koolhaas, J. M., de Boer, S. F., Coppens, C. M., & Buwalda, B. (2010). Neuroendocrinology of coping styles: Towards understanding the biology of individual variation. *Frontiers in Neuroendocrinology, 31,* 307–321.

Koski, S. E. (2011). Social personality traits in chimpanzees: Temporal stability and structure of behaviourally assessed personality traits in three captive populations. *Behavioral Ecology and Sociobiology, 65,* 2161–2174.

Kuczaj, S. A., & Eskelinen, H. C. (2014). Why do dolphins play? *Animal Behavior and Cognition, 1,* 113–127.

Kuczaj, S. A., & Frick, E. E. (2015). What do dolphins mean when they open their mouth? Poster session presented at the meeting for The Society for Marine Mammalogy, San Francisco, CA.

Kuczaj, S. A., Frick, E. E., Jones, B. L., Lea, J. S., Beecham, D., & Schnöller, F. (2015a). Underwater observations of dolphin reactions to a distressed conspecific. *Learning & Behavior, 43,* 289–300.

Kuczaj, S. A., Gory, J. D., & Xitco, M. J. (2009). How intelligent are dolphins? A partial answer based on their ability to plan their behavior when confronted with novel problems. *The Japanese Journal of Animal Psychology, 59,* 99–115.

Kuczaj, S. A., & Highfill, L. E. (2005). Dolphin play: Evidence for cooperation and culture? *Behavioral and Brain Sciences, 28,* 705–706.

Kuczaj, S. A., Highfill, L. E., & Byerly, H. C. (2012a). The importance of considering context in the assessment of personality characteristics: Evidence from ratings of dolphin personality. *International Journal of Comparative Psychology, 25,* 309–329.

Kuczaj, S. A., & Horback, K. M. (2013). Play and emotion. In S. A. Kuczaj & S. Watanabe (Eds.), *Emotions of animals and humans* (pp. 87–112). Japan: Springer.

Kuczaj, S. A., & Makecha, R. (2008). The role of play in the evolution and ontogeny of contextually flexible communication. In D. K. Oller & U. Griebel (Eds.), *Evolution of communicative flexibility, complexity, creativity, and adaptability in human and animal communication* (pp. 254–277). Cambridge, MA: MIT Press.

Kuczaj, S. A., Makecha, R., Trone, M., Paulos, R. D., & Ramos, J. A. (2006). Role of peers in cultural innovation and cultural transmission: Evidence from the play of dolphin calves. *International Journal of Comparative Psychology, 19,* 223–240.

Kuczaj, S. A., & Walker, R. T. (2006). How do dolphins solve problems. In E. A. Wasserman & T. R. Zentall (Eds.), *Comparative cognition: Experimental explorations of animal intelligence* (pp. 580–600). New York, NY: Oxford University Press.

Kuczaj, S. A., Winship, K. A., & Eskelinen, H. C. (2015b). Can bottlenose dolphins (*Tursiops truncatus*) cooperate when solving a novel task? *Animal Cognition, 18,* 543–550.

Kuczaj, S. A., & Yeater, D. B. (2006). Dolphin imitation: Who, what, when, and why? *Aquatic Mammals, 32,* 413–422.

Kuczaj, S. A., Yeater, D. B., & Highfill, L. E. (2012b). How selective is social learning in dolphins. *International Journal of Comparative Psychology, 25,* 221–236.

Kuppens, P. (2005). Interpersonal determinants of trait anger: Low agreeableness, perceived low social esteem, and the amplifying role of the importance attached to social relationships. *Personality and Individual Differences, 38,* 13–23.

Lee, P. C., & Moss, C. J. (2012). Wild female African elephants (*Loxodonta africana*) exhibit personality traits of leadership and social integration. *Journal of Comparative Psychology, 126,* 224–232.

Locurto, C. (2007). Individual differences and animal personality. *Comparative Cognition & Behavior Reviews, 2,* 67–78.

Lopes, M., Borger-Turner, J. L., Eskelinen, H. C., & Kuczaj, S. A. (2016). The influence of age, sex, and social affiliation on the responses of bottlenose dolphins (*Tursiops truncatus*) to a novel stimulus over time. *Animal Behavior and Cognition, 3,* 32–45.

Lusseau, D., & Newman, M. E. (2004). Identifying the role that animals play in their social networks.

Proceedings of the Royal Society of London B: Biological Sciences, 271, 477–481.

Mann, J. (1997). Individual differences in bottlenose dolphin infants. *Family Systems, 4,* 34–48.

Mashburn, K. L., & Atkinson, S. (2004). Evaluation of adrenal function in serum and feces of Steller sea lions (*Eumetopias jubatus*): Influences of molt, gender, sample storage, and age on glucocorticoid metabolism. *General and Comparative Endocrinology, 136,* 371–381.

Massen, J. M., & Koski, S. E. (2014). Chimps of a feather sit together: Chimpanzee friendships are based on homophily in personality. *Evolution and Human Behavior, 35,* 1–8.

Massen, J. J. M., Sterck, E. H. M., & de Vos, H. (2010). A review of close social associations in animals and humans: Functions and mechanisms of friendship. *Behaviour, 147,* 1379–1412.

Matzel, L. D., Han, Y. R., Grossman, H., Karnik, M. S., Patel, D., Scott, N., et al. (2003). Individual differences in the expression of a "general" learning ability in mice. *The Journal of Neuroscience, 23,* 6423–6433.

Matzel, L. D., Townsend, D. A., Grossman, H., Han, Y. R., Hale, G., Zappulla, M., et al. (2006). Exploration in outbred mice covaries with general learning abilities irrespective of stress reactivity, emotionality, and physical attributes. *Neurobiology of Learning and Memory, 86,* 228–240.

Möller, L. M., Beheregaray, L. B., Allen, S. J., & Robert, H. G. (2006). Association patterns and kinship in female Indo-Pacific bottlenose dolphins (*Tursiops aduncus*) of southeastern Australia. *Behavioural Ecology and Sociobiology, 61,* 109–117.

Morton, F. B., Weiss, A., Buchanan-Smith, H. M., & Lee, P. C. (2015). Capuchin monkeys with similar personalities have higher-quality relationships independent of age, sex, kinship, and rank. *Animal Behaviour, 105,* 163–171.

Ortiz, R. M., & Worthy, G. A. (2000). Effects of capture on adrenal steroid and vasopressin concentrations in free-ranging bottlenose dolphins (*Tursiops truncatus*). *Comparative Biochemistry and Physiology Part A: Molecular & Integrative Physiology, 125,* 317–324.

Park, K. J., Sohn, H., An, Y. R., Moon, D. Y., Choi, S. G., & An, D. H. (2013). An unusual case of care-giving behavior in wild long-beaked common dolphins (*Delphinus capensis*) in the East Sea. *Marine Mammal Science, 29,* 508–514.

Paul, E. S., Harding, E. J., & Mendl, M. (2005). Measuring emotional processes in animals: The utility of a cognitive approach. *Neuroscience and Biobehavioral Reviews, 29,* 469–491.

Pavlov, I. P. (1906). The scientific investigation of the psychical faculties or processes in the higher animals. *Science, 24,* 613–619.

Pavlov, I. P. (1941). The Conditioned Reflex. In W. H. Gantt (Ed.), *lectures on conditioned reflexes* (Vol. 2, pp. 166–185). Conditioned Reflexes and Psychiatry New York, NY: International Publishers.

Range, F., Bugnyar, T., Schlögl, C., & Kotrschal, K. (2006). Individual and sex differences in learning abilities of ravens. *Behavioural Processes, 73,* 100–106.

Samuels, A., & Gifford, T. (1997). A quantitative assessment of dominance relations among bottlenose dolphins. *Marine Mammal Science, 13,* 70–99.

Scott, E. M., Mann, J., Watson-Capps, J. J., Sargeant, B. L., & Connor, R. C. (2005). Aggression in bottlenose dolphins: Evidence for sexual coercion, male–male competition, and female tolerance through analysis of tooth-rake marks and behaviour. *Behaviour, 142,* 21–44.

Sih, A., Bell, A. A., & Johnson, J. C. (2004). Behavioural syndromes: An ecological and evolutionary overview. *Trends in Ecology & Evolution, 19,* 372–377.

Springer, M. S., Stanhope, M. J., Madson, O., & de Jong, W. W. (2004). Molecules consolidate the placental mammal tree. *Trends in Ecology & Evolution, 19,* 430–438.

Stamps, J. A., & Groothuis, T. G. G. (2010a). Developmental perspectives on personality: Implications for ecological and evolutionary studies of individual differences. *Philosophical Transactions of the Royal Society of London B: Biological Sciences, 365,* 4029–4041.

Stamps, J. A., & Groothuis, T. G. G. (2010b). The development of animal personality: Relevance, concepts and perspectives. *Biological Reviews, 85,* 301–325.

Suomi, S. J. (1997). Early determinants of behaviour: Evidence from primate studies. *British Medical Bulletin, 53,* 170–184.

Svartberg, K., & Forkman, B. (2002). Personality traits in the domestic dog (*Canis familiaris*). *Applied Animal Behaviour Science, 79,* 133–155.

Twiss, S. D., Cairns, C., Culloch, R. M., Richards, S. A., & Pomeroy, P. P. (2012a). Variation in female grey seal (*Halichoerus grypus*) reproductive performance correlates to proactive–reactive behavioural types. *PLoS ONE, 7,* e49598.

Twiss, S. D., Culloch, R., & Pomeroy, P. P. (2012b). An in-field experimental test of pinniped behavioral types. *Marine Mammal Science, 28,* 280–294.

Twiss, S. D., & Franklin, J. (2010). Individually consistent behavioural patterns in wild, breeding male grey seals (*Halichoerus grypus*). *Aquatic Mammals, 36,* 234–238.

Twiss, S. D., Pomeroy, P. P., & Anderson, S. S. (1994). Dispersion and site fidelity of breeding male grey seals (*Halichoerus grypus*) on North Rona, Scotland. *Journal of Zoology, 233,* 683–693.

Uher, J., & Asendorpf, J. B. (2008). Personality assessment in the Great Apes: Comparing ecologically valid behavior measures, behavior ratings, and adjective ratings. *Journal of Research in Personality, 42,* 821–838.

Vazire, S., & Gosling, S. D. (2004). Personality and temperament. In M. Bekoff (Ed.), *Encyclopedia of animal behavior* (Vol. 2, pp. 818–822). Westport, CT: Greenwood Press.

Vazire, S., Gosling, S. D., Dickey, A. S., & Schapiro, S. J. (2007). Measuring personality in nonhuman animals. In R. W. Robins, C. R. Fraley, & R. F. Krueger (Eds.), *Handbook of research methods in personality psychology* (pp. 190–206). New York, NY: The Guilford Press.

Veit, F., & Bojanowski, E. (1996). Behaviours accompanying a change in the dominance hierarchy of bottlenose dolphins with respect to adult males. *European Research on Ceteceans, 9,* 202–204.

Verbeek, M. E., Boon, A., & Drent, P. J. (1996). Exploration, aggressive behaviour, and dominance in pair-wise confrontations of juvenile male great tits. *Behaviour, 133,* 945–963.

Verbeek, M. E., De Goede, P., Drent, P. J., & Wiepkema, P. R. (1999). Individual behavioral characteristics and dominance in aviary groups of Great Tits. *Behaviour, 136,* 23–48.

Weinstein, T. A., & Capitanio, J. P. (2008). Individual differences in infant temperament predict social relationships of yearling rhesus monkeys, (*Macaca mulatta*). *Animal Behaviour, 76,* 455–465.

Weinstein, T. A., & Capitanio, J. P. (2012). Longitudinal stability of friendships in rhesus monkeys (*Macaca mulatta*): Individual- and relationship-level effects. *Journal of Comparative Psychology, 126,* 97–108.

Wilson, D. S., Clark, A. B., Coleman, K., & Dearstyne, T. (1994). Shyness and boldness in humans and other animals. *Trends in Ecology & Evolution, 9,* 442–446.

Wiszniewski, J., Brown, C., & Möller, L. M. (2012). Complex patterns of male alliance formation in a dolphin social network. *Journal of Mammalogy, 93,* 239–250.

Yasui, S., Konno, A., Tanaka, M., Idani, G., Ludwig, A., Leickfeldt, D., et al. (2013). Personality assessment and its association with genetic factors with captive Asian and African elephants. *Zoo Biology, 32,* 70–78.

Yeater, D. B., & Kuczaj, S. A. (2010). Observational learning in wild and captive dolphins. *International Journal of Comparative Psychology, 23,* 379–385.

Zillig, L. M. P., Hemenover, S. H., & Dienstbier, R. A. (2002). What do we assess when we assess a Big 5 trait? A content analysis of the affective, behavioral, and cognitive processes represented in Big 5 personality inventories. *Personality and Social Psychology Bulletin, 28,* 847–858.

Part IV

Applications of the Study of Nonhuman Personality

Individual Differences in Nonhuman Animals: Examining Boredom, Curiosity, and Creativity

13

Malin K. Lilley, Stan A. Kuczaj and Deirdre B. Yeater

Abstract

Personality psychology has traditionally focused on the study of individual differences in human cognition and behavior. More recently, the topic of individual differences in nonhuman animal behavior has featured more prominently in biology and psychology research. The study of individual differences in nonhuman animals has important implications for ecology, conservation, comparative psychology, agriculture, and the care of animals in zoological facilities. Individual differences in animal boredom will be examined in this chapter to highlight the importance of studying variation in boredom proneness and coping styles in nonhuman animals. The negative affective state of boredom is adaptive because it serves as motivation for an individual to re-engage with the environment, a process that can involve curiosity and creativity. Future research on animal boredom should investigate the behavioral and physiological correlates of boredom at both a species and an individual level in order to expand the existing literature and contribute to the future of animal welfare.

Defining "Individual Differences"

Humans are characterized by a number of traits that comprise each individual's personality. One model of human personality, known as the five-factor model or big five, posits that people tend to differ in how open to new experiences, agreeable, neurotic, extroverted, and conscientious they are (Digman 1990). In addition to a general factor of intelligence (Hopkins et al. 2014), humans and some other apes, also vary in clusters of skills that belong to particular domains, such as social and physical reasoning (Herrmann et al. 2010; Vonk and Povinelli 2011). Human personality psychology is an established field that aims to address several key questions: how much variation in traits is attributable to genetic influence and other factors?

M.K. Lilley (✉) · S.A. Kuczaj
The University of Southern Mississippi, Hattiesburg, MS, USA
e-mail: malin.lilley@gmail.com

D.B. Yeater
Sacred Heart University, Fairfield, CT, USA

J. Vonk et al. (eds.), *Personality in Nonhuman Animals*,
DOI 10.1007/978-3-319-59300-5_13

How are traits correlated with one another, with behaviors, and with other outcomes, and how did these traits develop over ontogenetic and evolutionary time?

These questions can also be explored when studying the behavior of nonhuman animals, provided that variation in behavior, cognitive tendencies, and/or affect does exist in a given species. One of the earliest arguments for the existence of animal personality came from Darwin (1871/1964). He believed that consistent individual differences were present in animals, and suggested that these behavioral traits could evolve in the same way as physical traits. Darwin (1872) also suggested that behaviors may be inherited based on acquired characteristics. The extent to which traits vary within individuals of a species likely depends upon variability within the environment. For instance, some animals that live in complex social societies experience more variability in their social and physical environment with shifts in dominance, social bonds, and partner preferences. Therefore, more socially complex species are more likely to exhibit individual variation in behavioral and personality traits. The study of individual differences in animals will likely be most fruitful when directed toward animals that live in changing social and physical environments (cf.,Sol, Griffin, and Bartomeus 2012).

In the mid-twentieth century, behaviorism and ethology led scientists to focus less on the existence of consistent individual differences and to focus more on average behaviors. Behaviorism assumed that animals were inherently similar and differed only as a result of environmental influences. During this time, emphasis was placed on collecting large sample sizes and calculating an average of behaviors, instead of examining individual differences (e.g., Chamove et al. 1972, studied 168 monkeys). There was also an emphasis on species-specific behaviors such as fixed action patterns (e.g., Thelen and Farish 1977). Later, the focus began to shift back to the study of consistent individual differences (e.g., Adamec 1975; Buirski et al. 1978; Huntingford et al. 2010; Stevenson-Hinde and Zunz 1978). After these studies, the topic of animal personality fell out of favor as anthropomorphism became taboo (Kennedy 1992, Chapter 1).

A review by Gosling (2001) reintroduced the notion of animal personality in the scientific field. A surge of studies on many species, including dolphins (*Tursiops truncatus*) (Kuczaj et al. 2012) and chimpanzees (*Pan troglodytes*, Weiss and King 2015), followed Gosling's review, and established that variation in behavior exists between individuals in the same population. After establishing that variation in behavior exists, the particular behaviors for which variation exists must be determined. For example, pumpkinseed sunfish (*Lepomis gibbosus*) vary in a personality trait commonly referred to as 'boldness' (Coleman and Wilson 1998), which has since been examined in many species (Réale et al. 2007). Variations in physical and social cognitive reasoning have also been reported in a smaller number of species, and mostly in nonhuman primates (e.g., Vonk and Povinelli 2011). Not all species vary in all behaviors or traits. Some species are more variable in certain domains than others, providing the example that socially complex species may vary more in social reasoning compared to species that lack complex social lives (Gartner, this volume).

Individual variation in a behavior or trait has been discussed in the literature using several terms, including "personality," "temperament," "behavioral profiling," "behavioral syndromes," "coping style," and "individual differences" (Sih et al. 2004; Tetley and O'Hara 2012). Often these terms are used interchangeably. "Personality," originally and most frequently used to describe stable and consistent differences in human behavior, has also been used in the literature to describe behavioral variation in nonhuman animals. "Temperament" has been frequently used to describe the biologically determined variation in human behavior, especially when referring to infants, and has also been used in reference to nonhuman animal behavior to avoid using "personality" and all of its anthropomorphic connotations (Gosling 2001). Some established dimensions of human personality, such as conscientiousness, might not exist in all other species, which is why "behavioral profiles" or "individual differences" are terms that can be more readily applied to nonhumans, without implying which dimensions can be found in those species.

Behavioral ecologists have also characterized individual variation in behavior using the term "behavioral syndromes." Sih et al. (2004, p. 372) define "behavioral syndrome" as "a suite of correlated behaviors reflecting between-individual consistency in behavior across multiple (two or more) situations." Each individual can be labeled as having a "behavioral type," such as being more or less aggressive. Sih and Bell (2008) explain that behavioral ecology frequently uses the term "behavioral syndrome" because it is more inclusive and general than "temperament," "personality," or "coping style," and behavioral ecologists are concerned with studying how correlations between different variables are related to evolutionary ecology. For an in-depth discussion of terminology definitions, see Réale et al. (2007). Terminology is perhaps one of the most difficult obstacles in developing precise, well-defined discussions of phenomena that can be useful for a comparison between similar concepts reported in different species.

The term "individual differences" will be used in this chapter to refer to variation between individuals in emotional dispositions, psychological states, reactions to stressful conditions, cognitive skills, behavioral patterns, social interactions, and other traits in both humans and nonhumans alike. Individual differences must exist within members of a given species and be consistent within individuals over time and across situations and contexts (Pervin and John 1997). For example, some pumpkinseed sunfish repeatedly avoid novel objects, whereas others are more likely to investigate them (Wilson et al. 1993). Likewise, some chimpanzees consistently solve cognitive tasks more quickly than conspecifics with similar experience (Vonk and Povinelli 2011).

Individual Differences: Why Do They Matter?

From an evolutionary perspective, natural selection acts upon variation in a trait within a population and may aid in species or population adaptation (Wilson 1998). Individuals higher or lower in a trait may have better fitness in different contexts; thus, variation within a population improves the chances that at least some individuals will survive and reproduce (Dall et al. 2004). For example, a bolder animal will be quicker to approach and capitalize on a novel food source, which could improve that individual's chances of survival in times of limited food resources. On the other hand, if resources are not scarce and predators are high in number, a bolder animal may easily become prey.

In a review of the literature, Call (2015) noted that boldness is related to creativity in animals. He reported, for example, that wild cavies (*Cavia aperea*) who were bolder and produced innovative behavior more quickly were slower at tasks that required reversal learning (e.g., Guenther et al. 2013). Call (2015) proposed that it is best for a particular population if there are some individuals that are quicker to innovate, some individuals that are more likely to copy the behavior of successful individuals, and some individuals that are quicker to adapt to a changing environment because each of these traits would facilitate different survival rates in different contexts. For example, wild cavies who are faster innovators may have higher survival rates in stable environmental conditions because they may be more successful at finding optimal food sources. In contrast, the wild cavies that are faster at reversal learning might have better survival rates in changing environmental conditions because they may adapt to the new conditions faster.

Variation in other traits can also be associated with fitness outcomes (Darwin 1871/1964). For example, more aggressive animals may have a better chance of survival and reproduction in circumstances where aggression provides protection from threats. In contrast, more aggressive animals are less socially connected with group members, which could decrease fitness in species where social affiliation is critical to survival and reproduction.

The above examples are quite general; however, research on individual differences has revealed similar trait dimensions, including Extraversion, Neuroticism, and Agreeableness that appear in several species (Gosling 2001). Current issues in nonhuman animal personality research include the lack of consistent trait definitions, lack of equivalent assessment methods, and the fact that differences in species-specific behavioral repertoires make cross-species comparisons challenging (Gosling 2001). These difficulties are also shared by researchers in the broader field of comparative psychology. Whereas some researchers have argued that tests of cognitive abilities lack internal and external validity, due to a lack of species-representative subjects and non-equivalent assessment methods (Boesch 2007, 2008), others argue that internal validity threats are controlled for experimentally and subjects are not a biased sample of the representative species (Tomasello and Call 2007). Both viewpoints should also be taken into consideration while designing cross-species comparison studies of personality.

Why Study Individual Differences?

Despite the challenges of studying individual differences in nonhuman animals, Gosling (2001) argued that animal personality research may inform how human personality is influenced by biological mechanisms, genetics, and the environment, and also how personality changes, develops, and is connected to health. King et al. (2008) and Weiss and King (2015) directly compared individual differences in different primate species in order to explore the developmental and evolutionary origins of personality. These comparisons are often difficult between taxa in other nonhuman animals due to the lack of equivalent behavioral repertoires. For example, comparisons of primate species may use traits such as "sympathetic" and "helpful" as descriptors of the factor "agreeableness" (King et al. 2008), though these traits, and possibly "agreeableness," too, may not be applicable to the behavior of distantly related species, such as zebrafish. Studies on differences within individual species can also be informative with regard to species-typical behavior and variation within and between populations of a particular species.

Individual differences in bottlenose dolphin (*Tursiops truncatus*) behavior have been reported in a range of contexts, including feeding, maternal care, group movements, and interactions with human caretakers (Highfill and Kuczaj 2010). Learning more about individual differences in dolphins will further understanding of dolphin behavior and development between and within wild populations. Additionally, Highfill and Kuczaj (2010) suggest that a better understanding of individual differences can aid human caretakers in providing effective environmental enrichment, social groupings, and breeding programs.

A review of animal personality ratings as an animal management tool found that, although only a small number of studies on animal personality have actually looked at the implications for animal welfare, some found clear implications for improving breeding success, pair breeding success, and the cohesion of social groups (Tetley and O'Hara 2012). Additionally, Tetley and O'Hara (2012) present evidence that the success and ease of events such as transfers and introductions to new habitats or social groups could be predicted by using personality profiles for each animal. Research on a greater variety of species and in different contexts is needed to explore the application of personality research to animal management programs. For example, different personality traits can predict whether an animal would fare well as a program animal in public education programs, or as an assistance animal, but also can help predict which animals might be most likely to survive reintroductions to natural habitats (see also Watters, Bermner-Harrison, and Powell, this volume). The principle that an animal's personality can affect its behavior in various circumstances is important for both animals in a captive setting and animals in the wild. Although not often discussed in research on personality, wild animals often need to adapt to changing environmental conditions or social situations and personality may play a role in their ability to adapt, although plasticity in behavior can also be important for survival (Found and St. Clair 2016).

Despite there being room for further exploration of these traits and behavioral tendencies in many species, this chapter will focus on the question of how individual differences in nonhuman animals are examined with regard to the understudied topic of boredom and its relation to curiosity and creativity. Individual differences in humans' proneness to boredom, styles of dealing with boredom, and outlets for relieving boredom have only recently been investigated (Daniels et al. 2015). In animals, the majority of researches related to boredom have been conducted in captive settings, due to control over environmental conditions and the efforts of the caretakers to reduce boredom of the animals in human care. Additional research can establish species-specific measures for individual differences in an animal's tendency to become bored, style of boredom, means of coping with boredom, and strategies for relieving boredom.

What Is Boredom?

Many children complain that they are bored at least once over summer vacation. They say, "there is nothing to do." Adults in the child's life usually respond, "there is plenty to do!" and proceed to list numerous activities. In this situation, it is not that the children actually have nothing to do, but that none of the options holds the children's interest. Irritation and restlessness set in and are part of a negative affective state that the children wish to avoid, thus increasing the intensity of their search for a sufficiently interesting activity. In this situation, children experience a mental state of boredom that fosters a cognitive and behavioral process.

Broadly speaking, an organism may experience boredom when it is unable to engage its attention with its external or internal environment (Eastwood et al. 2012). Boredom, a negative affective state, functions to push the organism from a state of habituation with the present situation so that the organism engages its attention with some stimulus. There is thus an adaptive process associated with boredom that starts with an organism's inability to engage with the environment and ends with the re-engagement of attention, the latter of which may take the form of curiosity, exploration, contemplation or play. If re-engagement does not occur, an individual may experience prolonged boredom, which can result in negative outcomes, such as learned helplessness and stereotypic behavior (Wemelsfelder 1984). Learned helplessness occurs when an organism no longer attempts to avoid an aversive situation due to the inability to control its environment and is often likened to depression found in humans (Seligman 1972).

Individual Differences in Human Boredom

Boredom has been studied to a limited degree in humans. Vodanovich (2003) reviewed measures of different types of boredom, including job boredom, leisure boredom, and personality measures associated with boredom. Across these studies, there was no consensus on a definition for boredom and each of the measures relied on self-report of the subjects. The important take away from this review is that boredom can be found in a number of contexts and can be caused by a lack of appropriate stimulation in several facets of an individual's life. The Boredom Proneness Scale (Farmer and Sundberg 1986) assesses how likely participants are to become bored and therefore may be useful as a comparison for individual differences in boredom proneness that are found in nonhuman animals. Daniels et al. (2015) established a link between causes of boredom and the manner in which people attempt to cope with boredom. One group of people reported via a questionnaire that, when a situation becomes boring, they try to avoid the boring situation. The same people who try to avoid boing situations were also likely to report becoming bored easily and believed that becoming bored was part of their personality. In contrast, another group of people reported that when a situation becomes boring they attempt to make the situation more interesting or make an effort to re-focus their attention.

The literature on boredom in humans has established that there are differences between individuals' propensity of becoming bored and furthermore, that individuals may have different coping mechanisms when they experience boredom (Daniels et al. 2015; Goetz et al. 2014; Vodanovich 2003). Although people have different styles of coping with boredom, there are different types of boredom as well (Goetz et al. 2014). The types of boredom found in high school and university students were indifferent (slightly positive valence, very low arousal; relaxed, withdrawn), calibrating (slightly negative valence and low arousal; uncertain, receptive to change/distraction), searching (slightly negative valence and medium arousal; restless, active pursuit of change/distraction), reactant (high levels of negative valence and relatively high levels of arousal; highly reactant, motivated to leave the situation for specific alternatives), and apathetic (high level of negative valence and very low arousal; similar to learned helplessness or depression) boredom. The type of boredom varied within individuals, suggesting that the strategies one would take to resolve boredom would depend on the type of boredom that one was experiencing.

Humans report that boredom is associated with a negative affective state (Fureix and Meagher 2015); however, recent research points to positive outcomes associated with boredom, such as increased creativity (Mann and Cadman 2014). Boredom can also be adaptive because it motivates an organism to re-engage its attention with new thoughts or stimuli (Bench and Lench 2013). We put forward the idea that the transition from boredom to arousal is a process that involves curiosity and creativity. Curiosity, the engagement of the brain's attentional system, is what helps end a period of boredom and focus attention on some external or internal stimulus.

Eastwood et al. (2012) calls for further investigation of the behavioral and physiological markers of boredom in humans. A systematic investigation into nonverbal correlates of boredom could help determine what categories of behavior are indicative of boredom. Learning more about boredom in humans and the individual differences associated with it will facilitate the study of boredom in nonhuman animals.

Animal Boredom

Identifying what boredom might look like in an animal that is unable to engage its attention with the environment is a challenging task. Fureix and Meagher (2015) address this issue by discussing behavioral correlates of boredom in humans and how those may be compared to the behavior of nonhuman animals. A strong emphasis should be placed on the fact that boredom in nonhuman animals is not well understood, and it is an easy mistake to attribute a behavioral state, such as inactivity, to a mental state, such as boredom (Fureix and Meagher 2015). In reality, an inactive animal might not be bored, but instead could be basking in the sun for pleasure or resting while it digests a meal. Conversely, an active animal may not necessarily be engaged with its environment, but instead may be exhibiting stereotypic behavior. Thus, inactivity and activity may be associated with cognitive engagement or boredom in different circumstances. Fureix and Meagher (2015) suggest that inactivity in various contexts should be considered when attempting to explore animals' cognitive engagement in relation to animal welfare.

Although activity levels alone are not reliable indicators of mental engagement, the behavior of an animal in different contexts can point to what behaviors are associated with positive affect and what behaviors are associated with negative affect, including boredom (Fureix and Meagher 2015). One of the first to argue that boredom in animals can be studied scientifically, Wemelsfelder (1984) proposed that boredom was the result of environmental deprivation and that animals often perform abnormal behaviors in circumstances of environmental deprivation in an attempt to fulfill certain behavioral needs. Some animals may perform behaviors in the absence of appropriate stimuli, such as copulation attempts with inanimate objects. Although this type of abnormal behavior does not necessarily reflect a lack of mental engagement, it may reflect a lack of appropriate environmental stimulation. Other animals may repetitively pace an enclosure or excessively groom themselves in an attempt to relieve boredom (Wemelsfelder 1984).

Wemelsfelder (1984) discussed a means for studying animal boredom by comparing levels of stereotypic behavior, frequency of behaviors performed in the absence of appropriate stimuli, general apathy, and immobility over long periods of time to a baseline of species-specific behavior in a stimulating environment. Although boredom may be assessed in an animal by comparing its behavior to species-typical behavior, individual differences in baseline activity levels should also be considered when attempting to determine whether an animal is bored (Wemelsfelder 1984). A less active animal could be content with its present circumstances, resulting in it resting for a longer period of time each day, whereas a more active animal could be content with its circumstances but moving about much more.

Meagher and Mason (2012) studied mink raised in enriched versus impoverished conditions with the goal of assessing the effects of environmental enrichment on the animals' signs of boredom, anhedonia, and apathy. Based on the behavior of humans in each of these mental states, Meagher and Mason (2012) wrote that bored animals seek out interaction with all stimuli, whereas anhedonic individuals decrease interaction with previously pleasurable stimuli, and apathetic animals decrease interaction with all stimuli. Their results suggested that the mink raised in impoverished conditions were bored because the animals showed increased interaction with all stimuli when compared to the animals raised in the enriched condition. Meagher and Mason (2012) also reported that increased interaction with stimuli was associated with periods of lying motionless while awake but was not associated with increased stereotypic behavior. This study highlights the need to operationally define boredom. By specifying that bored animals will seek out interaction with both novel and familiar stimuli as bored humans would, Meagher and Mason (2012) can argue that impoverished conditions contribute to boredom. Additionally, looking at the behaviors correlated with boredom, such as lying still while awake, is important for assessing boredom in other species, as each species may have different behaviors associated with boredom.

Although different species have different behavioral repertoires, humans and nonhuman animals may evaluate their environments based on similar principles. Veissier et al. (2009) proposed that sheep (*Ovis aries*) evaluate their environment based on the suddenness, predictability, familiarity, consistency, and the extent to which the animal has control, similar to the way that humans do. The authors argue that sheep also experience emotions, including boredom, which results from predictability and invariability. Though this might be the case for many animals, this model does not necessarily account for the idea that an animal can be bored in an environment that contains novel stimuli, which may occur due to differences both within and between species. Likewise, an animal with a less variable environment may not experience boredom in a stable environment or could even alter its behavior in order to create variability in the environment. For example, Kuczaj et al. (2006) discuss bubble play in bottlenose dolphin calves. Bubble play occurs when an individual expels air underwater or uses a body part to create a bubble beneath the water's surface and sometimes manipulates the bubbles with the rostrum or another part of the body (Kuczaj et al. 2006). One calf created challenges for itself by releasing a different number of bubbles at different depths with the apparent goal of popping all of them before they reached the surface. Once the calf was able to pop them all under a certain set of conditions, it modified its swim pattern, the depth of bubble release, and the number of bubbles released to make the game more challenging. Kuczaj et al. (1998) discussed how some captive orcas (*Orcinus orca*) appeared to enrich their environment by using fish as bait to attract seagulls into their habitat. These scenarios suggest that animals can create their own forms of environmental enrichment by altering their behavior. The authors point out that enrichment that creates environmental variability and fosters some behavioral processes may be more beneficial than enrichment that is focused solely on providing rewards. For example, the whales that played with the gulls rarely ate or even killed them, but instead seemed to make a game of the activity. Thus, the whales engaged in the activity, not with the goal of obtaining food, but instead as a means to vary behavior and to engage in some behavioral process (Kuczaj et al. 1998). The above examples of cetaceans playing with elements in their environment highlight the need to study individual differences both in how likely it is that an individual might become bored and how an individual might resolve that boredom.

Individual Differences in Animal Boredom

Many zoological facilities attempt to reduce pacing and other stereotypic behavior by introducing environmental enrichment, including novel objects, foraging challenges, or natural elements, into enclosures. For many animals, a successful enrichment program reduces the frequency of stereotypic behaviors and increases the frequency of naturalistic behaviors (Tarou and Bashaw 2007). However, not all animals respond equally to all types of enrichment. For example, when researchers placed mink (*Neovison vison*) in a barren enclosure, some mink paced the enclosure in a repetitive pattern, while others rested in the corner (Dallaire et al. 2012). The animals that exhibited the most stereotypic behavior prior to enrichment became immobile and withdrawn when enrichment was introduced (Dallaire et al. 2012). Perhaps these individuals were exhibiting learned helplessness and did not respond to the novel stimuli for this reason, or were highly neophobic and feared the new unfamiliar stimuli. In contrast, individual mink exhibiting less stereotypic behavior prior to being placed in an enriched environment, showed an increased interaction with novel stimuli and an increase in activity levels. As in humans, individual differences in nonhuman animals can influence how individuals respond to different situations and react to boredom. Therefore, it is important to examine individual differences in animal boredom and coping strategies (Dallaire et al. 2012; Eskelinen et al. 2015; Lopes et al. 2016). The improved welfare of animals in zoological facilities has been the main focus of the research on animal boredom until now, and, although animal welfare is an important outcome, research on animal boredom is also important to the broader field of psychology, the subfields of comparative psychology, and animal behavior and cognition.

Environmental enrichment research has been conducted with many different species and focuses on methods of enrichment. For example, research on stereotypical behavior in wombats (*Lasiorhinus latifrons*) that live in Australian zoos has found that the animals spent very little time foraging (Hogan and Tribe 2007). When provided with foraging opportunities, the animals spent more time foraging, which elicited a broader range of behaviors; however, the amount of stereotypic behavior was not reduced (Hogan et al. 2010). Similarly, a study of gorillas (*Gorilla gorilla gorilla*) found increased activity levels and time spent foraging upon the introduction of different kinds of environmental enrichment, but there were no significant differences in abnormal behavior between the conditions (Charmoy et al. 2015). Additionally, research on polar bears (*Ursus maritimus*) has found that the addition of unfamiliar odors to the habitat interrupted walking bouts and increased investigatory behaviors, such as sniffing, but did not eliminate the stereotypic pacing (Wechsler 1992). A study of harbor seals (*Phoca vitulina concolor*) and gray seals (*Halichoerus grypus*) examined stereotypic behavior with a variety of enrichment conditions. Random swim patterns and exploratory behavior increased and stereotypic behavior decreased for all subjects but one (Hunter et al. 2002). Individual seals interacted with different stimuli to different degrees, as some seemed to favor one enrichment device over the others- evidence that it is important to plan enrichment that is beneficial to all animals. Interestingly, one seal displayed an increase in stereotypic behaviors upon introduction of environmental enrichment. Hunter et al. (2002) suggested that this could be due to the seal's subordinate position or reflect differences in behavioral tendencies between individuals.

A meta-analysis of studies found that the introduction of environmental enrichment into the housing of animals living in zoological facilities reduced the percent of time animals spent performing stereotypical behavior by an average of 50% (Swaisgood and Shepardson 2006). Most environmental enrichment conditions in those studies involved the addition of novel objects, variable feeding methods, training, or more naturalistic habitats. Because many aspects of environmental enrichment were added in combinations or at the same time, the meta-analysis was not able to determine which methods reduced stereotypies most effectively. Moreover, the presence of stereotypies was never completely abolished in any of these studies. Because no one intervention eliminated stereotypies, further research needs to be conducted, paying attention to species-specific ecology, behavior, and even individual experience and differences (Swaisgood and Shepardson 2006). With additional research, a parallel may be established between the individual differences of animals and humans in coping with boredom, the types of boredom, and strategies for resolving boredom. Mellen and Sevenich MacPhee (2001) suggest placing a strong emphasis on an individual animal's history and experiences both when setting outcome goals for environmental enrichment, and also when deciding what enrichment strategies to use.

In the studies examining the effectiveness of environmental enrichment, the authors sometimes interpreted an animal's interaction with a novel object as a sign that the animals were less bored when interacting with that object. Caution should be used in stating that the animals interacting with the objects were bored prior to the object's introduction. Although the animal may have been bored, the animal could also have been already engaged with the environment but found the novel object to be more interesting, or was aroused by the introduction of the object. Meagher and Mason (2012) clarified that an animal should show increased interaction with all stimuli in order to be considered to have been bored; however, this definition should also be used with caution, as some neophobic animals might be bored, but also fear at least some of the novel stimuli and therefore do not interact with all of them. Additionally, caution should be used in concluding that all animals exhibiting stereotypies are bored. Stereotypies can be caused not only by a lack of sensory stimulation, but also by frustrated motivations resulting from the urge to perform specific behaviors, few environmental challenges, ample free time, and stressful conditions (Swaisgood and Shepardson 2006). For example, rhesus macaques (*Macaca mulatta*) that performed self-injurious behavior were less likely to respond behaviorally to a human intruder entering the room, compared to those that did not perform self-injurious behavior (Peterson et al. 2016). The macaques were likely performing the self-injurious behavior due to a cause other than boredom, as according to Meagher and Mason (2012), they would have shown an increase in response to the human intruder if they had been bored. Self-injurious behavior may be the result of stressful conditions, an unlikely environment in which boredom would develop.

Coping style has been reported as a factor in determining if an individual develops stereotypies or experiences learned helplessness when placed in a barren environment (Ijichi et al. 2013). Performing stereotypic behaviors may be an outlet for some individuals and function to reduce boredom or stress. For example, some people may bite their nails in a stressful situation to relieve tension, while others may try to think of another topic or attempt to solve the problem causing the stress. Nonhuman animals may have different ways in which they respond to stress as well. Neurobiological research supports that animals, both with and without stereotypic behavior, have similar physiological measures of stress, and thus, have established different coping mechanisms. Specifically, proactive individuals have higher levels of dopamine and higher activity levels, but lower levels of stress hormones. Conversely, reactive animals do not appear to be outwardly stressed because they have low dopamine and activity levels, but have a high neuroendocrine response (Ijichi et al. 2013). Stereotypical behavior as a proactive coping mechanism may help to relieve stress, as compared to animals that become withdrawn and disengaged; however, there is no definitive answer on whether stereotypies help to reduce stress in all animals (Würbel et al. 2006).

Other measures of personality are reported to interact with the presence of stereotypy as well. Orange-winged parrots (*Amazona amazonica*) rated high in extraversion-like traits were less likely to display stereotypic behavior, but those high in neuroticism-like traits were more likely to engage in feather-damaging behavior (Cussen and Mench 2015). Striped mice (*Rhabdomys dilectus*) that were categorized as bold on three different behavioral measures displayed more stereotypic behaviors compared to conspecifics that were not categorized as bold (Joshi and Pillay 2016). Similarly, rhesus macaques (*Macaca mulatta*) classified as active in response to a human intruder, rated as low on gentle temperament, and that frequently interacted with novel objects were more likely to exhibit motor stereotypies (Gottlieb et al. 2013). Overall, coping strategies and personality are associated with differential prevalence of stereotypic behavior and can be indicative of how likely an individual is to find that a particular situation is not engaging enough.

An additional factor reducing stereotypic behavior, and possibly boredom, in captive animals is the ability of animals to have some degree of choice or control over their environment (Markowitz and Aday 1998; Owen et al. 2005; Ross 2006). When provided with a choice of exhibit enclosures to occupy, giant pandas (*Ailuropoda melanoleuca*) had lower stress hormone levels and exhibited less stereotypic behavior, including pacing, scratching, and behaviors directed at the door that was between the enclosure areas (Owen et al. 2005). Similarly, two polar bears showed reduced stereotypic behavior and an increase in social play when given free access between the indoor and outdoor habitat areas (Ross 2006). Simply being given a choice of which habitat area to occupy resulted in improved welfare in both giant pandas and polar bears (Markowitz and Aday 1998; Owen et al. 2005; Ross 2006). Positive reinforcement training could also be a method of providing an animal with a choice of which behaviors to perform (Laule et al. 2003). For example, training an animal in a zoological or laboratory setting to voluntarily participate in veterinary procedures, such as blood draws, reduces the stress and anxiety the animal would experience compared to the stress of being forcibly restrained for the same procedure. Training can also provide an animal with other choices and opportunities for behavioral diversity and cognitive engagement (Pryor 2015).

What is clear from both the meta-analyses and literature reviews of environmental enrichment studies is that many zoological facilities have an enrichment program focused on Setting Goals, Planning, Implementing, Documenting, Evaluating, and Readjusting (SPIDER) based on the needs of the individual animals (Mellen and Sevenich MacPhee 2001). The acronym serves as a guide for caretakers to provide effective environmental enrichment for animals in their care. For example, if an animal ignores a particular form of enrichment or the animal appears to be frightened, documenting, evaluating, and readjusting are critical steps in finding effective enrichment. Simply adding a novel object to the animal's habitat each day may not be a positive change in the animal's life. Alligood and Leighty (2015) noted that although publications on environmental enrichment are increasingly prevalent, more emphasis should be placed on single-subject experimental designs. This endorsement is consistent with the idea that individual differences influence what type of enrichment is effective for a particular individual. Additionally, evaluation and readjustment of enrichment strategies are two important components of providing effective environmental enrichment that are often overlooked. Although most animals in zoological facilities are provided daily enrichment, the animals' response and the effectiveness of the enrichment are not always documented and used to inform future enrichment strategies. Evaluating enrichment effectiveness in terms of goals for each animal may improve animal welfare in zoological and laboratory settings. For example, a study of common squirrel monkeys (*Saimiri sciureus*) found that different forms of enrichment, such as food presentations, toys, and walkways, resulted in more positive or negative welfare indicators depending on the individual (Izzo et al. 2011). Positive responses to novelty and rates of aggression were predictors of welfare indicators (Izzo et al. 2011). An example of individual-focused research found that environmental enrichment tailored to specific individual macaques (*Macaca* spp.) was effective in decreasing stereotypic behavior in all study subjects and also decreased stress hormone levels in seven out of nine study subjects (Cannon et al. 2016). Such positive results of individual-focused enrichment strategies will hopefully encourage others to conduct and report on similar studies.

The Future of Boredom Research

Most research on animal boredom pertains to environmental enrichment and stereotypical behavior in zoological facilities. Although this research is important and should be promoted, in order to better understand animal boredom, further research must be conducted to distinguish boredom from other psychological states and to better understand the behavioral and physiological signs of boredom. Other topics to investigate include what factors contribute to individual differences in animal boredom, how individuals experience boredom and try to resolve it, and how humans can help to reduce boredom in animals that are in agricultural or zoological settings. This research will be useful for understanding personality in nonhuman animals and for informing enrichment programs.

Research in its initial stages on animal boredom has established that impoverished living conditions are one cause of underdevelopment, stereotypic behavior, and learned helplessness, suggesting that the construct of boredom exists for nonhuman animals (Dallaire et al. 2012; Mason et al. 2007; Meagher and Mason 2012; Wemelsfelder et al. 2000). Additionally, there is evidence that there are individual differences in boredom because animals do not react in the same ways when placed in the same conditions and provided with the same enrichment (Biondi, Bó, andVassallo 2010; Dallaire et al. 2012; Eskelinen et al. 2015; Lopes et al. 2016), though when this enrichment is tailored to individuals' behavioral needs, stereotypies and stress levels can be reduced (e.g., Cannon et al. 2016). Future research should consider using Meagher and Mason's (2012) paradigm of introducing novel and familiar stimuli to samples of other species and tailor environmental enrichment to the behavioral needs of individual animals, as did Cannon et al. (2016).

A survey of 60 zoological facilities revealed that, although staff at most facilities believed that enrichment is important for animal well-being, time and resource constraints often kept enrichment efforts to a small amount of time each day (Hoy et al. 2010). For example, 88.8% of staff surveyed reported an average of 1.5 h of their work day was spent on enrichment efforts (Hoy et al. 2010). Thus, staff at zoological facilities have little spare time, and so it is important that they are able to spend the majority of this time on the most effective enrichment possible. In order to do this, they must know what the most effective type of enrichment is for each animal. Often the most effective enrichment consists of objects that capture the animals' curiosity.

Curiosity in Nonhuman Animals

Curiosity is the motivational force that moves an individual from a state of boredom to a state of re-engagement with the environment. Curiosity is a well-explored concept in humans. Numerous questionnaires have been designed to measure the extent to which a person is interested in a novel perceptual stimulus or is eager to participate in new experiences (see, e.g., Collins et al. 2004).

Research on animal personality has also examined individual differences on the shy-bold continuum, which ranges from neophilia (literally "new loving") to neophobia (literally "new fearing"). The trait of boldness is similar to the five-factor model trait of "extraversion" (Goldberg 1990, 1993). Open field tests and novel object exploration have been used to characterize individuals of many species, including birds, (Biondi et al. 2010), rats (Dellu et al. 1993), and fish (Wilson et al. 1994), as being more bold or more shy compared to conspecifics. Individual differences in curiosity cannot be reduced to whether an individual behaves more or less boldly upon the introduction of a novel object; however, because the shy–bold continuum has been used so extensively to characterize individual differences in the exploration of novel objects, it will be used in this chapter as one approach to exploring curiosity.

Reader (2015) discusses how an individual's tendency to explore is influenced by experience and environmental conditions. Multiple factors that contribute to individual differences in curiosity include differences in genetic predispositions (Sih et al. 2004), positive feedback from behavior (Morand-Ferron and Giraldeau 2010), and social cues (Giraldeau and Dubois 2008). Although it is likely that many factors contribute to the exploratory tendencies of a particular individual, Reader (2015) discusses how the payoffs of exploring a new environment versus the exploitation of a familiar environment can maintain inter-individual variation in a population and influence an individual's behavior in a specific situation. For example, Reader discusses how using a previously known food source may save time and energy that would be necessary to explore for a new food source; however, the familiar food source could be depleted and, in this situation, the better option would be to explore for a food source.

An important distinction in the literature is the difference between search and exploration. Although boredom may facilitate both, there is a lack of research on how animals use search and exploration in different contexts. One can imagine that a bored child might explore a new playground to relieve boredom or search for a particular toy that has proven to relieve boredom in the past. A child may choose to search for a particular toy or explore a new environment depending on past experience and the child's motivation to relieve or avoid boredom. In a parallel situation, a bored animal may explore a new environment without seeking a particular stimulus or may search for specific sources of novelty, such as unfamiliar objects, a strategy which relieved boredom in the past.

Individual animals may also have different strategies for exploratory behavior as well. Some animals may have flexible strategies for exploratory behavior, whereas others may be more committed to one strategy, such as allowing a conspecific to explore the novelty first. Kuczaj (2017) uses the phrase "watchful cautious" to describe dolphins that peered over the "shoulder" of a conspecific to investigate a novel bubble-ring machine. Although these animals were exhibiting neophilic behavior, they allowed bolder animals to interact with the unfamiliar stimulus first, thereby giving more cautious individuals an opportunity to learn about the stimulus without having to initiate the interaction themselves. The watchful cautious approach would be an advantageous strategy if unknown objects might turn out to be harmful. The fact that the cautious animals exhibited signs of curiosity demonstrates the importance of curiosity for the behavioral repertoire of a social group. Reader (2015) also emphasizes that social influences should not be ignored when examining individual differences in curiosity.

Kuczaj (2017) uses the example of dolphins creating "games" with bubbles to explain the important role of curiosity in innovation and creativity. Kuczaj argues that individual differences in curiosity partially account for individual differences in creativity and innovation because, without the driving force of curiosity, animals are less likely to be creative and innovative. A creative animal is one that modifies its own behavior to produce novelty in the environment, while innovations occur when the behavior of a creative act serves a useful purpose and is adopted by others (Kuczaj 2017). The dolphins that were the most curious and investigated a bubble-producing machine went on to creatively interact with the bubbles, using various body parts to pop them. Although bubble-popping is not necessarily adaptive, other behaviors such as innovation in foraging could provide the animal with a fitness advantage.

Not only is it important to consider individual differences from an evolutionary and developmental perspective, it is also important to consider specific traits associated with search and exploration. Whereas individuals are often characterized as being more shy or bold overall, Wilson et al. (1994) argue that this trait can also be considered domain-specific. That is, in a social situation, an animal may behave boldly, but when considering novel environmental stimuli, the same individual could be classified as shy. Kuczaj et al. (2012) have also argued that dolphin personality can vary across contexts so that an individual could be high in openness regarding environmental stimuli, but low in openness with respect to humans; however, it is possible that, instead of measuring a single trait in two different contexts, these were measures of two distinct traits. This adds to the discussion of environmental enrichment as a change that may not be positive for individuals who are wary of new objects. In terms of animal management decisions, it is possible that these same individuals would respond more positively to the introduction of a new conspecific if they tended to be bold when considering social interactions, a prediction that could not be made from simply characterizing that individual's neophilia in response to a novel object.

Early experience can significantly impact how an individual will interact with novelty. Rats (*Rattus norvegicus*) placed in impoverished conditions during development, solitary and without stimulus objects, interacted with novel objects significantly less frequently as adults (Renner 1987). Rats kept in enriched conditions displayed more frequent and diverse interactions with objects and were faster to explore the novel area compared to the rats kept in impoverished conditions (Renner 1987). Although the significance of early experience should not be underestimated, factors such as age can contribute to differences in curiosity-related behaviors as well. Juvenile rats increase exploratory behavior and object manipulation as they approach adulthood (Renner et al. 1992), while chimpanzees and humans generally decrease in the traits Openness and Extraversion with age (King et al. 2008). Parallels have been suggested between individual differences in exploration of novelty in rats and sensation-seeking in humans (Dellu et al. 1993). Some rats are prone to risk-taking and exploration of novelty that appears independent of experience, just as humans that are high in the sensation-seeking trait prefer novel or varied situations (Dellu et al. 1993).

Research on birds has helped to explain that part of the reason why exploration would be beneficial for an individual is that exploration is correlated with better problem-solving abilities. It seems that age differences and individual differences in curiosity both play a role in exploratory behavior and its association with problem-solving. In the neotropical raptor, Chimango Caracara (*Milvago chimango*), less neophobic juveniles solved a novel problem faster than more neophobic juveniles (Biondi et al. 2010). Further informing this area of research, a study of common myna birds (*Sturnus tristis*) found that individuals with higher levels of exploration and lower levels of neophobia were more likely to exhibit motor and consumer innovations (Sol et al. 2012). Sol et al. (2012) were able to conclude that neophobia is a consistent individual trait in the common myna and that this trait is inversely related to innovation and problem-solving abilities.

The authors of a review article on avian personality call for more comparative work to study the evolution of animal personality, but caution that behavioral profiles cannot always be compared across taxa (Groothuis and Carere 2005). For example, research has found that wild spotted hyenas (*Crocuta crocuta*) were more successful on a novel problem-solving task if they were less neophobic and exhibited higher behavioral diversity when trying to solve the problem (Benson-Amran and Holekamp 2012). However, because this research was conducted in a wild population, it is unclear if interaction with the experimental device was most enticing for the bolder or more curious animals or if previously bored animals were most interested in trying to solve the task. The disparity in methods for assessing personality traits and different behavioral profiles of various species highlights the need for attention to methods used in other species when designing new studies. For example, frequency counts of specific behaviors are not easily compared across species (Wemelsfelder 1984). A more effective approach for assessing boredom in various species should include behavioral measures, such as activity levels, and physiological measures, such as heart rate or hormone levels, for comparison across species. For example, even if minimal behavioral changes occur, examining hormone levels could reveal changes in how individuals respond to environmental enrichment (e.g., Cannon et al. 2016). Not only would this approach help explain the origins of animal personality, it would also help to complete a picture of what behaviors accompany curiosity in many species. Being able to describe curiosity from a behavioral and physiological perspective would also allow for further study of individual differences in boredom.

Creativity, Play, and Problem-Solving

When an organism is able to re-engage its attention, the organism is considered to no longer be bored (Eastwood et al. 2012). This means that the organism has taken interest in some stimulus or is curious about something. However, boredom can result in more than an increase in curiosity. Individual differences have been documented in behaviors that are the result of curiosity, including creativity, problem-solving, play, and social interactions.

Experiencing boredom can result in increased creativity. When human participants were instructed to complete a monotonous task (Mann and Cadman 2014), they showed increases in the number of uses they could name for an object. Though research has not yet made the link between boredom and creativity in nonhuman animals, nonhuman animals do show individual differences in tendencies to be creative. For example, birds vary in their ability to innovate behavior and solve problems both within and between age groups (Biondi et al. 2010).

Individual differences in dolphins' ability to vary their behavior were reported by Kuczaj and Eskelinen (2014a). When instructed to vary their behavior by performing a behavior that was different from previous behaviors they had performed during that session, three dolphins differed in the average number of successive behaviors they could perform with this restriction and also varied the different energy levels at which they performed the behaviors. For example, the younger male performed more behaviors in succession without repeating one and also performed higher energy behaviors compared to the two older male subjects. Having used a similar training task, Pryor (2015) recounted that training an animal to behave "creatively" or at least to vary its behavior on command, resulted in a broader range of behaviors even when training sessions were not taking place, and that this reduced stereotypic behavior. This variation occurred spontaneously and seemed to reflect a more innovative and flexible behavioral repertoire after this "creativity" training took place.

Kuczaj (2017) provides a review of studies supporting the argument that individual differences in traits such as the shy–bold continuum influence an individual's tendency to be creative. Animals that are more innovative also tend to be more bold, curious, and neophilic. Kuczaj (2017) also discusses individual differences in an animal's tendency to invent new behaviors and learn new behaviors from others, providing the example of orcas (*Orcinus orca*) baiting live seagulls from Kuczaj and Walker (2012). Some whales were observed to vary their strategies or adopted new strategies that were more difficult and required more attempts to be successful. Although many different foraging strategies have been observed in wild orcas, the observations of captive-living orcas catching seagulls concretely demonstrates this species' flexibility and adaptability in foraging strategies (Baird 2000).

Play behavior usually takes place only when an organism's basic needs are met (Burghardt 2014). If an individual is hungry, injured, or in danger, there is no time for play. If basic needs are met though, an organism is then in a situation where multiple behaviors, including play, boredom, rest, or a sequence of these states, could occur. Play has been observed in many species and currently many theories on the evolutionary origins and adaptive value of play have been proposed, including that play may relieve excessive metabolic energy, help with development, and promote physical and cognitive maintenance (Burghardt 2014). In the same review, Burghardt (2014) also suggests that play might relieve stress and boredom and may develop where there is a lack of stimulation. Although this is not the only reason play may occur, the link between boredom and play has only begun to be explored.

Kuczaj and Eskelinen (2014b) propose that dolphins play to develop social skills, cognitive and motor skills, and to challenge themselves. In an instance where a dolphin consistently increases the complexity of its play, it is because the dolphin is no longer satisfied with the former level of stimulation and thus makes its play more interesting for its own benefit. Wild and captive dolphins exhibit individual variation in the frequency and types of objects with which they play (Greene et al. 2011). As mentioned previously, Kuczaj et al. (2012) argued that context should be considered when measuring personality, as some dolphins are more open to experiences with objects, while others are more open to experiences with humans or with conspecifics. Social interactions are often a part of play and may be a source of stimulation an individual seeks if it is bored. The research on dolphin play and personality suggests that individual animals have preferences in play partners and vary in their tendency to seek conspecifics (Kuczaj et al. 2012; Kuczaj and Eskelinen 2014b). These personality differences are likely to influence the type and frequency of play behavior.

Conclusion

Individual differences in personality have been reported in several species (Gosling 2001). Although between-species comparisons are sometimes difficult, characterizing individuals on a trait such as how bold one animal is compared to conspecifics has been done in many species. Being able to assess variation of a trait within a particular species and understanding how that affects the behavior and fitness of individuals leads to a better understanding of each species, is beneficial for the study of comparative psychology, and can aid caretakers in animal welfare.

Research has established that humans exhibit individual differences in causes, coping methods, and outcomes of boredom. The existing literature on animal boredom supports that there are individual differences in the behavior of animals when experiencing various environmental conditions, ranging from impoverished to enriched. Measures related to boredom in nonhuman animals may parallel those found in humans, and so their study may expand what is known about boredom as a psychological state in humans and, of course, other species. Curiosity and creativity are both topics that are essential in understanding how boredom can be alleviated and how the cycle of boredom to mental engagement regularly occurs to maintain an organism's well-being. Future research will help explore boredom from a comparative perspective, such that the causes and outcomes of boredom are better understood. This research should focus on describing boredom from a physiological and behavioral perspective, describing individual differences in the causes of boredom, coping mechanisms, and outcomes, and in developing environmental enrichment that is tailored to the needs of individual animals.

References

Adamec, R. (1975). Behavioral and epileptic determinants of predatory attack behavior in the cat. *Canadian Journal of Neurological Sciences, 2,* 457–466.

Alligood, C., & Leighty, K. (2015). Putting the "E" in SPIDER: Evolving trends in the evaluation of environmental enrichment efficacy in zoological settings. *Animal Behavior and Cognition, 2,* 200–217.

Baird, R. W. (2000). The killer whale. In H. Whitehead & J. Mann (Eds.), *Cetacean societies: Field studies of dolphins and whales* (pp. 127–153). Chicago: University of Chicago Press.

Bench, S., & Lench, H. (2013). On the function of boredom. *Behavioral Sciences, 3,* 459–472. doi:10.3390/bs3030459

Benson-Amran, S., & Holekamp, K. (2012). Innovative problem solving by wild spotted hyenas. *Proceedings of the Royal Society B: Biological Sciences, 279,* 4087–4095.

Biondi, L. M., Bó, M. S., & Vassallo, A. I. (2010). Inter-individual and age differences in exploration, neophobia and problem-solving ability in a Neotropical raptor (*Milvago chimango*). *Animal Cognition, 13,* 701–710. doi:10.1007/s10071-010-0319-8

Boesch, C. (2007). What makes us human (*Homo sapiens*)? The challenge of cognitive cross-species comparison. *Journal of Comparative Psychology, 121,* 227–240.

Boesch, C. (2008). Taking development and ecology seriously when comparing cognition: Reply to Tomasello and Call (2008). *Journal of Comparative Psychology, 122,* 453–455.

Buirski, P., Plutchik, R., & Kellerman, H. (1978). Sex differences, dominance, and personality in the chimpanzee. *Animal Behaviour, 26,* 123–129.

Burghardt, G. (2014). A brief glimpse at the long evolutionary history of play. *Animal Behavior and Cognition, 1,* 90–98.

Call, J. (2015). Conservatism versus innovation: The great ape story. In A. Kaufman & J. Kaufman (Eds.), *Animal creativity and innovation* (pp. 397–414). New York: Elsevier.

Cannon, T., Heistermann, M., Hankison, S., Hockings, K., & McLennan, M. (2016). Tailored enrichment strategies and stereotypic behavior in captive individually housed macaques (*Macaca spp.*). Journal of Applied. *Animal Welfare.* doi:10.1080/10888705.2015.1126785

Chamove, A. S., Eysenck, H. J., & Harlow, H. F. (1972). Personality in monkeys: Factor analyses of rhesus social behavior. *Quarterly Journal of Experimental Psychology, 24,* 496–504.

Charmoy, K., Sullivan, T., & Miller, L. J. (2015). Impact of different forms of environmental enrichment on foraging and activity levels in gorillas (*Gorilla gorilla gorilla*). *Animal Behavior and Cognition, 2,* 233–240.

Coleman, K., & Wilson, D. (1998). Shyness and boldness in pumpkinseed sunfish: individual differences are context-specific. *Animal Behavior, 56,* 927–936.

Collins, R. P., Litman, J. A., & Spielberger, C. D. (2004). The measurement of perceptual curiosity. *Personality and Individual Differences, 36,* 1127–1141. doi:10.1016/S0191-8869(03)00205-8

Cussen, V., & Mench, J. (2015). The relationship between personality dimensions and resiliency to environmental stress in orange-winged Amazon parrots (*Amazona amazonica*), as indicated by the development of abnormal behaviors. *PLoS ONE, 10,* 1–11.

Daniels, L., Tze, V., & Goetz, T. (2015). Examining boredom: Different causes for different coping profiles. *Learning and Individual Differences, 37,* 255–261.

Dall, S., Houston, A., & McNamara, J. (2004). The behavioural ecology of personality: Consistent individual differences from an adaptive perspective. *Ecology Letters, 7,* 734–739.

Dallaire, J. A., Meagher, R. K., & Mason, G. J. (2012). Individual differences in stereotypic behaviour predict individual differences in the nature and degree of enrichment use in caged American mink. *Applied Animal Behaviour Science, 142,* 98–108. doi:10.1016/j.applanim.2012.09.012

Darwin, C. (1871/1964). *The descent of man, and selection in relation to sex.* London: John Murray.

Darwin, C. (1872). *The expressions of the emotions in man and animals.* London: John Murray.

Dellu, F., Mayo, W., Piazza, P. V. V., Le Moal, M., & Simon, H. (1993). Individual differences in behavioral responses to novelty in rats. Possible relationship with the sensation-seeking trait in man. *Personality and Individual Differences, 15,* 411–418. doi:10.1016/0191-8869(93)90069-F

Digman, J. M. (1990). Personality structure: Emergence of the five-factor model. *Annual Review of Psychology, 41,* 417–440.

Eastwood, J. D., Frischen, A., Fenske, M. J., & Smilek, D. (2012). The unengaged mind: Defining boredom in terms of attention. *Perspectives on Psychological Science, 7,* 482–495. doi:10.1177/1745691612456044

Eskelinen, H. C., Winship, K. A., & Borger-Turner, J. L. (2015). Sex, age, and individual differences in bottlenose dolphins (*Tursiops truncatus*) in response to environmental enrichment. *Animal Behavior and Cognition, 2,* 241–253. doi:10.12966/abc.08.04.2015

Farmer, R., & Sundberg, N. (1986). Boredom proneness—The development and correlates of a new scale. *Journal of Personality Assessment, 50,* 4–17. doi:10.1207/s15327752jpa5001_2

Found, R., & St. Clair, C. (2016). Behavioural syndromes predict loss of migration in wild elk. *Animal Behaviour, 115,* 35–46.

Fureix, C., & Meagher, R. K. (2015). What can inactivity (in its various forms) reveal about affective states in nonhuman animals? A review. *Applied Animal Behaviour Science, 171,* 8–24. doi:10.1016/j.applanim.2015.08.036

Giraldeau, L.-A., & Dubois, F. (2008). Social foraging and the study of exploitative behavior. *Advances in the Study of Behavior, 38,* 59–104.

Goetz, T., Frenzel, A., Hall, N., Nett, U., Pekrun, R., & Lipnevich, A. (2014). Types of boredom: An experience sampling approach. *Motivation and Emotion, 38,* 401–419.

Goldberg, L. R. (1990). An alternative "description of personality" : The big-five factor structure. *Journal of Personality and Social Psychology, 59,* 1216–1229.

Goldberg, L. R. (1993). The structure of phenotypic personality traits. *American Psychologist, 48,* 26–34.

Gosling, S. (2001). From mice to men: What can we learn about personality from animal research? *Psychological Bulletin, 127,* 45–86.

Gottlieb, D., Capitanio, J., & McCowan, B. (2013). Risk factors for stereotypic behavior and self-biting in Rhesus macaques (*Macaca mulatta*): Animal's history, current environment, and personality. *American Journal of Primatology, 75,* 995–1008.

Greene, W., Melillo-Sweeting, K., & Dudzinski, K. (2011). Comparing object play in captive and wild dolphins. *International Journal of Comparative Psychology, 24,* 292–306.

Groothuis, T. G. G., & Carere, C. (2005). Avian personalities: Characterization and epigenesis. *Neuroscience and Biobehavioral Reviews, 29,* 137–150. doi:10.1016/j.neubiorev.2004.06.010

Guenther, A., Brust, V., Dersen, M., & Trillmich, F. (2013). Learning and personality types are related in cavies (*Cavia aperea*). *Journal of Comparative Psychology, 128,* 1–8.

Herrmann, E., Hernández-Lloreda, M. V., Call, J., Hare, B., & Tomasello, M. (2010). The structure of individual differences in the cognitive abilities of children and chimpanzees. *Psychological Science, 21,* 102–110.

Highfill, L., & Kuczaj, S. (2010). How studies of wild and captive dolphins contribute to our understanding of individual differences and personality. *International Journal of Comparative Psychology, 23,* 269–277.

Hogan, L. A., Johnston, S. D., Lisle, A., Horsup, A. B., Janssen, T., & Phillips, C. J. C. (2010). Stereotypies and environmental enrichment in captive southern hairy-nosed wombats, *Lasiorhinus latifrons*. *Applied Animal Behaviour Science, 126,* 85–95. doi:10.1016/j.applanim.2010.05.009

Hogan, L. A., & Tribe, A. (2007). Prevalence and cause of stereotypic behaviour in common wombats (*Vombatus ursinus*) residing in Australian zoos. *Applied Animal Behaviour Science, 105,* 180–191. doi:10.1016/j.applanim.2006.06.006

Hopkins, W., Russel, J., & Schaeffer, J. (2014). Chimpanzee intelligence is heritable. *Current Biology, 24,* 1649–1652.

Hoy, J. M., Murray, P. J., & Tribe, A. (2010). Thirty years later: Enrichment practices for captive mammals. *Zoo Biology, 29,* 303–316. doi:10.1002/zoo.20254

Hunter, S. A., Bay, M. S., Martin, M. L., & Hatfield, J. S. (2002). Behavioral effects of environmental enrichment on harbor seals (*Phoca vitulina concolor*) and gray seals (*Halichoerus grypus*). *Zoo Biology, 21,* 375–387. doi:10.1002/zoo.10042

Huntingford, F., Andrew, G., Mackenzie, S., Morera, D., Coyle, S., Pilarczyk, M., et al. (2010). Coping strategies in a strongly schooling fish, the common carp *Cyprinus carpio*. *Journal of Fish Biology, 76,* 1576–1591.

Ijichi, C., Collins, L., & Elwood, R. (2013). Evidence for the role of personality in stereotypy predisposition. *Animal Behaviour, 85,* 1145–1151.

Izzo, G. N., Bashaw, M. J., & Campbell, J. B. (2011). Enrichment and individual differences affect welfare indicators in squirrel monkeys (*Saimiri sciureus*). *Journal of Comparative Psychology, 125,* 347.

Joshi, S., & Pillay, N. (2016). Association between personality and stereotypic behavior in the African striped mouse *Rhabdomys dilectus*. *Applied Animal Behaviour Science, 174,* 154–161.

Kennedy, J. (1992). *The new antrhopomorphism*. Cambridge: Cambridge University Press.

King, J., Weiss, A., & Sisco, M. (2008). Aping humans-age and sex effects in chimpanzee (*Pan troglodytes*) and human (*Homo sapiens*) personality. *Journal of Comparative Psychology, 122,* 418–427.

Kuczaj, S.A. II (2017). Animal creativity and innovation. In J. Call, G. Burghardt, I. Pepperberg, C. Snowdon, & T. Zentall (Eds.), *APA handbook of comparative psychology: Perception, learning, and cognition*. Washington, DC, US: American Psychological Association.

Kuczaj, S., & Eskelinen, H. (2014a). The "creative dolphin" revisited: What do dolphins do when asked to vary their behavior? *Animal Behavior and Cognition, 1,* 66–77.

Kuczaj, S. A., & Eskelinen, H. C. (2014b). Why do Dolphins Play? *Animal Behavior and Cognition, 2,* 113. doi:10.12966/abc.05.03.2014

Kuczaj, S., Highfill, L., & Byerly, H. (2012). The importance of considering context in the assessment of personality characteristics: Evidence from ratings of dolphin personality. *International Journal of Comparative Psychology, 25,* 309–329.

Kuczaj, S., Lacinak, T., Garver, A., & Scarpuzzi, M. (1998). Can animals enrich their own environment? In V. Hare & K. Worley (Eds.), *Proceedings of the third international conference on environmental enrichment* (pp. 168–170). San Diego, CA: Shape of Enrichment Inc.

Kuczaj, S., Makecha, R., Trone, M., Paulos, R., & Ramos, J. (2006). Role of peers in cultural innovation and cultural transmission: Evidence from the play of dolphin calves. *International Journal of Comparative Psychology, 19,* 223–240.

Kuczaj, S. A., II, & Walker, R. T. (2012). Dolphin problem solving. In T. Zentall & E. Wasserman (Eds.), *The Oxford handbook of comparative cognition* (pp. 736–756). London: Oxford University Press.

Laule, G., Bloomsmith, M., & Schapiro, S. (2003). The use of positive reinforcement training techniques to enhance the care, management, and welfare of primates in the laboratory. *Journal of Applied Animal Welfare Science, 6,* 163–173.

Lopes, M., Borger-Turner, J., Eskelinen, H., & Kuczaj, S. (2016). The influence of age, sex, and social affiliation on the responses of bottlenose dolphins (*Tursiops truncatus*) to a novel stimulus over time. *Animal Behavior and Cognition, 3,* 32–45.

Mann, S., & Cadman, R. (2014). Does being bored make us more creative? *Creativity Research Journal, 26,* 165–173. doi:10.1080/10400419.2014.901073

Markowitz, H., & Aday, C. (1998). Power for captive animals: Contingencies and nature. In D. J. Shepherdson, J. D. Mellen, & M. Hutchins (Eds.), *Second nature: Environmental enrichment for captive animals* (pp. 47–58). Washington, DC: Smithsonian Institution Press.

Mason, G., Clubb, R., Latham, N., & Vickery, S. (2007). Why and how should we use environmental enrichment to tackle stereotypic behaviour? *Applied Animal Behaviour Science, 102,* 163–188. doi:10.1016/j.applanim.2006.05.041

Meagher, R. K., & Mason, G. J. (2012). Environmental enrichment reduces signs of boredom in caged mink. *PLoS ONE.* doi:10.1371/journal.pone.0049180

Mellen, J., & Sevenich MacPhee, M. (2001). Philosophy of environmental enrichment: Past, present, and future. *Zoo Biology, 20,* 211–226.

Morand-Ferron, J., & Giraldeau, L. A. (2010). Learning behaviorally stable solutions to producer-scrounger games. *Behavioral Ecology, 21,* 343–348.

Owen, M., Swaisgood, R., Czekala, N., & Lindburg, D. (2005). Enclosure choice and well-being in giant pandas: Is it all about control? *Zoo Biology, 24,* 475–481.

Peterson, E., Worlein, J., Lee, G., Dettmer, A., Varner, E., & Novak, M. (2016). Rhesus macaques (*Macaca mulatta*) with self-injurious behavior show less behavioral anxiety during the human intruder test. *American Journal of Primatology, 79,* 1–8.

Pervin, L., & John, O. P. (1997). *Personality: Theory and research* (7th ed.). New York: Wiley.

Pryor, K. (2015). Creating creative animals. In A. Kaufman & J. Kaufman (Eds.), *Animal creativity and innovation* (pp. 483–498). New York: Elsevier.

Reader, S. M. (2015). Causes of individual differences in animal exploration and search. *Topics in Cognitive Science.* doi:10.1111/tops.12148

Réale, D., Reader, S. M., Sol, D., McDougall, P. T., & Dingemanse, N. J. (2007). Integrating animal temperament within ecology and evolution. *Biological Reviews, 82,* 291–318.

Renner, M. J. (1987). Experience-dependent changes in exploratory behavior in the adult rat (*Rattus norvegicus*): Overall activity level and interactions with objects. *Journal of Comparative Psychology, 101,* 94–100. doi:10.1037//0735-7036.101.1.94

Renner, M. J., Bennett, A. J., & White, J. C. (1992). Age and sex as factors influencing spontaneous exploration and object investigation by preadult rats (*Rattus norvegicus*). *Journal of Comparative Psychology, 106,* 217–227. doi:10.1037/0735-7036.106.3.217

Ross, S. (2006). Issues of choice and control in the behaviour of a pair of captive polar bears (*Ursus maritimus*). *Behavioural Processes, 73,* 117–120.

Seligman, M. (1972). Learned helplessness. *Annual Review of Medicine, 23,* 407–412. doi:10.1146/annurev.me.23.020172.002203

Sih, A., & Bell, A. (2008). Insights for behavioral ecology from behavioral syndromes. *Advances in the Study of Behavior, 38,* 227.

Sih, A., Bell, A., Johnson, J., & Ziemba, R. (2004). Behavioral syndromes: An ecological and evolutionary overview. *Trends in Ecology and Evolution, 19,* 372–378.

Sol, D., Griffin, A. S., & Bartomeus, I. (2012). Consumer and motor innovation in the common myna: The role of motivation and emotional responses. *Animal Behaviour, 83,* 179–188. doi:10.1016/j.anbehav.2011.10.024

Stevenson-Hinde, J., & Zunz, M. (1978). Subjective assessment of individual rhesus monkeys. *Primates, 19,* 473–482.

Swaisgood, R., & Shepardson, D. (2006). Environmental enrichment as a stratedgy for mitigating stereotypies in zoo animals: A literature review and meta-analysis. In G. Mason & J. Rushen (Eds.), *Stereotypic animal behaviour: Fundamentals and applications to welfare.* Wallingford: CAB International.

Tarou, L., & Bashaw, M. (2007). Maximizing the effectiveness of environmental enrichment: Suggestions from the experimental analysis of behavior. *Applied Animal Behaviour Science, 102,* 189–204.

Tetley, C., & O'Hara, S. (2012). Ratings of animal personality as a tool for improving the breeding, management and welfare of zoo animals. *Animal Welfare, 21,* 463–476.

Thelen, E., & Farish, D. (1977). An analysis of the grooming behavior of wild and mutant strains of Bracon Hebet or (*Braconidae: Hymenoptera*). *Behaviour, 62,* 70–102.

Tomasello, M., & Call, J. (2007). Assessing the validity of ape-human comparisons: A reply to Boesch (2007) *Journal of Comparative Psychology, 122,* 449–452.

Veissier, I., Boissy, A., Désiré, L., & Greiveldinger, L. (2009). Animals' emotions: Studies in sheep using appraisal theories. *Animal Welfare, 18,* 347–354.

Vodanovich, S. (2003). Psychometric measures of boredom: A review of the literature. *The Journal of Psychology: Interdisciplinary and Applied, 137,* 569–595.

Vonk, J., & Povinelli, D. (2011). Individual differences in long-term cognitive testing in a group of chimpanzees. *International Journal of Comparative Psychology, 24,* 137–167.

Wechsler, B. (1992). Stereotypies and attentiveness to novel stimuli: A test in polar bears. *Applied Animal Behaviour Science, 33,* 381–388. doi:10.1016/S0168-1591(05)80074-7

Weiss, A., & King, J. (2015). Great ape origins of personality maturation and sex differences: A study of orangutans and chimpanzees. *Journal of Personality and Social Psychology, 108,* 648–664.

Wemelsfelder, F. (1984). Animal Boredom : Is a scientific study of the subjective experiences of animals possible ? In M. W. Fox & L. D. Mickley (Eds.), *Advances in animal welfare science, vol. 85* (pp. 115–154). Washington, DC: The Humane Society of the United States.

Wemelsfelder, F., Haskell, M., Mendl, M., Calvert, S., & Lawrence, A. (2000). Diversity of behaviour during novel object tests is reduced in pigs housed in substrate-impoverished conditions. *Animal Behaviour, 60,* 385–394. doi:10.1006/anbe.2000.1466

Wilson, D. (1998). Adaptive individual differences within single populations. *Philosophical Transactions of the Royal Society B: Biological Sciences, 353,* 199–205.

Wilson, D., Clark, A., Coleman, K., & Dearstyne, T. (1994). Shyness and boldness in humans and other animals. *Trends in Ecology and Evolution, 9,* 442–446.

Wilson, D. S., Coleman, K., Clark, A. B., & Biederman, L. (1993). Shy-bold continuum in pumpkinseed sunfish (*Lepomis gibbosus*): An ecological study of a psychological trait. *Journal of Comparative Psychology, 107,* 250.

Würbel, H., Bergeron, R., & Cabib, S. (2006). The coping hypothesis of stereotypic behaviour. Box 3, Chapter 1. In G. Mason & J. Rushen (Eds.), *Stereotypic animal behaviour: Fundamentals and applications to welfare* (pp. 14–15). Wallingford: CAB International.

Phenotype Management: An Inclusive Framework for Supporting Individuals' Contributions to Conservation Populations

14

Jason V. Watters, Samantha Bremner-Harrison and David M. Powell

Abstract

There are generally two types of animal populations managed for conservation purposes, in situ and ex situ. The management goals for each type of population differ and this drives the manner by which the populations experience selection. Population members of different behavioral types may respond to the same stimuli in varying ways, generating potentially supportive effects for achieving conservation goals. The cumulative impact of observable phenotypic or behavioral variation predicts the potential of meeting population goals. Phenotype management is a conservation strategy that employs understanding of the varied outcomes for individuals in developing the potential for successful conservation populations. To employ phenotype management, it is useful to consider the environmental factors that drive the expression of varied behavioral types and life history trajectories. Diversity of habitat and developmental circumstance may be crucial to generating phenotypically diverse populations. Ex situ populations may be spread across numerous locations as meta-populations and members of these populations may experience a diversity of husbandry protocols, social groupings, and climates—resulting in population level behavioral diversity. A focus on habitat heterogeneity and in situ habitat restoration may support phenotypic diversity in populations of concern.

Introduction

This chapter develops a unified framework for supporting the contributions of individual animals to conservation populations. We recognize two fundamentally different types of conservation populations—in situ and ex situ populations. The two types of populations differ both in form

J.V. Watters (✉)
San Francisco Zoological Society, San Francisco, CA, USA
e-mail: jasonw@sfzoo.org

S. Bremner-Harrison
Nottingham Trent University, Nottingham, UK

D.M. Powell
Wildlife Conservation Society, Bronx Zoo, New York, NY, USA

J. Vonk et al. (eds.), *Personality in Nonhuman Animals*,
DOI 10.1007/978-3-319-59300-5_14

and function as well as with regard to the apparent selection pressures they face. In the grand scheme of conservation work, these two types of populations play different roles. The members of in situ populations are fostered to fill the ecological niche of their ancestral populations and are generally maintained in the same or similar areas to where the ancestral population was originally found. The members of ex situ populations are often distributed across several locations as meta-populations—though not always. Metaphorically, these populations serve as genetic banks—savings accounts for the possible infusion of genes or individuals back into wild populations. "Withdrawals" from these accounts are generally infrequent and rarely deplete the entire "fund," leaving many animals in ex situ populations to play other roles—the most common being teacher and breeder. Teacher animals are those maintained under human care at zoos and other public conservation institutions, but some of these animals may also be breeders during their lives. These animals are often described as ambassadors for theirs or other species. It is our intention here to recognize the varied purposes and attempt to provide an inclusive view of individuals' roles in these populations and how they support conservation outcomes. The core of the concept we develop is that variant phenotypes interact with their world in different ways and that this leads to situation-dependent payoffs. We also recognize that development, life history, and selection pressures play a role in the expression of individual phenotypes and attempt to suggest future areas of study and application where conservation programs may benefit from assessing and directing the competence of individuals for the role they will play in conservation populations. We will focus primarily on behavioral phenotypes—personalities—because they are relatively easy to assess and likely play a role in the manner in which circumstances are perceived by the individual and, thus, in individual decision-making as well as group function.

Phenotype Management—Connecting the Concepts Across Conservation Populations

Watters et al. (2003) initially described "phenotype management" in the context of habitat restoration. They referred to the then theoretical process as that of supporting phenotypic diversity in populations targeted for habitat restoration by ensuring opportunities for alternate developmental pathways with habitat heterogeneity. They also indicated that habitat heterogeneity could promote facultative behavior switching. Watters et al. (2003) provided a simple theoretical treatment that suggested that—for an environment with a given carrying capacity—when environmental change was unpredictable, a diversity of phenotypes in the population supported a higher probability of population survival than a more homogenous population. Reciprocally, these authors indicated that when there is certainty about environmental fluctuation, focusing on a subset of phenotypes could result in the largest sustained population size. They also pointed out that a population's effective size was likely to be maximized when it contained individuals that used different approaches to gain mates such as the alternative mating tactics males that many species employ. They indicated that consideration of whether alternative reproductive tactics are expressed as a result of developmental processes or are a facultative response to the environment will help to maximize the number of breeders in populations of concern. Thus, by ensuring that a high proportion of potential breeders in a population actually reproduce, maximum genetic diversity is retained.

The original argument for phenotype management was based on the observation that population members of different phenotypes occupy varied niche spaces (sensu Bolnick et al. 2003). This separation of the way that individuals interact with their environment—at either a large or a fine scale—serves an important role in population function as well as population resilience to environmental change. For example, many species in the family Salmonidae have populations that express numerous life history trajectories. These trajectories vary not only in age to maturity but also in whether an individual becomes anadromous or remains in the natal stream as well as the duration of time that individuals on different trajectories are exposed to any one of the different environments (Hodge et al. 2016). The decision to follow one trajectory or another is generally understood to be based on an individual's assessment of its body condition —at some "decision time" (Thorpe 1994). A single population of *Oncorhynchus mykiss* (known collectively as rainbow trout and steelhead trout) in California's Klamath River Basin expresses as many as 38 different life histories (Hodge et al. 2016). The result of such phenotypic diversity in the same population is that deleterious—as well as beneficial—effects of environmental perturbations are unevenly distributed across the population. Thus, supporting the ability for a population to retain or develop a diversity of life histories is viewed as a useful conservation strategy (Emel and Bonett 2011; Greene et al. 2010). This sort of diversity is maintained in restored environments that include a diversity of habitats even when some of those habitats support only small proportions of the population (Watters et al. 2003).

With other colleagues, Watters has broadened the concept of phenotype management. Watters and Meehan (2007) combine ex situ populations with in situ ones to suggest that captive-reared animals can be managed to promote the expression of varied phenotypes and that doing so will facilitate the release of broadly diverse phenotypic groups from ex situ populations to in situ programs. They follow on the argument of Watters et al. (2003) that an unpredictable world warrants the development of diverse populations, rather than populations with an abridged phenotypic portfolio. They also indicate that developmental processes play a large role in the expression of phenotypes and suggest that environmental enrichment can be used to drive the expression of variant phenotypes in ex situ populations. Recommendations to assess the behavioral type of pre-release animals for the purpose of generating diverse groups for release are increasing in conservation practice (Bremner-Harrison et al. 2004; Dunston et al. 2016).

Carroll and Watters (2008) describe the concept of phenotypic potential—that set of phenotypes that can be expressed by a population's set of genotypes across all possible environments—regardless of their apparent fitness in a given environment. Carroll and Watters (2008) also suggest that in the era of fast acting ecological problems such as invasive species and abrupt environmental change, it might make little sense to try to "reset" in situ populations to their former state. Instead, it may be more logical to move forward, developing conservation populations capable of evolutionary change and providing the substrate for short-term adaptive plasticity that can, in turn, promote long-term genetic adaptation of conservation populations—even if it means that those populations do not ultimately look as their ancestral ones did. Thus, phenotype management may also include promoting within-population evolution of never before seen phenotypes that perform well enough to support population persistence in a changing world (Reed et al. 2011).

Watters and Powell (2012) reconsidered captive population management programs with the suggestion that animals' behavioral phenotypes could be assessed to place them into positions in captive populations where they would be most successful. Similar to the notion that novel phenotypes could help to support in situ population persistence, it should be recognized that phenotypic expression in captive animals is possibly dissimilar from the range of phenotypes observed in their wild counterparts. Nevertheless, Watters and Powell suggest that the diversity of phenotypes found in ex situ populations could support the numerous roles that members of these populations fill. They also suggested that assessing and understanding the situation-specific consequences of behavioral type for individuals in captive populations would help to support the welfare of these individuals—ultimately supporting overall population goals of maintaining genetic and behavioral diversity. The root of this concept is again that varied phenotypes function in different ways in different settings. A growing literature recognizes the potential for considering behavioral types in the planning of ex situ populations (Fox and Millam 2014; Kaumanns and Singh 2015) as well as the differential animal welfare outcomes that similar situations can have for variant types (Coleman 2012; Franks et al. 2013, 2014; Herrelko et al. 2012; Shepherdson et al. 2013; Torgerson-White and Bennett 2014) and the efficacy of different individuals in the "teaching" role (Luebke et al. 2016).

In this chapter, we use the term *phenotype management* to refer to the process of managing conservation populations with consideration of individuals' contributions to the population goals based on observable phenotypic characteristics or behavioral phenotype. This consideration requires understanding the outcomes for different individuals in different environmental circumstances. It also benefits from understanding the factors that drive the development of alternative phenotypes and how these are influenced because active phenotype management requires manipulation of the frequencies of phenotypes in populations. Thus, a complete view of phenotype management will consider both in situ and ex situ populations as well as the factors that drive the expression of varied phenotypes, including life history trajectories. As stated earlier, we focus primarily on behavioral phenotypes, although we will develop a general hypothesis that relates personality with life history trajectory in developing individuals. Although we appreciate that each individual has its own unique phenotype, we will focus our discussion on general phenotypes—such as shy and bold individuals—where straightforward assessments can reliably group individuals into functionally similar types or place them along a continuum.

Factors that Affect the Frequencies of Phenotypes in Populations

As previously mentioned, life history variation can lead to substantial differentiation in population structure. Numerous species have the capacity to vary their developmental trajectory based upon experienced conditions or perceived states (Stearns and Koella 1986). A general example is the animal that matures earlier than the average member of its population as a result of enhanced growth and physiological development. An alternative example is the animal that retains juvenile status for an extended period in order to extend growth and reach maturity at a larger than average size. Life history decisions can be based on size, perceived danger and numerous other states or conditions. We suggest that there might be an important link between behavioral phenotype and life history trajectory that to date remains relatively unstudied; that is, that the behavioral tendency of a given individual may promote its development along a certain trajectory. For example, individuals that are more socially interactive may in turn be less likely to avoid agonistic interactions than less socially interactive individuals. As a result, perhaps more socially interactive individuals are more likely to be territory holders and gain the benefits associated with holding a territory, including allowing faster growth and perhaps quicker development. In such a way, personality may drive the expression of alternative life histories or vice versa (Biro and Stamps 2008; Wolf et al. 2007). For example, in coho salmon (*Oncorynchus kisutch*), there are two alternative male types. The types act consistently differently during mating and each is the product of an alternative life history trajectory (Watters 2005). Thus, it might be that the result of a specific life history trajectory is a specific behavioral type. It is also possible, though, that certain behavioral types are more likely to develop along one trajectory than others as one tactic is associated with rapid early growth and likely territory holding. Either way, the environment must be structured to support the successful expression of alternative phenotypes.

Understanding the range of phenotypes that can be expressed by a population—its phenotypic potential—is a key to phenotype management. We consider two primary means by which phenotypic potential can be observed. The first approach is essentially a norm of reaction type of experiment. Here, one would observe the phenotypes that are expressed in varied environments or rearing conditions by individuals of the same genetic or family line. This approach is likely to provide insight to the range of phenotypes that will develop under predicted or existing conditions. An alternative approach is to observe the range of phenotypes expressed under conditions where selection is relaxed or highly scattered across the population. This situation is perhaps more realistic than one might at first imagine when considering the selection pressures (or lack of them) faced by ex situ populations.

One goal of many ex situ population managers is to reduce the selection placed on a population. To this end, much effort is put forward to reduce disease, provide nutritional support, and ensure breeding to capture as much of the initial founding stock's genetic variation as possible—in some cases even if that stock contains genetic anomalies (Ralls et al. 2000). Typical breeding programs in zoos aim to minimize genetic adaptation to captivity that might occur through artificial selection and result in loss of genetic variation in the population (Ballou and Lacy 1995) because maintenance of maximum genetic diversity minimizes inbreeding (Crnokrak and Roff 1999; Keller and Waller 2002; Ralls et al. 1979) and provides a buffer against future environmental fluctuations (Falconer 1981; Keller et al. 1994; Markert et al. 2010; Willi et al. 2006). It is, of course, impossible to remove all forms of selection, yet these efforts are aimed at essentially doing just that by generally pairing individuals for breeding based on their relatedness to the entire population (mean kinship, Ballou and Lacy 1995). Individuals of low mean kinship are paired for breeding to produce offspring based on demographic, genetic, and space considerations. In addition, many ex situ populations, such as those that are cooperatively managed within accredited zoos and aquariums, are composed of individuals that are held in relatively small numbers at numerous locations. For example, Bennett et al. (2015) studied 45 okapis (*Okapia johnstoni*) housed at 16 separate institutions. These animals were a part of the Association of Zoos and Aquariums Species Survival Plan© for okapi and are collaboratively managed as a connected meta-population. They represented nearly one half of this meta-population. The animals in this study were in climate areas ranging from the west to the east coast of the United States and from as far south as San Diego to as far north as Chicago. Indeed, even though captive populations most certainly do experience some form of selection, because of this distribution of "population members" across institutions—each in different climates and with varied animal husbandry regimes, selection is likely to be scattered and not focused in any clear direction. In addition, the fact that animals are regularly transferred from one institution to another to fulfill breeding recommendations means that long-term location-based selection is unlikely. In sum, it is likely that the efforts to minimize selection on these populations outweigh the scattered and short-term selective forces that these populations experience. Relaxed selection in captive rearing environments has been shown to result in increased behavioral variation over generations (McPhee 2004). Environments that minimize selection may also increase behavioral variation over the course of development in a single generation (Lee and Berejikian 2008). Thus, relaxing selection may provide insight to a population's phenotypic potential in both evolutionary and developmental time.

Phenotype Management as Applied to Ex Situ Populations for Conservation

Ex situ populations can broadly be managed in two, non-mutually exclusive ways: management for exhibition to the public (teachers) and management for propagation (breeders). Phenotype management has applications in both scenarios, and this section will explore how explicit consideration of phenotype could provide measurable progress toward achieving the goals of ex situ management programs.

In the conservation arena we are mainly managing populations of whole individuals. As such, our consideration of phenotype is largely restricted to an individual animal's consistent behavioral characteristics or tendencies, what others have referred to as personality (e.g., Powell and Gartner 2010) or behavioral types (Sih et al. 2004). Whole individuals are units that we can readily observe and assess in terms of both welfare and contribution to conservation goals. That said, there may very well be scenarios in which phenotype management in ex situ populations could extend to anatomy or physiology. For example, should it become known that the individuals most resistant to temperature stress, i.e., through a variety of physiological attributes they possess but which vary across individuals, are most likely to survive reintroduction, then phenotype management could be used to identify and propagate individuals that are more likely to survive temperature stress in the early stages of life, post-release. Similarly, one might try to identify individual animals that fare the best in urban, public zoological parks and breed these individuals preferentially. However, the danger remains that consideration of only a narrow range of phenotypic characteristics could inadvertently select for other traits that could threaten *population* viability in the long term (Sih and Watters 2005). For example, it has long been known that active zoo animals are more interesting to zoo guests (Bitgood et al. 1988; Margulis et al. 2003; Myers et al. 2004; Powell and Bullock 2014). However, Luebke et al. (2016) demonstrated that animals who make eye contact with zoo-goers also support visitor interest—suggesting that focusing only on active animals could result in loss of otherwise good educator animals from zoo populations if the propensity to make eye contact does not covary with activity.

Ex Situ Animals as Educators

Professional zoos and aquaria, hereafter collectively referred to as zoos, devote significant resources to developing and maintaining self-sustaining populations of animals for exhibition to the public (Gusset and Dick 2011). This is largely because it is no longer considered acceptable to continually obtain animals from the wild when captive sources are available, and in many cases it has become logistically or legislatively extremely difficult to bring some kinds of animals into ex situ management. Zoos are therefore obligated to manage animal populations for the long term. This means that one of the most fundamental goals for animals in zoos is breeding to sustain their species, and successful breeding is nearly always predicated upon excellent welfare.

For zoos to deliver on their missions of generating conservation support and educating the public (in addition to providing recreation for the public), the animals they exhibit should be good teachers as well as good breeders. By this we mean that they should demonstrate natural behaviors, they should appear to be thriving because the public cares about their welfare, and they should engage the public to foster a conservation ethic and facilitate biological literacy. Abundant research demonstrates that animals in zoos can inspire positive affective outcomes in zoo visitors (Grajal et al. 2016; Luebke et al. 2016; Powell and Bullock 2014) but animals that perform abnormal behaviors or do not appear healthy can also raise concerns in visitors (Godinez et al. 2013; Miller 2012), which presumably evokes negative affect.

What do we know about how personality/phenotype impacts breeding performance in captive wildlife? The data are sparse, but studies have demonstrated that certain personality traits or behavioral types (e.g., active, investigatory individuals) can be used to predict or promote successful breeding. For example, female giant pandas that are more shy exhibit less socio-sexual behavior (Powell et al. 2008). Similarly, bolder male pandas make more attempts at social interaction during the breeding season. For black rhinos, we have learned that breeding pairs in which the female is more dominant than the male have a higher chance of breeding successfully (Carlstead et al. 1999). Impacts of personality on reproductive success have also been studied in cheetah (Wielebnowski 1999), maned wolves (Carlstead and Kleiman 1998), primates (Maestripieri 1993), and birds (Fox and Millam 2014). Many more such studies should be conducted as the number of species under care in zoos is vast. Over time, a larger body of knowledge could reveal a few "rules of thumb" that might have multispecies application in terms of behavioral types or personality traits that are generally associated with reproductive success. Based on what we know now, we might hypothesize that behavioral types or personalities characterized as nervous, fearful, or shy around conspecifics are less likely to be successful breeders and could be less successful at rearing offspring in managed environments. These individuals might benefit from relatively "private" or off-exhibit breeding areas as well as an array of potential mates to choose from. Animals that are characterized as calm, confident, (appropriately) bold, and interested in their environment might be more likely to engage in successful courtship, mating, and rearing of young.

To ask what kinds of personality traits in zoo animals make them good "teachers," we have to ask zoo visitors about their experiences and correlate what they say with what animals do or do not do. Zoos are largely venues for informal learning; a smaller percentage of zoo-goers participate in structured educational zoo programs. Visitors are most often coming to a zoo on a recreational motivation (Morgan and Hodgkinson 1999; Reade and Waran 1996; Tofield et al. 2003), though educational interests are also cited, particularly with regards to trips to the zoo with children (Tofield et al. 2003).

Research has shown that emotion and learning are often related (Damasio 1994). Affective arousal is associated with increased attention and willingness to learn (Krapp et al. 1992; Pekrun 1992) as well as retention of facts (Buchanan and Lovallo 2001). Thus, it is thought that positive emotional experiences facilitate free-choice learning (Tofield et al. 2003). A large body of research demonstrates that viewing healthy animals at zoos results in positive affective responses (Luebke et al. 2016; Powell and Bullock 2014 and references therein).

Are there specific animal traits that draw the public in and elicit these positive emotional responses more powerfully? Several non-behavioral traits affect how well an animal species endears humans to them (see Powell and Bullock 2014 for discussion), but some behavioral characteristics have also proven important at capturing visitors' emotions. First, in addition to activity (Bitgood et al. 1988; Luebke et al. 2016; Margulis et al. 2003; Myers et al. 2004; Powell and Bullock 2014;), which we discussed earlier, behavioral diversity exhibited by the animals is also correlated with positive emotional arousal (Anderson et al. 2003). An exception to this general finding may occur when visitors see animals actively engaging in real or perceived abnormal behaviors, like pacing (Altman 1998; Reade and Waran 1996; Tunnicliffe 1995, but see Luebke et al. 2016). To date, the data suggest that generally active animals drive this positive affective response in the zoo-goer. Luebke et al. (2016) found that several types of activity: eating, moving around, interacting with conspecifics, and surprisingly, pacing, were associated with stronger positive emotional arousal.

Additional behavioral traits that have been identified more recently are willingness to come into proximity with and/or make eye contact with visitors. These behaviors facilitate the visitor experiencing an "encounter" and perhaps feeling a connection with the animal (Powell and Bullock 2014). Two recent studies have demonstrated that the extent to which visitors report having made eye contact with the animal or having had an encounter with the animal are positively correlated with positive emotional responses. Powell and Bullock (2014) surveyed visitors after seeing three carnivore exhibits at the Bronx Zoo and asked them to report on the positive emotions they experienced and to rate how strongly they made eye contact with the animals and how strong of an encounter they had with the animals. Both eye contact and encounter strength were strongly correlated with emotions. Luebke et al. (2016) reported on similar studies at three zoos focusing on four other species of mammals and also found that these experience variables, eye contact and sense of "encounter," had a strong impact on emotional arousal. They also demonstrated that the effect of up close encounters (which included perceived encounter strength and eye contact) had a stronger effect on positive affect than animal behavioral diversity observed at the exhibits. Taken together with findings on animal activity and behavioral diversity, these results suggest that there may be more than one behavioral type that is capable of supporting the role of "educator" in ex situ animal populations. Bold, active, or social animals may fill this role well. Indeed, a new hypothesis might suggest that shy and vigilant animals, while likely to spend more time in distant proximity from public areas, may also be individuals that are prone to making protracted eye contact with zoo-goers.

Another category of animals in zoos is those animals used in encounters with guests during narrated presentations, meet-and-greets, touch tanks, etc. We know of no published studies of the personality characteristics of animals that make certain individuals, or even certain species, more successful at educating the public in encounters where visitors can get close to and perhaps make contact with animals. We hypothesize that, like other good animal teachers, animals successful in this role will be calm, confident animals that are exploratory and interested in novel stimuli (e.g., unfamiliar classrooms, scents of unfamiliar people). It is also the case that many of these individuals are trained to fulfill this role from a very young age. Thus, there is likely a developmental component that focuses the expression of the personality factors associated with success in this role. Studies of animals that appear to thrive in this role versus those that do not and are reassigned to live in traditional exhibits or serve as breeders would be very informative. Such studies would focus on the behavioral types of individuals who are successful and unsuccessful in each role and also on the developmental circumstances each individual may have experienced.

Because animals of various personalities are likely to perceive similar circumstances in different ways, it is possible that animal welfare varies for animals of different behavioral types housed in the same conditions (Lilley et al. this volume; Tetley and O'Hara 2012; Watters and Powell 2012). The studies mentioned earlier that indicate that breeding success for exhibit animals varies with personality shed some light on this notion but the field specifically examining animal personality and welfare of this group of animals remains in need of substantial development. One cross-sectional study of a large sample of orangutans housed in numerous zoos found relationships between personality factors and subjective assessments of animal well-being (Weiss et al. 2006). The researchers followed up these individuals to determine that those orangutans that were ranked higher on a measure of subjective well-being—termed "happiness" by the researchers—lived longer lives (Weiss et al. 2011). Future work in the area of exhibit animal welfare should investigate interactions between personality type and specific components of the zoo environment, including space, habitat complexity, social group structure and husbandry regimes.

Crossover Conservation Populations: From in Situ to Ex Situ and Back

Occasionally, it becomes necessary to take wild animals into captive environments in order to attempt to avoid an extinction event. One example of such a program is the captive breeding program that supports the ongoing growth of the wild population of the Mexican wolf (*Canis lupus baileyi*) (Hedrick and Fredrickson 2008). When animals are taken into human care for the purpose of propagation for eventual reintroduction, housing, husbandry, and interactions with humans may be managed in a way dramatically different from other exhibit animals in zoos. And even within these populations of animals, the goals for individuals differ and thus the applicability and form of phenotype management will vary as well.

Reintroduction programs are typically multi-year initiatives. This is because often the species in question is drastically reduced in numbers and, left alone, its population is not viable. Typically, individuals are brought into captivity and used as founder stock for a breeding population. Whether their first, second, or later generation offspring are used as additional breeders or reintroduced will vary with program characteristics and species biology, but generally, the goal is to get animals back into the wild as soon as possible to avoid the behavior changes that appear to accompany long-term captive propagation (Williams and Hoffman 2009). It would be fair to say that in these scenarios or programs, breeders and individuals slated for release are managed with the goal of being as "hands-off" as possible. Obviously for successful propagation the environmental conditions in captivity must be right for reproduction and the animals must feel comfortable enough to do so.

The propagation for reintroduction scenario can present a conflict in terms of application of phenotype management. To the extent that animal personality traits are heritable (Bouchard and Loehlin 2001; Brent et al. 2014; Dingemanse et al. 2002; Weiss et al. 2000), the captive environment might select for a different range of behavioral types than is present in the founder stock. There may be a set of personality traits or behavioral types that thrive and reproduce better than others in propagation facilities, wherein living environments, and possibly husbandry practices are more standardized than is the case in zoos. These successful individuals might tolerate human presence more, be less tense or frightful of unfamiliar stimuli, and they might be more likely to form relationships with or dependence on humans. If the need to produce animals for reintroduction is urgent, it might be argued that focus be placed on those "good breeders" in the propagation program and breed them as much as possible to get offspring on the ground quickly for release with the caveat that the reintroduced generation might be less genetically diverse—and therefore possibly less personality diverse—than the founder base. There may be trade-offs between behavioral types that breed well in captivity and types that survive and/or reproduce best in a reintroduction program. Humans are one of the primary reasons for the death of captive bred individuals when they are reintroduced (Jule et al. 2008). However in some cases, animals that are most comfortable around humans may be the best ex situ breeders. To address this potential conflict, it might be instructive to consider propagation programs for reintroduction in phases. In the first phase, the animals may have extensive exposure to human caretakers so that optimal breeding protocols can be identified. For example, breeding introductions are closely observed and studied to promote compatibility, artificial nest or birthing structures are tested for preference by the animals, and mother–infant interaction is carefully documented to promote survivorship. Once a formula for successful propagation is developed, the next phase would involve incorporating the lessons learned from phase one and eliminating the need for human exposure to the extent possible. This could involve moving animals to larger enclosures, providing food/water remotely, or observing animals only via surveillance cameras. The hope would be that removal of the human interaction element of the program would not compromise successful propagation.

As previously discussed, typical long-term breeding programs in zoos aim to reduce selection and maintain as much natural genetic diversity as possible. Thus, if these programs are successful at meeting these goals, the offspring produced in typical zoo breeding programs are possibly more diverse in terms of personality characteristics than their wild counterparts. However, in some cases, it may take more hands-on management and manipulation to get various behavioral types to breed well in captive environments, and a variety of types of environments may have to be constructed. Current animal care practice supports the development of diverse environments in zoos. Conversely, in a propagation for reintroduction scenario, the choice could arise between rapidly producing large numbers of offspring with a narrow range of behavioral types that have been maintained under less management intervention and taking more time and hands-on manipulation to produce an equivalently sized cohort of animals for release that have a broader range of personality traits. Given our current understanding of behavioral types and their ecological and group functions (Wolf and Weissing 2012), we believe that it is always important to ensure some degree of diversity in release groups. Which types and their frequencies in the release population is an area in need of further research.

Behavioral Type Diversity Is Necessary for Release Success

Because the ultimate goal of captive propagation for reintroduction programs is returning animals to the wild, eventually, animals from these programs will be released. At this point, these ex situ populations morph into either in situ populations or unsuccessful (dead) release groups. Of course, to re-establish a population requires that released animals reproduce, but a cornerstone of re-establishment is survival. Here, groups of released animals from captive propagation programs are similar to wild born animals that are translocated for conservation purposes.

Traditionally, in situ conservation measures that involve the intentional movement of animals through conservation translocations (see IUCN 2013 for definitions) have focused on genetic and ecological characteristics when selecting source animals (IUCN 1998). However, in more recent decades there has been a shift of focus with behavioral suitability or *competence* being given more consideration. Along with this focus on behavioral competence, a few studies demonstrate the importance of functional social groups. For example, Shier (2006) found that in translocations of prairie dogs (*Cynomys ludovicianus*) where animals were trapped and relocated with family group members, reproduction and survivorship following release were higher than when animals were trapped and relocated with no regard for familial relationships. Shier and Swaisgood (2012) later demonstrated that Stephen's kangaroo rats (*Dipodmys stephensi*), a species considered territorial and solitary, also had higher survival rates and reproductive success when translocated with familiar kangaroo rats (former neighbors from the trapping site) than unfamiliar kangaroo rats. Both of these studies suggest that familiarity with social group partners increases translocation success. However, if we can assume that intact family groups or stable territorial arrays of animals include individuals of varying personality or behavioral type then we might argue that phenotype management that produces an array of behavioral types for reintroduction would be preferable for facilitating post-release establishment, survival, and reproduction.

Investigations into the impact of personality or behavioral type on release success of reintroduced, captive-bred animals have demonstrated relationships between behavioral type and survival (Sinn et al. 2014) post-release movements (Bremner-Harrison et al. 2004, 2013; Stratton 2015), social structure (Dunston et al. 2016) and reproduction (Bremner-Harrison et al. 2013). These findings suggest a real need for inclusion of behavioral type as a reintroduction candidate selection criterion, a suggestion that has been put forward by both academics and conservation practitioners (e.g., Bremner-Harrison et al. 2004; McDougall et al. 2006; Watters et al. 2003).

These initial investigations highlighted a clear link between optimal behavioral type for particular environment conditions, both when looked at from an applied reintroduction perspective and from a wider personality and ecological perspective. Evolutionarily, this makes sense because individuals adapted for certain ecological or environmental conditions show the prevalent behavioral types. However, these studies present only a "snapshot" view of which behavioral type is optimal for specific environmental conditions, or for a particular factor studied, e.g., survival or reproductive output, at the time of the assessment. A broader view to take is to ask: What is required for long-term population sustainability when environmental conditions fluctuate? How do we look beyond the snapshot of what we know now to predict the range of phenotypes that will support long-term population persistence? Ensuring that there is the means for adaptation to release site conditions, a fluctuating environment, or to stochastic events requires sufficient diversity to increase likelihood of population success. There are three key arguments that support this call for diversity of behavioral phenotypes when constructing populations for conservation, whether they be founders for breeding or release.

Genetic diversity has long been recognized as a vital component for maximizing chances of success in conservation populations (see Chaps.

11–13 of Ewan et al. 2012 for an overview), and, as such, is an important recommendation in the updated IUCN Reintroduction Guidelines (2013) developed by the IUCN Reintroduction Specialist Group. However, several factors lead to the suggestion that diversity of behavioral type may be as important as genetic diversity when considering movement of animals. First, a review by Smith and Blumstein (2008) explored the existence of trade-offs in personality. A main finding of this meta-analysis was the relationship between the behavioral trait boldness and the fitness correlates of survival and reproduction. Bolder individuals may have a greater likelihood of mortality through predation, but if surviving, may show higher levels of reproductive output, thus leading to population growth. Their shyer counterparts, however, tend to demonstrate greater survival due to reduced risk-taking behavior, but have lower reproductive output. The trade-off between these particular behavioral traits has ramifications for reintroduction success where the primary goals, particularly in the early stages of a conservation program, are for founder animals to survive and reproduce within the area of release, thus establishing a sustainable population.

Second, the persistence of diversity in behavioral type within in situ populations suggests that variant types each have similar adaptive value. Many studies have demonstrated genetic heritability of personality (Laine and van Oers, this volume; van Oers et al. 2005), which, in a stable environment might suggest genetic selection for an optimal personality type and a long-term loss of variation. However, fluctuation of ecological drivers in environments across both seasons and time results in variable selection pressures. Thus, variation in behavioral type at the population level allows for optimal environmental exploitation of shifting conditions. This is evidenced through differences within populations in behaviors such as feeding or vigilance (e.g., Favreau et al. 2014), exploration (Dingemanse et al. 2012), or offspring survival (Dingemanse et al. 2004). In addition, the growing literature on social group composition suggests that diversity in group members' behavioral types plays a fundamental role in overall group fitness (Sih and Watters 2005; Pruitt and Riechert 2011).

Finally, at the individual level, whereas an animal's behavioral type is considered consistent over time and contexts (Réale et al. 2000), there is growing recognition of the impact of consistent variation or plasticity within an individual's personality type (Biro and Adriaenssens 2013). While previously explored from ecological and evolutionary perspectives, theories have converged to advance understanding of the adaptive nature of plasticity (Dingemanse et al. 2010). Plasticity at the population level serves as a further driver to ensure variability within a population, thus allowing a diversity of strategies to cope with fluctuating resources.

In combination, personality related trade-offs, persistence of diverse populations, and consistency of within individual plasticity strongly support recommendations that incorporating personality into founding groups of individuals for conservation measures may help to maximize the likelihood of success. A trial translocation by Stratton (2015, PhD thesis) tested this theory through an experimental release program that used wood mice as a model species. Mice were trapped in the wild, and tested for activity and boldness, first in the wild at the point of capture and again in captivity. Mice were placed in one of three release groups (all shy, all bold, and mixed) and released at three sites with closely matched environmental conditions. Post-release monitoring indicated that the mixed group, i.e., higher diversity of behavioral types, had the lowest rates of mortality and dispersal, thus showing greater survival and site retention and meeting these two fundamental goals of conservation translocation projects (Stamps and Swaisgood 2007). Although this study had some limitations, for example, there were restrictions on moving resident mice at the release site and extrapolating results beyond one model study species should be done with caution, the results of this study do support the general recommendation for providing diversity in founding release groups. The results warrant further studies that examine the conditions that affect optimal frequencies of alternative types in release groups.

Habitat, the Crucial Ingredient for Phenotype Success

Above, we discussed the need for variety of the experience of individuals in founding populations to maintain or allow behavioral diversity to develop. However, by what means can this be achieved? Typically, endangered animals have been maintained in captivity under a "one size fits all" regime, whereby all individuals in a group at a particular facility are maintained under similar conditions. Where environmental conditions are limited there is little potential for the ontogeny or expression of a range of behavioral phenotypes, therefore the resultant range of behavioral types is similarly limited. Where environments are enriched to create diverse sub-habitats the opportunities for expression of a broad range of phenotypes is greatly expanded (Watters and Meehan 2007). The observation that different behavioral types are expressed between and within habitats in wild populations supports the case for creating habitats for captive animals that will encourage development of diverse behavioral phenotypes.

There are a growing number of studies that have investigated the variability expressed in personality traits between populations in differing environmental conditions. For example, Brown et al. (2005) assessed boldness in eight populations of tropical poeciliid from streams with differing levels of predation risk. Across all sites, fish that originated from areas with high predation risk showed higher boldness than those from areas with low predation risk. Boldness and fish length were related with smaller fish emerging to forage earlier, thus scoring higher on the boldness continuum; however, results of previous lab-based studies (Brown and Braithwaite 2004) suggest that this difference is the result of smaller fish having higher metabolic needs rather than a function of environment. Dingemanse et al. (2012) demonstrated consistent average differences in personality and plasticity both between and within four populations of great tit, *Parus major*, sourced from sites with differing environmental conditions. Assessed from a within-population perspective, Favreau et al. (2014) explored both ecological and behavioral hypotheses to determine effects of habitat and social conditions on eastern gray kangaroos, *Macropus giganteus*. Increased similarities in behavior between individuals in groups were found in relation to ecological conditions, such as decreased patch-richness, and social conditions, such as when in large group sizes, but greater consistent individual differences in behavior were observed when patch-richness was high or group sizes were small. Consistent individual differences were also observed relative to reproductive state and vigilance/foraging and in behavioral plasticity, indicating a variety of individual strategies. Another example employed a suggestion made by Watters et al. (2003) to vary the habitat structure of facultative phenotype switching pupfish. There, Gumm et al. (2011) were able to increase the population size of endangered Leon Springs pupfish (*Cyprinodon bovinus*) by modifying the habitat to support more males that employed a territory holding rather than a satellite mating tactic. This simple change required modifying the habitat to increase the number of shallow breeding areas. Moreover, prior to the habitat modifications, pupfish eggs were predated at very high rates by the also endangered Pecos gambusia (*Gambusia nobilis*), and so Gumm and her colleagues were able to mitigate this negative interaction between endangered species with these simple habitat modifications, too.

The examples above indicate that providing complex habitats facilitates the development and expression of behavioral phenotypes as individuals respond to divergent environmental factors. This is particularly desirable if animals are to be used for conservation purposes or where behavioral diversity is to be maintained over time, similar to the retention goal of 90% of genetic diversity over 100 years in a captive breeding program (Frankham et al. 2002). However, in captivity, spatial constraints may limit the ability to produce and maintain a variety of environmental conditions. There may be greater opportunities for spatial and enclosure variation for animals reared in facilities purpose built for breeding and release programs, such as the

National Black-Footed Ferret Conservation Center in Colorado, USA, or the U.S National Park Service breeding facilities for California Channel Island fox on the California Channel Islands, USA. However, for programs that exist within zoos, there may be difficulties in flexibility within physical resources to provide a variety of habitat types or environmental conditions for multiple groups of the same species. As described above, within institutions, the majority of species are held in relatively small numbers as one group or in a single exhibit. This limits the possibilities for habitat variation in a single location. But, because long-term captive breeding populations are often spread over sometimes as many as dozens of institutions, it may be possible to take a coordinated approach in producing a range of environmental conditions to facilitate development of a variety of behavioral phenotypes, and then drawing on multiple institutions to produce a behaviorally heterogenous group of individuals for release.

One aspect of within-population studies of behavioral variation that has not been thoroughly studied is the impact of microhabitats, whereby a variety of habitat types are contained within the overall habitat, or, in the captive instance, the animal's enclosure. If access to these microhabitats is unrestricted, individuals are likely to exploit niches within their environment that are suited to their personality type. An example is provided by Kobler et al. (2011), whereby more consistently aggressive fish within one stretch of river were found in areas of reduced complexity and structure (in terms of openness of habitat and degree of presence of reeds, branch jams, tree roots, stones and water plants), compared to less-aggressive individuals who were repeatedly located in areas of higher complexity and structure. Kobler et al. (2011) speculated that fish that have a propensity to be more aggressive may choose to occupy the less complex microhabitat as the difficulty in defending it provides an under-utilized food resource, a potentially risky trade-off. Alternatively, fish that choose the less complex habitat adapt to becoming more aggressive. This could be relevant to captive-reared populations where niche-management through enclosure design, enclosure conditions and implementation of varied enrichment could be utilized to ensure that a range of behavioral phenotypes develop within a captive group.

A note of caution is provided by Hensley et al. (2012), however. Giant sea anemones (*Condylactis gigantea*) demonstrated a significant association between personality type and habitat differences along a continuous habitat gradient created by a combination of density and blade height of seagrass and water depth. The authors hypothesize that this congregation at a preferred habitat type may, over time, lead to segregation of personality types with an ultimate risk of speciation. Although reproductive isolation is highly unlikely to occur in a managed captive breeding program taking place over a range of environmentally controlled habitats across which animals may be transferred, or indeed through the means of artificial insemination, it bears thinking about when breeding for conservation and release, whereas the aim may be to create founding groups with high levels of diversity, the group still has to operate as a functionally cohesive population.

We have discussed above the ability of populations of individuals to comprise a diverse range of behavioral phenotypes despite being under similar environmental conditions. However, it is worth noting that the environmental or habitat constraints present do place a restriction on the range of behavioral phenotypes expressed within that population (Watters et al. 2003). Multiple studies illustrate that a population within a particular habitat expresses a population level mean for a given behavioral trait, with deviation around that mean representing consistent individual variation (Carrol and Watters 2008). The level of variation expressed within a population depends on genetic diversity and environmental complexity, which interact in the expression of behavioral types (Watters and Meehan 2007), the success of which determines the genetic diversity of the next generation (Watters et al. 2003). Therefore, within ex situ populations there will be levels of behavioral type that are not adaptive to given environmental conditions at certain points in time, e.g., excessively bold individuals in areas of high predation threat or excessively shy individuals in areas of high mate or resource competition. As a result, maladaptive behavioral types will be selected against and either not expressed, or lost from the population if selected against persistently.

Analogously, it is often assumed that individuals within the same captive-bred population will be subject to similar environmental and genetic selection pressures, with a population mean being expressed around the optimal behavior type for the particular habitat conditions. However, a single captive metapopulation's conditions can be very diverse across locations—for example public facilities with high levels of visitors such as zoos, compared to closed captive breeding facilities that are purpose built to cater for one species only, This situation then may result in divergent optimal behavior types and the ranges expressed around the means. When populations are maintained in captivity over time for conservation breeding, there is a risk of directional selection occurring whereby the genetic diversity and diversity of behavioral phenotypes may be lost or reduced in the population. This risk is likely to be highest in populations that are maintained in single breeding centers where all individuals experience a similar environment. However, while an assessment of mice held in captivity for differing periods of time demonstrated a reduction in predator avoidance behavior, it also revealed an increase in behavioral variance for those mice bred across a greater number of generations (McPhee 2004), whereas this increase in phenotypic variation was interpreted as a result of relaxed selection having potentially detrimental impacts for reintroduction programs, it may prove to be of benefit for these programs. Osborne and Seddon (2012, p. 75) discuss the importance of habitat for reintroduction. Among the eight components they list as impacting reintroduction success they include that "A habitat's suitability and its characteristics vary across the species range." Although we recognize the impact of habitat differences on behavioral type, we are nowhere near being able to appropriately match individuals, or even populations displaying a particular mean type, with habitats. Therefore, further to our earlier arguments on diversity, releasing individuals with a wide variance of behavioral types into occupied or empty habitats may increase the likelihood of success. Furthermore, although it may appear logical that animals that are returned to the point of source following a period in captivity would have a higher likelihood of comprising the requisite range of behavioral types, this is not necessarily the case. Between the time of population decline and release, habitat and environmental conditions at the source/release site may have changed significantly since the species was last present either through local evolution or restoration measures (Osborne and Seddon 2012). Thus, releasing a population that shows increased variance around the mean from the original founding population of captive animals may increase rather than constrain the likelihood of success.

Conclusion

We have argued for greater consideration of diversity of behavioral type in the management of various types of populations to realize greater success toward conservation goals. Although a strong theoretical and empirical background for this line of thinking has developed over the last 15 years, still much remains to be done to effectively put these concepts into action in the conservation arena. More progress will come with explicit hypothesis testing regarding the success or failure of different behavioral types in applied in situ and ex situ settings. Success in propagation at zoos or dedicated breeding centers needs to be measured against behavioral type and related to environmental and husbandry circumstances. Multiple releases of animals in reintroduction programs also provide an empirical framework for testing best practices for expanding the role of phenotype management in situ.

References

Altman, J. D. (1998). Animal activity and visitor learning at the zoo. *Anthrozoös, 11*(1), 12–21.

Anderson, U. S., Kelling, A. S., Pressley-Keough, R., Bloomsmith, M. A., & Maple, T. L. (2003). Enhancing the zoo visitor's experience by public animal training and oral interpretation at an otter exhibit. *Environment and behavior, 35*(6), 826–841.

Ballou, J. D., & Lacy, R. C. (1995). Identifying genetically important individuals for management of genetic variation in pedigreed populations. In *Population Management for Survival and Recovery: Analytical methods and strategies in small population conservation* (76–111). New York, NY: Columbia University Press.

Bennett, C., Torgerson-White, L., Fripp, D., Watters, J., & Petric, A. (2015). A multi-institutional assessment of factors influencing locomotion and pacing in captive okapis (*Okapia johnstoni*). *Journal of Applied Animal Welfare Science, 18,* S43–S61. doi:10.1080/10888705.2015.1075835

Biro, P. A., & Stamps, J. A. (2008). Are animal personality traits linked to life-history productivity? *Trends in Ecology & Evolution, 23*(7), 361–368. doi:10.1016/j.tree.2008.04.003

Bitgood, S., Patterson, D., & Benefield, A. (1988). Exhibit design and visitor behavior empirical relationships. *Environment and behavior, 20*(4), 474–491.

Bolnick, D. I., Svanback, R., Fordyce, J. A., Yang, L. H., Davis, J. M., Hulsey, C. D., et al. (2003). The ecology of individuals: Incidence and implications of individual specialization. *American Naturalist, 161*(1), 1–28. doi:10.1086/343878

Bouchard, T. J., & Loehlin, J. C. (2001). Genes, evolution, and personality. *Behavior Genetics, 31*(3), 243–273.

Bremner-Harrison, S., Cypher, B. L., & Harrison, S. W. (2013). An investigation into the effect of individual personality on reintroduction success, examples from three North American fox species: swift fox, California Channel Island fox and San Joaquin kit fox. In: P. S. SOORAE, (Ed.), *Global re-introduction perspectives: 2013. Further case studies from around the globe* (152–158). Abu Dhabi, UAE: IUCN/SSC Re-introduction Specialist Group and Environment Agency.

Bremner-Harrison, S., Prodohl, P. A., & Elwood, R. W. (2004). Behavioural trait assessment as a release criterion: Boldness predicts early death in a reintroduction programme of captive-bred swift fox (*Vulpes velox*). *Animal Conservation, 7,* 313–320. doi:10.1017/s1367943004001490

Brent, L. J. N., Semple, S., MacLarnon, A., Ruiz-Lambides, A., Gonzalez-Martinez, J., & Platt, M. L. (2014). Personality traits in rhesus macaques (*Macaca mulatta*) are heritable but do not predict reproductive output. *International Journal of Primatology, 35*(1), 188–209.

Brown, C., & Braithwaite, V. A. (2004). Size matters: A test of boldness in eight populations of the poeciliid Brachyraphis episcopi. *Animal Behaviour, 68,* 1325–1329. doi:10.1016/j.anbehav.2004.04.004

Brown, C., Jones, F., & Braithwaite, V. (2005). In situ examination of boldness-shyness traits in the tropical poeciliid, Brachyraphis episcopi. *Animal Behaviour, 70,* 1003–1009. doi:10.1016/j.anbehav.2004.12.022

Carlstead, K., Mellen, J., & Kleiman, D. G. (1999). Black rhinoceros (*Diceros bicornis*) in US zoos: I. Individual behavior profiles and their relationship to breeding success. *Zoo Biology, 18*(1), 17–34. doi:10.1002/(sici)1098-2361(1999)18:1<17::aid-zoo4>3.0.co;2-k

Carlstead, K., & Kleiman, D. G. (1998). Behavior profiles of individual animals: A new tool for multi-institutional studies. *In Proceedings of the American Zoo and Aquarium Association Meeting* (39–43).

Carroll, S. P., & Watters, J. V. (2008). Managing phenotypic variability with genetic and environmental heterogeneity: Adaptation as a first principle of conservation practice. In *Conservation biology: Evolution in Action* (181–198). Oxford, NY: Oxford University Press.

Coleman, K. (2012). Individual differences in temperament and behavioral management practices for nonhuman primates. *Applied Animal Behaviour Science, 137*(3), 106–113.

Crnokrak, P., & Roff, D. A. (1999). Inbreeding depression in the wild. *Heredity, 83,* 260–270. doi:10.1038/sj.hdy.6885530

Damasio, A. R. (1994). Descartes' error: Emotion, rationality and the human brain. New York, Putnam.

Dingemanse, N. J., Kazem, A. J., Réale, D., & Wright, J. (2010). Behavioural reaction norms: animal personality meets individual plasticity. *Trends in Ecology & Evolution, 25*(2), 81–89.

Dingemanse, N. J., Both, C., Drent, P. J., & Tinbergen, J. M. (2004). Fitness consequences of avian personalities in a fluctuating environment. *Proceedings of the Royal Society of London, Series B: Biological Sciences, 271,* 847–852.

Dingemanse, N. J., Both, C., Drent, P. J., Van Oers, K., & Van Noordwijk, A. J. (2002). Repeatability and heritability of exploratory behaviour in great tits from the wild. *Animal Behaviour, 64,* 929–938.

Dingemanse, N. J., Bouwman, K. M., van de Pol, M., van Overveld, T., Patrick, S. C., Matthysen, E., et al. (2012). Variation in personality and behavioural plasticity across four populations of the great tit *Parus major*. *Journal of Animal Ecology, 81*(1), 116–126. doi:10.1111/j.1365-2656.2011.01877.x

Dunston, E. J., Abell, J., Doyle, R. E., Evershed, M., & Freire, R. (2016). Exploring African lion (*Panthera leo*) behavioural phenotypes: Individual differences and correlations between sociality, boldness and behaviour. *Journal of Ethology, 34*(3), 277–290.

Emel, S. L., & Bonett, R. M. (2011). Considering alternative life history modes and genetic divergence in conservation: A case study of the Oklahoma salamander. *Conservation Genetics, 12*(5), 1243–1259. doi:10.1007/s10592-011-0226-9

Ewan, J. G., Armstrong, D.P., Parker, K.A., & Seddon, P. J. (Eds.), (2012). *Reintroduction biology: Integrating-science and management.* Hoboken, NJ: Wiley-Blackwell.

Falconer, D. (1981). *An introduction to population genetics*. London: Longman.

Favreau, F. R., Goldizen, A. W., Fritz, H., Blomberg, S. P., Best, E. C., & Pays, O. (2014). Within-population differences in personality and plasticity in the trade-off between vigilance and foraging in kangaroos. *Animal Behaviour, 92,* 175–184. doi:10.1016/j.anbehav.2014.04.003

Fox, R. A., & Millam, J. R. (2014). Personality traits of pair members predict pair compatibility and reproductive success in a socially monogamous parrot breeding in captivity. *Zoo Biology, 33*(3), 166–172. doi:10.1002/zoo.21121

Frankham, R., Briscoe, D. A., & Ballou, J. D. (2002). *Introduction to conservation genetics*. Cambridge university press.

Franks, B., Higgins, E. T., & Champagne, F. A. (2014). A theoretically based model of rat personality with implications for welfare. *PLoS ONE, 9*(4), 9. doi:10.1371/journal.pone.0095135

Franks, B., Reiss, D., Cole, P., Friedrich, V., Thompson, N., & Higgins, E. T. (2013). Predicting how individuals approach enrichment: Regulatory focus in cotton-top tamarins (*Sanguinus oedipus*). *Zoo Biology, 32,* 427–435. doi:10.1002/zoo.21075

Godinez, A. M., Fernandez, E. J., & Morrissey, K. (2013). Visitor behaviors and perceptions of jaguar activities. *Anthrozoos, 26,* 613–619.

Grajal, A., Luebke, J. F., Clayton, S., Kelly, L. A. D., Karazsia, B. T., Saunders, C. D., … & Mann, M.E. (2016). The relationship between affective connections to animals and proenvironmental behaviors. *Conservation Biology,* 31(2), 322–330.

Greene, C. M., Hall, J. E., Guilbault, K. R., & Quinn, T. P. (2010). Improved viability of populations with diverse life-history portfolios. *Biology Letters, 6,* 382–386. doi:10.1098/rsbl.2009.0780

Gumm, J. M., Snekser, J. L., Leese, J. M., Little, K. P., Leiser, J. K., Imhoff, V. E., et al. (2011). Management of interactions between endangered species using habitat restoration. *Biological Conservation, 144,* 2171–2176. doi:10.1016/j.biocon.2011.05.006

Gusset, M., & Dick, G. (2011). *WAZA magazine 12: Towards sustainable population management*. Gland: WAZA Executive Office.

Hedrick, P. W., & Fredrickson, R. J. (2008). Captive breeding and the reintroduction of Mexican and red wolves. *Molecular Ecology, 17,* 344–350. doi:10.1111/j.1365-294X.2007.03400.x

Hensley, N. M., Cook, T. C., Lang, M., Petelle, M. B., & Blumstein, D. T. (2012). Personality and habitat segregation in giant sea anemones (Condylactis gigantea). *Journal of Experimental Marine Biology and Ecology, 426,* 1–4.

Herrelko, E. S., Vick, S. J., & Buchanan-Smith, H. M. (2012). Cognitive research in zoo-housed chimpanzees: Influence of personality and impact on welfare. *American Journal of Primatology, 74,* 828–840. doi:10.1002/ajp.22036

Hodge, B. W., Wilzbach, M. A., Duffy, W. G., Quinones, R. M., & Hobbs, J. A. (2016). Life history diversity in Klamath River steelhead. *Transactions of the American Fisheries Society, 145*(2), 227–238. doi:10.1080/00028487.2015.1111257

IUCN (2013). *Guidelines for reintroductions and other conservation translocations, Version 1.0.* Gland, Switzerland: IUCN Species Survival Commission.

IUCN (1998). *Guidelines for Re-introductions. IUCN/SSC Re-introduction Specialist Group*. Gland, Switzerland: IUCN.

Jule, K. R., Leaver, L. A., & Lea, S. E. G. (2008). The effects of captive experience on reintroduction survival in carnivores: A review and analysis. *Biological Conservation, 141*(2), 355–363. doi:10.1016/j.biocon.2007.11.007

Kaumanns, W., & Singh, M. (2015). Towards more individual-based and fitness-oriented captive mammal population management. *Current Science, 109,* 1121–1129.

Keller, L. F., Arcese, P., Smith, J. N. M., Hochachka, W. M., & Stearns, S. C. (1994). Selection against inbred song sparrows during a natural-population bottleneck. *Nature, 372,* 356–357. doi:10.1038/372356a0

Keller, L. F., & Waller, D. M. (2002). Inbreeding effects in wild populations. *Trends in Ecology & Evolution, 17,* 230–241. doi:10.1016/s0169-5347(02)02489-8

Kobler, A., Maes, G. E., Humblet, Y., Volckaert, F. A. M., & Eens, M. (2011). Temperament traits and microhabitat use in bullhead, *Cottus perifretum*: Fish associated with complex habitats are less aggressive. *Behaviour, 148,* 603–625. doi:10.1163/000579511x572035

Krapp, A., Hidi, S., & Renninger K.A. (1992). Interest, learning and development. In *The role of interest in learning and development* (3–25). Hillsdale, NJ: Erlbaum.

Lee, J. S. F., & Berejikian, B. A. (2008). Effects of the rearing environment on average behaviour and behavioural variation in steelhead. *Journal of Fish Biology, 72,* 1736–1749. doi:10.1111/j.1095-8649.2008.01848.x

Luebke, J. F., Watters, J. V., Packer, J., Miller, L. J., & Powell, D. M. (2016). Zoo visitors' affective responses to observing animal behaviors. *Visitor Studies, 19*(1), 60–76. doi:10.1080/10645578.2016.1144028

Maestripieri, D. (1993). Maternal anxiety in rhesus macaques (*Macaca mulatta*). 1. Measurement of anxiety and identification of anxiety-eliciting situations. *Ethology, 95,* 19–31.

Margulis, S. W., Hoyos, C., & Anderson, M. (2003). Effect of felid activity on zoo visitor interest. *Zoo Biology, 22,* 587–599.

Markert, J. A., Champlin, D. M., Gutjahr-Gobell, R., Grear, J. S., Kuhn, A., McGreevy, T. J., et al. (2010). Population genetic diversity and fitness in multiple environments. *BMC Evolutionary Biology, 10,* 13. doi:10.1186/1471-2148-10-205

McDougall, P. T., Reale, D., Sol, D., & Reader, S. M. (2006). Wildlife conservation and animal temperament: Causes and consequences of evolutionary change for captive, reintroduced, and wild populations. *Animal Conservation, 9,* 39–48. doi:10.1111/j.1469-1795.2005.00004.x

McPhee, M. E. (2004). Generations in captivity increases behavioral variance: Considerations for captive breeding and reintroduction programs. *Biological Conservation, 115,* 71–77.

Miller, L. J. (2012). Visitor reaction to pacing behavior: Influence on the perception of animal care and interest in supporting zoological institutions. *Zoo Biology, 31,* 242–248.

Morgan, J. M., & Hodgkinson, M. (1999). The motivation and social orientation of visitors attending a contemporary zoological park. *Environment and Behavior, 31,* 227–239.

Myers, O. E., Saunders, C. D., & Birjulin, A. A. (2004). Emotional dimensions of watching zoo animals: An experience sampling study building on insights from psychology. *Curator: The Museum Journal, 47,* 299–321.

Osborne, P. E., & Seddon, P. J. (2012). Selecting suitable habitats for reintroductions: Variation, change and the role of species distribution modelling. In *Reintroduction biology: Integrating science and management* (73–104). Hoboken, NJ: Blackwell Publishing Limited.

Pekrun, R. (1992). The impact of emotions on learning and achievement: Towards a theory of cognitive/motivational mediators. *Applied Psychology, 41*(4), 359–376.

Powell, D. M., & Bullock, E. V. W. (2014). Evaluation of factors affecting emotional responses in zoo visitors and the impact of emotion on conservation mindedness. *Anthrozoos, 27,* 389–405. doi:10.2752/175303714x13903827488042

Powell, D. M., & Gartner, M. C. (2010). Applications of personality to the management and conservation ofnonhuman animals. In *From genes to animal behavior* (185–199). Berlin: Springer.

Powell, D., Hong, L., Carlstead, K., Kleiman, D., Zhang, H., Zhang, G., et al. (2008). Relationships between temperament, husbandry, management, and socio-sexual behavior in captive male and female giant pandas (*Ailuropoda melanoleuca*). *Acta Zoologica Sinica, 54,* 169–175.

Pruitt, J. N., & Riechert, S. E. (2011). How within-group behavioural variation and task efficiency enhance fitness in a social group. *Proceedings of the Royal Society B-Biological Sciences, 278,* 1209–1215. doi:10.1098/rspb.2010.1700

Ralls, K., Ballou, J. D., Rideout, B. A., & Frankham, R. (2000). Genetic management of chondrodystrophy in California condors. *Animal Conservation, 3,* 145–153. doi:10.1111/j.1469-1795.2000.tb00239.x

Ralls, K., Brugger, K., & Ballou, J. (1979). Inbreeding and juvenile mortality in small populations of ungulates. *Science, 206,* 1101–1103.

Reade, L. S., & Waran, N. K. (1996). The modern zoo: How do people perceive zoo animals? *Applied Animal Behaviour Science, 47,* 109–118.

Reed, T. E., Schindler, D. E., & Waples, R. S. (2011). Interacting effects of phenotypic plasticity and evolution on population persistence in a changing climate. *Conservation Biology, 25,* 56–63. doi:10.1111/j.1523-1739.2010.01552.x

Réale, D., Gallant, B. Y., Leblanc, M., & Festa-Bianchet, M. (2000). Consistency of temperament in bighorn ewesand correlates with behaviour and life history. *Animal Behaviour, 60*(5), 589–597.

Shepherdson, D., Lewis, K. D., Carlstead, K., Bauman, J., & Perrin, N. (2013). Individual and environmental factors associated with stereotypic behavior and fecal glucocorticoid metabolite levels in zoo housed polar bears. *Applied Animal Behaviour Science, 147,* 268–277. doi:10.1016/j.applanim.2013.01.001

Shier, D. M. (2006). Effect of family support on the success of translocated black-tailed prairie dogs. *Conservation Biology, 20,* 1780–1790. doi:10.1111/j.1523-1739.2006.00512.x

Shier, D. M., & Swaisgood, R. R. (2012). Fitness costs of neighborhood disruption in translocations of a solitary mammal. *Conservation Biology, 26,* 116–123. doi:10.1111/j.1523-1739.2011.01748.x

Sih, A., & Watters, J. V. (2005). The mix matters: Behavioural types and group dynamics in water striders.*Behaviour, 142* (9–10), 1417–1431.

Sih, A., Bell, A., & Johnson, J. C. (2004). Behavioral syndromes: An ecological and evolutionary overview. *Trends in Ecology & Evolution, 19,* 372–378. doi:10.1016/j.tree.2004.04.009

Sinn, D. L., Cawthen, L., Jones, S. M., Pukk, C., & Jones, M. E. (2014). Boldness towards novelty and translocation success in captive-raised, orphaned Tasmanian devils. *Zoo Biology, 33,* 36–48.

Smith, B. R., & Blumstein, D. T. (2008). Fitness consequences of personality: A meta-analysis. *Behavioral Ecology, 19,* 448–455. doi:10.1093/beheco/arm144

Stamps, J. A., & Swaisgood, R. R. (2007). Someplace like home: experience, habitat selection and conservationbiology. *Applied Animal Behaviour Science*, 102(3), 392–409.

Stearns, S. C., & Koella, J. C. (1986). The evolution of phenotypic plasticity in life-history traits—Predictions of reaction norms for age and size at maturity. *Evolution, 40,* 893–913. doi:10.2307/2408752

Stratton, T. D. (2015). *Use of personality to improve reintroduction success: the effects of behavioural variation within release groups* (Doctoral dissertation, Nottingham Trent University).

Tetley, C., & O'Hara, S. (2012). Ratings of animal personality as a tool for improving the breeding, management and welfare of zoo mammals. *Animal Welfare-The UFAW Journal, 21,* 463.

Thorpe, J. E. (1994). Salmonid flexibility—Responses to environmental extremes. *Transactions of the American Fisheries Society, 123,* 606–612.

Tofield, S., Coll, R. K., Vyle, B., & Bolstad, R. (2003). Zoos as a source of free choice learning. *Research in Science & Technological Education, 21,* 67–99.

Torgerson-White, L., & Bennett, C. (2014). Rating methodology, personality axes, and behavioral plasticity: A case study in African lions. *Animal Behavior and Cognition, 1,* 230–248.

Tunnicliffe, S. D. (1995). *Talking about animals: Studies of young children visiting zoos, a museum and a farm.* Dissertation submitted to King's College, University of London.

Van Oers, K., De Jong, G., Van Noordwijk, A. J., Kempenaers, B., & Drent, P. J. (2005). Contribution of genetics to the study of animal personalities: a review of case studies. *Behaviour*, 142(9–10), 1185–1206.

Watters, J. V. (2005). Can alternative male tactics 'fighter' and 'sneaker' be considered 'coercer' and 'cooperator' in coho salmon? *Animal Behaviour, 70* (5), 1055–1062.

Watters, J. V., Lema, S. C., & Nevitt, G. A. (2003). Phenotype management: A new approach to habitat restoration. *Biological Conservation, 112,* 435–445. doi:10.1016/s0006-3207(02)00343-9

Watters, J. V., & Meehan, C. L. (2007). Different strokes: Can managing behavioral types increase post-release success? *Applied Animal Behaviour Science, 102,* 364–379. doi:10.1016/j.applanim.2006.05.036

Watters, J. V., & Powell, D. M. (2012). Measuring animal personality for use in population management in zoos: Suggested methods and rationale. *Zoo Biology, 31,* 1–12. doi:10.1002/zoo.20379

Weiss, A., Adams, M. J., & King, J. E. (2011). Happy orang-utans live longer lives. *Biology Letters, 7,* 872–874. doi:10.1098/rsbl.2011.0543

Weiss, A., King, J. E., & Figueredo, A. J. (2000). The heritability of personality factors in chimpanzees (*Pan troglodytes*). *Behavior Genetics, 30,* 213–221.

Weiss, A., King, J. E., & Perkins, L. (2006). Personality and subjective well-being in orangutans (*Pongo pygmaeus* and *Pongo abelii*). *Journal of Personality and Social Psychology, 90,* 501–511. doi:10.1037/0022-3514.90.3.501

Wielebnowski, N. C. (1999). Behavioral differences as predictors of breeding status in captive cheetahs. *Zoo Biology, 18,* 335–349.

Willi, Y., Van Buskirk, J., & Hoffmann, A. A. (2006). Limits to the adaptive potential of small populations. *Annual Review of Ecology, Evolution, and Systematics, 37,* 433–458.

Williams, S. E., & Hoffman, E. A. (2009). Minimizing genetic adaptation in captive breeding programs: A review. *Biological Conservation, 142,* 2388–2400. doi:10.1016/j.biocon.2009.05.034

Wolf, M., van Doorn, G. S., Leimar, O., & Weissing, F. J. (2007). Life-history trade-offs favour the evolution of animal personalities. *Nature, 447,* 581–584. doi:10.1038/nature05835

Wolf, M., & Weissing, F. J. (2012). Animal personalities: Consequences for ecology and evolution. *Trends in Ecology & Evolution, 27,* 452–461.

15 You Are What You Eat: The Interplay Between Animal Personality and Foraging Ecology

Sandra M. Troxell-Smith and Valentina S.A. Mella

Abstract

Foraging animals face a constant dilemma—balancing the need to acquire food without putting themselves at risk. Individuals within a population vary in personality, and as a consequence they differ in how they perceive, interpret, and respond to foraging choices. Despite the inherent necessity for animals to acquire food, and thus direct link to fitness, the interplay between foraging and animal personality has received little attention. In this chapter, we first summarize studies that have integrated animal personality and foraging decisions in wild and captive populations, then highlight research deficiencies and suggest future research opportunities to unite the study of personality and foraging ecology. We also introduce and discuss patch-use theory as a specific research methodology that may help to facilitate future assessment of individual foraging differences. We believe that further investigation into the influence of personality on individual foraging decisions is key to understanding a valuable and largely understudied aspect of overall animal success and the ecological consequences of personality.

Differences in behavior between individuals of the same population have long been noted (e.g., Clark and Ehlinger 1987; Huntingford 1976; van Oortmerssen and Bakker 1981; Wilson 1998). Yet, these differences have often been left unexplained, or attributed simply to statistical noise (Dall et al. 2004; Ruckstuhl and Neuhaus 2002). Recent research suggests that such differences are stable between individuals,

S.M. Troxell-Smith (✉)
Department of Biological Sciences,
University of Illinois at Chicago,
845 W Taylor St (M/C 066), Chicago,
IL 60607, USA
e-mail: sandra.troxell@gmail.com

V.S.A. Mella
School of Life and Environmental Sciences,
The University of Sydney, Sydney,
NSW 2006, Australia

J. Vonk et al. (eds.), *Personality in Nonhuman Animals*,
DOI 10.1007/978-3-319-59300-5_15

representing specific personality traits (Réale et al. 2007). As a consequence, individuals in a population may vary consistently in their responses to the environment (Réale et al. 2007).

Animal personality has been described as consistent inter-individual differences in behavior across time and contexts (Réale et al. 2007), and has recently received increasing attention regarding its influence on the choices of individual animals. Animal personality can be quantified by measuring behavior in standardized conditions and ranking individuals based on their responses (Réale et al. 2007; but see Tetley and O'Hara 2012 for discussion of preferential use of observer/keeper ratings to evaluate personality in zoo-housed species). A behavioral response is considered stable, and therefore deemed to represent a personality trait, when the rank order of individuals is maintained over time, though its absolute value might change (Stamps and Groothuis 2010). Typical personality traits include boldness, exploration, activity, and aggressiveness (Réale et al. 2007).

A behavioral syndrome is generated when individuals within a population vary in a suite of correlated behaviors across functional contexts (Garamszegi et al. 2007). For example, cross-context correlations (e.g., correlations between two measures of the same personality trait quantified in different contexts) may reveal that some individuals are consistently bolder across a range of different situations, and cross-behavior correlations (e.g., correlations between two distinct personality traits) may reveal that some individuals are, for instance, jointly bolder and more aggressive, while others are always shy and less aggressive (see reviews in Bell 2007; Sih et al. 2004a, b, 2012). Some individuals may present a 'proactive' behavioral style (i.e., coping style; Koolhaas et al. 1999): consistently more aggressive, more exploratory, more active, and bolder than 'reactive' individuals, which in turn are less competitive, less exploratory, more passive, and more shy.

Such personality traits can be adaptive and maintained if individuals at the extreme ends of the spectrum have higher fitness than those that are in the middle (Wilson 1998). Thus, personality traits are inherently linked to individual fitness (Réale et al. 2007; Smith and Blumstein 2007) and have important implications for ecological activities. Bold red squirrels (*Tamiasciurus hudsonicus*) occupy a wider range of habitats of varying quality than their shy counterparts (Boon et al. 2008). More exploratory female great tits (*Parus major*) disperse farther upon fledging than less exploratory females (Dingemanse et al. 2003); however, less exploratory females were found to have a higher nest success and larger fledglings (Both et al. 2005). Bold bighorn sheep (*Ovis canadensis*) ewes are less likely to be predated upon by cougars (*Puma concolor*) than shy individuals (Réale and Festa-Bianchet 2003), whereas bold swift fox (*Vulpes velox*) experienced greater mortality when reintroduced into the wild compared to shy individuals (Bremner-Harrison et al. 2004).

Despite the increasing interest in the ecological implications of animal personality, few studies have linked individual personality to inter-individual differences in foraging ecology and decision-making. This is unfortunate; foraging is one of the key behaviors that influence fitness (Dukas 1998). Therefore, investigating and understanding how individual animals make foraging decisions is vital to understanding overall individual fitness.

Foraging Ecology and the Giving-Up Density Framework

Foraging ecology provides a highly useful conceptual framework with which to understand individual animal behavior and decision-making. Optimal foraging theory investigates how animals balance the costs and benefits associated with making foraging decisions (MacArthur and Pianka 1966). Resources are not ubiquitous throughout the environment; instead they typically occur in patches (MacArthur and Pianka 1966). Patch-use theory examines how animals make decisions on when to approach and leave a particular patch. Charnov's (1976) Marginal Value Theorem predicts that an animal will leave a patch when the resource payoff falls below the average rate of return for the surrounding environment. How a forager interprets the resource payoff will depend on how it perceives the reward from the resource, and the costs associated with foraging in that patch. Brown (1988) extended Marginal Value Theorem by identifying three costs associated with foraging: metabolic expenditure from searching for and processing food in addition to thermoregulatory costs (C); missed opportunity costs (MOC, i.e., other non-foraging related activities the animal can perform); and perceived or realized exposure to forms of risk or discomfort (P). These costs can be quantified in the equation: $H= C + MOC + P$, where (H) represents the quitting harvest rate. When an animal's quitting harvest rate (H) no longer exceeds the sum of the other components, the benefits of foraging in the patch no longer exceed the costs, and the forager should leave the patch (Brown 1988).

Unfortunately, the quitting harvest rate (H) can be challenging to measure directly in natural foraging situations. It can be difficult to estimate how a forager will perceive the quality of the food resources in a patch, and estimating the natural food density in the environment prior to a foraging bout is also problematic. Providing an animal with experimental food patches, on the other hand, creates an opportunity for more controlled foraging scenarios and manipulations (Brown 1988). An experimental food patch consists of a container, a known quantity of food appealing to the species of interest, and an inedible substrate. The inedible substrate is necessary to ensure that each piece of food is slightly harder to get than the last (Brown 1988). By measuring the amount of food left over in a patch once the animal stops foraging (the Giving-Up Density or GUD), one gains a surrogate for the quitting harvest rate, which is ultimately a reflection of how an animal perceives the costs of foraging in that environment (Brown 1988). Since its inception in 1988, patch-use and GUD methodology has been widely utilized to investigate how foragers perceive a variety of foraging costs (see Bedoya-Perez et al. 2013). The use of GUD provides a simple and useful measure of foraging efficiency, which can be used to add an invaluable insight into the role of personality in the foraging process of animals.

In general, optimal foraging theory assumes that all individuals should behave in the same (optimal) way; the study of personality shows instead that individuals do not necessarily behave in a single optimal way, but exhibit behavioral differences that reflect alternative adaptive strategies. A recent review presented the idea that personality and individual specialization in the foraging context may covary, and that the first would drive the latter by promoting differences in individual foraging behavior (Toscano et al. 2016). Patch-use and GUD studies have acknowledged the existence of individual variation in foraging outcomes, but direct links to personality have been understudied (but see Mella et al. 2015). For example, Troxell-Smith et al. (2016) used GUDs to compare the foraging abilities and aptitudes of both domestic and wild house mice (*Mus musculus*) in a laboratory setting. Whereas the domestic mice foraged more effectively overall compared to wild mice, significant individual variation in resulting foraging effort (GUDs) within each mouse strain was also discovered that could not be attributed to differences in physiological factors alone. Similarly, another recent study used GUDs to determine the exhibit preferences (or landscape of comfort) of zoo-housed okapi (*Okapia johnstoni*). Individuals varied greatly in their preferences for, and overall use of, specific foraging locations throughout the space, demonstrating that each animal perceived the same exhibit space differently (Troxell-Smith et al. 2017). While individual animal personality was not directly assessed in either study, the presence of individual variation does suggest that the differences in foraging effort may reflect personality differences.

Other factors not related to personality may also affect individual foraging decisions. For example, individual variation in performance in patch-use studies has been previously attributed to the physiological state of the forager, including satiation level, developmental stage, and reproductive state (Bedoya-Perez et al. 2013). For example, individuals that are well-fed will tend to avoid risky foraging situations (leave higher GUDs) compared to those that are food-deprived (Altendorf et al. 2001; Brown et al. 1992; Lima 1998), demonstrating that the value of the food for a hungry animal is high enough to outweigh the risk of predation (Berger-Tal et al. 2010; Sánchez et al. 2008). Another study found that individuals with high ectoparasite loads spend less time foraging and more time engaged in parasite removal compared to those with lower ectoparasite loads, resulting in higher GUDs (Raveh et al. 2011; Webster et al. 2007). Likewise, immunochallenged individuals are willing to forage more and to take more foraging risks than individuals that are not immunochallenged (Schwanz et al. 2012).

The abovementioned physiological factors can easily be controlled in captive experiments. However, individual personality may bias foraging decisions above and beyond a forager's physiological state. In fact, differences in physiological traits may be linked to both individual personality and foraging demands. For example, highly active common mud crabs (*Panopeus herbstii*, Toscano and Monaco 2015), and bold mealworm beetles (*Tenebrio molitor*, Krams et al. 2013) have higher resting metabolic rates compared to their less active and less bold counterparts. As a consequence, these individuals may require increased foraging activity to sustain and support their greater energetic demands ('increased intake model;' Careau et al. 2008). Energetically expensive behaviors require higher metabolic rates to be supported (Biro and Stamps 2008), and thus, individuals with bold or highly active/exploratory personality traits may also require increased foraging activity to sustain greater energetic demands (Careau and Garland 2012). Studies utilizing a patch-use/GUD paradigm to quantify individual foraging effort can provide valuable insight into the connections between physiology, personality, and foraging decisions.

Personality and Foraging Ecology: Studies in Wild Species

It is well established that differences in the foraging strategies of individuals that are not attributable to differences in age, sex, size, or environmental effects exist in the wild (e.g., Baird et al. 1992; Werner and Sherry 1987). For example, sea otters (*Enhydra lutris*) from the same population show extreme inter-individual variation in diet composition, which persists throughout life and cannot be explained by environmental factors (Estes et al. 2003). Similar inter-individual differences in foraging behavior have been observed in other free-living populations, such as Brünnich's guillemots (*Uria lomvia*, Woo et al. 2008) and northern gannets (*Morus bassanus*, Patrick and Weimerskirch 2014). A study by Potier et al. (2015) showed that the apparent flexibility in foraging behavior of great cormorants (previously explored only at the population level), is actually due to a wide range of individual behavior repeatability, suggesting the existence of different individual foraging profiles. However, research to date has primarily evaluated personality in individuals temporarily brought into captivity, and only a few studies have directly explored the effects of personality on foraging behavior in wild species.

The existence of different personalities in a population implies that individuals vary consistently in where and how they forage in their natural environment, generating variation in the use of feeding opportunities. These differences in foraging strategies are adaptive and likely arise to reduce competition for resources. Bold mud crabs (*Panopeus herbstii*) (as assessed by behavioral responses to cues of predation) inhabit subtidal parts of reefs, where foraging opportunities are better, whereas shy crabs tend to inhabit the safe intertidal parts of the reef, where exposure to predators is low (Griffen et al. 2012). Fast-exploring free-ranging great tits (*Parus major*), that were tested in a novel environment, were found to travel farther distances in search of food than slow-exploring birds when challenged by a sudden removal in food supply at feeders in the wild (van Overveld and Matthysen 2010). Slow-exploring individuals continued to visit known food patches even following food removal, while fast-exploring individuals quickly sought new patches (Arvidsson and Matthysen 2016). Similarly, exploratory tendency in blue tits (*Cyanistes caeruleus*) is linked to their ability of finding new feeding sites in the wild (Herborn et al. 2010). Foraging success in a social context can also be highly influenced by personality. Following screening for boldness using responses to a simulated predation event, social groups consisting of both bold and shy guppies (*Poecilia reticulata*) approached a novel feeder more quickly, and had overall greater foraging success, than groups of either just bold or just shy individuals (Dyer et al. 2009). So the personality of individuals in a population can be related to different strategies employed in the foraging context and, therefore, to consistent individual differences in food resource use.

The consistent variation in foraging behavior observed in wild individuals of the same population is in some instances shaped early in life. General activity and space use in a novel environment of wild juvenile brook charr (*Salvelinus fontinalis*), which determine a proactive and a reactive personality type, are associated with the food searching tactics (i.e., active search or sit-and-wait, respectively) that are adopted and maintained as adults (Wilson and McLaughlin 2007). A study on foraging skills of juvenile meerkats (*Suricata suricatta*) found that early-life foraging efficiency of individuals is maintained later in life. Individuals that were willing to find food on their own as pups had higher immediate foraging efficiency than individuals that were begging for food from helpers (Thornton 2008). As a consequence, 'beggars' had lower body conditions as adults. However, when the body weight of these individuals was manipulated through long-term food provisioning, foraging efficiency for 'beggars' remained surprisingly unvaried. This seems to indicate that the two foraging strategies adopted by juvenile meerkats (i.e., self-feeding and begging) may be permanent and hence reflect consistent differences maintained during the life of individuals.

If wild individuals with different personalities differ in their foraging decisions and efficiency, this can result in differential foraging outcomes, with possible important fitness consequences. Hence, consistent individual differences in foraging behavior can, in some instances, seemingly represent a constraint preventing the achievement of an optimal strategy. So why do they exist? Patrick and Weimerskirch (2014) showed that differences in foraging habits of wild black browed albatross (*Thalassarche melanophrys*) are related to individual differences in personality. Bird personality was assessed using responses to a novel object. It was then discovered that bolder birds feed in shallow waters, where competitive interactions are substantial, whereas shy individuals forage in deep areas of the ocean. Although shy males always have higher breeding success, fitness benefits for females depend on year quality, with bold females having higher fitness in poor quality years, when competition for food is high. Therefore, the fitness benefit of different personality types changes depending on environmental conditions. As these conditions vary between years, different personality types will be favored in different years. This explains why different personality types are maintained within the same population.

In the context of wild foraging animals faced with the trade-off between food and fear, differences between individuals of the same population suggest that personality may provide alternative ways of solving the foraging problem. The first study to provide evidence of how personality affects foraging decisions of individuals relating to both food quality and predation risk in a natural context was conducted on a generalist herbivore, the common brushtail possum (*Trichosurus vulpecula*). Mella et al. (2015) quantified the GUDs of free-ranging brushtail possums using artificial feeding patches in the wild, in response to altered food quality and level of predation risk. Brushtail possums were offered a choice between nontoxic food at risky patches (ground feeders) paired with increasingly more toxic food at safe patches (above-ground feeders). Results showed that boldness (quantified in captivity via open-field tests) influences foraging in response to food toxins and predation risk, and that these two costs are perceived differently by individuals with different personalities, so that each will perceive its own individualized landscape of food and fear (as suggested by McArthur et al. 2014). As a consequence, individuals' foraging behavior depended on their personalities, and this resulted in diverse foraging outcomes between bold and shy animals. Bold individuals spread their foraging time among nontoxic feeders, using both risky and safe ones, while shy animals mainly concentrated their foraging at safe, nontoxic patches. In general, shy possums had higher foraging efficiency (foraging time per food consumed) than bold possums. At risky-feeders, shy possums achieved the same foraging outcome (GUD) as bold possums, but at safe, nontoxic feeders, they harvested more food (lower GUD) than bold individuals. The most likely explanation for these results is that shy possums explore foraging patches quickly (but efficiently) to avoid competition with bolder conspecifics (Mella et al. 2015). Therefore, individuals from the same population with different personalities may use alternative (but equally successful) strategies to deal with the trade-off between food quality and predation risk. The lack of literature linking personality with foraging behavior highlights the need of more studies on wild animals to understand the extent of personality-related variation in foraging behavior.

Personality and Foraging Ecology: Studies in Captive Species

Animals housed in captivity (including domestic, laboratory, and zoo animals) provide a unique, and largely untapped, opportunity to concurrently study personality and foraging decisions. Daily situations and random events are more easily regulated and controlled for in captivity compared to similar situations in the wild, and there is an added benefit of always knowing where your study species and individuals are located. Moreover, for species living in captivity, understanding how individuals vary in their response to foraging opportunities can have important implications for the assessment, improvement, and overall understanding of animal welfare. Several studies have taken advantage of the control captivity provides when conducting personality studies, and have investigated the impacts of personality on various aspects of the lives of animals. However, research directly linking animal personality to foraging decisions has received far less attention.

The few studies that have investigated both personality and foraging ecology in captive animals have largely focused on how differences in personality (particularly measures of boldness) influence foraging decisions in a social context. For instance, Bergvall et al. (2011) measured three personality traits (boldness, dominance, and flexibility) in tame fallow deer (*Dama dama*) using reactions to novel objects. They further investigated how these traits influenced foraging decisions in three foraging scenarios: familiar food was presented in a familiar context, familiar food presented in a novel context, and novel food presented in a familiar context. Results demonstrate that bold individuals were more likely to consume novel food in familiar situations, and familiar food in novel situations compared to shy individuals. Such results demonstrate that personality influences neophobic responses in the foraging context, which may be adaptive in an environment with fluctuating resources.

Boldness has also been shown to influence the foraging tactics of domestic barnacle geese (*Branta leucopsis*). Kurvers et al. (2010) identified individual geese as bold, shy or intermediate based on reactions to novel object tests. In a social producer–scrounger paradigm, groups of two individuals (one bold, one shy or two intermediate) were given the opportunity to find food patches (producing) or join patches (scrounging). Shy individuals were more likely to scrounge than bold individuals, whereas bold individuals tended to spend more time producing than shy individuals. These results demonstrate that choice of foraging tactics in social scenarios can be directly linked to individual personality. In an additional study, Kurvers et al. (2012) found that bold geese were more willing to explore food patches located away from conspecifics, whereas shy individuals took longer to approach food patches when far from conspecifics, or when foraging alone, suggesting that personality influences sensitivity to social foraging cues.

Similar effects of boldness and shyness during social and competitive foraging has also been quantified in domestic sheep (*Ovis aries*, Michelena et al. 2009, 2010). Individual sheep were first classified as bold or shy as a result of their willingness to leave a social group to explore a novel environment. In a subsequent social foraging experiment, bold individuals split into subgroups at smaller group numbers than shy individuals, demonstrating that boldness influences willingness to move away from conspecifics and explore new areas (Michelena et al. 2009). Furthermore, the coexistence of bold and shy individuals within a population can influence collective decision-making while foraging. A model based on experimental data collected on sheep suggests that different foraging tactics may arise within a social group mainly because of differences in conspecific social attraction (Michelena et al. 2010). Support for this theory also comes from a study on pumpkinseed fish (*Lepomis gibbosus*). Bold individuals, which fed more quickly and to a greater degree than shy individuals, were less likely to swim in close proximity to other fish, suggesting that foraging tactics in social species may be related to individual personality traits (Wilson et al. 1993).

A recent study in laboratory-housed common marmosets (*Callithrix jacchus*) suggests that both boldness and exploration influence foraging decisions in a captive context (Šlipogor et al. 2016). Marmoset personality was established by individual responses to a series of five novel object/novel situation tests. When presented with a valuable food in a novel and therefore potentially threatening situation, bold marmosets spent more time in proximity to the threatening stimuli to gain access to a food reward, whereas shy individuals spent more time vigilant and took longer to approach the threatening stimulus. Furthermore, when provided with access to a novel food item, more exploratory marmosets approached the novel food resources faster than their 'avoidant' counterparts (Šlipogor et al. 2016), further substantiating these personality traits (i.e., boldness and exploration) as important drivers in foraging decisions.

As highlighted above, most studies on animal personality and foraging ecology in captive animals to date have been performed on domestic and laboratory species. Zoo populations, however, also present an opportunity to investigate individual differences in foraging. As with domestic and laboratory animals, zoo conditions are highly controlled, and provide an opportunity to easily observe animals in a variety of situations over time. Moreover, quantification of animal personality can have a direct impact on care and welfare (Tetley and O'Hara 2012; Watters and Powell 2011), as it can directly influence how individuals respond to new environments, environmental change, and interactions with both conspecifics and individuals of other species (Powell and Gartner 2011). As a result, studies investigating personality in zoos are not novel. However, existing studies have focused on relating individual differences in behavior to aspects of animal physiology, such as reproductive success (Powell and Svoke 2008; Razal et al. 2016; Wielebnowski 1999) and stress responses (Grand et al. 2012; Shepherdson et al. 1993). Additional studies have investigated how personality influences housing success (Kuhar et al. 2006), and general aspects of welfare, well-being, and husbandry (Baker and Pullen 2013; Gartner et al. 2016; Powell and Gartner 2011; Schaefer and Steklis 2014). Still, little consideration has been given to how animal personality influences other day-to-day aspects of life for zoo animals, particularly in relation to foraging choices. In fact, foraging ecology has received little attention from zoo researchers, save for studies focused on the efficacy of foraging enrichment (i.e., providing captive animals with additional opportunities and devices designed to increase foraging times, Young 2003; but see Rapaport 1998). This lack of research may be due to the assumption that foraging choices in zoos are already too limited to study effectively, as animals typically receive a far more limited dietary selection than experienced in the wild (Newberry 1995), and in some cases may be fed to satiety. Also, the factors that largely motivate foraging decisions in the wild (e.g., predation risk, required search and handling time, toxicity of diet, etc.) are largely reduced or absent in captive situations (McPhee and Carlstead 2010; Mogerman 2011), thus reducing the time and energy investment required for zoo animals to find food.

Nevertheless, understanding how zoo animals approach and respond to foraging challenges has the potential to provide valuable information about how these captive animals interpret their environment. Moreover, integrating foraging ecology and personality to better understand how individuals differ in their interpretation of foraging challenges can greatly contribute to individual-based welfare decisions. For example, for socially foraging species, boldness (Rudin and Briffa 2012), exploration (Favati et al. 2014) and aggressiveness (Wilson et al. 2013) have all been found to influence position in dominance hierarchies, which can determine and possibly limit access to food (Gende and Quinn 2004; Hansen and Closs 2005). Overall willingness to forage in social groups (González-Bernal et al. 2014; Michelena et al. 2009), and explore and utilize novel foraging locations (van Overveld and Matthysen 2010) have also been linked to personality traits, which could ultimately impact utilization of zoo exhibit space. As ensuring proper nutrition and caloric intake is a fundamental aspect of captive animal husbandry and care, conducting studies that integrate personality and foraging success in zoos can aid in informing care and management decisions.

Conclusion

In this chapter we have reviewed the existing, albeit limited, literature relating personality to foraging decisions in wild and captive populations, and have highlighted the largely untapped research potential of investigating the effects of animal personality on foraging decisions. Foraging ecology provides a useful framework to study and understand how personality variation may be maintained in populations. Furthermore, utilizing patch-use methodology to explore these connections can facilitate the investigation of relationships between animal personality and individual foraging strategies and outcomes. We therefore urge researchers to pursue more studies combining foraging ecology and personality to deepen our understanding of how animals decide where, and how much, to forage in captive and wild scenarios. We recommend continuing to investigate personality and foraging in the wild (e.g., Mella et al. 2016), as ecological validations of captive personality studies are still scarce (but see Boon et al. 2008; Dammhahn and Almeling 2012; Dingemanse et al. 2003; Mella et al. 2015). This step is crucial in advancing our knowledge of the role of personality traits in foraging ecology in natural contexts. Similarly, we encourage increased utilization of captive, and especially zoo, populations in studies incorporating foraging ecology and personality. Such studies will not only provide controlled and easily managed research scenarios for investigators, but the results gained have the potential to drastically improve animal health and welfare.

References

Altendorf, K. B., Laundré, J. W., Lopez Gonzalez, C. A., & Brown, J. S. (2001). Assessing effects of predation risk on foraging behavior of mule deer. *Journal of Mammalogy, 82,* 430–439.

Arvidsson, L. K., & Matthysen, E. (2016). Individual differences in foraging decisions: Information-gathering strategies or flexibility? *Behavioral Ecology, 27,* 1353–1361.

Baird, R. W., Abrams, P. A., & Dill, L. M. (1992). Possible indirect interactions between transient and resident killer whales: Implications for the evolution of foraging specializations in the genus *Orcinus*. *Oecologia, 89,* 125–132.

Baker, K., & Pullen, P. K. (2013). The impact of housing and husbandry on the personality of cheetah (*Acinonyx jubatus*). *Journal of Zoo and Aquarium Research, 1,* 35–40.

Bedoya-Perez, M. A., Carthey, A. J. R., Mella, V. S. A., McArthur, C., & Banks, P. B. (2013). A practical guide to avoid giving up on giving-up densities. *Behavioral Ecology and Sociobiology, 67,* 1541–1553.

Bell, A. M. (2007). Future directions in behavioural syndromes research. *Proceedings of the Royal Society of London B: Biological Sciences, 274,* 755–761.

Berger-Tal, O., Mukherjee, S., Kotler, B. P., & Brown, J. S. (2010). State of emergency: Behavior of gerbils is affected by the hunger state of their predators. *Ecology, 91,* 593–600.

Bergvall, U. A., Schäpers, A., Kjellander, P., & Weiss, A. (2011). Personality and foraging decisions in fallow deer, *Dama dama*. *Animal Behaviour, 81,* 101–112.

Biro, P. A., & Stamps, J. A. (2008). Are animal personality traits linked to life-history productivity? *Trends in Ecology & Evolution, 23,* 361–368.

Boon, A., Reale, D., & Boutin, S. (2008). Personality, habitat use, and their consequences for survival in North American red squirrels Tamiasciurus hudsonicus. *Oikos, 117,* 1321–1328.

Both, C., Dingemanse, N. J., Drent, P. J., & Tinbergen, J. M. (2005). Pairs of extreme avian personalities have highest reproductive success. *Journal of Animal Ecology, 74,* 667–674.

Bremner-Harrison, S., Prodohl, P. A., & Elwood, R. W. (2004). Behavioural trait assessment as a release criterion: Boldness predicts early death in a reintroduction programme of captive-bred swift fox (*Vulpes velox*). *Animal Conservation, 7,* 313–320.

Brown, J. (1988). Patch use as an indicator of habitat preference, predation risk, and competition. *Behavioral Ecology and Sociobiology, 22,* 37–47.

Brown, J. S., Morgan, R. A., & Dow, B. D. (1992). Patch use under prédation risk: A test with fox squirrels, Sciurus niger. *Annales Zoologici Fennici, 29,* 311–318.

Careau, V., & Garland, T., Jr. (2012). Performance, personality, and energetics: Correlation, causation, and mechanism. *Physiological and Biochemical Zoology, 85,* 543–571.

Careau, V., Thomas, D., Humphries, M., & Réale, D. (2008). Energy metabolism and animal personality. *Oikos, 117,* 641–653.

Charnov, E. L. (1976). Optimal foraging, the marginal value theorem. *Theoretical Population Biology, 9,* 129–136.

Clark, A. B., & Ehlinger, T. J. (1987). Pattern and adaptation in individual behavioral differences. In P. P. G. Bateson & P. Klopfer (Eds.), *Perspectives in Ethology* (Vol. 7, pp. 1–47). Springer US.

Dall, S. R. X., Houston, A. I., & McNamara, J. M. (2004). The behavioural ecology of personality: Consistent individual differences from an adaptive perspective. *Ecology Letters, 7,* 734–739.

Dammhahn, M., & Almeling, L. (2012). Is risk taking during foraging a personality trait? A field test for cross-context consistency in boldness. *Animal Behaviour, 84,* 1131–1139.

Dingemanse, N. J., Both, C., van Noordwijk, A. J., Rutten, A. L., & Drent, P. J. (2003). Natal dispersal and personalities in great tits (*Parus major*). *Proceedings of the Royal Society of London B: Biological Sciences, 270,* 741–747.

Dukas, R. (1998). Evolutionary ecology of learning. In R. Dukas (Ed.), *Cognitive ecology: The evolutionary ecology of information processing and decision making* (pp. 129–164). University of Chicago Press.

Dyer, J. R. G., Croft, D. P., Morrell, L. J., & Krause, J. (2009). Shoal composition determines foraging success in the guppy. *Behavioral Ecology, 20,* 165–171.

Estes, J. A., Riedman, M. L., Staedler, M. M., Tinker, M. T., & Lyon, B. E. (2003). Individual variation in prey selection by sea otters: Patterns, causes and implications. *Journal of Animal Ecology, 72,* 144–155.

Favati, A., Leimar, O., & Løvlie, H. (2014). Personality predicts social dominance in male domestic fowl. *PLoS ONE, 9,* 1–7.

Garamszegi, L. Z., Erritzøe, J., & Møller, A. P. (2007). Feeding inovations and parasitism in birds. *Biological Journal of the Linnean Society, 90,* 441–455.

Gartner, M. C., Powell, D. M., & Weiss, A. (2016). Comparison of subjective well being and personality assessments in the Clouded Leopard (*Neofelis nebulosa*), Snow Leopard (*Panthera uncia*), and African Lion (*Panthera leo*). *Journal of Applied Animal Welfare Science, 19,* 294–302.

Gende, S. M., & Quinn, T. P. (2004). The relative importance of prey density and social dominance in determining energy intake by bears feeding on Pacific salmon. *Canadian Journal of Zoology, 82,* 75–85.

González-Bernal, E., Brown, G. P., & Shine, R. (2014). Invasive cane toads: Social facilitation depends upon an individual's personality. *PLoS ONE, 9,* e102880.

Grand, A. P., Kuhar, C. W., Leighty, K. A., Bettinger, T. L., & Laudenslager, M. L. (2012). Using personality ratings and cortisol to characterize individual differences in African Elephants (Loxodonta africana). *Applied Animal Behaviour Science, 142,* 69–75.

Griffen, B. D., Toscano, B. J., & Gatto, J. (2012). The role of individual behavior type in mediating indirect interactions. *Ecology, 93,* 1935–1943.

Hansen, E. A., & Closs, G. P. (2005). Diel activity and home range size in relation to food supply in a drift-feeding stream fish. *Behavioral Ecology, 16,* 640–648.

Herborn, K. A., Macleod, R., Miles, W. T. S., Schofield, A. N. B., Alexander, L., & Arnold, K. E. (2010). Personality in captivity reflects personality in the wild. *Animal Behaviour, 79,* 835–843.

Huntingford, F. A. (1976). The relationship between anti-predator behaviour and aggression among conspecifics in the three-spined stickleback, *Gasterosteus aculeatus*. *Animal Behaviour, 24,* 245–260.

Koolhaas, J. M., Korte, S. M., De Boer, S. F., Van Der Vegt, B. J., Van Reenen, C. G., Hopster, H., et al. (1999). Coping styles in animals: Current status in behavior and stress- physiology. *Neuroscience and Biobehavioral Reviews, 23,* 925–935.

Krams, I., Kivleniece, I., Kuusik, A., Krama, T., Freeberg, T. M., Mänd, R., et al. (2013). Predation selects for low resting metabolic rate and consistent individual differences in anti-predator behavior in a beetle. *Acta Ethologica, 16,* 163–172.

Kuhar, C. W., Stoinski, T. S., Lukas, K. E., & Maple, T. L. (2006). Gorilla Behavior Index revisited: Age, housing and behavior. *Applied Animal Behaviour Science, 96,* 315–326.

Kurvers, R. H. J. M., Nolet, B. A., Prins, H. H. T., Ydenberg, R. C., & Van Oers, K. (2012). Boldness affects foraging decisions in barnacle geese: An experimental approach. *Behavioral Ecology, 23,* 1155–1161.

Kurvers, R. H. J. M., Prins, H. H. T., van Wieren, S. E., van Oers, K., Nolet, B. A., & Ydenberg, R. C. (2010). The effect of personality on social foraging: Shy barnacle geese scrounge more. *Proceedings of the Royal Society of London B: Biological Sciences, 277,* 601–608.

Lima, S. L. (1998). Stress and decision making under the risk of predation: Recent developments from behavioral, reproductive, and ecological perspectives. *Advances in the Study of Behavior, 27,* 215–290.

MacArthur, R., & Pianka, E. (1966). On optimal use of a patchy environment. *American Naturalist, 100,* 603–609.

McArthur, C., Banks, P. B., Boonstra, R., & Forbey, J. S. (2014). The dilemma of foraging herbivores: Dealing with food and fear. *Oecologia, 176,* 677–689.

McPhee, M. E., & Carlstead, K. (2010). The importance of maintaining natural behaviors in captive mammals. In D. G. Kleiman, K. V. Thompson, & C. Kirk Baer (Eds.), *Wild mammals in captivity: Principles and techniques for zoo management* (2nd Ed., pp. 303–313). University of Chicago Press.

Mella, V. S. A., Krucler, J., Hawkins, J., Herath, A. P. H. M., Johnstone, K. C., Troxell-Smith, S. M., et al. (2016). Effective field-based methods to quantify personality in brushtail possus (*Trichosurus vulpecula*). *Wildlife Research, 43,* 332–340.

Mella, V. S. A., Ward, A. J. W., Banks, P. B., & McArthur, C. (2015). Personality affects the foraging response of a mammalian herbivore to the dual costs of food and fear. *Oecologia, 177,* 293–303.

Michelena, P., Jeanson, R., Deneubourg, J.-L., & Sibbald, A. M. (2010). Personality and collective decision-making in foraging herbivores. *Proceedings of the Royal Society of London B: Biological Sciences, 277,* 1093–1099.

Michelena, P., Sibbald, A. M., Erhard, H. W., & McLeod, J. E. (2009). Effects of group size and personality on social foraging: The distribution of sheep across patches. *Behavioral Ecology, 20,* 145–152.

Mogerman, J. E. E. H. (2011). Zoo foraging ecology: Patch use and giving-up densities as tools for animal care. Ph. D, dissertation, University of Illinois at Chicago.

Newberry, R. C. (1995). Environmental enrichment: Increasing the biological relevance of captive environments. *Applied Animal Behaviour Science, 44,* 229–243.

Patrick, S. C., & Weimerskirch, H. (2014). Personality, foraging and fitness consequences in a long lived seabird. *PLoS ONE, 9,* e87269.

Potier, S., Carpentier, A., Grémillet, D., Leroy, B., & Lescroël, A. (2015). Individual repeatability of foraging behaviour in a marine predator, the great cormorant, *Phalacrocorax carbo. Animal Behaviour, 103,* 83–90.

Powell, D. M., & Gartner, M. C. (2011). Applications of personality to the management and conservation of nonhuman animals. In M. Inoue-Murayama, S. Kawamura, & A. Weiss (Eds.), *From genes to animal behavior* (pp. 185–199). Springer Japan.

Powell, D. M., & Svoke, J. T. (2008). Novel environmental enrichment may provide a tool for rapid assessment of animal personality: A case study with giant pandas (*Ailuropoda melanoleuca*). *Journal of Applied Animal Welfare Science, 11,* 301–318.

Rapaport, L. (1998). Optimal foraging theory predicts effects of environmental enrichment in a group of adult golden lion tamarins. *Zoo Biology, 17,* 231–244.

Raveh, A., Kotler, B. P., Abramsky, Z., & Krasnov, B. R. (2011). Driven to distraction: Detecting the hidden costs of flea parasitism through foraging behaviour in gerbils. *Ecology Letters, 14,* 47–51.

Razal, C. B., Pisacane, C. B., & Miller, L. J. (2016). Multifaceted approach to personality assessment in cheetahs (*Acinonyx jubatus*). *Animal Behavior and Cognition, 3,* 22–31.

Réale, D., & Festa-Bianchet, M. (2003). Predator-induced natural selection on temperament in bighorn ewes. *Animal Behaviour, 65,* 463–470.

Réale, D., Reader, S. M., Sol, D., McDougall, P. T., & Dingemanse, N. J. (2007). Integrating animal temperament within ecology and evolution. *Biological Reviews of the Cambridge Philosophical Society, 82,* 291–318.

Ruckstuhl, K. E., & Neuhaus, P. (2002). Sexual segregation in ungulates: A comparative test of three hypotheses. *Biology Review, 77,* 77–96.

Rudin, F. S., & Briffa, M. (2012). Is boldness a resource-holding potential trait? Fighting prowess and changes in startle response in the sea anemone, *Actinia equina. Proceedings of the Royal Society of London B: Biological Sciences, 279,* 1904–1910.

Sánchez, F., Korine, C., Kotler, B. P., & Pinshow, B. (2008). Ethanol concentration in food and body condition affect foraging behavior in Egyptian fruit bats (*Rousettus aegyptiacus*). *Naturwissenschaften, 95,* 561–567.

Schaefer, S. A., & Steklis, H. D. (2014). Personality and subjective well-being in captive male western lowland gorillas living in bachelor groups. *American Journal of Primatology, 76,* 879–889.

Schwanz, L. E., Previtali, M. A., Gomes-Solecki, M., Brisson, D., & Ostfeld, R. S. (2012). Immunochallenge reduces risk sensitivity during foraging in white-footed mice. *Animal Behaviour, 83,* 155–161.

Shepherdson, D. J., Carlstead, K., Mellen, J. D., & Seidensticker, J. (1993). The influence of food presentation on the behavior of small cats in confined environments. *Zoo Biology, 12,* 203–216.

Sih, A., Bell, A., & Johnson, J. C. (2004a). Behavioral syndromes: An ecological and evolutionary overview. *Trends in Ecology & Evolution, 19,* 372–378.

Sih, A., Bell, A. M., Johnson, J. C., & Ziemba, R. E. (2004b). Behavioral syndromes: An integrative overview. *The Quarterly Review of Biology, 79,* 241–277.

Sih, A., Cote, J., Evans, M., Fogarty, S., & Pruitt, J. (2012). Ecological implications of behavioural syndromes. *Ecology Letters, 15,* 278–289.

Šlipogor, V., Gunhold-de Oliveira, T., Tadić, Z., Massen, J. J. M., & Bugnyar, T. (2016). Consistent inter-individual differences in common marmosets (*Callithrix jacchus*) in boldness–shyness, stress–activity, and exploration–avoidance. *American Journal of Primatology, 78,* 961–973.

Smith, B. R., & Blumstein, D. T. (2007). Fitness consequences of personality: A meta-analysis. *Behavioral Ecology, 19,* 448–455.

Stamps, J. A., & Groothuis, T. G. G. (2010). Developmental perspectives on personality: Implications for ecological and evolutionary studies of individual differences. *Philosophical Transactions of the Royal Society of London. Series B, Biological Sciences, 365,* 4029–4041.

Tetley, C. L., & O'Hara, S. J. (2012). Ratings of animal personality as a tool for improving the breeding, management and welfare of zoo mammals. *Animal Welfare, 21,* 463–476.

Thornton, A. (2008). Early body condition, time budgets and the acquisition of foraging skills in meerkats. *Animal Behaviour, 75,* 951–962.

Toscano, B. J., Gownaris, N. J., Heerhartz, S. M., & Monaco, C. J. (2016). Personality, foraging behavior and specialization: Integrating behavioral and food web ecology at the individual level. *Oecologia, 182,* 55–69.

Toscano, B. J., & Monaco, C. J. (2015). Testing for relationships between individual crab behavior and metabolic rate across ecological contexts. *Behavioral Ecology and Sociobiology, 69,* 1343–1351.

Troxell-Smith, S. M., Tutka, M. J., Albergo, J. M., Balu, D., Brown, J. S., & Leonard, J. P. (2016). Foraging decisions in wild versus domestic *Mus musculus*: What does life in the lab select for? *Behavioural Processes, 122,* 43–50.

Troxell-Smith, S. M., Watters, J. V., Whelan, C. J., & Brown, J. S. (2017). Zoo foraging ecology: Preference and welfare assessment of two okapi (*Okapia johnstoni*) at the Brookfield Zoo. *Animal Behavior and Cognition, 4,* 187–199.

van Oortmerssen, G. A., & Bakker, T. C. M. (1981). Artificial selection for short and long attack latencies in wild *Mus musculus* domesticus. *Behavior Genetics, 11,* 115–126.

van Overveld, T., & Matthysen, E. (2010). Personality predicts spatial responses to food manipulations in free-ranging great tits (*Parus major*). *Biology Letters, 6,* 187–190.

Watters, J. V., & Powell, D. M. (2011). Measuring animal personality for use in population management in zoos: Suggested methods and rationale. *Zoo Biology, 29,* 1–12.

Webster, S. J., Dill, L. M., & Butterworth, K. (2007). The effect of sea lice infestation on the salinity preference and energetic expenditure of juvenile pink salmon (*Oncorhynchus gorbuscha*). *Canadian Journal of Fisheries and Aquatic Sciences, 64,* 672–680.

Werner, T. K., & Sherry, T. W. (1987). Behavioral feeding specialization in *Pinaroloxias inornata*, the "Darwin's Finch" of Cocos Island, Costa Rica. *Proceedings of the National Academy of Sciences of the United States of America, 84,* 5506–5510.

Wielebnowski, N. (1999). Behavioral differences as predictors of breeding status in captive cheetahs. *Zoo Biology, 18,* 335–349.

Wilson, A. D. M., & McLaughlin, R. L. (2007). Behavioural syndromes in brook charr, *Salvelinus fontinalis*: Prey-search in the field corresponds with space use in novel laboratory situations. *Animal Behaviour, 74,* 689–698.

Wilson, A. J., Grimmer, A., & Rosenthal, G. G. (2013). Causes and consequences of contest outcome: Aggressiveness, dominance and growth in the sheepshead swordtail, *Xiphophorus birchmanni*. *Behavioral Ecology and Sociobiology, 67,* 1151–1161.

Wilson, D. S. (1998). Adaptive individual differences within single populations. *Philosophical Transactions of the Royal Society B: Biological Sciences, 353,* 199–205.

Wilson, D. S., Coleman, K., Clark, A. B., & Biederman, L. (1993). Shy-bold continuum in pumpkinseed sunfish (*Lepomis gibbosus*): An ecological study of a psychological trait. *Journal of Comparative Psychology, 107,* 250–260.

Woo, K. J., Elliott, K. H., Davidson, M., Gaston, A. J., & Davoren, G. K. (2008). Individual specialization in diet by a generalist marine predator reflects specialization in foraging behaviour. *Journal of Animal Ecology, 77,* 1082–1091.

Young, R. J. (2003). Food and foraging enrichment. In *Environmental enrichment for captive animals* (pp. 85–106). Blackwell Publishing.

Applications of Research in Nonhuman Animal Personality

16

Carly Loyer and James Ha

Abstract

Nonhuman animal personality research has increased dramatically in recent years and encompasses a vast range of species, traits examined, and methods of measurement. To make use of this research, it is necessary to understand how various approaches interact with and complement each other, as well as examine the challenges posed for researchers in this field. In this chapter we describe a variety of ways nonhuman animal research has been or has the potential to be applied to current problems faced by both nonhuman animals and humans. The potential applications for this field of research are as far-reaching as the approaches used to examine nonhuman animal personality.

Introduction

At different points in history it has been declared that only humans have emotions, use tools, or have a concept of "self", only to have researchers discover at least one exception to each of these rules shortly after they began looking for them. Charles Darwin's book "The Expression of Emotions in Man and Animals" documents multiple comparisons between human and nonhuman emotional responses. Rats (*Rattus norvegicus*) and dogs (*Canis lupus familiaris*) have been found to have specific ways of breathing or vocalizing that are akin to the human expression of laughter (Panksepp 2000, 2007; Panksepp and Burgdorf 2003; Simonet et al. 2005). New Caledonian crows (*Corvus moneduloides*) do not only use tools, they *create* them, modifying the sticks they use as "grub extractors", to improve their usefulness (Hunt and Gray 2004; Weir et al. 2002). Recognition of "self" in the ubiquitous mirror task has been observed in species ranging from chimpanzees (*Pan troglodytes*, Gallup 1970) to elephants (*Elephas maximus*, Plotnik et al. 2006). As each of these distinctions has been refuted, the line between humans and the rest of the animal kingdom has been blurred further.

C. Loyer · J. Ha (✉)
Department of Psychology, University of Washington, Box 351525, Seattle, WA 98195-1525, USA
e-mail: jcha@u.washington.edu

J. Vonk et al. (eds.), *Personality in Nonhuman Animals*,
DOI 10.1007/978-3-319-59300-5_16

Although nonhuman animals (henceforth: NHA) are used as models for humancentric psychological and medical treatment problems ranging from post-traumatic stress disorder, parental attachment, and autism to cancer treatment and medical emergencies, until the last few decades it was generally considered overly anthropomorphic to focus on NHA personality. Even fundamental statistical designs focused on group means rather than individual differences. However, research in the field of individual differences in NHAs has recently (in the last 20 years or so) transitioned from taboo to popularity. With this recent surge in NHA personality research, it has become clear that nonhuman animals do, in fact, display consistent individual differences in their patterns of behavior across contexts and time. To put it bluntly, they have "personality". Other chapters of this book outline the analogues between human and NHA personality and give the reader a taste of the "basic science" discoveries that NHA personality research has uncovered in terms of the evolutionary and biological underpinnings of individual differences. The goal of this chapter is to explain *why* it is important that NHAs have personality, and to outline some of the practical applications of research in this area.

Beyond the interest in individual differences for its own sake, a majority of behavioral research is focused on predicting behavior. Most research looks at average increases or decreases in behavior across a sample when the independent variable changes, e.g., average increase in singing by song sparrows (*Melospiza melodia*) as a function of the season, presence of potential mates, or parasite load. In cases like these, individual differences are seen as statistical "noise" obscuring the population-level effects in which the researcher is actually interested (a problem also noted by Réale et al. 2007). Nonetheless, a better understanding of that noise allows a researcher to better control for it, improving the validity and generalizability of the population-level results, and decreasing the sample sizes required for statistical significance, which is of both practical and ethical value.

Within-Species Applications

Although a large portion of research conducted on NHAs is for "basic science", and NHA personality certainly has plenty to offer in terms of basic discoveries, improved understanding of individual differences has enormous potential to benefit not just the species being studied, but the subjects themselves. For NHAs in the care of humans, we have a responsibility and often a vested interest in providing living conditions that maximize their normal range of behavior and well-being, both physically and emotionally. From an ethical standpoint, given that we choose to keep NHAs in captivity for entertainment, companionship, for research, to assist in our daily lives, and to provide food, we have an equal duty to ameliorate the potential negative effects of that captivity. Standards of animal care in private homes, in zoos and aquariums, and in research laboratories have improved drastically in the last century and a half (Gauthier and Griffin 2005). We have gone from providing the bare minimum in terms of shelter and food to designing elaborate, naturalistic enclosures, providing opportunities for NHAs to use their basic hunting and foraging skills, and arranging social housing when it is species appropriate. These changes reflect a major shift not only in our concern for NHAs' physical health, but also toward our responsibility to maximize their psychological well-being. Scientists have repeatedly demonstrated that providing these types of environmental and social enrichment can increase activity and may improve medical and/or reproductive outcomes (Johansson and Ohlsson 1996; Meagher et al. 2014; Passineau et al. 2001). However, NHA caretakers have long observed that environmental enrichment or deprivation do not affect all NHAs' behavior equally.

Environment/Personality Interactions

In collaboration with the Woodland Park Zoo, we investigated behavioral differences in three elephants (two were Asian elephants, *Elephas maxima*, and one was African, *Loxodonta*) that had been sharing the same space for the majority of a thirty-plus year span. Bamboo, Chai, and Watoto (the African elephant) shared their enclosure at the Woodland Park Zoo, but though all three had similar high standards of care and enrichment opportunities, their keepers noticed behavior patterns distinct enough that the zoo wanted to better understand. Specifically, Chai had developed a very distinctive rocking stereotypy that both intrigued and concerned zoo visitors. Bamboo and Watoto, on the other hand, showed very little rocking behavior. The zoo had been incorporating more and more enrichment in the form of toys, food puzzles, naturalistic browse, and training with the keepers, and wanted to determine whether those efforts had had an impact on the stereotypical behavior of any or all of the elephants. They recruited researchers at the University of Washington to oversee data collection and analysis of the elephants' behavior throughout the day. Using focal scan sampling, researchers catalogued what the elephants did at various times of day for 4 years. The results were consistent with staff and volunteer observations: Chai spent a greater proportion of her time engaged in stereotypic behavior than did Bamboo or Watoto, and the *type* of stereotypy she engaged in was different than that engaged in by Bamboo and Watoto.

Examining the elephants' range of behaviors provided additional information regarding the activity that Chai's rocking behavior replaced. During the times of day when Chai was most likely to be seen rocking, Bamboo and Watoto dramatically increased their independent foraging activity. The perimeter of the elephant enclosure was intentionally planted with live, edible foliage, which the elephants regularly snagged and devoured as part of their enrichment program. During the hour or so leading up to a regular feeding, Bamboo and Watoto would often snack on the plants around the perimeter, while Chai was more likely to spend that time in anticipatory rocking. This information suggested that providing Chai with more foraging opportunities, or "training" her to exploit existing opportunities as the other elephants were already doing, might be an effective intervention in reducing her rocking behavior (Loyer et al. 2013). The results of this study were promising: across the observation years (and correlated with increased enrichment opportunities) there was a decrease in stereotypy across all of the elephants. This study supports the thinking that enrichment, or added environmental complexity, can reduce the frequency and duration of stereotypic behaviors, and that NHAs have consistent individual differences in their responses to environmental enrichment, suggesting that enrichment may need to be targeted to an NHA's personality to have maximum impact.

Effect of Environmental Enrichment

One simple explanation for differences in the efficacy of environmental enrichment is that some NHAs may simply be too afraid to interact with the enrichment provided. This would explain why, when new structures and methods of food delivery are available, some individuals show a substantial increase in activity while others show no change. Walker and Mason (2012) tested this hypothesis by first assessing the degree of neophobia (fear of novel objects or situations) in female mice (*Mus musculus*) by placing novel objects into their home cages and calculating how long it took them to make contact with the object. Then they gave the mice-free access to an enriched cage environment that included a running wheel, objects to chew on, nesting materials, complex surfaces for the mice to climb on and around, and a variety of manipulable toys. The researchers measured the use of the enriched environment by the amount of food consumed in the enriched cage compared to their standard laboratory cages, and by the quantity of two "consumable" forms of enrichment (a cardboard planter pot, which the mice shredded, and a length of string that they could pull into the cage to chew or use for nesting materials). They found that mice with higher levels of neophobia (those who took longer to interact with a novel object in their cage) tended to eat less food in the enriched cage and to consume less of the available enrichment. These results support the hypothesis that enrichment items may actually be frightening for individuals that are more fearful. They also beg the question of whether neophobia might be indicative of a general personality trait of fearfulness. In such a case, we would predict that neophobic individuals would show extra caution across multiple scenarios, for example, a greater response to an environmental stressor or increased vigilance under threat of predation. Alternatively, if no relationship (or even an opposite relationship) was found between neophobia and behavior in other contexts, it opens an entirely new line of inquiry as we try to explain why (evolutionarily, biologically, or ontogenetically) an individual might be fearful in one context and not in others. Occasionally studies like this deliver reasonably straightforward results, but it is far more common for researchers to find a complex relationship between behavior and environment.

The University of California at Davis keeps a colony of orange-winged Amazon parrots (*Amazona amazonica*) for research and has conducted multiple behavioral studies with them that examine the relationship between environmental enrichment and personality. Previous research at the university demonstrated that parrots reared in enriched environments tend to be less neophobic than those reared in barren or completely stable environments (Cussen and Mench 2015). To test whether reduction in neophobia is due to the enrichment items themselves or the consistent exposure to novelty, Fox and Millam (2007) compared neophobia (measured as the latency to eat something tasty from a feeder with a new object hanging nearby) before, during, and after orange-winged Amazon parrots were exposed to 11 weeks of either stable environmental enrichment objects (the low novelty condition), or rotating enrichment items (the high novelty condition). On a group level, the high novelty treatment had a greater impact on neophobia than the low novelty treatment. However, for individuals that were extremely neophobic prior to the treatment, their neophobia actually *increased* in the high novelty condition compared to those in the low novelty condition. These results indicate that *exposure to novelty is only an effective treatment for neophobia if that exposure is not too severe for the individual being exposed.* The implications for human behavioral treatments of fear, anxiety, and neophobia-related conditions are immediately obvious.

Response to Barren Environments

The previous two studies examined how enriched environments can improve potentially negative *personality traits and how personality traits affect NHA responses to enrichment.* At the opposite end of the spectrum, scientists have studied how NHAs vary in their response to barren environments based on personality. One common metric of welfare used in zoos and laboratories is the amount of time animals spend engaged in repetitive behaviors called stereotypies. These are behaviors like the rocking we previously described in Chai the elephant, or the famous pacing of caged predators (Clubb and Mason 2003; Mason 2010; Clubb and Vickery 2006). There are quantifiable differences between conspecific individuals housed in similar conditions or even in the same enclosures when it comes to the degree and type of stereotypic behavior they exhibit (Cussen and Mench 2015; Loyer et al. 2013), health outcomes (Cavigelli 2005), and social relationships (Weinstein and Capitanio 2008).

Cussen and Mench (2015) looked at how feather plucking (a common problem behavior in captive psittacines) and locomotor stereotypies were affected by 20 weeks of barren housing in orange-winged Amazon parrots that had previously been rated on what the authors labeled "extraversion" and "neuroticism". These ratings were obtained by having two people familiar with all of the birds assign a value from 0 to 7 based on how well an adjective described their personality. Birds that were considered high on the factor they labeled neuroticism were aggressive, eccentric, excitable, fearful, inflexible, high-strung, shy, and were not affiliative, tame, or confident. Birds that were considered highly extraverted were active, bold, impulsive, intelligent, cooperative, persistent, and were not cautious or lazy. Prior to enrichment deprivation, the researchers collected behavioral data that was repeated at the end of deprivation and after they had reintroduced enrichment.

There were some fairly universal changes in behavior: when deprived of enrichment, birds spent more time preening (cleaning and straightening out their feathers with their beaks), more time engaged in locomotor stereotypies, and more time generally active during the barren housing treatment than in either the baseline or the re-enriched conditions. Additionally, the birds' feather condition deteriorated significantly during the barren condition, indicating feather plucking and over-preening (and although they returned to baseline levels of preening after enrichment was reintroduced, the birds' feather condition did not improve even after the enrichment was returned). While the changes described above were found in nearly all birds, there were also significant differences that were predicted by individual parrots' personality ratings. Birds that were rated as more "neurotic" had poorer feather condition, even though they did not spend any more time preening than their less neurotic counterparts. Birds that were higher on extraversion showed a smaller increase in stereotypy after the barren environment and after re-enrichment: The deprivation had a smaller impact on their stereotypy than birds that were less extraverted. This difference in sensitivity to stressful environments based on personality has been found in multiple species, and although the welfare implications should be sufficient motivation to continue research in this line, research performed on domestic pigs (*Sus scrofa domesticus*) at an experimental farm at a university in the Netherlands offers additional arguments for understanding how personality and environment interact.

Bolhuis et al. (2006) investigated how different rearing environments as well as housing environment later in life impacted domestic pigs on the experimental farm at Wageningen University. They looked at behavior as well as health to determine how personality or coping style interacted with the environment. Pigs' personalities were assessed using the "Backtest" in the first few weeks of life (see also Horback this volume). At 10 days old, experimenters held the piglets on their backs for 60s and counted the number of times they struggled during the restraint. The test was repeated at 17 days old, and the escape attempts were added across the two tests in an effort to get a more global picture of the pigs' behavior. They categorized piglets as high resisting (HR) or low resisting (LR), and then studied the responses of both groups of pigs to two housing environments: barren floors versus straw bedding. The researchers manipulated whether pigs had access to straw during rearing (from weaning until 10 weeks old), and later during the "finishing phase" (the period between 10 and 22 weeks old, when they are slaughtered). One-quarter of the pigs spent the entire time with straw bedding available, another quarter spent the entire time on barren substrate, and the remaining pigs started with either barren floors or straw and were switched at 10 weeks onto the opposite flooring. During the study, they looked at the pigs' behavior, overall health, and weight gain, and found complicated interactions between rearing environment, later housing environment, and coping style. HR pigs tended to be more aggressive than LR pigs, and the LR pigs tended to be more sensitive to the environment and its changes: while HR pigs in barren environments showed no difference in the amount of time they spent chewing/biting at their penmates based on their rearing environment, LR pigs reared in barren environments spent more time chewing on their penmates, especially if they transitioned to enriched housing during their finishing phase. LR pigs also played more on straw bedding than on barren floors, particularly if they were also reared on straw.

These findings on behavioral differences have welfare implications for livestock raised for human consumption and should be sufficient on their own to warrant further investigation. However, the researchers also found that personality and environment interacted when it came to health and weight in the pigs, which has immediate practical implications for farming protocols. There were significant differences in the average daily weight gain in HR pigs raised in enriched environments: If they stayed in an enriched environment, they gained about 10% more weight during the finishing phase than if they were switched to barren floors. They also found that LR pigs had a significant decrease in the occurrence of gastric lesions when they spent the second half of their lives on straw bedding. Although these results present a substantial argument in favor of environmental enrichment, they also highlight the fact that individual differences in behavior mean different ways of expressing stress, different degrees of response to particular stressors, and a very clear need for better understanding of each of these facets of NHA personality. They also suggest that selective breeding for personality or behavior in addition to physical attributes has the potential to improve productivity on farms with minimal cost. While this research was conducted with pigs, other studies examining livestock coping styles/behavioral syndromes/personalities have found similar variation in behavior that could inform husbandry and breeding practices to both improve animal welfare and increase efficiency and productivity on farms (Lansade et al. 2014; Müller and Schrader 2005). Because research in NHA personality is relatively new, scientists have much more work to do in testing the generalizability of findings from one species to the next. Still, there is sufficient evidence to indicate that some basic personality traits are comparable even across taxa (Gosling 2001; Gosling and John 1999).

Across-Species Applications

An example of research in one species having implications for many others comes from a study by Ferland et al. (2014). Ferland and her colleagues investigated the complex relationship between an individual's sensitivity to novelty, and the effect of repeated exposure to that novelty on impulsive behavior in Long-Evans rats (*R. norvegicus*). In rats, one way to measure motor impulsivity is by looking at their ability to essentially "wait their turn" in a task where they have to choose the correct hole out of five options in order to earn a sugar pellet. Between trials, the rats are required to wait a whopping 5 s before a light indicates which hole is the correct choice for the next trial. Poking their nose into any of the holes before the 5 s is up is an example of motor impulsivity, and it turns out that in *some* rats, you can drastically increase the number of premature responses by repeatedly exposing them to a stimulating, enriched environment. Researchers assigned half of their rats to the enriched environment condition, and for 16 days would place them in a large, enriched cage for an hour prior to testing while the other half remained in their normal housing during that hour. Some rats began nose-poking prematurely far more often at the end of the stimulating environment condition than they had during baseline testing, so the researchers repeated the experiment with a second cohort of rats. This time, at the start of the experiment they used infrared beams to measure the rats' amount of locomotor activity in a novel environment for 60 min prior to beginning the task training. Because not all rats in the enriched environment condition had changed their proportion of premature responses, the experimenters were trying to identify any pre-existing behavioral differences between rats that were impacted by the environmental enrichment and those who were not. They found a significant difference in activity between rats whose premature responses increased drastically in the stimulating environment condition and those who did not: Rats that were more sensitive to the extra stimulation were also more active in the first 40 min in a novel environment, but by the end of their 60 min, they were no more active than the less-sensitive rats. Thus, a consistent personality trait appeared to govern both their exploratory behavior in novel environments, and their impulsivity in response to environmental enrichment.

The implications of this research (and other experiments like it) are numerous. It demonstrates the complicated interaction between NHA personality and environment (hyperactivity may be related to an increased sensitivity to novel stimuli, which may result in an increase in impulsive behavior, which, in humans, can be considered part of the personality itself!). If this experiment could be applied to species used for assisting humans, it may help inform our decisions about how to more effectively train animals, like dogs, or which individual animals to use for particular tasks.

Selection and Evaluation of Animals for Specific Tasks: Working Animals

Consider, for a moment, the behavioral requirements of a guide dog tasked with helping a blind person maneuver around a busy metropolitan area. Impulsivity is an enormously risky trait when it might result in a handler being yanked into traffic, and it is one reason that dogs "fail out" of the training (Burrows et al. 2008). If some dogs, like rats, are likely to become more impulsive after prolonged exposure to a highly stimulating environment (such as a big city), and if those dogs can be identified before extensive time and money have been spent preparing them to be guide dogs, guide dog organizations (and the individuals they benefit) could save a substantial investment that might have otherwise been wasted. A few studies have examined different behaviors and personality assessment techniques to determine their efficacy in predicting success of canine candidates, with varied outcomes (see also Fratkin, this volume).

The earlier canine candidates can be identified, and the sooner unsuitable dogs can be eliminated from the program, the more efficiently working-dog programs can function. Breeding and training working dogs is enormously expensive, ranging from around $19,000 for a police dog (http://www.nationalpolicedogfoundation.org) to $42,000 for a seeing-eye dog (http://www.guidedogsofamerica.org). In addition to the financial cost, training for these programs constitutes a serious investment of time (an estimated 18–24 months for service or assistance dogs; http://servicedogcentral.org; http://www.assistancedogsinternational.org). Therefore, the cost of training an individual dog that is later ejected from the program due to unsuitable behavior is immense. On the other hand, if the methods used for early rejection are unreliable, excellent candidates will be removed from the program before they have a chance to demonstrate their suitability. Goddard and Beilharz (1986) tested Labrador puppies that were purpose-bred to be guide dogs to determine how early important behavioral and personality differences could be identified to help with selection of both breeding stock and dogs that would be successful in the training program. They found that consistent differences in fearfulness (one of the most important factors in a dog's success as a guide dog) could be detected as early as 8 weeks of age, but that tests to predict adult fearfulness were more accurate as the dog's age increased. Based on their research, Goddard and Beilharz suggest that genetic selection for fear in breeding stock is most accurate in adulthood, which is crucial information for working-dog breeding programs. Many programs neuter males that will be sent through training when they are adolescents, which means that dogs that may have potentially valuable genetic contributions are eliminated from the breeding pool before their potential can be adequately assessed. Wilsson and Sundgren (1998) were interested in this problem and tested whether tests conducted on 8-week-old puppies could adequately predict adult behavior. They found that the "… correspondence of puppy test results to performance at adult age was negligible" (p. 1).

This type of NHA personality research, aimed at determining which personality traits stabilize at different stages ontogenetically, can have immediate practical impacts in addition to feeding back into broader theories of personality development. Additionally, research that helps identify the most accurate and efficient method(s) and age(s) for assessing NHA personality can improve our selection process for purposeful breeding in species we live and work with. These results should be beneficial not only for working canines, but for livestock and companion animals as well.

In an effort to produce results comparable across species, multiple NHA studies have imitated methods used in human research by having observers familiar with the NHAs being assessed rate the degree to which adjectives apply to the individuals (Gosling 1998) or by asking about individual responses to specific events (Hsu and Serpell 2003). Unfortunately, these methods depend on the availability of familiar observers (which is sometimes impossible) and introduce a risk of subjectivity and of anthropomorphizing NHA behavior. Thus, behavioral coding is often used instead, although this brings its own set of limitations, one of the most serious being the heavy time and resource investment required. To run a single individual through a battery of tasks designed to assess multiple facets of personality can require hours, even if these tasks are conducted in a single environment. Many researchers avoid this using extremely simple assessment tools that measure behavior differences on one or perhaps two axes (Amy et al. 2012; Brommer et al. this volume; Korsten et al. 2010; Sinn et al. 2006). This simplification of personality to one or two dimensions is useful for establishing that individual differences exist, but does not feed back into broader theory about how personality and environment interact as it tests extremely limited slices of each. In the same vein of critique, behavioral assays are often implemented in a laboratory setting, which calls into question the generalizability to related behaviors in the wild or in real-life circumstances, or worse, whether the tasks themselves are sufficiently biologically relevant to allow interpretation of the results.

The degree to which the limitations of these techniques hobble their practicality depends on their purpose. If the goal is to breed more docile animals to improve the ease of handling them in captivity, it is probably sufficient to perform simple assays by well-trained observers who need not know the animals individually. Dmitri Belyaev's famous work with silver foxes (*Vulpes vulpes*) in Russia may be the best proof of this. Using a simple two-part assay, Belyaev and his colleagues selected individuals that showed minimal signs of fear or aggression toward humans and bred them. Within 10 generations, almost 20% of the selectively bred population were what Belyaev labeled "domestic elite", meaning that, when interacting with humans, they would whimper, sniff the experimenters' hands, and lick them much like domestic dogs do (Trut 1999). In this case, the researchers were interested in domestication and in improving ease of handling, while looking at the behavioral, genetic, and physiological side effects of this artificial selection. This experiment has provided important information for basic science about the genetics of behavior and offers a starting point for those interested in selecting animals for captivity or livestock based on improving ease of handling. However, the subjects of this research were bred and raised specifically for this project and therefore had fulfilled their duties simply by demonstrating more affiliation with humans. In more applied settings, the personality or behavioral tasks required of NHAs are often far more complex, and the humanitarian and economic costs of improper breeding and selection of may be far more severe. As a result, more complicated personality or behavior measures are often required. Determining the best method of measurement for each purpose is one subset of applied NHA personality, and again, the domestic dog provides an opportune species for investigation.

In an experiment that examined multiple ways of evaluating working-dog behavior, Rooney et al. (2007) compared the ratings of dogs' handlers/trainers, experienced trainers who were unfamiliar with the subject dogs, and scientists, as well as objective measures obtained during an assessment completed at the end of a training program at the Defence Animal Centre in the UK. The trainers rated the dogs throughout their 10-week training program. At the end of this program, the dogs were run through a series of search tasks that were video recorded and rated independently by scientists and experienced military dog trainers who were unfamiliar with the subjects. These independent observers rated the dogs on similar (but not identical) characteristics to those assessed by their trainers. Finally, the recordings of the search tasks were analyzed for behavior measures thought to be important for a dog's success as a working detection dog. The outcomes of each of these methods of assessing the dogs' behavior were compared. Ideally, if each method is equally valid, the agreement between the methods would be high, which is what they found: The trainers who had worked with the dogs for 10 weeks and knew them rated their abilities similarly, the independent scientists and unfamiliar trainers who rated the dogs based solely on the video recordings rated the dogs very similar to one another, and overall, the trainers' ratings and independent scientist and trainer ratings for the dogs were correlated with each other and with the objective measures. This type of research is crucial to our ability to identify the best way to measure NHA's personality and working potential. If it is equally valid to use people familiar with an individual to rate their overall personality or behavior, versus people familiar with the species and the tasks required for the circumstances, or to use ethological measures that can be assessed by someone who need not be familiar with the species or the tasks as long as they can be trained to measure the crucial criteria, it allows us to (a) pick the best measurement method based on time, cost, and personnel available and (b) more easily compare outcomes across studies that utilize different methods to assess personality and behavior.

Selection of Companion Animals

In addition to the humancentric importance of using effective behavior and personality assessments for working dogs, most NHAs kept as pets stand to benefit from improving the tests used to assess their performance and personality. Behavior assessments are used in animal shelters and at rescues to determine the right type of home environment for the NHAs, as well as to identify NHAs that present a public safety threat. A test that is supposed to measure a dog's bite risk (the chance that a dog will bite a human or another NHA) necessarily has a major impact on the dog's life: with limited space and resources in shelters, dogs are often euthanized based on behavior rather than medical necessity. In these cases, knowing which (if any) assay can accurately predict bite risk could potentially save NHA lives while improving safety for the general public.

Currently, most behavioral assessments used by animal shelters, breeders, or rescue organizations have not been scientifically validated. Furthermore, they tend to be blunt instruments aimed at predicting bite risk (Taylor and Mills 2006), as opposed to understanding the nuances of behavior that may promote or degrade their relationship with prospective adopters. Tests of common assessments show a limited ability to differentiate dogs with a known history of aggression from those who have no aggressive history (Bennett et al. 2012; Bollen and Horowitz 2008; Paroz et al. 2008), calling into question their utility as a tool for determining adoptability and placing dogs with appropriate families. Given the stakes of these assessments, this is an area of applied research in NHA personality that is in need of far more attention.

Human Applications

The studies outlined previously have covered a range of applied research in NHA personality that is useful for the sake of understanding NHA personality and behavior in itself. In some cases, it is useful for ethical reasons and can benefit NHAs themselves. In others, it is primarily beneficial to the humans living and working with these NHAs, and in still other circumstances, it is beneficial to NHAs and humans alike. Besides the utility of understanding NHA personality for its own sake, NHAs have been used as models for human behavior and medicine essentially since the beginning of scientific enquiry.

We use NHAs as models for humans when it is impractical or unethical to conduct experiments on humans. It is not uncommon to deliberately breed or genetically engineer animals to specifically exhibit a medical or behavioral disorder, to identify factors that might affect that disorder's presentation, or to test treatments. Some strains of laboratory rats and mice are more likely to show fearful behavior (Higley et al.

1991), mice have been bred for their propensity to display stereotypies (Muehlmann et al. 2012) so that they can be used as models for autism treatment, and we examined the relationship between exploratory behavior or neophilia and genetics in species ranging from vervet monkeys (*Chlorocebus pygerythrus*) (Bailey et al. 2007) to great tits (*Parus major*) (Fidler et al. 2007; Korsten et al. 2010). With NHAs, the possibility of selective breeding combined with decreased time between generations allows researchers to get answers faster than they would with human subjects. The availability of species with social lives that are similarly complex to those of humans allows for research on interactions between different personality types, and NHA models are the only experimental option when testing the effect of major stressors on personality development. In humans, personality has been linked to differential outcomes in a variety of areas, including substance use and abuse (Cloninger et al. 1988; Kotov et al. 2010), academic performance (Chamorro-Premuzic and Furnham 2003; Komarraju et al. 2009), and relationship outcomes and satisfaction (Shiota and Levenson 2007; Vohs et al. 2011). Thus, if we are to use NHAs to understand human behavior and medical issues, it is critical that we at least attempt to account for the individual differences that might affect the conclusions we are attempting to make.

Conclusion

Applications of NHA personality research are numerous, and span multiple fields from conservation (see Troxell-Smith this volume; Watters et al. this volume), captive animal welfare (see Horback this volume; Lilley, Kuczaj, and Yeater this volume), and the safety of the general public, to medicine and evolutionary biology (Brommer et al. this volume; Van Oers and Laine this volume), and the list goes on. Given the vast number of areas and species in which this research can be applied, perhaps it is no surprise that we have yet to come up with a universal method of assessment, and indeed, that universality is probably not only difficult (if not impossible) to achieve, but inadvisable. Depending on the specific aspects of an individual's personality or behavior that need to be understood, assays must be tailored to be appropriately broad or specific to the task. Perhaps one way to consider this challenge is by comparing NHA personality tests to aptitude tests, much like many high-school students take to determine "what they should be when they grow up." The goal of these tests is to ensure that the right individuals end up in the right vocations, but there are two ways one might get to the same point: (1) Develop a general assessment that examines behavior in a variety of contexts and try to use this assessment to direct individuals to the best task, or (2) develop an assessment that is geared at differentiating candidates' aptitude at a specific task with a simple "yes or no" answer at the end. Each approach has its advantages and disadvantages. Whereas a broader assessment is better able to examine how facets of behavior and personality interact, it may necessarily sacrifice the ability to predict behavior in specific circumstances as the result of its broader perspective. Alternatively, the simpler assessment that examines small pieces of behavior under limited conditions may be more accurate at predicting behavior in similar circumstances in the future, but may, in isolation, tell us less about the broader construct of personality. However, these slices of information, when combined, can help us piece together a better understanding of personality (or individual differences, or coping styles, or behavioral phenotype, or whatever one chooses to call it), in both human and nonhuman animals. NHA personality research has enormous potential to improve the lives of all animals, human, and otherwise.

References

Amy, M., van Oers, K., & Naguib, M. (2012). Worms under cover: Relationships between performance in learning tasks and personality in great tits (*Parus major*). *Animal Cognition, 15,* 763–770. doi:10.1007/s10071-012-0500-3.

Bailey, J. N., Breidenthal, S. E., Jorgensen, M. J., McCracken, J. T., & Fairbanks, L. A. (2007). The association of DRD4 and novelty seeking is found in a nonhuman primate model. *Psychiatric Genetics, 17,* 23–27. doi:10.1097/YPG.0b013e32801140f2.

Bennett, S. L., Litster, A., Weng, H. Y., Walker, S. L., & Luescher, A. U. (2012). Investigating behavior assessment instruments to predict aggression in dogs. *Applied Animal Behaviour Science, 141,* 139–148. doi:10.1016/j.applanim.2012.08.005.

Bolhuis, J. E., Schouten, W. G. P., Schrama, J. W., & Wiegant, V. M. (2006). Effects of rearing and housing environment on behaviour and performance of pigs with different coping characteristics. *Applied Animal Behaviour Science, 101,* 68–85. doi:10.1016/j.applanim.2006.01.001.

Bollen, K. S., & Horowitz, J. (2008). Behavioral evaluation and demographic information in the assessment of aggressiveness in shelter dogs. *Applied Animal Behaviour Science, 112,* 120–135. doi:10.1016/j.applanim.2007.07.007.

Burrows, K. E., Adams, C. L., & Millman, S. T. (2008). Factors affecting behavior and welfare of service dogs for children with autism spectrum disorder. *Journal of Applied Animal Welfare Science: JAAWS, 11,* 42–62. doi:10.1080/10888700701555550.

Cavigelli, S. (2005). Animal Personality and Health. *Behaviour, 142,* 1223–1244. doi:10.1163/156853905774539355.

Chamorro-Premuzic, T., & Furnham, A. (2003). Personality predicts academic performance: Evidence from two longitudinal university samples. *Journal of Research in Personality, 37,* 319–338. doi:10.1016/S0092-6566(02)00578-0.

Cloninger, C. R., Sigvardsson, S., & Bohman, M. (1988). Childhood personality predicts alcohol abuse in young adults. *Alcohol Clinical and Experimental Research, 12*(4), 494–505.

Clubb, R., & Mason, G. (2003). Captivity effects on wide-ranging carnivores. *Nature, 425,* 473–474.

Clubb, R., & Vickery, S. (2006). Locomotory stereotypies in carnivores: Does pacing stem from hunting, ranging, or frustrated escape? In G. Mason & J. Rushen (Eds.), *Stereotypic animal behaviour: Fundamentals and applications to welfare* (2nd ed., pp. 58–85). Cambridge, MA: CABI.

Cussen, V. A., & Mench, J. A. (2015). The relationship between personality dimensions and resiliency to environmental stress in orange-winged amazon parrots (*Amazona amazonica*), as indicated by the development of abnormal behaviors. *PLoS ONE, 10,* e0126170. doi:10.1371/journal.pone.0126170.

Ferland, J. M. N., Zeeb, F. D., Yu, K., Kaur, S., Taves, M. D., & Winstanley, C. A. (2014). Greater sensitivity to novelty in rats is associated with increased motor impulsivity following repeated exposure to a stimulating environment: Implications for the etiology of impulse control deficits. *European Journal of Neuroscience, 40,* 3746–3756. doi:10.1111/ejn.12748.

Fidler, A. E., van Oers, K., Drent, P. J., Kuhn, S., Mueller, J. C., & Kempenaers, B. (2007). Drd4 gene polymorphisms are associated with personality variation in a passerine bird. *Proceedings Biological Sciences/The Royal Society, 274,* 1685–1691. doi:10.1098/rspb.2007.0337.

Fox, R., & Millam, J. (2007). Novelty and individual differences influence neophobia in orange-winged Amazon parrots (Amazona amazonica). *Applied Animal Behaviour Science, 104,* 107–115. doi:10.1016/j.applanim.2006.04.033.

Gauthier, C., & Griffin, G. (2005). Landmarks in animal-based research and key moral. *Revue Scientifique et Technique (International Office of Epizoonotics)* (2005).

Goddard, M. E., & Beilharz, R. G. (1986). Early prediction of adult behaviour in potential guide dogs. *Applied Animal Behaviour Science, 15,* 247–260. doi:10.1016/0168-1591(86)90095-X.

Gosling, S. D. (1998). Personality dimensions in spotted hyenas (*Crocuta crocuta*). *Journal of Comparative Psychology, 112,* 107–118. doi:10.1037/0735-7036.112.2.107.

Gosling, S. D. (2001). From mice to men: What can we learn about personality from animal research? *Psychological Bulletin, 127,* 45–86. doi:10.1037/0033-2909.127.1.45.

Gosling, S. D., & John, O. P. (1999). Personality dimensions in nonhuman animals: A cross-species review. *Current Directions in Psychological Science, 8,* 69–75. doi:10.1111/1467-8721.00017.

Higley, J. D., Hasert, M. F., Suomi, S. J., & Linnoila, M. (1991). Nonhuman primate model of alcohol abuse: Effects of early experience, personality, and stress on alcohol consumption. *Proceedings of the National Academy of Sciences of the United States of America, 88,* 7261–7265. doi:10.2307/2357662.

Hsu, Y., & Serpell, J. A. (2003). Development and validation of a questionnaire for measuring behavior and temperament traits in pet dogs. *Journal of the American Veterinary Medical Association, 223,* 1293–1300. doi:10.2460/javma.2003.223.1293.

Hunt, G. R., & Gray, R. D. (2004). Direct observations of pandanus-tool manufacture and use by a New Caledonian crow (*Corvus moneduloides*). *Animal Cognition, 7,* 14–120. doi:10.1007/s10071-003-0200-0.

Johansson, B. B., & Ohlsson, A. L. (1996). Environment, social interaction, and physical activity as determinants of functional outcome after cerebral infarction in the rat. *Experimental Neurology, 139,* 322–327. doi:10.1006/exnr.1996.0106.

Komarraju, M., Karau, S. J., & Schmeck, R. R. (2009). Role of the Big Five personality traits in predicting college students' academic motivation and achievement. *Learning and Individual Differences, 19,* 47–52. doi:10.1016/j.lindif.2008.07.001.

Korsten, P., Mueller, J. C., Hermannstädter, C., Bouwman, K. M., Dingemanse, N. J., Drent, P. J., et al. (2010). Association between DRD4 gene polymorphism and personality variation in great tits: A test across four wild populations. *Molecular Ecology, 19,* 832–843. doi:10.1111/j.1365-294X.2009.04518.x.

Kotov, R., Gamez, W., Schmidt, F., & Watson, D. (2010). Linking "big" personality traits to anxiety, depressive, and substance use disorders: A meta-analysis. *Psychological Bulletin, 136,* 768–821. doi:10.1037/a0020327.

Lansade, L., Valenchon, M., Foury, A., Neveux, C., Cole, S. W., Layé, S., et al. (2014). Behavioral and transcriptomic fingerprints of an enriched environment in horses (*Equus caballus*). *PLoS ONE, 9,* 1–20. doi:10.1371/journal.pone.0114384.

Loyer, C., Ha, J. C., Fernandez, E., & Hawkes, N. (2013, July). *Zoo elephant behavior patterns and effects of alternative feeding strategies*. Poster session presented at the meeting of the Animal Behavior Society, Boulder, CO.

Mason, G. J. (2010). Species differences in responses to captivity: Stress, welfare and the comparative method. *Trends in Ecology and Evolution, 25,* 713–721. doi:10.1016/j.tree.2010.08.011.

Meagher, R. K., Dallaire, J. A., Campbell, D. L. M., Ross, M., Møller, S. H., Hansen, S. W., et al. (2014). Benefits of a ball and chain: Simple environmental enrichments improve welfare and reproductive success in farmed American mink (*Neovison vison*). *PLoS ONE*. doi:10.1371/journal.pone.0110589.

Muehlmann, A. M., Edington, G., Mihalik, A. C., Buchwald, Z., Koppuzha, D., Korah, M., et al. (2012). Further characterization of repetitive behavior in C58 mice: Developmental trajectory and effects of environmental enrichment. *Behavioural Brain Research, 235,* 143–149. doi:10.1016/j.bbr.2012.07.041.

Muller, R., & Schrader, L. (2005). Behavioural consistency during social separation and personality in dairy cows. *Behaviour, 142,* 1289–1306.

Panksepp, J. (2000). The riddle of laughter: Neural and psychoevolutionary underpinnings of joy. *Current Directions in Psychological Science, 9,* 183–186. doi:10.1111/1467-8721.00090.

Panksepp, J. (2007). Neuroevolutionary sources of laughter and social joy: Modeling primal human laughter in laboratory rats. *Behavioural Brain Research, 182,* 231–244. doi:10.1016/j.bbr.2007.02.015.

Panksepp, J., & Burgdorf, J. (2003). "Laughing" rats and the evolutionary antecedents of human joy? *Physiology and Behavior, 79,* 533–547. doi:10.1016/S0031-9384(03)00159-8.

Paroz, C., Gebhardt-Henrich, S. G., & Steiger, A. (2008). Reliability and validity of behaviour tests in Hovawart dogs. *Applied Animal Behaviour Science, 115,* 67–81. doi:10.1016/j.applanim.2008.04.010.

Passineau, M. J., Green, E. J., & Dietrich, W. D. (2001). Therapeutic effects of environmental enrichment on cognitive function and tissue integrity following severe traumatic brain injury in rats. *Experimental Neurology, 168,* 373–384. doi:10.1006/exnr.2000.7623.

Plotnik, J. M., de Waal, F. B. M., & Reiss, D. (2006). Self-recognition in an Asian elephant. *Proceedings of the National Academy of Sciences of the United States of America, 103,* 17053–17057. doi:10.1073/pnas.0608062103.

Réale, D., Reader, S. M., Sol, D., Mcdougall, P. T., & Dingemanse, N. J. (2007). Integrating animal temperament within ecology and evolution. *Biological Reviews, 82,* 291–318. doi:10.1111/j.1469-185X.2007.00010.x.

Rooney, N. J., Gaines, S. A., Bradshaw, J. W. S., & Penman, S. (2007). Validation of a method for assessing the ability of trainee specialist search dogs. *Applied Animal Behaviour Science, 103,* 90–104. doi:10.1016/j.applanim.2006.03.016.

Shiota, M. N., & Levenson, R. W. (2007). Birds of a feather don't always fly farthest: Similarity in Big Five personality predicts more negative marital satisfaction trajectories in long-term marriages. *Psychology and Aging, 22,* 666–675. doi:10.1037/0882-7974.22.4.666.

Simonet, P., Versteeg, D., & Storie, D. (2005). Dog laughter: Recorded playback reduces stress related behaviour in shelter dogs. In N. J. Clum, S. Silver, & P. Thomas (Eds.), *Proceedings of the 7th International Conference on Environmental Enrichment*. New York City, NY: Wildlife Conservation Society.

Sinn, D. L., Apiolaza, L. A., & Moltschaniwskyj, N. A. (2006). Heritability and fitness-related consequences of squid personality traits. *Journal of Evolutionary Biology, 19,* 1437–1447. doi:10.1111/j.1420-9101.2006.01136.x.

Taylor, K. D., & Mills, D. S. (2006). The development and assessment of temperament tests for adult companion dogs. *Journal of Veterinary Behavior: Clinical Applications and Research, 1,* 94–108. doi:10.1016/j.jveb.2006.09.002.

Trut, L. N. (1999). Early canid domestication: The farm-fox experiment. *American Scientist, 87,* 1–4.

Vohs, K. D., Finkenauer, C., & Baumeister, R. F. (2011). The sum of friends' and lovers' self-control scores predicts relationship quality. *Social Psychological and Personality Science, 2,* 138–145. doi:10.1177/1948550610385710.

Walker, M. D., & Mason, G. (2012). Female C57BL/6 mice show consistent individual differences in spontaneous interaction with environmental enrichment that are predicted by neophobia. *Behavioural Brain Research, 227,* 508–513. doi:10.1016/j.bbr.2011.09.032.

Weinstein, T. A. R., & Capitanio, J. P. (2008). Individual differences in infant temperament predict social relationships of yearling rhesus monkeys, *Macaca mulatta*. *Animal Behaviour, 76,* 455–465. doi:10.1016/j.anbehav.2008.01.024.

Weir, A. A. S., Chappell, J., & Kacelnik, A. (2002). Shaping of hooks in new caledonian crows. *Science, 297,* 5583.

Wilsson, E., & Sundgren, P. E. (1998). Behaviour test for eight-week old puppies—Heritabilities of tested behaviour traits and its correspondence to later behaviour. *Applied Animal Behaviour Science, 58,* 151–162. doi:10.1016/S0168-1591(97)00093-2.

Index

Note: Page numbers followed by *f* and *t* indicate figures and tables respectively

A

Activity, 120, 170, 244
  associated with dog personality, 216–217*t*
  names of dimensions, in structure of dog personality, 208*t*
  to quantify traits in adult zebrafish, 117*t*
  personality framework, 207*t*
Age, 218
  and antipredator behavior, 162*f*
  associated with dog personality, 216–217*t*
  correlations across contextual gradient, 91*f*
  genotype-age interaction, 81
  in swine personality research, 195–199*t*
Agency, 42, 47, 49
Aggression, 7, 119, 207*t*, 244
  associated with dog personality, 216–217*t*
  names of dimensions, in structure of dog personality, 208*t*
  personality framework, 207*t*
  to quantify traits in adult zebrafish, 117*t*
American Society for the Prevention of Cruelty to Animals (ASPCA), 49
Animal boredom
  bubble play, 263
  coping style, 265
  environmental deprivation, 263
  environmental enrichment research, 264, 265
  inactivity and activity, 262
  meta-analysis, 265
  neurobiological research, 266
  positive reinforcement training, 266
  self-injurious behavior, 265
  species-specific behavior, 263
  stereotypical behavior, 264–266
  zoological facilities, 264
Animal personality
  distinction between quantitative and molecular genetics, 57
  field of
    evidence for, 136–137
    juvenile bull shark, 137
    large free ranging sharks, behavioral assays in, 138
    personality and individual differences, 136
  genetic influence, theory of, 56
  molecular genetics of
    association mapping, 64–65
    biological mechanisms, 60
    candidate gene approach, 61
    CRISPR-Cas method, 60
    genome-wide QTL approaches, 63
    linkage-based mapping, 63–64
    pleiotropy, 65–66
    QTL, 60
    resource, 60
    SNP, 60–61
  non-additive genetic effects, 57
  polygenic trait, 56–57
  quantitative genetics, 57–59
  social behavior in, 47–49
  structure of
    circular profile, 45, 45*f*
    extraversion and agreeableness, 44
    factor-analytic models, 44–45
    five-factor model, 44
    rating and coding approaches, 46
    vector method, 45–46
  trait correlations, 59
Animal personality research
  behavior rating scale for young chimpanzees, 5–6
  constitutional differences and functional disturbances, 4–5
  early expressions, 4
  emotional and non-emotional groups, rats, 6–7
  emotions profile index, 8
    cross-species method, 10
    dolphins, Kellerman's interpretation of, 9
    inter-rater reliability, 10
    issues, 9
    olive baboons, 9
  Five-Factor Model plus Dominance, 12
  Madingley questionnaire, 11–13
  monkeys, personality in
    biological basis and social–behavioral, 11
    rhesus social behavior, 10
  recent history of, 13–15
  temperament in chimpanzees, 7–8
Animal welfare, 186, 264, 267, 312, 316
  ongoing research, 199
Association mapping, 64–65 *See also* Genome-wide association study (GWAS)

J. Vonk et al. (eds.), *Personality in Nonhuman Animals*,
DOI 10.1007/978-3-319-59300-5

B

Balancing selection, 80–81
negative frequency-dependent selection, 81
Behavioral ecology
behavioral plasticity
between-individual variation, 90
character-state approach, 89–90
cross-context correlations, 90
individual differences, 86
ontogenetic classes, 90
reaction norm approach, 87
types, 86–87
behavioral syndrome. *See* Behavioral syndrome
biological perspective, 74
evolutionary quantitative genetic perspective
balancing selection, 80–81
disruptive selection, 80
evolutionary stasis, 79
genotype-environment/genotype-age interaction, 81
mutation–selection balance, 80
sexual antagonistic selection, 80
sexual selection, 80
personality in
additive-genetic variance, 78
consistency, 75
individual-specific deviation, 74
phenotypic variance, 75, 76
quantitative genetics, 76, 78
repeatability, 75
residual deviation,74
psychological perspective, 74
repeatable behavior*See also* Repeatable behavior
negative frequency-dependence, 86
stable state distribution and feedback loops, 82–86
trade-offs, 82
theoretical framework, 74
Behavioral syndrome, 296
BVN, 92
coping styles, 95
correlational selection, 93, 94
individual-level correlation, 93
multidimensionality, 91
phenotypic correlation., 92, 93
physiology, aspects of, 91
plasticity and, 97–100
POLS hypothesis, 96, 97, 97*t*
residual covariance, 92
Big Five Factor*See also* Five Factor Model (FFM), 75, 112
dimensions, 40–41, 44, 46, 206
agreeableness, 40
conscientiousness, 40
extraversion. *See* Extraversion
neuroticism. *See* Neuroticism
openness. *See* Openness
framework, 12, 21, 44, 45
individual differences, 257
Bivariate normal (BVN) distribution, 92
Boldness
to quantify traits in adult zebrafish, 117*t*
repeatable behavior*See also* Repeatable behavior, 75
and shyness, 116–118
QTL analysis, 118
tasks used to quantify traits in adult zebrafish, 117*t*
Bold–shy personality*See also* Boldness, 245
Bonobo, 32
assertiveness, 30, 31, 33
wild, 34
Bottlenose dolphins, 141, 240
dominance hierarchies, 244
fission–fusion social structure, 238
personality ratings of, 239
Breed, 213–215
breed differences, 214*t*
modern dog breeding, 213
significant personality differences, 215
working roles for people, 215

C

Candidate gene approach, 61
DRD4 gene, 62
MAO-A, 62
polymorphisms, 63
Capuchin monkeys, 28
assertiveness in, 30
personality effects in, 248
Cetaceans. *See* Bottlenose dolphins
Chase, 188, 194*t*
names of dimensions, in structure of dog personality, 208*t*
Chimpanzee Personality Questionnaire (CPQ), 21
Chimpanzees, 226–227
temperament in, 7–8
young, behavior rating scale for, 5–6
Coding method, 46, 242, 243
Communion, 42–49
Comparative psychology, 3, 10, 14–15, 55, 260
Complementarity, 47
alternative model, 48
applications, 49
interpersonal complementarity, 48
model of, 47*f*
Confidence, 33, 210, 231
names of dimensions, in structure of dog personality, 208*t*
Copy-number variants (CNVs), 122
Corticosteroids, 246–247
Corticotropin-releasing hormone (CRH), 120, 121
Crawford, Meredith P., 5, 6
Crocodilians, 154, 178
Cross-species method, 10

D

Disruptive selection, 80
Distraction, 206, 262

names of dimensions, in structure of dog personality, 208*t*
Dog Mentality Assessment, 207
Dog personality
assessing, method for
convergent validity, 210–211
discriminant validity, 211
genetic and non-genetic effects, 208
inter-observer reliability, 210
observational tests, 209
predictive validity, 211–212
ratings, 209
test battery, 209
test–retest reliability, 210
biological factors
age, 218
breed, 213–215
environmental factors, 218–219
heredity, 212–213
morphology, 215
reproductive status, 217–218
sex, 215
structure of
Big Five, 206
dimensions, 206, 207*t*
Dog Mentality Assessment, 207
Dolphins, 8, 9, 131
affiliative behavior, 248
bottlenose dolphins, 141, 240
dominance hierarchies, 244
fission–fusion social structure, 238
personality ratings of, 239
cognitive abilities, 238
individual differences in bold–shy dimension, 246
Kellerman's interpretation of, 9
open mouth, 243
phrases and activity
consistency over time, 239
watchful cautious, 268
play behavior, 270–271
social cooperation, 146
Dopamine receptor D4 (*DRD4*) gene, 62, 63, 66, 114
Drive, 209
names of dimensions, in structure of dog personality, 208*t*

E

Emotions profile index (EPI), 8
cross-species method, 10
dolphins, Kellerman's interpretation of, 9
inter-rater reliability, 10
issues, 9
olive baboons, 9
Excitability, 206, 207, 214*t*
names of dimensions, in structure of dog personality, 208*t*
Exploration
exploration-avoidance, 118
to quantify traits in adult zebrafish, 117*t*
Exploratory factor analysis (EFA), 188
Extraversion, 22, 24, 25, 29, 31, 121, 268
Big Five dimension, 40
names of dimensions, in structure of dog personality, 208*t*
surgency, 13

F

Fear(fulness), 11
associated with dog personality, 216–217*t*
cheetahs, tense–fearful, 230, 233
names of dimensions, in structure of dog personality, 208*t*
personality framework, 207*t*
Felid personality, implications
big cats, 226
coat color, 229
domestic cat, 225–226
dominance factor, 227
FeLV, prevalence of, 230
FIV, 230
pharmacological treatments, 232
rearing environment, 229
welfare and conservation, 232–233
well-being and neuroticism, 230–231
wild*vs.* domestic cats, 227–228
Feline immunodeficiency virus (FIV), 230
Feline leukemia virus (FeLV), 230
Fine mapping, 63
Five Factor Model (FFM), 24, 112, 113, 114, 121, 191
See also Big Five Factor
individual differences, 257
origins of 54 traits comprising Hominoid Personality Questionnaire, 22*t*, 28
Plus Dominance (FFM + D) in chimpanzee personality, 12–13
Focus, names of dimensions, in structure of dog personality, 208*t*
Foraging ecology
behavioral response, 296
behavioral syndrome, 296
bold bighorn sheep, 296
captive species, 300–302
experimental food patch, 297
GUD, 297–300
inedible substrate, 297
natural food density, 297
optimal foraging theory, 296, 297
patch-use theory, 296, 297
personality traits, 296
physiological factors, 298
proactive behavioral style, 296
wild species, 298–300
Friendliness, 5, 210, 230, 231, 245
names of dimensions, in structure of dog personality, 208*t*

G
Genome-wide association study (GWAS), 61, 64–65, 67
Genotype-environment/genotype-age interaction, 81
Giving-up density (GUD), 297–300
Glucocorticoid receptors (GR), 121
Gorilla, 32, 33
  individual differences, 264
  nature of relationship, 47, 232
  personality factor, 233
Great apes, 32, 131, 226
Grey seals, 240–241, 242, 244

H
High resisting (HR) piglets, 311
  in swine personality research, 195*t*, 198*t*
Hominoid personality questionnaire (HPQ)
  assumptions and non-assumptions
    behavioral codings, 28
    behavior-descriptive verb items, 28
    Big Five/Five-Factor Model, 28
    crab-eating macaques, 28, 29
    human observations, 28
    neurophysiological and molecular genetic studies, 29
  evaluation, research tradition
    crab-eating macaques, 31
    gorillas, 33
    partial female dominance, 32
    personality sex differences, 31, 32
    reproductive success, 32
  habits
    factors/components, 26–27
    heritability estimates, 26
    ICCs, 25
    interrater reliability, 25
  origins
    Big Five/Five-Factor Model, 21, 24
    chimpanzee behavior, 21
    conscientiousness domains, 24
    curious and conventional, 23
    heritability findings, 21, 22
    neuroticism facets, 23
    orangutan personality structure, 24
    trait descriptive adjectives, 21
    zoo-housed chimpanzees, 24
Hostile dominant behavior, 48
Human boredom, 261–262
Hypothalamic pituitary adrenal (HPA) axis, 120, 121, 186, 230, 246, 247
Hypothalamus–pituitary–interrenal (HPI) axis, 121

I
Impulsive-disinhibited personality (IDP), 121–122
  cluster B personality disorders, 121
  copy-number variants, 122
  SIRPB1 candidate gene for, 122
Individual differences
  agreeableness, 260
  animal personality ratings, 260
  behavioral ecologists, 259
  behaviorism and ethology, 258
  boredom
    animal boredom, 262–267
    boredom research, future of, 267
    human boredom, 261–262
  creativity, 270
  Five-Factor Model/Big Five, 257
  human personality psychology, 257
  nonhuman animals, curiosity in
    environmental stimuli, 269
    food source, 268
    neophobia, 268
    problem-solving abilities, 269
    shy-bold continuum, 268
    watchful cautious approach, 268
  personality, 258
  play behavior, 271
  survival rates, 259
  temperament, 258
  traits and behavioral tendencies, 261
Individual optimization, 74
  behavioral ecology theory, 81
  repeatable behaviors, 82
Inter-observer reliability, 210
Interpersonal circumplex
  animal personality, social behavior in, 47–49
    "arrogant-calculating" behavior, 48
    dynamic "field-regulatory" system, 47
    "friendly dominant" behavior, 48
    "friendly submissive" behavior, 48
    "hostile dominant" behavior, 48
    "hostile submissive" behavior, 48
    interpersonal circumplex, 49
    interpersonal complementarity, 47
    "Meet Your Match" program, 49
  animal personality, structure of
    circular profile, 45, 45*f*
    extraversion and agreeableness, 44
    factor-analytic models, 44–45
    Five-Factor Model, 44
    rating and coding approaches, 46
    vector method, 45–46
  Big Five personality dimensions, 40, 41
  implications, 41
  lexical approach, 40
  limitation, 41
  humans
    agency dimension, 42
    anthropology, 42
    circular taxonomy, 43
    evolutionary psychology, 42
    interpersonal circle, 42
    social behavior, 42
    sociology, 42
Intraclass correlation coefficients (ICCs), 25, 29, 75

K

Kangaroo rats, 286, 287
Kangaroos, 288

L

Laboratory strains, 120
Linkage-based mapping, 63–64
  DRD4 gene, 63
  F2-cross designs, 64
  pedigree information, 63
  personality and animal well-being, 63
  predictive markers, 63
  SNP markers, 64
Livestock, 186, 311–312, 314
  QTL for behavior traits in, 63
Lizards
  common lizard, 174
  heritability and development, 173
  hormone-dependent personality trait expression, 173–174
  male personality and female predation risk, 175
  measuring behavioral traits, 170
  morphology, correlated variation in, 176–177
  Namibian rock agama, 174
  personality and pace-of-life syndrome, 175–176
  personality traits/behavioral syndromes, 167, 168–169*t*
  population density and fitness, 174
  statistical analysis, 171–172
Low resisting (LR) piglets, 311
  in swine personality research, 195*t*, 198*t*

M

Madingley questionnaire, 11–13
Marine mammals
  bottlenose dolphins, 238
  defining personality for
    consistency over time, 239
    temperament, 238
  friendships, 247–248
  learning, 245–246
  physiology, 246–247
  research
    bottlenose dolphins, 240
    cross-species comparisons, 241–242
    grey seals, 240–241
    refining methods and techniques, 242–243
  social rank and dominance, 243–243
Monkeys, 4
  biological basis and social–behavioral activity, 11–12
  capuchin monkeys, 28–30, 33
  common squirrel monkeys, 267
  cynomolgus monkeys, 231
  dominant monkeys, 9
  factor analysis of rhesus social behavior, 10
  Old World monkeys, 33
  vervet monkeys, 316
Monoamine oxidase A (MAO-A), 62
Motivation, 154
  to feed, 84
  names of dimensions, in structure of dog personality, 208*t*
Mutation–selection balance, 80

N

Namibian rock agama, 174
Negative frequency-dependent selection, 81
Neophobia, 245, 268, 269, 309, 310
Neuroticism, 230, 231, 266, 310
  Big Five dimension, 40, 41
  Five-Factor Model neuroticism domain, 23
Nonhuman animal (NHA) personality
  across-species applications
    behavioral coding, 314
    companion animals, selection of, 315
    domestic elite, 314
    human applications, 315–316
    Long-Evans rats, 312
    working-dog behavior, 314, 315
  animal care, 308
  basic science, 308
  environment/personality interactions
    barren environments, 310–312
    environmental enrichment, effect of, 309–310
    staff and volunteer observations, 309
  fundamental statistical designs, 308
  rats and dog, 307
Nonhuman primates, 14
Non-invasive methods, 247

O

Obedience, 206
  names of dimensions, in structure of dog personality, 208*t*
Openness, 13, 21, 23, 24, 41, 269
  Big Five dimensions, 40
  in cognitive enrichment, 232
  and learning, 245
  origins of 54 traits comprising Hominoid Personality Questionnaire, 22*t*
Orangutan, 24, 32
  personality, 23
  Personality Questionnaire, 23

P

Pace-of-life syndrome (POLS), 96, 97, 97*t*, 99
Pathological five model (PFM), 113
Pavlov, Ivan, 4, 245
Personality
  animal personality. *See* Animal personality
  dog personality. *See* Dog personality
  felid personality. *See* Felid personality, implications
  polygenic trait, 56–57

Phenotype management
behavioral type diversity, release success
genetic diversity, 286, 287
investigations, 286
mice, 287
plasticity, 287
post-release, 287
conservation, ex situ populations for, 281–282
conservation populations
animals' behavioral phenotypes, 279
decision time, 279
habitat heterogeneity, 278
in situ and ex situ populations, 280
life history decisions, 280, 281
meta-population, 281
ex situ animals, educators
breeding performance, 282
contact and encounter strength, 283
positive emotional arousal, 283
socio-sexual behavior, 282
zoos, 282
in situ to ex situ
captive breeding program, 284
good breeders, 285
human interaction, 285
reintroduction programs, 284–285
predation risk, 288
reproductive isolation, 289
Phylogenetic lineages, 227
Pinnipeds. *See* Grey seals
Play behaviors, 246, 271
Playfulness, 206, 207
names of dimensions, in structure of dog personality, 208*t*
Pleiotropy, 65–66, 95, 186
Polygenic trait, 80
personality as, 56–57
Predation, 134
intensity variation, 158
Predation risk, 81–83, 86–88
female predation risk, 175
in shark, 134
Predation threat, 290, 310
test, 187
Primates, 30, 32, 172, 226, 228, 282
nonhuman, 14, 20, 30, 155, 248, 258
non-primates, 23, 227
Principal components analysis (PCA), 171, 172, 188, 191

Q

Quantitative trait loci (QTL), 60
QTL mapping. *See* Linkage-based mapping

R

Random regression phenotypic models (RRPMs), 88, 89
Random regression animal model (RRAM), 88, 89
Rats, 269, 307
aggressive male rats, 246
emotional rats, 7
fearful behavior, 315
kangaroo rats, 286
Long-Evans rats, 312
non-emotional rats, 7
Roman high-avoidance (RHA) rats, 114
Roman low-avoidance (RLA) rats, 114
strains of, 7
Reaction norm approach
contextual gradient, 87
genotype*vs.* environment interaction, 88
RRAM, 88, 89
RRPM, 88, 89
Reactivity, 12, 245
cortisol reactivity, 247
emotional reactivity, 21
/fearfulness, 213, 215, 218
associated with dog personality, 216–217*t*
personality framework, 207*t*
Repeatable behavior
negative frequency-dependence, 86
stable state distribution and feedback loops, 82–86
empirical evidence, 85
negative feedbacks loops, 83
performance model, 84
positive feedbacks, 82–83
risk-taking behavior, 83, 84
RRV, 83
state-dependent safety, 85
trade-offs, 82
Reptile behavior, personality and individuality in
anthropomorphic traits, 156
crocodilians, studies of, 178
lizards
common lizard, 174
heritability and development, 173
hormone-dependent personality trait expression, 173–174
male personality and female predation risk, 175
measuring behavioral traits, 170
morphology, correlated variation in, 176–177
Namibian rock agama, 174
personality and pace-of-life syndrome, 175–176
personality traits/behavioral syndromes, 167, 168–169*t*
population density and fitness, 174
statistical analysis, 171–172
nonavian reptiles, 154
psychological propensities, 155
reptilian taxa, 154
snakes, individuality and antipredator behavior in, 156–158
behavioral consistency across temperatures, 164
chemical prey extracts, 165
experiential effects, 161–163
heritability of, 165–166

hognose snakes, 157
individual consistency over time, 158–161
inter-individual variation, fitness consequences of, 166–167
long-term developmental trait consistency, 161
moving finger trial, 158
natricine snakes, 157, 161
nonmoving trial, 158
non-natricine snakes, 164
open-field exploration, 164–165
sociality, 167
variety of, 157
watersnake species, 157
tongue-vomeronasal system, 156
turtles, studies of, 177–178
Research tradition, 20, 29
evaluating, 30–33
biologically meaningful relations, 30
bullying, 32
conscientiousness, 33
partial female dominance, 32
personality sex differences, 31
personality structures, 30
social competence, 32
Residual reproductive value (RRV), 83–85
Responsiveness to training, 206, 213, 218, 220
associated with dog personality, 216–217*t*
personality framework, 207*t*
Risk-taking behavior, 65, 83, 84, 173
Roman high-avoidance (RHA) rats, 114
Roman low-avoidance (RLA) rats, 114

S
Sex, 215, 217
characteristics associated with dog personality, 216*t*
Sex determination, 179
Sexual antagonistic selection, 80
Sexual selection, 80–81
Sharks
Chondrichthyans, 130
experimental challenges and constraints, 134–135
personality of
acoustic telemetry studies, 134
catsharks,131
juvenile lemon sharks, 133, 133*f*, 134
personality tests, 132
risks and benefits, 134
semi-captive environments, 131
social network analyses, 131
slow growth and reproduction rates, 130
study on shark personality
core areas, use of, 138, 140
elasmobranchs field studies, 138, 139*t*
individual differences, 141
investigating personality, 141
juvenile lemon sharks, 143
Notorynchus cepedianus, 142
personality traits, 143
predation risk, 142
shark social behavior, 142
tools for, 143–144
studying personality, relevance and feasibility of, 130
Single-nucleotide polymorphism (SNP), 60–61
gene, 62
marker, 64
Snakes, individuality and antipredator behavior in, 156–158
behavioral consistency across temperatures, 164
chemical prey extracts, 165
experiential effects, 161–163
heritability of, 165–166
hognose snakes, 157
individual consistency over time, 158–161
inter-individual variation, fitness consequences of, 166–167
long-term developmental trait consistency, 161
moving finger trial, 158
natricine snakes, 157, 161
nonmoving trial, 158
non-natricine snakes, 164
open-field exploration, 164–165
sociality, 167
variety of, 157
watersnake species, 157
Sociability, 119–120
names of dimensions, in structure of dog personality, 208*t*
personality framework, 207*t*
to quantify traits in adult zebrafish, 117*t*
Sounders, 189
Submissiveness, 43, 206, 210
personality framework, 207*t*
Swine, personality in
biological welfare, 199
domestic farm animals
assessing personality traits, 186–187
environmental and social stress for, 186
pigs, ethograms for, 187
pleiotropy, 186
test–retest reliability, 186
domestic swine welfare, 190–191
implications, 199–200
natural behavior, 189–190
personality data analysis
calculating trait scores, 188
coping styles, 189
PCA, 188
psychological welfare, 199
research, 191–192
back-test, 193–194
novel object test, 193
open field test, 192–193
response to humans, 194–195
social challenge, 193

T
Temperament, 4, 7, 155, 238, 258
  in chimpanzees, 7–8
Temperament and Character Inventory (TCI), 112
Test–retest reliability, 6, 186, 210
Top predator, 141
Trainability, 206, 207, 211, 212, 218
  breed differences, 214*t*
  names of dimensions, in structure of dog personality, 208*t*
Turtles, 177–178
  boldness, 177
  compete for single food item, 178

U
Ushering, 135

V
Vector method, 45–46
Vervet monkeys, 316

W
Watchful cautious approach, 268
Wild-type animals, 121
Working-dog behavior, 314, 315

Y
Young chimpanzees, behavior rating scale for, 5–6

Z
Zebrafish
  activity, 120
  aggression, 119
  Casper, 116
  exploration-avoidance, 118
  FFM, 112, 113
  genetic differences, 114
  genetic models, 114
  human personality and personality disorders
    anxiety disorders, 122
    depressive disorders, 120–121
    IDP, 121–122
  motivated agent, 111
  neurobehavioral model, 115, 123
  nonhuman species, 114
  personality traits, ontogeny of, 116
  RHA and RLA, 114
  rodent lines, 115
  rodent models, 115
  shyness–boldness, 116–118
  sociability, 119–120
  social actor, 111

FSC
www.fsc.org
MIX
Papier | Fördert
gute Waldnutzung
FSC® C083411

Kolibri 360 GmbH
Ferdinand-Jühlke-Straße 7
99095 Erfurt, Deutschland
produktsicherheit@kolibri360.de